Solar Collectors and Panels, Theory and Applications

edited by

Reccab M. Ochieng

Solar Collectors and Panels, Theory and Applications
Edited by Reccab M. Ochieng

Published by InTech
Janeza Trdine 9, 51000 Rijeka, Croatia

Publishing Process Manager Iva Lipovic

Technical Editor Teodora Smiljanic

Cover Designer InTech Design Team

Additional hard copies can be obtained from orders@intechopen.com

Solar Collectors and Panels, Theory and Applications, Edited by Reccab M. Ochieng
p. cm.
ISBN 978-953-307-142-8

For EU product safety concerns:
IN TECH d.o.o., Prolaz Marije Krucifikse Kozulić 3, 51000 Rijeka, Croatia,
info@intechopen.com or visit our website at intechopen.com.

Meet the editor

Reccab M. Ochieng became an associate professor of Physics in the Department of Physics and Materials Science at Maseno University in early 2010. Prior to that he spent over 22 years as a lecturer, senior lecturer and researcher in various public Universities in Kenya and the southern African country of Zambia where he taught and carried out research in the Department of Physics at the University of Zambia from 2003 to 2005 and then from 2006 to 2008. He received his Ph.D. from Maseno University, Kenya. Though his main area of research in the recent past has been in new and renewable energy, his early background was in Theoretical Plasma Physics, specifically Strongly Coupled Plasma Physics and Mathematical Physics in which he holds an MSc. from the University of Nairobi, Kenya. He later ventured into experimental Physics in which he did a master degree in High Temperature Superconductivity and Electronics at the University of Arkansas in USA before embarking later on in new and renewable energy research at Maseno University.

Contents

Preface

The title of this book was specifically chosen to encompass all the chapters presented in the book. Solar collectors and panels have become household and industrial items being used for power production, heating, cooling and are even being used for outer space research because of their environmentally friendly nature.

The reason for writing this book was to put together some material which are related in a way and can give those interested in the field of renewable energy a quick start but can also provide detailed information on what is going on in the dynamic area of solar collectors and panels research. The book provides a quick read for experts, researchers as well as novices in the field of solar collectors and panels research, technology, applications, theory and trends in research. The book covers the use of solar panels applications in detail, ranging from lighting to use in solar vehicles.

Theory and applications have been discussed as well, with a view of giving an in depth knowledge to the reader.

The efforts of many researchers in the area of renewable energy, specifically solar collector panels and technology have helped create a wealth of information that cannot be complete encompassed in such a book. This book is therefore an attempt to bridge the gap of lack of information. The first 16 chapters deal with solar panels including solar photovoltaics applications and chemistry spans.

Solar thermal energy system which basically include concentrators and collectors are presented in the last 4 chapters.

Hopefully researchers, engineers, users, policy makers and implementers of solar collectors and panels to whom this book is dedicated will strive towards building a better society by using the knowledge discussed for the preservation and improvement of our environment.

Editor

Reccab M. Ochieng
Department of Physics and Materials Science,
Maseno University, P.O. Box 333, Maseno 40105,
Kenya

1

Internal Lighting by Solar Collectors and Optical Fibres

P. Sansoni, D. Fontani, F. Francini, L. Mercatelli,
D. Jafrancesco and E. Sani, D. Ferruzzi
CNR-INO National Institute of Optics, Largo E. Fermi, 6, 50125 Firenze, Italy

1. Introduction

Sunlight concentration on small surfaces is widely studied [1-3], experimented and mostly applied to photovoltaic power generation [4-6]. More rarely these solar collectors are coupled to optical fibres [7-9], with the advantage of always having a circular absorber shape. On the contrary the photovoltaic (PV) cell is typically squared and therefore it requires a secondary optical system to reshape the image and to improve the light distribution uniformity.

The introduction of optical concentrators, especially high concentration systems, has two positive effects: it reduces the area of expensive solar cells and it increases their efficiency. The main reasons for this development are enhanced efficiency of CPV (concentrating photovoltaic) systems due to new solar cells, improved size of PV installations and increasing interest in alternative technologies, both due to government incentives and to the poor Silicon availability. In general it can be assumed that an improvement in the volume of the collection system reduces the costs, given that the system provides a higher production of energy.

This chapter presents optical systems exploiting the sunlight using optical collectors and fibres to illuminate building interiors. In particular Sect. 5 describes in details a solar plant demonstrator installed to provide illumination of museum showcases [10]. This daytime lighting system was developed from design to production, installation and testing in working conditions. The light focused by the solar collector can be used either for direct illumination or to accumulate power [11] for lighting at times when there is no sunlight. The first function is obtained coupling an optical fibre to a solar collector. The second consists in focusing the solar light on a PV cell, which converts the light into electrical energy. This function has been suggested by the long closing times, typical of the museum. Due to the fact that the internal illumination was not required for hours or entire days (museum closure day), during these periods the PV cells can exploit the solar light.

A solar collector with optimised features and collection performance was specifically designed for mass production to reduce costs. The evolution of solar concentrators for fibre coupling [12-13] is discussed in Sections 2-3, with theoretical and experimental comparisons based on optical tests [14] performed on the realised samples. The field tests, with direct exposition to the sun, required to design and built suitable mechanical systems to support and move the concentrators: examples of tracking systems [15-16] are reported in Sect. 4.

Besides museum lighting required an alternative illumination source that can be used when sunlight is not available. This was realised by employing novel LEDs (Light Emitting Diodes) with very low power consumption. But several fundamental requirements must be met. First, illuminance levels are dictated by the need to control deterioration of the exhibit items. Moreover, the LED light must replicate light from the sun. Finally, light uniformity and colour are important. Light hue and colour balance were examined using photometric and colorimetric measurements [17-20] to define the suitable filters for plastic fibre and LED light. Additional tests were carried out on lens production to evaluate collection efficiency and image size, examine optical treatments and experiment the effects of external agents and UV exposure. Section 5 illustrates all mentioned aspects analysed during the development of the museum lighting plant.

2. Optical design of fibre-coupled collectors

Optical systems for sunlight exploitation have been optically designed and tested in our laboratory since 1997 [12-14, 10]. They are modular devices including solar collectors, optical fibres and mechanical and electronic systems for sun tracking. The main element of the device is the sunlight concentrator coupled to an optical fibre for power transportation to the utilisation point. Our first theoretical studies and practical experimentation of an optical system for sunlight collection were developed for a European project. The selected optical fibres were single fibres made of quartz, which are characterized by extremely reduced losses. The collector was optically designed to be coupled to an optical fibre with core diameter 0.6mm and numerical aperture NA=0.48 (angular semi-aperture 28.7°). Several optical projects, with increasing complexity, were designed for the concentrator and some of them were realized. The optical configurations included Mangin, Parabolic and Cassegrain collectors: Table 1 summarizes their main optical characteristics; in particular the last column reports the material chosen for the realisation.

Label	Collector name	Primary mirror	Secondary mirror	Other optical elements	Material
A1	Standard Mangin	spherical	flat		Glass
A2	Modified Mangin I	spherical	spherical		Glass
A3	Modified Mangin II	spherical	spherical	correction lens	Glass
B1	Parabolic	parabolic	flat		Glass
B2	Parabolic with lens	parabolic	flat	spherical correction lens	Aluminium
C1	Conic Cassegrain CCM	elliptic	spherical		Quartz, Plastic

Table 1. Optical layouts of collectors coupled to quartz fibre Ø=0.6mm.

The efficiency obtained by each collector coupled to a 0.6mm quartz fibre was theoretically estimated and experimentally measured. This comparative study showed that the Cassegrain collector C1, especially designed for the European project, was the most compact and performed the best light collection. The Mangin configuration represented the best trade-off between collection efficiency and cost, since these collectors have only spherical surfaces that are easier to be optically manufactured. Concerning the mechanical alignment between concentrators and optical fibre, for Mangin and Parabolic collectors this problem was more complex to solve than for the Cassegrain optics because they used a secondary mirror that is physically separated from the primary mirror (it is realized on the protective window). This alignment difficulty increased if an array of concentrators was used. Hence the modular unit was constituted by a tile mounting four collectors and each collector was coupled to a single quartz fibre.

Starting from the fibre-matching requirements of image diameter 0.6mm and *NA*=0.48, we developed the optical systems in Table 1. Then we estimated the illuminated area on the focal plane of each solar collector in order to compare the sun image to the fibre diameter. The image is mainly created by the spot within upper and lower rays, but it also includes the root mean square spot diagram at maximum field 0.25 deg (indicated in the following as "rms spot"). The rms spot size usually denotes the root mean square of the spot radial size and it gives information on the rays spread and thus on the geometrical aberration of the optical system. The diameter of this area should be within 0.6mm to obtain the best fibre-coupling.

Class A contains the solar collectors where all surfaces are spherical. These configurations present small image diameter and rms spot over 0.1mm. In class B there are systems with parabolic mirror, while the conic Cassegrain represents class C. Collectors of classes B and C are characterised by a very good image quality. These layouts are more complicated and expensive to be realised than the others, but in the best case the collected power is about three times higher with respect to A1. A3 is a trade-off between the two collector types since all its surfaces are spherical but the image quality is quite good. The enter pupil diameter of the solar collector varies from 40mm to 71mm and it has been optimised in the spectral range between 400nm and 1400nm of wavelength, for a 0.5 deg field of view. An exemplification of the images pertaining to the six optical systems under consideration is shown in Fig. 1. Within the fibre core (plotted as dashed circle) the central yellow spot indicates the sun image, while the two lateral dashed spots represent the rms spots.

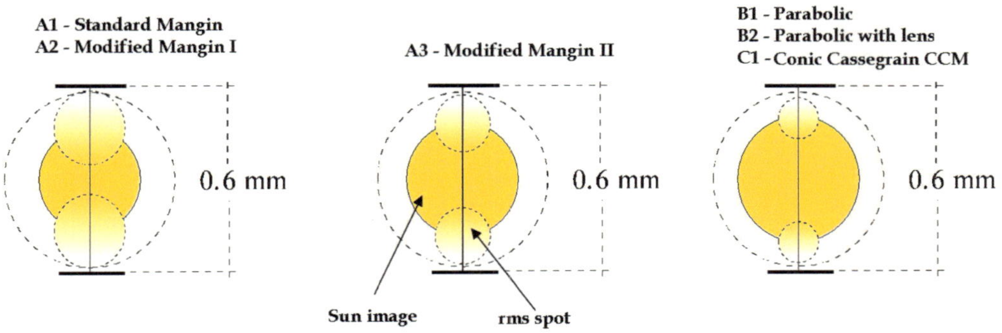

Fig. 1. Schematic view of images formed by the six collectors.

In the next paragraphs of this section each of the six optical layouts is described in details: optical configuration and performance are examined and compared. Then in Sect. 3 some experimental measurements on realised samples of the designed collectors are analysed and compared with the theoretical estimations.

2.1 Mangin collectors

The Mangin system is composed of a glass meniscus, aluminised on the rear surface, with a first spherical mirror and a secondary mirror, which can be flat or spherical. The optical path between the two surfaces of the meniscus allows the control of the spherical aberration, which can be minimised adjusting their curvature radius.

Three different configurations (A1, A2, A3) were selected, optimising their optical parameters with the aim of reaching the largest enter pupil diameter (EPD) in order to collect the maximum of power. The layouts are depicted in Figures 2, while the optical projects are reported in Figures 3: A1 in (a), A2 in (b), A3 in (c). All optical designs for A, B, C collectors were developed using Zemax ray tracing software.

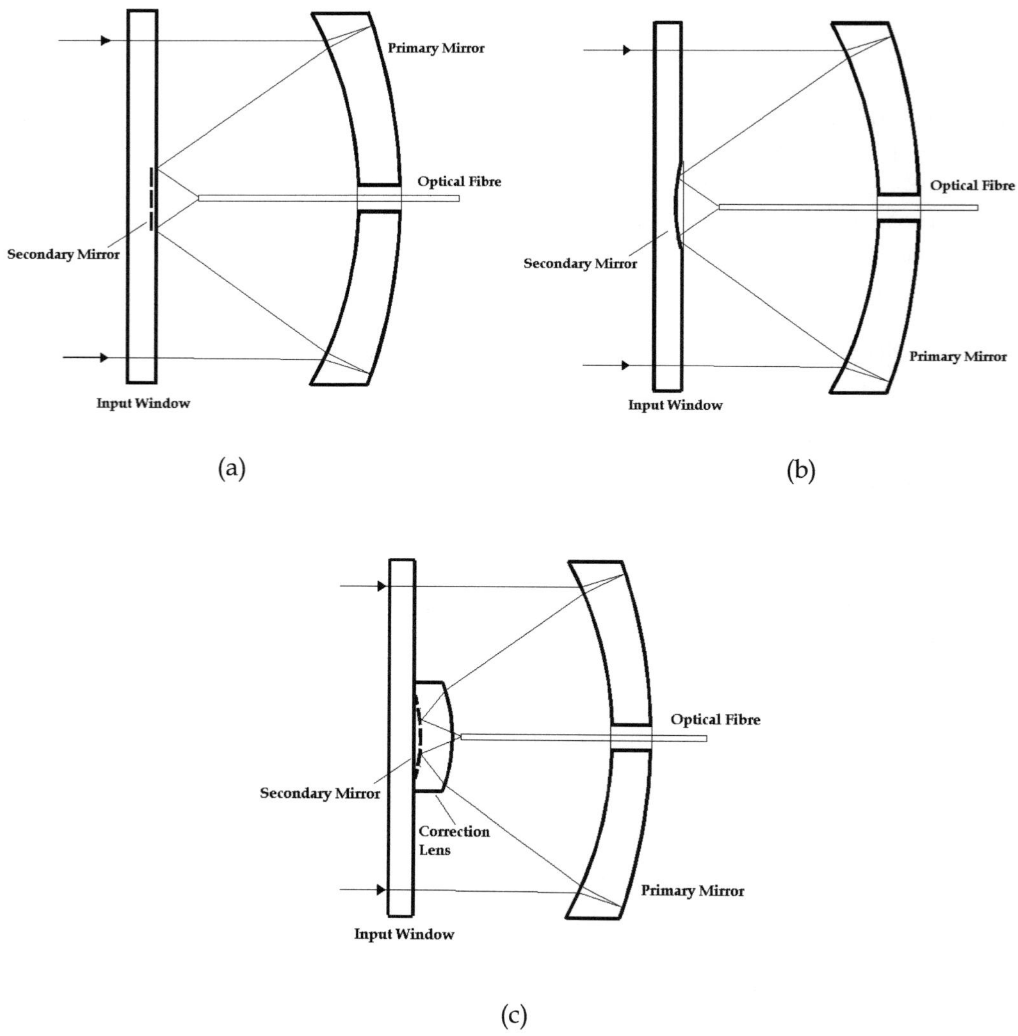

Fig. 2. (a) Layout of Mangin A1, (b) Layout of Mangin A2, (c) Layout of Mangin A3

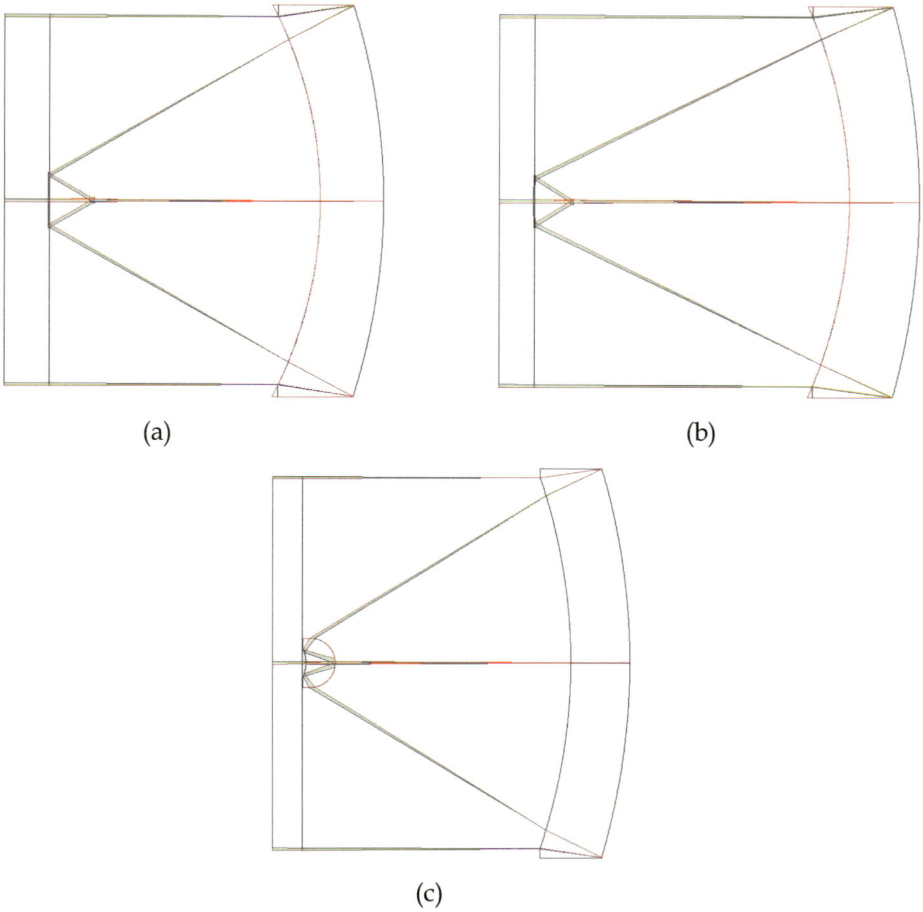

Fig. 3. (a) Zemax optical design of Mangin A1 (b) Zemax optical design of Mangin A2 (c) Zemax optical design of Mangin A3

They have the following optical features (the focal length f is at wavelength 580nm; $f/\#=f/EPD$ is the f number; θ is the maximum angle of rays with respect to the optical axis):

A1) spherical primary mirror; flat secondary mirror; maximum EPD=40mm; f=39.76mm, f/#=0.994 (θ=26.7°);

A2) spherical primary mirror; spherical secondary mirror; EPD=52mm, f=48mm, f/#=0.923 (θ=28.44°);

A3) spherical primary mirror; spherical secondary mirror with correction lens; EPD=62mm, f=58.1mm, f/#=0.937 (θ=28.5°).

Table 2 in Sect. 3.1 compares the main optical characteristics among all examined collectors of classes A, B, C.

Comparing the different layouts in class A, we can see that a more corrected system allows us to increase the enter pupil EPD and it yields to a higher collected power on the optical fibre. For A1 and A2 we can use Fused Silica to minimise the chromatic aberration; while for A3 we used SF1 glass for the primary mirror and PBM3 for the correction lens.

A1 is very cheap to realise, but the total spot diameter is small (0.513mm); while A2 allows to obtain a larger sun image (0.598mm). The total spot diameter for A3 is 0.594mm, but it is the most complicated: there is a correction lens on the surface of the secondary mirror (the rear surface of this lens was aluminised to avoid the crossing of two further surfaces). The differences among the sun images can be evidenced considering their profiles on a central line in the image plane. The curves for A1, A2 and A3 are reported in Fig. 4 plotting the average image profile versus the points along the image line.

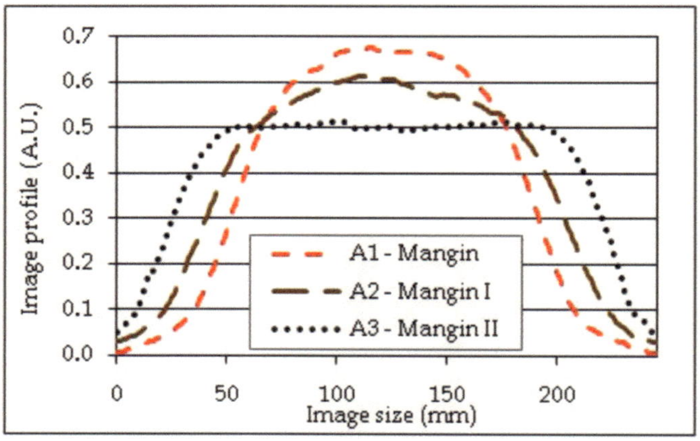

Fig. 4. Profiles of images for A1, A2, A3.

Standard Mangin A1 was realised in Glass in two versions: with EPD and *f* 40mm (Mangin40) or with EPD and *f* 60mm (Mangin60). The enlarged diameter of Mangin60 had the aim of improving the collected energy with respect to Mangin40. Modified Mangin A2 and A3 are more complicated to be realised, but the same performance level can be obtained by a standard Mangin of EPD 60mm. Thus Mangin60 represents a trade-off between A3 with EPD 62mm and A1. The manufacturing cost for the realisation of Mangin60 is low because of the traditional work to obtain spherical surfaces: a sample of Mangin60 is shown Fig. 5. For Mangin and Parabolic collectors the optical fibre is placed very close to the secondary mirror by means of a support crossing the collector in its central part, as Fig. 6 illustrates.

Fig. 5. Collector A1 (Mangin60) realized in glass.

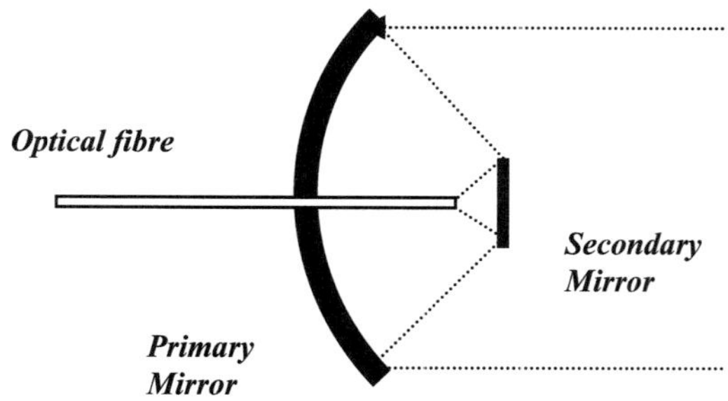

Fig. 6. Fibre position for Mangin and Parabolic collectors.

2.2 Parabolic collectors

The parabolic system is composed of two mirrors, the primary is parabolic and the secondary is flat, with or without correction lens: B1 and B2, respectively. The position of the secondary mirror is analogous of the corresponding Mangin layouts A1 and A3 shown in Figures 3a and 3c.

Figures 7a and 7b present the standard views of the optical designs for B1 and B2. To provide a more realistic view of B2, Fig. 7c presents a three-dimensional model of the Parabolic with lens.

The optical parameters are summarized in Tab. 2 of Sect. 3.1, but for B1 the maximum reachable EPD is 70mm, *f* is 65mm (at a wavelength of 580nm), f/# is 0.93 and the maximum axial angle θ is 28.3°; while for B2 EPD is 71.1mm, *f* is 65mm, f/# is 0.914 and θ=28.67°.

The optical quality of B1 is very good; the total spot diameter is 0.610 mm with a very high uniformity of light distribution, as shown by the image profiles in Fig. 8. It is useful to remind that a parabolic mirror is free from on-axis aberrations (spherical and chromatic), then the spot diagram for the on-axis field is exactly a point (if we do not consider higher order aberrations). In B2 an additional correction lens allows an improvement of the quality of this system, decreasing the off-axis aberrations (coma and astigmatism) and the total spot diameter results 0.598 mm. For B1 and B2 Fig. 8 compares the mean image profiles along a central image line. The figure evidences that the images of B1 and B2 are larger than those of the three Mangin collectors in Fig. 4.

B1 was realised by an Italian firm (SILO) as Glass parabolic mirror built by traditional optical working: EPD is 70mm and *f* is 65mm for the SILO parabolic mirror. A B2 sample was acquired among the commercial products available on the market, selecting the Aluminium Diamond turned parabolic mirror (by Advance - Coherent) with EPD=71.1mm and *f*=65mm.

The useful features of the parabolic layouts are the optimised optical parameters, which create high collection efficiency, combined to a large enter pupil, giving a wide effective area. The output power, analysed in Sect. 3, depends on both efficiency factor and effective area. Collectors B1 and B2 represent the best solutions optimising these two quantities.

2.3 Conic Cassegrain collector

The optical project of the Catadioptric Concentrator Monoblock (CCM) was developed with the aim of optimising the optical characteristics of the collector but also its compactness. In Tables 1 and 2 it is classified as conic Cassegrain and indicated as C1. The first surface is elliptic and the second one is spherical. The maximum EPD is 56mm, *f* is 55mm and f/# is 0.98. As for the previous collectors, C1 fulfils the fibre-matching requirements: the output

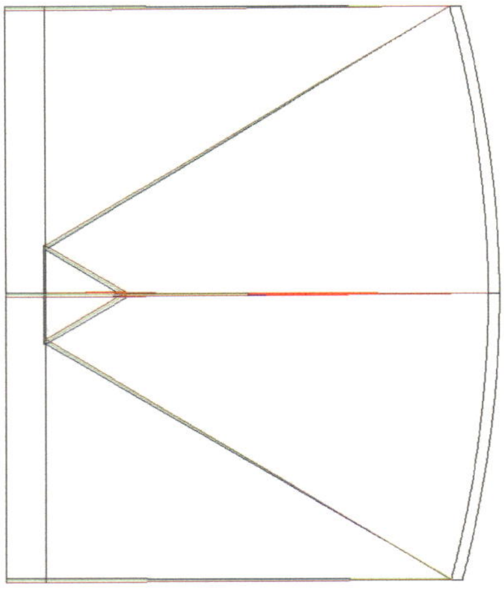

(a)

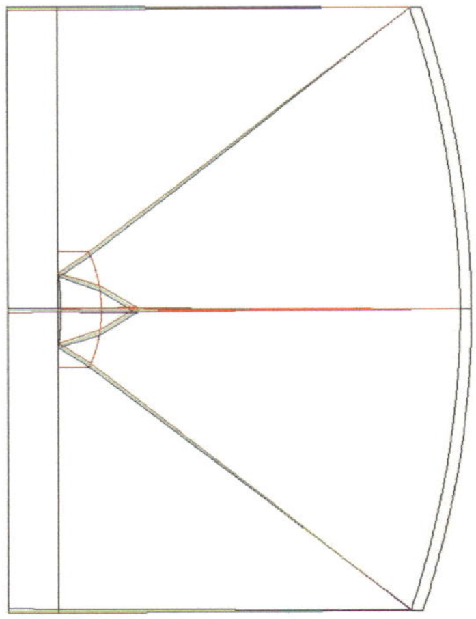

(b)

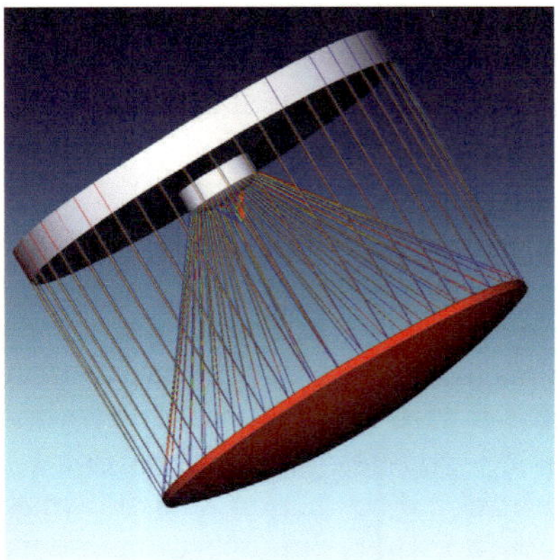

(c)

Fig. 7. (a) Zemax optical design of Parabolic B1. (b) Zemax optical design of Parabolic B2. (c) Zemax 3D model of Parabolic B2.

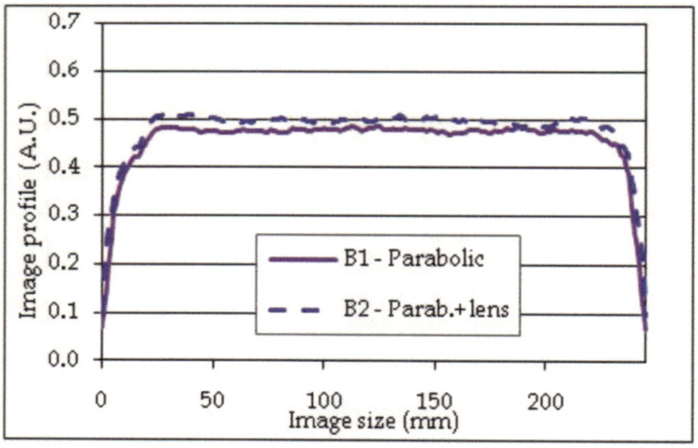

Fig. 8. Profiles of B1 and B2 images.

angle θ=26.98° is within the fibre acceptance angle (28.7°) and the total image diameter (0.534mm) is considerably shorter that the fibre size (0.6mm).
The optical working principle of this collector is well known: it consists of two optical elements in a coaxial configuration of Cassegrain type. The C1 scheme is presented in Fig. 9; while Fig.10a reports the optical design and Fig.10b a 3D model. The characteristics of C1 seemed to be particularly useful for fibre-coupling and the realisation procedure appeared to be innovative.
C1 was realised in a unique piece of quartz (see Sect. 3.3), obtaining an objective characterised by extremely reduced dimensions and great mechanical stability. The input surface is flat and its central part was optically worked to obtain the spherical surface of the secondary mirror. The primary mirror is aspherical and in its centre there is a non-aluminised flat zone, which represents the output surface. The crossing of the glass constituting the concentrator along its entire internal optical path contributes to compensate the spherical aberration of the mirrors and it allows correcting the aberrations of a primary mirror with ellipsoid shape. The internal path within the material introduces acceptable chromatic aberrations.
Successively some samples of CCM were realised in PMMA ((polymethylmethacrylate), reducing weight and realisation costs (see Sect. 3.3).
The image profile of C1 is compared with those of B2 and A3 in Fig. 11 showing that the C1 image is larger than A3 (the largest Mangin image) but considerably smaller with respect to B2 (the largest Paraboloid image). The optical quality of this system is good, but lower with respect to the quality of the two parabolic ones (see Tab. 2 in Sect. 3.1).
The main advantages of C1 are immediately visible comparing the layouts of the six collectors, the CCM thickness (25mm) is considerably shorter and, being a monoblock, it is easier to be mounted and aligned. These characteristics of the CCM are a fundamental advantage since the device is modular and it is supposed to incorporate several collectors in each tile (illustrated in Sect. 4).

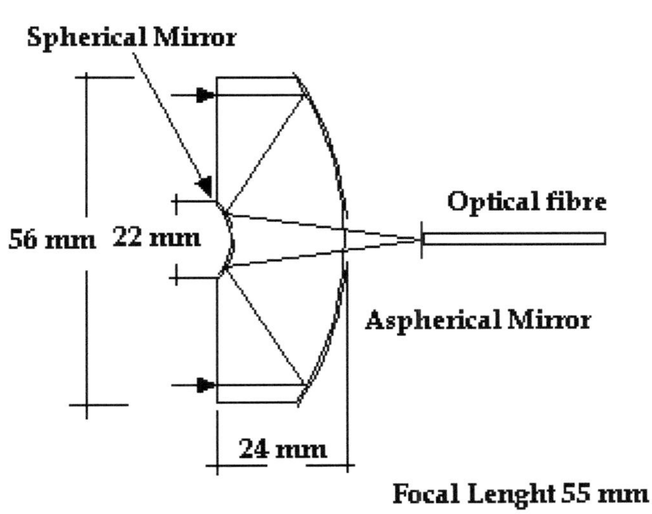

Fig. 9. Layout of CCM: collector C1.

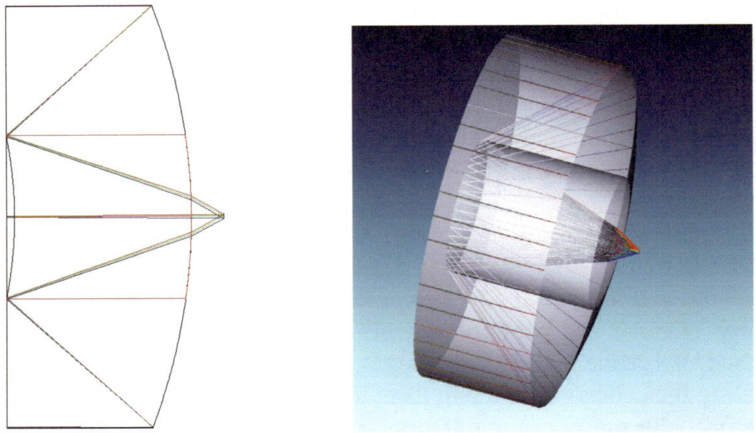

(a) (b)

Fig. 10. (a) Zemax optical design of conic Cassegrain C1. (b) Zemax 3D model of conic Cassegrain C1.

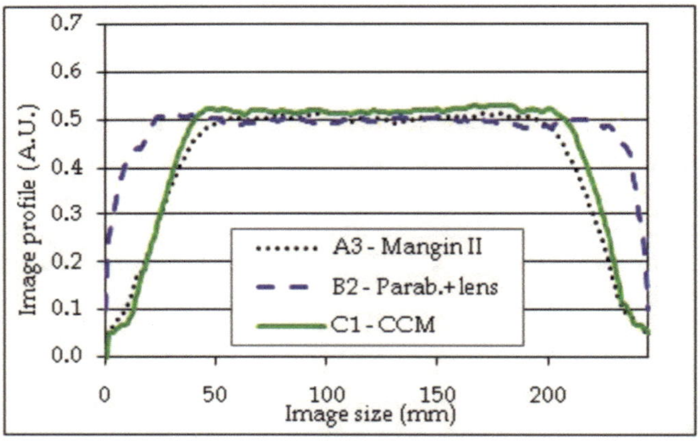

Fig. 11. Profile of C1 image, compared to B2 and A3 profiles.

3. Theoretical and experimental comparison of the collectors

3.1 Theoretical efficiency of collectors

Our modular device was composed of units, each of which consisted of a concentrator coupled to an optical fibre that transported the power to the utilisation point. Characteristics and performance of the six optical projects were theoretically estimated, calculating in particular the collection efficiency of each collector coupled to a 5m quartz fibre of size 0.6mm and NA=0.48. Table 2 summarises these optical features to compare the concentrators presented in Sect. 2.

Table 2 reports the power collection performance for the six concentrators of Tab. 1: the collector efficiency factor is the ratio output power / input power, while the total efficiency factor corresponds to the collector coupled to the fibre. The effective area is the surface effectively exposed to the sun and it was used to calculate the output power theoretically obtained at fibre end. In this output power calculation the total efficiency factor takes into

1	2	3	4	5	6	7	8	9	10	11
Layout	Max. enter pupil diam. (mm)	Focal length (mm)	f/# number	Image size (mm)	Rms spot diam. (mm)	Total spot size (5+6) (mm)	Collector efficiency factor	Total efficiency factor	Effective area (mm²)	Output power (W)
A1 - STANDARD MANGIN (M1 Spherical, M2 Flat)										
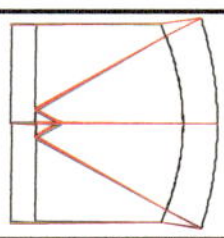	40.0	39.8	0.99	0.350	0.163	0.513	0.73	0.64	1230	0.677
A2 - MODIFIED MANGIN I (M1 Spherical, M2 Spherical)										
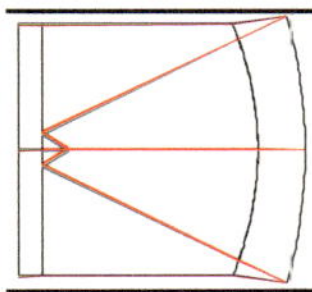	52.0	48.0	0.923	0.418	0.180	0.598	0.73	0.64	2085	1.147
A3 -MODIFIED MANGIN II (M1 Spherical, M2 Allum. Spher. Lens)										
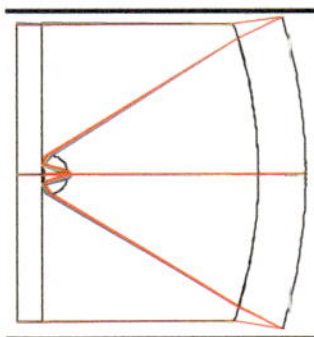	62.0	58.1	0.937	0.500	0.094	0.594	0.66	0.58	3004	1.498
B1 - PARABOLIC (M1 Parabolic, M2 Flat)										
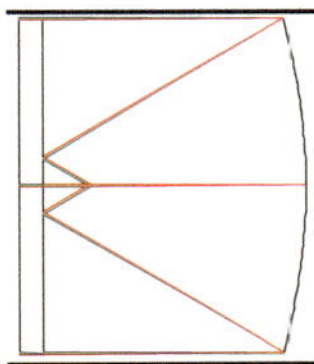	70.0	65.0	0.93	0.567	0.043	0.610	0.81	0.71	3732	2.279
B2 - PARABOLIC WITH LENS (M1 Parabolic, M2 Allum. Spher. Lens)										
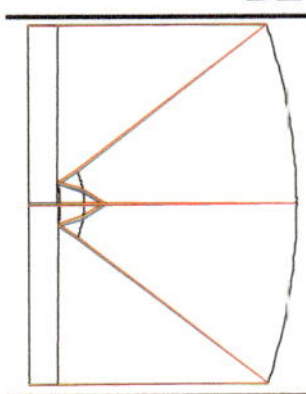	71.1	65.0	0.914	0.560	0.038	0.598	0.73	0.64	3909	2.152
C1 - CONIC CASSEGRAIN CCM (M1 Elliptic, M2 Spherical)										
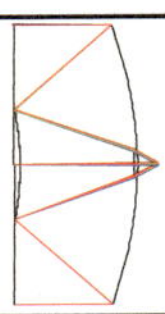	56.0	55.0	0.98	0.480	0.054	0.534	0.73	0.64	2082	1.146

Table 2. Collectors characteristics and sunlight collection performance.

account all losses of the collector-fibre system. Data in Table 2 were evaluated considering the following efficiency factors: 0.9 per mirror surface, 0.95 per glass surface, 0.97 for the 5m quartz fibre. The effect of the protective glass that should cover the whole tile of collectors was not taken into account in this table. The input power density value considered was

860 W/m^2, which is approximately the solar irradiance arriving on the Earth surface in the best conditions at a latitudes around 45°N. While the standard value of the irradiance outside the atmosphere is 1367 W/m^2.

A crucial parameter in this comparative table is the estimation of the final power obtained at the end of the optical fibre coupled to each collector. The highest output power values correspond to the two Parabolic B1 and B2, followed by Mangin A3 power value. C1 and A2 reach output power levels slightly lower than Mangin A3, but they are still better than A1. Even though the C1 image appears to be as good as those of B1 and B2, the output power obtained by the CCM does not reach the parabolic values. The advantage of the parabolic optical projects is that they combine high collection efficiency with wide effective area: they optimise these two quantities maximising the output power. The most compact optical layout is C1 that allows obtaining high power concentration with reduced aberrations. Table 2 shows that the rms spot of aberrations (lateral dashed spot in Fig. 1) is significantly reduced for C1, B1 and B2, with respect to all Mangin collectors.

In conclusion if the main purpose is to maximise the output power the parabolic collectors will be preferred. The CCM will be chosen if we privilege the compactness and the facility to be mounted and aligned. By a commercial point of view, if the main requirement is to have a collector easy to be realised at a competitive price the Mangin collector will be the better solution. On the base of the optical projects described in Sect. 2 and compared in this section, samples of all collectors were practically realised. The results of their experimental tests are reported in the next section.

3.2 Experimental tests on realised samples

Several samples of concentrators in Tab. 1 were realised as discussed in Sect. 2 and we experimentally measured the performance of the realised collectors coupled to an optical fibre. The experimentation included laboratory tests and field measurements with direct explosion to the sun. The tests in laboratory were performed using an optical system reproducing the solar divergence [14]; while the field tests require the employment of tracking systems to correctly orient the concentrators (discussed in Sect. 4).

Only B1 of Tab. 2 was available as commercial product, while the other samples were optically manufactured by Italian firms on the base of the optical designs in Sect. 2. The results for the collection efficiency, which is the ratio between the power measured at fibre end and the power arriving on the collector effective area, are reported in Table 3. The experimental values were measured in Florence (Italy) during the month of November between 11.30 to 12.30 AM.

Taking into account efficiency results, realisation costs and optical system compactness, we decided to extensively test in real conditions only Mangin60 and CCM. Field tests were repeated in all seasons of the year and at different hours during the day. The average output power measured at the end of a 5m quartz fibre was 0.80÷0.85 W for Mangin60 and 0.95÷1.05 W for CCM in Silica. The CCM realised in PMMA (with EPD=56mm, f=53.7 mm and effective area 2082mm^2) provided an output power of 0.80÷0.90 W, thus resulting slightly less efficient than the CCM in Silica. For the plastic realisation of C1 the lower performance is both due to the use of PMMA instead of Silica (quartz) and to the more precise optical manufacturing of the Silica C1. Beyond the reduced thickness, easy mounting and easy aligning, the C1 in PMMA presents a further advantage: it is lighter and cheaper than C1 in Silica. The tests evidenced that the drawbacks of Silica C1 are heaviness and

label	Collector	EPD(mm), f(mm), effective area (mm^2)	Theoretical efficiency at fibre end (Tab.2 col.9)	Lab. meas. of efficiency at fibre end	Field meas. of efficiency at fibre end
A1	Mangin 60 in Glass	60.0, 60.0, 2672	0.64	0.41	0.33
A1	Mangin 40 in Glass	40.0, 40.0, 1230	0.64	0.41	---
B1	Paraboloid in Glass	70.0, 65.0, 3732	0.71	0.39	0.31
B2	Paraboloid (commercial)	71.1, 65.0, 3909	0.64	0.32	0.33
C1	CCM in Quartz	56.0, 55.0, 2082	0.64	0.54	0.53

Table 3. Experimental efficiency of sunlight collection.

difficulties to realise the aspherical surface and obviously the very high cost for this realisation. The CCM in PMMA reduces both weight and costs still providing good collection efficiency.

It is useful to note that the obtained output power is in sunlight, whose luminous efficacy is much higher than that of electric lighting. The luminous efficacy is 70÷105 lm/W for direct sunlight and 110÷130 lm/W for diffuse skylight [21]; while for an incandescent lamp it is 10÷18 lm/W [22]. Hence the application to internal illumination exploits also the elevated luminous efficacy of solar light.

3.3 Plastic collector development

The successive step was the development of plastic optical components, which reduced the realization costs. Collector C1 was realized both in Silica (quartz) and in PMMA (plastic); it is important to note that the optical project of C1 ought to be redesigned for the realization

1	2	3	4	5	6	7	8	9	10	11
Layout and realisation	Max. enter pupil diam. (mm)	Focal length (mm)	f/# number	Image size (mm)	Rms spot diam. (mm)	Total spot size (5+6) (mm)	Collector efficiency factor	Total efficiency factor	Effective area (mm^2)	Output power (W)
C1 in Fused Silica	*56.0*	*55.0*	*0.98*	*0.480*	*0.054*	*0.534*	*0.73*	*0.64*	*2082*	*1.15*
C1 realised in Quartz	56.0	55	0.98	---	---	0.8	0.66	0.54	2082	0.97
C2 in PMMA (plastic)	*56.0*	*53.7*	*0.96*	*0.5*	*0.075*	*0.075*	*0.73*	*0.64*	*2082*	*1.10*
C2 realised in PMMA	56.0	54	0.96	---	---	---	---	---	2082	0.85

Table 4. Features and collection performance of Cassegrain CCM collectors.

in PMMA (polymethylmethacrylate). Table 4 summarises optical parameters and collection performance of conic Cassegrain CCM collectors: it compares two theoretical optical projects, C1 in Fused Silica and C2 in PMMA (*in italics*), to the CCM manufactured in quartz and plastic. The measurements on C1 realised in quartz were performed in laboratory and inside a hole of 0.6mm diameter there was 84% of the total energy in the focal plane. The output power for C2 realised in PMMA is the average of the values obtained in the field tests.

Two CCM samples are presented in Fig. 14: Fig. 14a shows a C1 in Silica, Fig. 14b a C2 in PMMA. The optical performance of the two components was comparable, but the weight was considerably reduced for the plastic optics. The work proceeded modifying the optical project of collector C2 for being mass produced and for being coupled to a larger optical fibre (with core diameter 1.2mm and *NA*=0.48). The final component was an aspherical lens in PMMA (described in Sect. 5.1) with much reduced width (14.9mm) and weight (24g) compared to the C1 in quartz (width 23.8mm, weight 136g).

(a) (b)

Fig. 14. (a) Cassegrain collector C1 realized in quartz. (b) Cassegrain collector C2 realized in plastic.

A further improvement was the use of plastic (or glass) fibre bundles instead of quartz single fibres. Considering the coupling to the aspherical lens in PMMA, the chosen value for the core diameter of the single fibre was 1.5mm, to take into account the spot enlargement and to facilitate the alignment. For the museum application to the illumination of several very large showcases (3m X 5m), requiring a large number of lighting terminations, fibre bundles were preferred. Optical fibre bundles made in plastic are very flexible, almost unbreakable and considerably cheaper than quartz fibre bundles, especially for a diameter of 1.5mm. Moreover for such a large core diameter a plastic fibre has extremely shorter bend radius with respect to a quartz fibre (as discussed in Sect. 5.4).

4. Sun tracking systems to support and orient the collectors

4.1 Sun tracking method

The sun tracking technique was studied [15], experimented and tested under working conditions. Suitable mechanical systems, to support and move the optical system, were

designed and built. The movements to align the optical collectors in the sun direction were performed in two directions by using an equatorial configuration: the directions being on the temporal axis and on the declination axis.
The methodology to track the sun position employed a double guiding system that uses two complementary procedures. The first one provides the preliminary orientation, then the second realises the fine positioning and adjustments. The first tracking system is of a passive type and drives the motors to correctly orient the collectors every day of the year. The second one is of a dynamic type and employs an optical pointing system. The core of this active tracking system is the sun pointer, which works as a double pinhole camera [16]. The pointer has two sensors, with decreasing field of view, that are used in sequence, improving the precision of the sun tracking. The system tracks the sun position with an angular precision higher than 0.1°. It is reliable and adaptable to all weather conditions and environmental variations. In the case of sun shading or temporary sun absence, the system provides a realignment of collectors in a few seconds. Furthermore, it is able to compensate for possible errors in the positioning of the device, which should be placed with the temporal axis parallel to the Earth's axis.
The solar collecting device is modular and its basic unit is a tile holding four concentrators. The tiles are mounted on a support whose orientation follows the sun position using a tracking system. This modular solar plant can be placed either on the roof or in the playground of a building.

4.2 The tile with 4 Mangins and 4 Cassegrains

Mangin and CCM Cassegrain concentrators were mounted on tiles of four collectors, as shown in Fig. 12a and 12b, respectively. The use of a small tile facilitates the alignment operations and improves the possibility of its massive reproduction, so the tile dimensions were 14cm x 14cm. Reduced size of the tile and system geometry makes it adaptable to the available space and to specific architectural requirements.

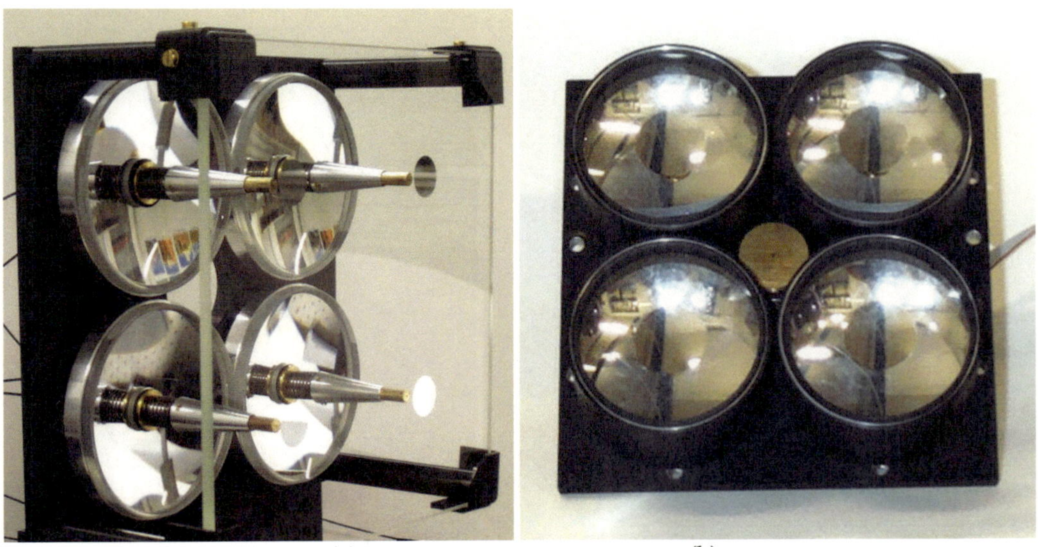

(a) (b)

Fig. 12. (a) Tile of 4 Mangins.(b) Tile of 4 CCM Cassegrains.

Fig. 12a presents the tile with 4 Mangin60 collectors made of glass. The housing for the single collector is a metallic support of diameter 64 mm, which also holds the mirror. The fibre holder with its focusing adjustment is placed in the centre of the metallic support. The secondary mirrors are realised by evaporated aluminium on the input window covering the four collectors; consequently the Mangin alignment is more difficult than for CCM.
The tile holding 4 CCM collectors made of quartz is shown in Fig. 12b. The fibre adjustments mechanism is placed in the rear part of the tile, behind the collector. The sensor driving the sun tracking system is visible in the centre of the tile in Fig 12b.
Both tiles had the same external dimensions, so they were exchangeable if they were mounted inside the same moving support.

4.3 Motorised frames holding 1 tile or 9 tiles

Suitable mechanical and electronic systems were developed to support and move the optics. The equatorial structure included two motors, one for the temporal axis and one for the declination axis. Optical collector movement was performed by a double guiding system: preliminary orientation by a passive system driving the motors to correctly orient the optics on a daily basis, then fine positioning and adjustments using an optical sun pointer. These electro-mechanical structures were necessary to perform field test, with direct exposition to the sun of the collectors. To test in operative conditions the realised collectors, two different moving frames were realised. The first (Frame A) supported a single tile of four collectors: in Fig. 13a it mounts Cassegrain C1 collectors. It includes the sun tracking system and two micro motors, which keep the tile aligned in the sun light direction. The second configuration (Frame B) contained 9 tiles in a larger frame that can contain 36 collectors: Fig. 13b shows the frame with Mangin collectors. The dimensions are for Frame A: 16cm x 24cm; and for Frame B: 60cm x 60cm. The tiles in Fig. 12a and in Fig. 12b are interchangeable, so they can be mounted on both frames.
The most interesting feature of these frames is their self-alignment capability.

(a)

(b)

Fig. 13. (a) Frame A testing C1 collectors. (b) Frame B testing Mangin A1 collectors.

5. The application to internal museum illumination

The possible applications of this innovative solar device were room illumination, water heating or energy supply for domestic devices. For this latter use a photovoltaic panel provides the conversion into electric energy, which can be stored for later use. However the most promising application of these systems appeared to be the illumination of buildings, in particular for artistic purposes such as in museums or in special uses requiring solar light (colour rendering and colour recognition). In these cases the low power level reached by the system is an advantage, because the lighting of artworks has illuminance restrictions for their conservation. On the other hand direct solar lighting has an additional value in museum exhibitions, since it improves the colour rendering of the exposed objects, and it is necessary for colour identification in industrial production, which is fundamental for example for fabric.

Hence our original system for sunlight collection was applied to internal lighting and in particular it was adapted to illuminate museum interiors [10]. It was developed from optical and electro-mechanical design to production, installation and testing in working conditions. The device included solar collectors, optical fibres, photovoltaic cells, mechanical and electronic systems for sun tracking. The optical collector was studied with comparative ray tracing analyses and experimental tests to optimise the optical configuration. Then the optical design was adapted to plastic component production to reduce costs. The final concentrator was a narrow and lightweight aspherical lens made of plastic. To provide illumination, the solar collector is coupled to a fibre bundle. For energy supply the sunlight is concentrated on a photovoltaic cell that converts the light into electric energy for utilization or storage. In addition the museum required an alternative light source for the case of sun absence, which was realised by low power consumption LEDs (Light Emitting Diodes).

A plant demonstrator was successfully installed in a Florentine museum, where it was developed and adjusted to illuminate the interior of large showcases with front size of length 5m and height 3m. Figure 14 presents a solar collecting device of the museum plant. The installation consisted in two groups of devices placed in two separated locations: on the museum roof, shown in Fig. 15a, and on the museum garden, shown in Fig. 15b.

Fig. 14. A device of the museum plant demonstrator.

(a) (b)

Fig. 15. (a) The installation on the museum roof. (b) The installation on the museum garden.

Collector samples were tested to assess collection performance; they also endured optical tests in operative conditions. Plastic and glass fibres were preferred to quartz fibres due to the difficulties of museum installation. Therefore the long fibre bundles represented the main cause of energetic losses, but some light can also be lost in an imprecise collector-fibre coupling. Fibre bundles transported the light inside the showcases realizing the lighting points, which were suitably distributed to maximize illumination uniformity. Appropriate filters, selected to achieve colour balance, were mounted on plastic fibre and LEDs. The museum experts indicated the correct illuminance levels, taking into account the recommendations of the International Council of Museum [19] for the exposed objects: they were basically weapons, armatures and metallic objects; the more fragile exhibit items were costumes and textiles. The result was a quite uniform lighting separately obtained by optical fibres or LEDs: both illuminations fulfilled illuminance equivalence and illuminance level requirements.

5.1 Collectors and fibres for the museum plant

For more than a decade our laboratory has studied sunlight collectors comparing the optical performance of different configurations to concentrate sunlight into an optical fibre [12-14]. Collection efficiency assessment and optical characterisation were carried out on samples of Paraboloids, Mangin collectors and a specially designed concentrator called Catadioptric Concentrator Monoblock (CCM). The Cassegrain collector CCM, presented in Sect. 2.3, was realised in quartz (Silica) and in plastic (polymethylmethacrylate PMMA): the plastic version has considerably reduced dimension, weight and cost, maintaining good collection efficiency.

Finally, to obtain the collector for the museum demonstrator, the optical project of the plastic CCM was adapted to provide collectors that could be mass produced, so allowing a significant reduction in cost. The result was a narrow and lightweight aspherical lens made of PMMA: Fig. 16 shows a sample. Figure 17 compares the optical project of the aspherical lens in PMMA, designed for coupling with a plastic fibre of core diameter 1.2mm, to the optical project of the Conic Cassegrain in PMMA (collector C2), designed for coupling to a quartz fibre of core diameter 0.6mm. The comparison evidences that the main effects of the adaptation of the optical design were to simplify the optical surfaces and to lengthen the focal distance. In the optical design of the plastic aspherical lens the focal distance is 54.5mm, the entrance pupil diameter is 55mm, the numerical aperture is 0.48 and the thickness is 14.9mm.

Fig. 16. A sample of the production of aspherical lenses in PMMA.

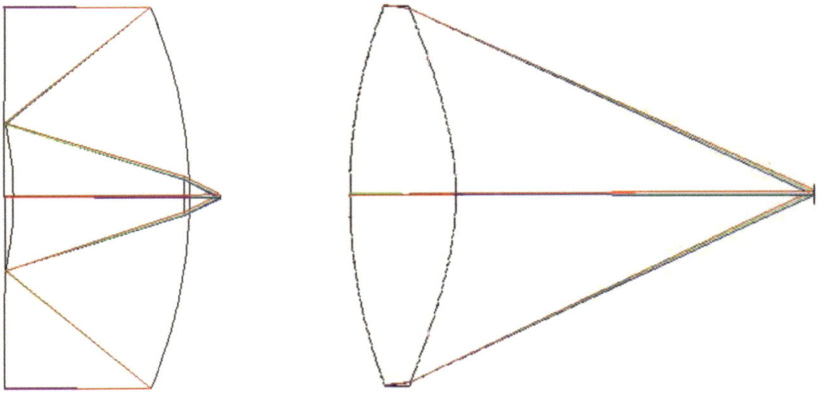

Fig. 17. Comparison of the optical designs: Cassegrain in PMMA (left) and aspherical lens in PMMA (right).

A first production of this plastic collector was realised for the museum installation. Tests were carried out on lens production to evaluate collection efficiency and image size (in Sect. 5.2), examine optical treatments and experiment the effects of external agents and UV exposure (in Sect. 5.3). The optical parameters, measured on randomly selected samples, presented extremely reduced standard deviations, confirming production homogeneity and reproducibility of lens fabrication process.

To enhance the collection efficiency the lenses had an anti-reflection treatment on both sides: the tests, repeated on the treated lenses, verified the expected performance improvement. The average collection efficiency E of the coated lenses was 98% and in addition the treatment reduced the E standard deviation. The features experimentally measured on the production samples were: mean focal length of 62.4mm, diameter of 55mm and width of 14.9mm.

The lenses were exposed to the external atmospheric agents to test the endurance of the treatment; then to simulate the operative conditions, the samples experienced an artificial Ultra-Violet irradiation, using a UV lamp. The most disturbing effect was water condensation, basically due to humidity and thermal excursion; while the effect of UV exposure was negligible on the lens spectrum. So the anti-reflection treatment lasted on the lenses, which maintained a very high optical quality with more than 95% of collection efficiency.
Each solar lens was coupled to a polymeric fibre bundle of length 30m, which had seven terminations, at its other extreme, arriving into the showcases. These terminations were optical fibres of core diameter 1.5mm and *NA*=0.48: they represented the actual lighting points, suitably arranged within the museum showcases, as discussed in Sect. 5.6. Fibre bundles were preferred to single fibre because the higher number of terminations allows illuminating the large showcases with more uniformly distributed light. The number of optical fibres in a bundle is defined by the circular geometry necessary to realise the bundle: 7 is the first possible fibre number after 1. Optical features of plastic and glass fibres are compared in Sect. 5.4, where the choice of plastic fibres is justified. For the specific application of museum lighting experts in this area suggested the use yellow-orange light: this aspect is discussed in Sect. 5.5.

5.2 Light collection and image size of the plastic lens

Sunlight collection efficiency was assessed on twenty samples of plastic collector production, randomly selected from available lenses. The optical tests were performed on a white light collimator, which reproduced the solar light divergence, examining collection efficiency E and focal distance f. The efficiency of sunlight collection [14] is measured as a ratio between the light focused within the nominal image and the light entering in the lens EPD (enter pupil diameter). The nominal image of the lens is the image obtained from a ray tracing simulation using the optical project of the lens: it had approximate diameter 1.2mm. The E values are obtained with a possible error of ±2%. The tests were addressed to verify the homogeneity of the plastic lens production and reproducibility of the lens fabrication process. The average focal length was 62.4mm with a standard deviation of 0.28mm.
The plastic lenses were treated on both faces with an anti-reflection coating to improve efficiency of sunlight collection. Measurements of f and E were repeated on the treated lenses and test results confirmed the expected improvement in collection efficiency values. The treated lenses reached an E value of 99%, with an average value of 98%; while the lenses without treatment reached 93%, with a mean value of 91%. But the more interesting result is that the anti-reflection treatment reduces the standard deviation of the E data from 1.3 to 0.58.
For focalization into a fibre it is useful to measure image spot dimension and light distribution within the focused image. The optical set-up for this image control included a white light source reproducing the solar divergence. The beam impinged on the tested lens and a detection system was located in its image plane. It consisted of a photo-detector combined with a multi-hole mask, with hole diameters Ø from 0.8 mm to 1.6 mm. Each hole of the mask acted as a spatial filter on the image. The mask was moved in front of the detector to measure the light corresponding to each hole. The light was measured by the current generated by the detector. The data for filtered images were compared to the data for unfiltered image. The results of filtered image light were relative measurements, expressed as a ratio with respect to the total light flux in the focal plane.
This image analysis verified that the lens production was homogeneous in this aspect. The filtered image light, expressed as a percentage, was: 95% (Ø = 1.2mm); 98% (Ø = 1.3mm);

99% (Ø = 1.5mm). The diameter 1.2mm corresponds to the nominal image of the PMMA lens. While the value 1.5mm corresponds to the core diameter of the optical fibre selected for the museum plant.

5.3 Plastic lens exposure to atmospheric agents and UV irradiation

The solar lens with anti-reflection coating was tested in various atmospheric conditions to estimate the endurance of the treatment. An obvious negative effect was the deposit of raindrops or dust on the upper lens surface. However, the most disturbing effect appeared to be water condensation on the lower lens surface, occurring in sunny weather or raining conditions, because it is basically due to moisture presence and thermal changes between night and day. The collection efficiency E was monitored during the first ten weeks of exposure to the atmosphere. E was measured every week before and after cleaning the lens, since there were water droplets on one or both lens surfaces. Without cleaning, E decreased from 97% to 93% in three weeks, then E fluctuated between 91% and 76%, with an average value of 83%. The measurements on the lens covered by droplets and moisture had very low reproducibility and reliability. The results for cleaned lenses showed a very slight decreasing trend during the ten testing weeks, with E going from 97% to 95%.

Another essential analysis for the anti-reflection treated lenses is the exposure to ultra-violet radiation. The plastic used to manufacture the lenses is PMMA (polymethylmethacrylate) and UV irradiation typically induces a colour variation from transparent to yellow in this material. The anti-reflection treatment on the PMMA lens should reduce this degradation effect. A test was performed with a continuous exposure to an UV lamp for several weeks. It was found by measuring lamp and sun emissions with the same power meter in the same spectral range that UV laboratory radiation corresponds to about three months of sunlight exposure. The measurements did not show evidence of colour variations in the plastic lens material: comparison of the spectral transmission (in the wavelength range 200nm to 700nm) measured before and after UV irradiation indicated the difference was within the experimental error of the spectrometer.

In conclusion the collection efficiency of coated lenses was mostly affected by moisture presence and thermal changes between night and day. However these are only test examples and preliminary analyses, aimed at estimating how the operating conditions affect the solar collector efficiency. It would be useful to perform more detailed tests to correlate each single element (temperature, moisture, cloud cover amount, wind speed, spectral range of the radiation, etc) to the deterioration of performance of the plastic lens.

5.4 Fibre selection and tests for the museum installation

The optical coupling between solar lens and fibre bundle was one of the major causes of energy losses in the whole optical system of the museum demonstrator. The focused light should be received by the fibre with the maximum optical coupling obtained in the best alignment condition. As regards the core diameter, we finally chose the value of 1.5mm, to take into account the spot enlargement and to facilitate the alignment.

Another important cause of losses was the absorption in the optical fibre, which basically depends on material, diameter and length of the fibre. The fibre bundles should be characterised by high performance as regards spectral transmission. Typical materials for optical fibre production are quartz, glass and plastic. Silica has very good light transmission, but it is expensive, especially for bundle production, and quartz fibres are very fragile and rigid. Glass fibres have a light attenuation higher than silica fibres, but they are considerably cheaper and more flexible, which is a fundamental advantage. Generally plastic is the preferred material to make fibres bundles, since it facilitates production and plastic fibre bundles are inexpensive, almost unbreakable and extremely flexible. In particular they have a bend radius of few centimetres for a fibre diameter of 1.5mm, while a silica fibre of the same diameter has a 900mm bend radius. Glass fibres are slightly more rigid than plastic ones, but they usually have lower transmission losses. Nevertheless an innovative plastic fibre bundle, realised in a polymeric mixture with an original composition, can reach a similar transmission performance to that of glass fibres.

For the museum installation, fibre bundles were selected in preference to a single fibre. Two fibre materials, glass and plastic, were considered by examining samples of fibre bundle with seven terminations: our samples of plastic fibres were produced by DGA (www.dga.it), while the samples of glass fibres were produced by 3M (www.3m.com). The sample of plastic fibre bundle had a single core diameter of 1.5mm and length 30m. The sample of the glass fibre bundle had a single core diameter 0.6mm and length 40m. To compare the optical performance of these two fibre types, measurements were carried out with sunlight and by analysing the illuminance at the fibre ends. These field tests examined the light transmitted by the seven terminations of the fibre bundle coupled to the plastic lens exposed to the sun. The use of the sun tracking system (in Sect. 4) is fundamental for performing these tests, because it keeps the lens in the sun's direction. The tests were performed at noontime, when the illuminance of the sunlight impinging on the demonstrator collectors was 950 lx to 1020 lx. Measurements were repeated with various sun conditions and on different days.

The illuminance obtained on the exposed object was measured at two reference distances: 50cm and 75cm. These lengths correspond to minimum and maximum distances between lighting points and exposed objects within the museum showcases. For the plastic fibre bundle the illuminance was 300 lx to 510 lx at 50cm and 150 lx to 270 lx at 75cm. The glass fibre bundle provides illuminance values of 340 lx to 560 lx at 50cm and 230 lx to 260 lx at 75cm. As seen from the results, the measurement values fluctuate during the test and it was found that they can vary even more between days and sun conditions. The final choice for the application of the museum plant was to employ polymeric fibre bundles.

5.5 Light level and colour suitable for museum illumination

The museum demonstrator employed a combination of solar light and other sources, represented by white LED with high emission levels at low supplying power (DGA product number 700001.31 "1W fixed LED gem", ref. www.dga.it). Museum object illumination has specific requirements on illuminance levels, light colour and light distribution uniformity. The first task was to reach a mean illuminance of 100÷120 lx, with the uniformity of light distribution being maximised within the showcases. The second task was the colorimetric equivalence between LED and fibre illumination. The third task was to obtain a yellow-orange colour. This section is devoted to photometric analyses and colour studies on the three light categories: sunlight guided by glass and plastic fibres and LED emission. The purpose was to minimise the colour difference between the three illumination categories by introducing suitable filters. The aspect of illuminance values is separately examined in Sect. 5.6, since they depend on the source distribution within the showcases.

A preliminary analysis compared the spectral components of the three illumination categories. Figure 18 presents the emission spectrum of the white LED and the illuminance spectrum of the sun after passing through glass and plastic, in the visible range. They were measured using a Minolta CS1000 spectrophotometer, which examined a *Spectralon* (LabSphere™) surface illuminated by the radiation under test. The LED light was located between 420nm and 700nm and it was characterised by two isolated peaks, while the light guided by fibres presented a more continuous spectrum. Glass fibres transmitted in the whole visible range and over 800nm in the infra red region. The transmission of plastic fibres lied within 380nm and 700nm, almost covering the whole visible range. The colour temperatures were 4294 °K for glass fibres, 7982 °K for plastic fibres and 5183 °K for the white LED, whilst the Colour Rendering Index was: 95.4 for glass fibres, 67.3 for plastic fibres and 72.8 for the LED. A visual comparison of the solar illumination transmitted by the two fibre types is shown in the photo of Fig. 19: plastic fibres supplied a blue illumination, while glass fibres provided a yellow lighting.

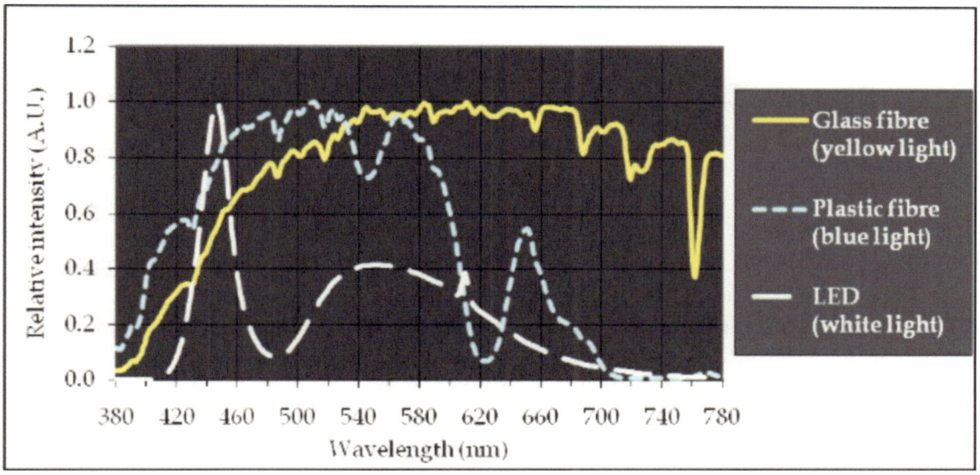

Fig. 18. Spectral comparison of the lighting using white LED, plastic and glass fibres.

Fig. 19. Visual comparison of the sunlight transmitted by plastic and glass fibres.

Glass fibre appeared to be more appropriate for obtaining the correct hue. Nevertheless for the museum installation we finally decided to use polymeric fibre bundles because they are almost unbreakable and easier for installation, owing to their very short bend radius. However, the light guided by polymeric fibre bundles required some filtering.

The introduction of filters was necessary to match colour requirements. The filters were chosen from the catalogue of *Supergel filters* produced by Rosco (www.rosco.com). The museum experts preferred the yellow hue of the light transmitted by glass fibres to the blue hue of the plastic fibre illumination. Therefore, the glass fibre light was taken as the reference for the colour matching, and filtering was used for the other two lighting categories. In addition to modifying the colour, the filter attenuated the light, thus reducing the illuminance obtained within the showcases.

The selection of suitable filters was performed on the basis of photometric tests between the three lighting categories. The scheme for Colour_Test_1, comparing Glass Fibre and LED lights, is reported in Fig. 20a; while Fig. 20b presents the scheme for Colour_Test_2, comparing Plastic Fibre and filtered LED lights.

In Colour_Test_1, the radiation guided by glass fibres represented the reference quantity. This glass fibre lighting was compared to the filtered LED emission. Spectral tests on the effect of a set of filters mounted on the LED sources individuated the filter (FILTER_L), which minimised the colour difference. In Colour_Test_2 the filtered LED illumination was considered as the reference. The comparison test was performed for the light guided by plastic fibres and the emission of LED with FILTER_L. The choice of the best filter (FILTER_F) for plastic fibres was made by testing several filters and finding the spectrum approaching the reference one.

The experimental set-up included two channels guiding the two types of radiation to be compared on two faces of a *Spectralon* cube. In front of the *Spectralon* cube a screen with a hole was positioned so that the observer, located at a suitable distance, had a view angle of 2° (*fovea* vision). For balancing the luminance, neutral filters were mounted on the two channel lights, thus facilitating the colour matching by the observer.

All tests were repeated with several different observers to obtain a preliminary selection of the most suitable filters. Then the final filter choice was made on the basis of the chromatic coordinates measured by the Minolta spectrophotometer CS1000.

The examined quantities were the chromatic coordinate (u',v') and the distance D on the (u',v') diagram: the results for the two colour tests are separately compared in Tables 5a, 5b, 6a and 6b. The criterion for selecting the optimum filter was the minimum distance between reference and filtered light. The (u',v') chromatic coordinates were preferred to the (x,y) coordinates since they appeared to be more linear. The 1976 (u',v') chromaticity diagram is significantly more uniform than the (x,y) diagram, yet it is still far from perfect. In fact in the (u',v') diagram the distance between two colour-points, in a quadratic calculation, is not rigorously correct because indistinguishable colours are included inside ellipses. However, the use of the distance between two colour-points is more correct in the (u',v')-system than in the (x,y)-system [17-18].

For Colour_Test_1, Table 5a examines the colour of LED emission and glass fibre lighting, both measured without filtering. The errors are < 1% for all quantities in Tables 5 and 6. The preliminary choice of FILTER_L was represented by filters #2 "Bastard Amber" and #304 "Pale Apricot" of the Rosco catalogue. The chromatic coordinates measured after the introduction of the proposed filters are compared in Table 5b, where filter #02 corresponded to the minimum distance on the chromaticity diagram.

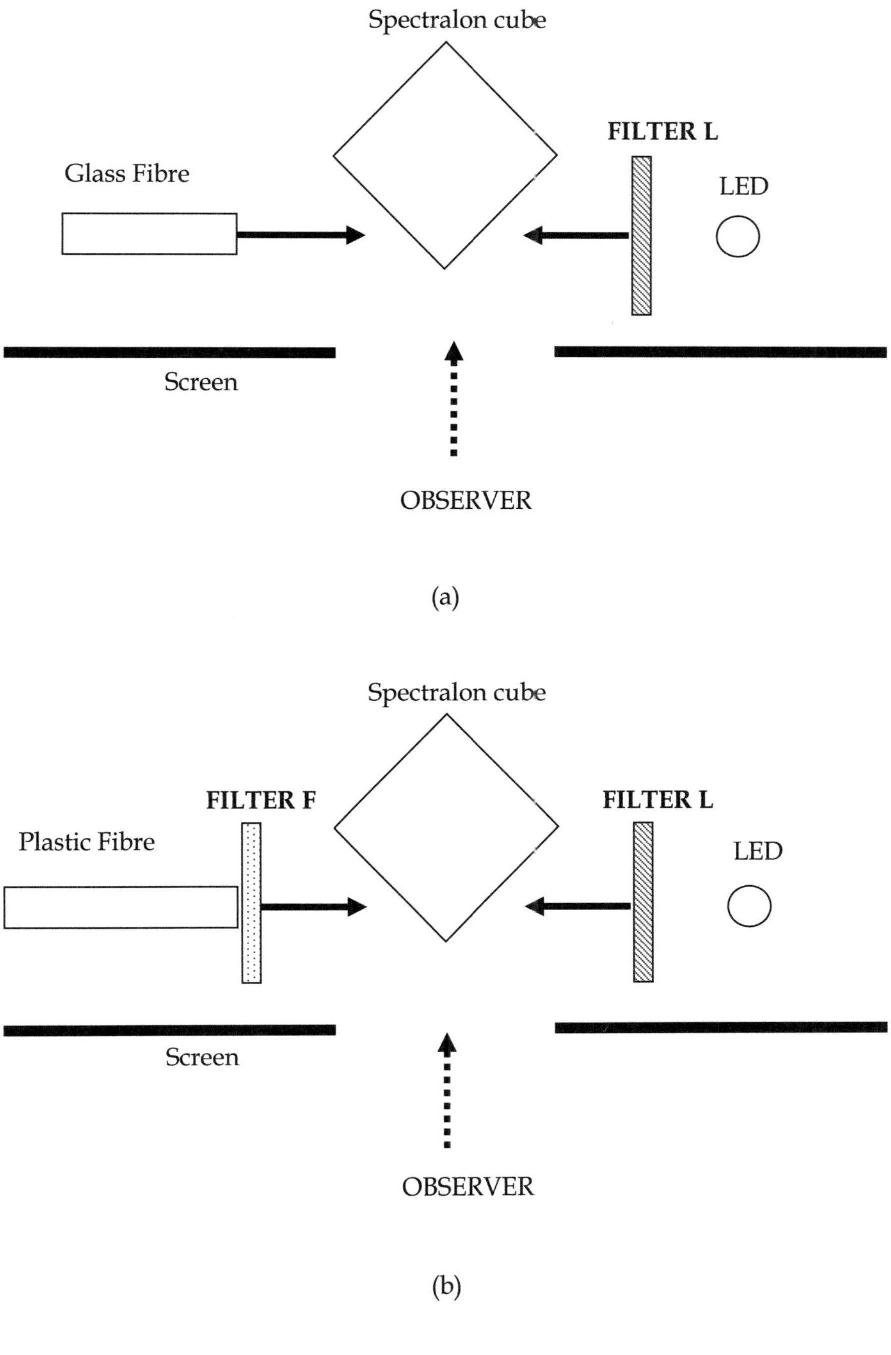

Fig. 20. (a) Set-up for Colour_Test_1 comparing Glass Fibre and LED lights. (b) Set-up for Colour_Test_2 comparing Plastic Fibre and filtered LED lights.

	u'	v'	D
Glass Optical Fibre	0.2206	0.5079	---
LED	0.2000	0.4915	0.0263

Table 5. (a) Colour_Test_1. Chromatic coordinates u'v' and distance D in the u'v' diagram for the lights before filtering.

	u'	v'	D
Glass Optical Fibre	0.2206	0.5079	---
LED + filter #02	0.2183	0.5083	0.0023
LED + filter #304	0.2226	0.5011	0.0071

Table 5. (b) Colour_Test_1. Chromatic coordinates u'v' and distance D in the u'v' diagram for the lights after filtering.

	u'	v'	D
LED + filter #02	0.2168	0.5073	---
Plastic Optical Fibre	0.1695	0.4845	0.0525

Table 6. (a) - Colour_Test_2. Chromatic coordinates u'v' and distance D in the u'v' diagram for the lights before filtering.

	u'	v'	D
LED + filter #02	0.2168	0.5073	---
Plastic Fibre + filter #03	0.2064	0.5065	0.0104
Plastic Fibre + filter #17	0.2274	0.5205	0.0169
Plastic Fibre + filter #317	0.2310	0.5236	0.0216

Table 6. (b) - Colour_Test_2. Chromatic coordinates u'v' and distance D in the u'v' diagram for the lights after filtering.

In Colour_Test_2, the light colour was measured with LED with filter #02 and on plastic fibre without a filter: Table 6a shows the results. Three possibilities for FILTER_F were identified in the Rosco catalogue: #3 "Dark Bastard Amber", #17 "Light Flame" and #317 "Apricot". Table 6a compares the chromatic coordinates measured with the possible FILTERS_F and the minimum distance *D* in the (u',v') diagram corresponded to filter #03.

Combining the results of both colour tests, it can be concluded that the nearest illumination colours were obtained by:

1. Light transmitted by glass optical fibres
2. Emission of LED with filter #02 "Bastard Amber"
3. Light guided by plastic fibres with filter #03 "Dark Bastard Amber"

5.6 Installation and validation of the museum plant demonstrator

A demonstrator of our solar collection system was installed in a prestigious museum in Florence to provide illumination inside several large showcases. The width of the showcases can be 5m or 2m, while the height is 3m. The photos of Fig. 21 present two 5m X 3m showcases: the pictures show the showcases before (left) and after (right) the installation of the solar lighting plant. The installation of the lighting terminations within the showcases was realised in the occasion of a re-styling of the exposition showcases, with displacement

Fig. 21. Two museum showcases without (left) and with (right) the internal lighting supplied by the installed solar plant.

of the shelves and consequent new arrangement of the exhibit items (particularly evident in the lower pictures). The museum plant demonstrator included two separated installations: five devices were placed on the museum roof (Fig. 15a) and four devices were located in the garden. The roof devices were devoted to supply internal illumination in a room of the museum; while the garden installation had didactic purposes.
Each device (in Fig. 14) included eight solar lenses (in Fig. 16), coupled to eight fibre bundles, each of which had seven fibre terminations. The plastic optical fibres transported the light, concentrated by the solar collectors, within the showcases realising the lighting points that are suitably distributed within the spaces to be lighted. The total number of lighting terminations was 5x8x7=280 (from 5 devices with 8 collectors each and 7 terminations in every fibre bundle).
Museum illumination had several fundamental requirements on: illuminance depending on the exhibit items; equivalence between the two lighting types (solar light and LED); light colour and uniformity. Lighting hue and colour balance have been examined in Sect. 5.5, where photometric and colorimetric measurements have determined the appropriate filters for LED emission and light guided by plastic fibres. The museum experts indicated 100÷120 lx as average illuminance required to light the showcase interior. This value took into account the illuminance levels recommended by the International Council of Museum [19-20]. The exhibition objects were basically weapons, armatures and metallic objects: items made of metal, stone and ceramic have no limits on maximum illuminance; but some exposed objects were made of leather or wood and others contained horn, bone or ivory and for these materials the illuminance limit is 150 lx. The more fragile exhibit items were costumes and textiles that should not receive illumination higher than 50 lx.
The two lighting configurations, with plastic fibres or LED, were separately estimated and practically experimented directly within the showcases to individuate the best arrangement of the lighting points. The vertical positioning of the lighting spots improved the light uniformity, with respect to the horizontal positioning. The total emission angle was about 120° for LED and around 60° for the plastic fibre (numerical aperture NA=0.48) thus the LED lighting achieved a higher distribution inside the showcases. On the other hand, fibre terminations could be orientated to maximise the uniformity of lighting distribution. The selected fibres disposition and LED arrangement fulfilled illuminance correspondence and illuminance level requirements. The illuminance measured on the showcase background resulted to be between 80 lx and 170 lx; the employed luxmeter had an error of 2% ±1 digit. The solar illuminance within the showcases obviously depended on the external sunlight irradiation, which presented daily and monthly variations. This effect introduced fluctuations in the solar illuminance provided by the fibres, but the illuminance variations were judged compatible with the requirements of museum lighting.

6. References

[1] Winston R. Light collection within the framework of geometrical optics. *J. Opt. Soc. Amer.* 60 (2), 245-247 (1970).

[2] Winston R, Minano J C, Benitez P. *Non-Imaging Optics. Optics and Photonics.* Elsevier Academic Press USA, 2005.

[3] Collares - Pereira M, Rabl A, Winston R. *Lens-mirror combinations with maximal concentration*. Applied Optics 16 (10), 2677-2683 (1977).
[4] Jenkins DG. *High-uniformity solar concentrators for photovoltaic systems*. Proc. SPIE 4446, 52-59, (2001).
[5] Luque A. *Solar cells and optics for photovoltaic concentration*. The Adam Hilger Series on Optics and Optoelectronics. Bristol and Philadelphia; ISBN 0-85274-106-5; 1989.
[6] Winston R, Goodman N B, Ignatius R, Wharton L. *Solid-dielectric compound parabolic concentrators: on their use with photovoltaic devices*. Applied Optics 15 (10), 2434-2436 (1976).
[7] Xiaohui Ning. *Three-dimensional ideal θ1/θ2 angular transformer and its uses in fiber optics*. Applied Optics 27 (19), 4126-4130 (1988).
[8] Cariou J M, Dugas J, Martin L. *Transport of Solar Power with Optical Fibres*. Solar Power 29 (5), 397-406 (1982).
[9] Liang D, Nunes Y, Monteiro L F, Monteiro M L F, Collares -Pereira M. *200W solar power delivery with optical fiber bundles*. SPIE Vol. 3139, 277-286 (1997).
[10] Sansoni P, Francini F, Fontani D, Mercatelli L, Jafrancesco D. *Indoor illumination by solar light collectors*. Lighting Res. & Technol. 40 (4), 323-332 (2008).
[11] *Solar Collectors, Power Storage and Materials*. Edited by Francis de Winter. The MIT press Cambridge, Massachusetts London ISBN 0-262-04104-9; 1991.
[12] Ciamberlini C, Francini F, Longobardi G, Piattelli M, Sansoni P. *Solar system for the exploitation of the whole collected energy*. Optics and Laser in Engineering 39 (2), 233-246 (2003).
[13] Fontani D, Francini F, Jafrancesco D, Longobardi G, Sansoni P. *Optical design and development of fibre coupled compact solar collectors*. Lighting Res. & Technol. 39 (1), 17-30 (2007).
[14] Fontani D, Francini F, Sansoni P. *Optical characterisation of solar collectors*. Optics and Laser in Engineering 45, 351-359 (2007).
[15] Fontani D, Sansoni P, Francini F, Jafrancesco D, Mercatelli L. *Sensors for sun pointing*. proceedings of WREC/WREN World Renewable Energy Congress / Network 2008, Editor A. Sayigh 2008 WREC, Glasgow - UK, 19-25 July 2008.
[16] Fontani D, Sansoni P, Francini F, Mercatelli L, Jafrancesco D. *A pinhole camera to track the sun position*. t5.1.O12, ISES Solar World Congress 2007, Beijing - China, 18-21 Sept. 2007.
[17] Wyszecki G, Stiles W S. *Color Science. Concepts and Methods. Quantitative Data and Formulae*. Second Edition A Wiley-Iterscience Publication, John Wiley and Sons Inc, New York; 1982.
[18] Y. Ohno, *CIE Fundamentals for Color Measurements*, Proc. IS&T NIP16 International Conference on Digital Printing Technologies, Vancouver, Canada, Oct. 15-20 2000: 540-545 (2000).
[19] Cuttle C. *Damage to museum objects due to light exposure*. Lighting Res. & Technol. 28 (1), 1-10 (1996).
[20] Castellini C, Cetica M, Farini A, Francini F, Sansoni P. *Dispositivo per il monitoraggio della radiazione ultravioletta e visibile in ambiente museale*. Colorimetria e Beni culturali - SIOF, atti dei convegni Firenze 1999 e Venezia 2000, 168-180 (2000).

[21] Littlefair P.J. *The luminous efficacy of daylight: a review* Lighting Res. & Technol., 17 (4), 162-182 (1985).
[22] EERE Information Centre (http://www1.eere.energy.gov/buildings/ssl/efficacy.html) of the U.S. Dept. of Energy - Energy Efficiency & Renewable Energy (EERE).

2

Photovoltaic Concentrators – Fundamentals, Applications, Market & Prospective

Andrea Antonini
CPower Srl
Italy

1. Introduction

The main obstacles for the photovoltaic energy to be competitive with standard energy sources are 3: the low efficiency, intended as low density of energy production for occupied area, the high cost of the constituting materials and the variability of the production which is correlated to the meteorological conditions.
While for the last point the solutions are related to technologies external to the PV, touching issues of grid management and distribution of solar plants, the first two issues are the aims of the PV research. One way investigated to improve the efficiency and reducing the costs is the concentrated photovoltaic (CPV); the light concentration allows higher efficiency for the cells' PV conversion and permits to replace large part of photoactive materials with cheaper components concentrating the light. Unfortunately, besides these advantages some limitations are present for the CPV too; the most evident are the necessity for the panel to be mounted on a sun tracker and the capacity to convert only the direct component of the sunlight; moreover, the reliability of the CPV systems has not yet been proofed in field for long time as for the standard PV, since this technology has achieved an industrial dimension only in the last years.
The photovoltaic concentrators spread on a large space of different possible configurations; there are concentrators with concentration factor from 2 to over 1000, there are CPV assemblies using silicon solar cells as well as using III-V semiconductors solar cells; there are CPV systems with one axis tracker as well as two axis tracker, and with different requirements on the pointing precision. All these different configurations have been developed from the first pioneer works in the '70s till the current commercial products, to find the best solutions for cost competitive solar energy.
The CPV industry is very different from that of other PVs; indeed, a CPV module or assembly is made of many components requiring high precision of mounting. So, the CPV sector appears like an hybrid between the microelectronic and the automotive industries. This possibility to derive large part of the automation necessary for medium-high volume of production from other well consolidated industrial field is an important advantage for the first assessment of CPV and an useful reference for the cost analysis of large productions.

2. Optics for concentrators

The optics for the Sun concentrators have been mostly developed during the last 30 years; the non-imaging optics, a branch of geometrical optics, has given a great contribution to the evolution of the shapes for solar light concentrators. For this application there isn't the concern to reconstruct images avoiding distortions, but the aim is to maximize the transfer of light flux from the first intercepting area of the concentrator, to the photovoltaic receiver. In this application, the light can be represented with sunrays, so the geometric optics is suitable to describe the optical properties of the concentrators.

Some optical parameters cover a substantial role in photovoltaic concentrators; the parameter are both geometrical, related to the ideal design of the parts, and physical, related to manufacturing issues and material choice.

The main geometrical parameters are:

- Concentration factor
- Acceptance angle

The main physical factors to consider in the optics for concentrators are:

- Light transmittance
- Light reflectance
- Light absorbance
- Dispersion
- YI (Yellowing Index)
- BRDF (Bidirectional Reflectance Distribution Function)
- BTDF (Bidirectional Transmission Distribution Function)

The BRDF is the Bidirectional Reflectance Distribution Function defined as the scattered radiance per unit incident irradiance; mathematically it's expressed as in Eq. (9).

$$BRDF(\theta_i,\varphi_i,\theta_S,\varphi_S)=\frac{dL_S(\theta_S,\varphi_S)}{dE_i(\theta_i,\varphi_i)} \tag{9}$$

Where θ_i, φ_i represent the angles of incidence for the incoming radiation, in spherical coordinates, while θ_s, φ_s are the angles indicating the scattering directions. L_s is the scattered radiance, while E_i is the incident irradiance. This optical property can become significant after the aging of the materials/surfaces, introducing unwanted light scattering at the reflector surfaces. The BTDF accounts for a detailed description of the scattering of the light through a transparent mean; usually, the parameter employed to describe the scattering of the light in transparent materials is the haze. The haze, defined as the ratio of the scattered light to the total light that get through a transparency, normally expressed as a percent, does not provide indication of the distribution of the light scattered (ASTM D 1003-97, 1997). Sometimes, this scattered light is not completely lost for CPV, but, however, the haze of a material is usually enough to estimate the optical performances useful for concentrators.

All these properties affect the optical efficiency of the solar concentrator, where the optical efficiency is usually defined as in (1):

$$\eta_{opt}=\frac{Irradiance\ @\ receiver\ surface}{Irradiance\ @\ entrance\ surface} \tag{1}$$

The aim of the optics designer is to maximize the optical efficiency, the concentration factor and the acceptance angle of the concentrator; moreover, for the photovoltaic application can be very important to consider other optical characteristics, like the spatial distribution of the irradiation onto the receiver surface and the light incidence angles distribution onto the

solar cells. Indeed, the PV devices usually work better with an even irradiation and with low incidence angles of the incident rays.
The geometrical concentration factor, defined as in (2), is a mere ratio of surfaces, which can growth indefinitely; however, to maintain an high efficiency, i.e. a maximal transfer of the incident energy flux of light, the concentration factor is constrained by the maximal light divergence of the incident rays.

$$C_{geom} = \frac{Entrance\ Area}{Receiver\ Area} \tag{2}$$

This constrain, obviously consistent with the second law of thermodynamic considering the Sun as heating body and the receiver (Smestad et al., 1990), is the sine brightness equation for ideal geometrical flux transfer; in its general form, with the receiver immersed in a material with refractive index n, this law is like in (3) for a 3D concentrator with axial symmetry. The θ_{in} represents the maximal incident angle for the incoming radiation respect to the normal direction at the entrance surface allowing for a maximal ray collection, while θ_{out} is the maximal angle for the rays at the receiver.

$$C_{max} = \frac{n^2 sin^2\theta_{out}}{sin^2\theta_{in}} \tag{3}$$

In fig.1 a schematic representation of a generic concentrator is sketched.

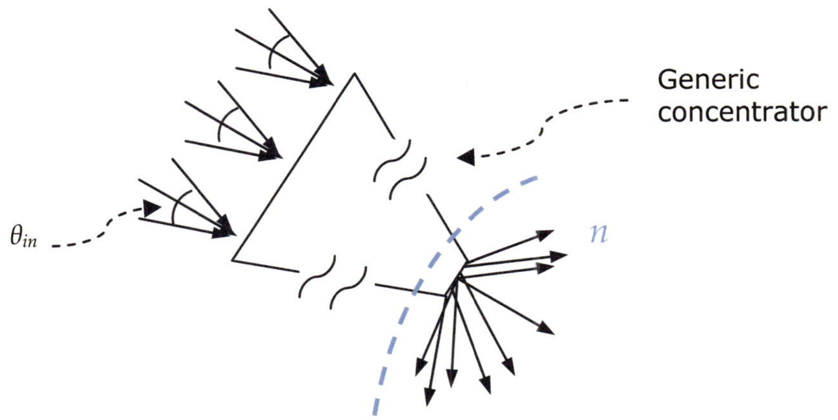

Fig. 1. Generic concentrator: the rays achieving the entrance with a maximal incident angle θ_{in} are collected to the exit aperture immersed in a means of refractive index n

Considering the maximal concentration achievable, the output angle is with $\theta_{out} = 90°$, so the theoretical max concentration becomes (4). For a solar concentrator with the receiver in air, i.e. with $\theta_{in}=0.27°$ and $n=1$, this value is 46000; this and even higher values using $n>1$ have been experimental obtained (Gleckman et al, 1989). The sunlight divergence, due to the non negligible dimension of the Sun, is determined by the Sun radius and the Sun-Earth distance.

$$C_{max} = \frac{n^2}{sin^2\theta_{in}} \tag{4}$$

For a linear concentrator the sine brightness equation is as (6), for an θ_{out}= $90°$; the demonstration is straightforward. Considering a radiance L, an ideal concentrator must conserve the flux (Φ_{in}= Φ_{out}) given by the radiance integrated onto the entry surface. For a linear concentrator, this flux becomes as in (5) and the concentration factor becomes (6). For a solar concentrator in air, it becomes about 200.

$$\Phi = \Phi_{in} = LA_{in}\int_0^{\theta_{in}} \cos(\theta)d\theta = \Phi_{out} = LA_{out}n\int_0^{\theta_{out}} \cos(\theta)d\theta \tag{5}$$

$$C_{max} = \frac{A_{in}}{A_{out}} = \frac{n}{sin(\theta_{in})} \tag{6}$$

In the CPV field, the acceptance angle is defined as the angle of incidence for the rays at which the optical efficiency of the concentrator achieves the 90% of its maximal value.
The two geometrical properties (optical efficiency and acceptance angle) of a light concentrator with defined concentration level are well represented with a graphic like in fig. (2), where the optical efficiency is plotted vs the incidence angle. The rectangular shaped dashed line with a side at the limit angle is the graph corresponding at an ideal concentrator; it collects at the exit surface all the rays with angle lower than the Θ_{max} defined by the theoretical limit. The other lines represent 2 possible characteristics of non-ideal concentrators; their acceptance angle can be determined in correspondence of the 90% of the optical efficiency.

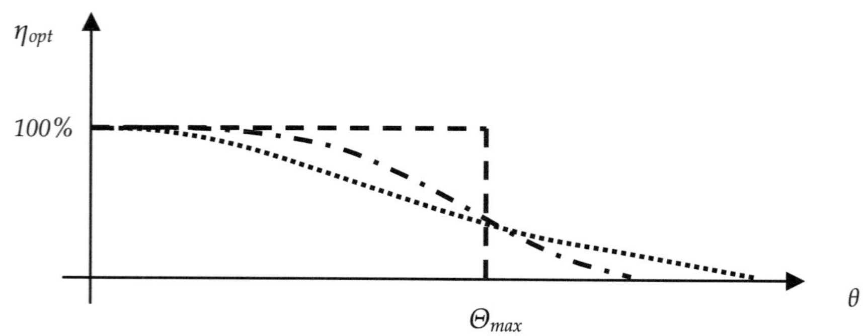

Fig. 2. Optical efficiency vs incident angle for solar concentrators: the rectangular shaped dotted line represents the characteristics of an ideal concentrator, while the others are for non ideal concentrating geometries

In the real applications, the concentrators have surfaces different from the geometrical ideals; this because the geometrical shapes allowing for the theoretically best results are limited and usually with complex structures or requiring special materials. These conditions are constrains for the cost competitiveness of the concentrators, so a trade-off between performances and cost must be achieved.
As previously indicated, the theoretically maximal concentration of an optical system is limited; an optical invariant, called Lagrange invariant or étendue, accounts for this relation

between concentration and angle of divergence consistently with the thermodynamic limits. It describes the integral of the area and the angular extends over which is set a radiation transfer, as in (7).

$$\acute{e}tendue = n^2 \iint \cos(\theta) dA d\Omega \quad (7)$$

Using this optical invariant is possible to derive (4,6) (Winston et al., 2005). Considering a bundle of rays, the étendue can be represented univocally as a volume in a phase space characterized by the cosine directions of the rays and their positions in the real space; a geometric concentrator works as an operator with the function to modify this volume; in this transformation the étendue must be conserved.

2.1 Design methods

The design of solar concentrators has different drivers respect to imaging optical elements. Indeed, the design goal here is to maximize the flux density, i.e. the irradiance, at the receiver. Different methods can be implemented to achieve this result (Winston et al., 2005); one of the most commons is the edge ray method. This is based on the assumption that the edge rays in the phase space, i.e. with higher incidence angle at the entrance boundaries of the concentrator, correspond at the extreme rays, in term of positions as well as angles, at the receiver too; the rays between the edge rays are collected to the receiver as well, supposing smoothing and optical active surfaces in continuous media for the concentrator. The first example of non-imaging concentrator obtained with this technique is the compound parabolic concentrator (CPC), as shown in fig. (3); a bundle of parallel rays with an angle respect to the CPC's axis of symmetry (which is the max angle of divergence for the collected rays), is focused onto a point at the exit area by the reflection on a parabolic surface; this point is on the edge of the exit of the concentrator. All the rays entering with lower angle of incidence are collected at the exit surface. This kind of concentrator allows for the maximal theoretical level of concentration for a linear collector, and it's almost ideal for the 3D case, with a surface obtained by revolution.

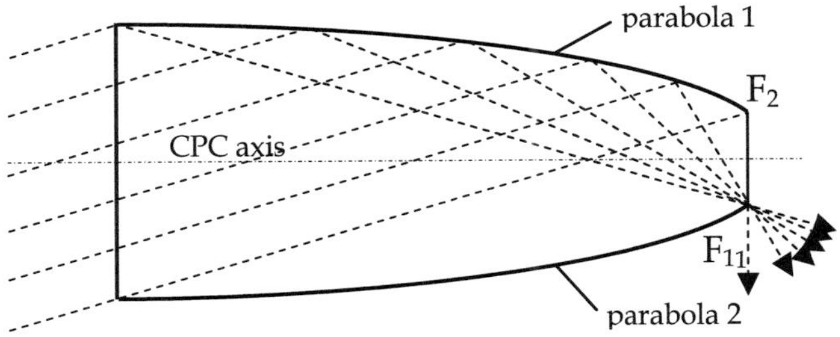

Fig. 3. Scheme of the edge ray method applied to a compound parabolic concentrator (CPC); the dotted arrows represents the incoming rays

Other methods have been developed since the 70's till today (flow line method, Tailored Edge Ray, Poisson bracket method, Simultaneous Multiple Surface, Point-source Differential Equation method) both analytical as well as numerical.

The design of solar concentrators must take into account many different aspects other than the geometrical optical efficiency and concentration levels; indeed, the physical optical properties former reported have to be considered, in order to achieve an effective high optical efficiency. Moreover, the concentrators should be as much compact as possible, deliver a suitable irradiance distribution at the receiver, allowing for cheap assembling and good thermal management of the system components. All these variables have enlarged the space of possible configurations for CPV optics and there is indeed a wide spectrum of real applications. Currently, most of them are based on Fresnel lenses for the primary optics; the Fresnel lenses are particular kind of lenses for which the dielectric transparent volume material is reduced at the minimum, as shown in fig.(4a), in order to reduce the mass, so the weight, as well as the light absorbance. Other solutions use the reflection of the light instead of the refraction to concentrate the light; the classical parabolic reflectors are used as well as more complex configuration in the form of cassegrain designs, as in fig. (4c); this optical design based on two reflections has the aim to achieve a compact structure, with the light focus behind the primary concentrator. The cassegrain structure is normally employed in telescopes, for the magnification of the far field objects, and, in its basic design for imaging optics, use a parabolic mirror reflecting toward a hyperbolic mirrored surface.

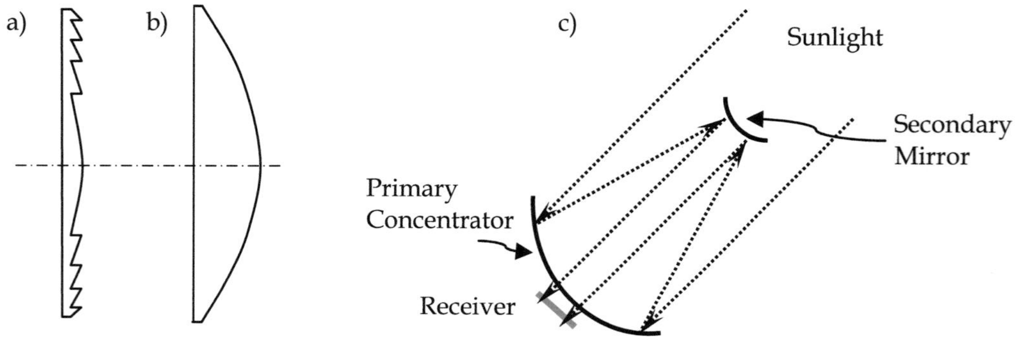

Fig. 4. Classical designs for photovoltaic concentrators: a) Fresnel lenses equivalent to the standard lens of b); c) schematic drawn of a cassegrain optics

The CPV optical systems are often composed of a primary concentrator with a secondary optical element (SOE); these secondary elements are usually joint to the photovoltaic cells and are employed to improve the concentration factor and the angular acceptance. Moreover, they are often used to increase the light uniformity on the receiver through multiple reflections with kaleidoscopic effect (Ries et al, 1997; Chen et al, 1963); in this latter case, to allow for good optical efficiency, the reflections must be associated to negligible losses. An optical phenomena used to achieve this result is the total internal reflection (TIR) effect; this is obtainable through the channeling of the light into transparent dielectric means shaped to allow the striking of the rays on their surfaces only with an angle lower than the limit angle Θ_c (8); this angle is a direct consequence of the Snell law, when the SOE is made of a material with dielectric index n_1, placed in a mean of dielectric index n_2. It works like a light pipe.

$$\Theta_c = Arcsin(\frac{n_2}{n_1}) \tag{8}$$

The shape of the secondary optics is directly related to the primary concentrator, because it works on the already deflected bundle of rays. So, a number of different designs for these components can be found. However, the most popular can be classified in few groups, like domed shapes, CPCs, truncated pyramids or cones (Victoria et al., 2009). Other original configurations can be found, depending on the requirements of every CPV manufacturer.

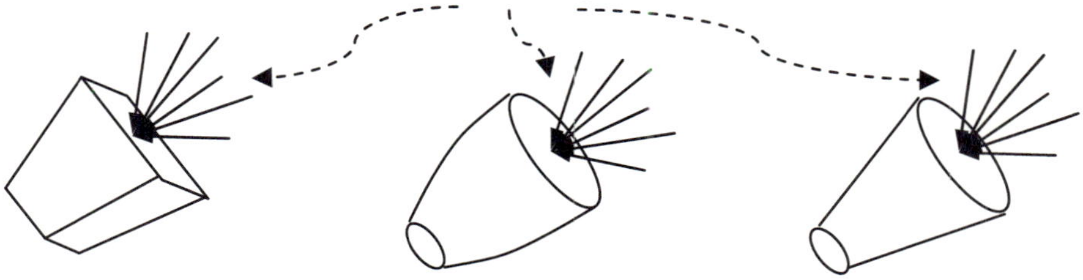

Fig. 5. Examples of geometries for simple secondary concentrators

Currently, powerful modern raytrace-based analysis tools for optics design are available; the majority of these software employ the Monte-Carlo method to solve the coupled integral differential equations used to calculate the illuminance distribution in 3D models (Dutton & Shao, 2010).

These software tools often allow for the accounting of physical parameters too, delivering very realistic estimations for optical performances.

Fig. 6. Cassegrain type optics for solar concentration arranged in modules by Solfocus Inc. (www.solfocus.com)

2.2 Other concepts

In order to maximize the conversion efficiency of the solar cells and of the complete concentrating system, some CPV designs act onto the spectral properties of the light together with the geometrical ones. Each photovoltaic materials has the best photovoltaic performances for wavelengths with energy slightly higher than the semiconductor bandgap. A splitting of the incoming light or the wavelength shifts are tricks used in dichroic and luminescent concentrators to try to increase the PV conversion efficiency.

2.2.1 Dichroic concentrators

The idea to split the sun spectrum in light beams and to drive theem toward different cells of selected material is not new. As well as the idea of concentrating the light, it can be realized in a number of different configurations; the constrains for its implementations are mainly related to the costs of these assemblies, considering that additional complexities are introduced; indeed, to split the solar spectrum, two physical ways are possible: dispersion through a transparent prism or reflection/transmission through dichroic filter working for light interference. The light is concentrated too, in order to reduce the costs of the cells dedicated to defined wavelengths. In these configurations the theoretical efficiency can achieve its maximal level, because each cell produces power in the best conditions of irradiation, without constrains of series electrical connections as happen for multijunction monolithic structures. Multi-cells arrays with a record efficiency of 43% have been fabricated (Green & Ho-Baillie, 2010) to demonstrate the feasibility of this approach.

2.2.2 Luminescent concentrators

The aforementioned solutions and methods to concentrate the light are not the only developed for photovoltaic applications. One important limitation of these designs is the necessity to use tracking structure to follow the sun. This constrain must not be considered always a limitation; indeed, especially for utility scale installations, tracking structure are used for standard flat plate modules too, in order to improve the energy harvesting, being always on the plane perpendicular to the sunrays. However, the possibility to use static photovoltaic concentrator able to capture also the diffuse radiation has been developed, using a different optical approach, not just the geometrical optics, but involving also some physical properties of particular material like the luminescence; these concentrators, named luminescent concentrators, are usually made of a flat plate of transparent material, with solar cells connected to the sides of the plate; inside the transparent material, luminescent particles like organic dyes or quantum dots are dispersed, absorbing part of the light spectrum and re-emitting light with shifted wavelengths, matching the spectral response of the cells. The re-emitted light is than guided toward the solar cells through the transparent mean, using the total internal reflection at the surface. The limiting point of this technology is the low efficiency achieved due to the losses in the different physical processes involved; it is currently in the order of 6-7% for record prototypes; moreover, the usual concentration for this kind of modules is in the order of 10-40 and the overall size of each luminescent concentrator, to avoid significant losses for light absorption from the transparent material, must be limited.
In fig.(7) a sketch describing the basic concept of these concentrator is reported.

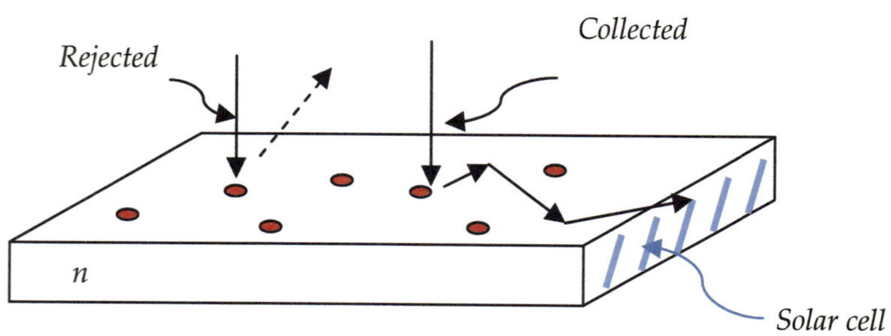

Fig. 7. Simplified drawn of a luminescent solar concentrator

3. Solar cells

The solar cells used in CPV are made with many different technologies, depending on the kind of used concentrator. In general, for low and medium concentration level, up to about

300 Suns, cells made of Silicon are still used; for higher concentrations, cells based on III-V semiconductors are usually employed; these latter cells allow for efficiency in the order of 40% and find their natural application under high concentration. Due to the high cost of the base materials and processes, these ultra-high efficiency cells found application for space satellites and for terrestrial concentrators. Thin film solar cells, in particularly made of CIS-CIGS, have given interesting results under concentration too (Ward et al, 2009), but, till now, no significant applications have been developed out of the laboratory scale.
The light concentration, through the increasing of the concentration of the minority carriers, improves the efficiency of the solar cells logarithmically. The produced current is linearly proportional to the irradiation level; because of the generated power is given by the product between the current and the voltage and the voltage increases logarithmically with the concentration level as in (9), the power increases in the mentioned super-linear way. In (9) *C* is the concentration level, while J_{ph_1sun} is the photo-generated current under one standard sun level of irradiation.

$$V_{oc} = \frac{AkT}{q} \ln\left(\frac{CJ_{ph_{1Sun}}}{J_0} + 1\right) \tag{9}$$

Where J_0 is the dark current of the diode and *A* is the ideality factor of the device.
An additional advantage for CPV cells is the performances reduction with the temperature, which is lower under concentrated light respect to the same effect under one Sun of irradiation, for the same kind of cell. This is true in general, for all semiconductor; in addition, III-V cells, often used in CPV, have a lower temperature coefficient than standard crystalline silicon solar cells. For example, the interdigited back contact silicon solar cells have a voltage temperature coefficient of about -1.78 mV/°C under one sun and of about -1.37 mV/°C at 250 suns (Yoon, 1994), while for GaAs from –2.4 mV/°C under one sun, to -1.12 mV/°C at 250 suns (Siefer, 2005). The dependence of the temperature coefficient with the concentration appears, in first approximation, with a logarithmic behaviour, as in (10); considering the V_{oc} as the voltage associated to the energy gap between the quasi-Fermi energy levels of the illuminated cell, as from fig.(8), this value is given by (11), where *C* is the concentration level, while *B* is a parameter dependent on various physical characteristics of the material.

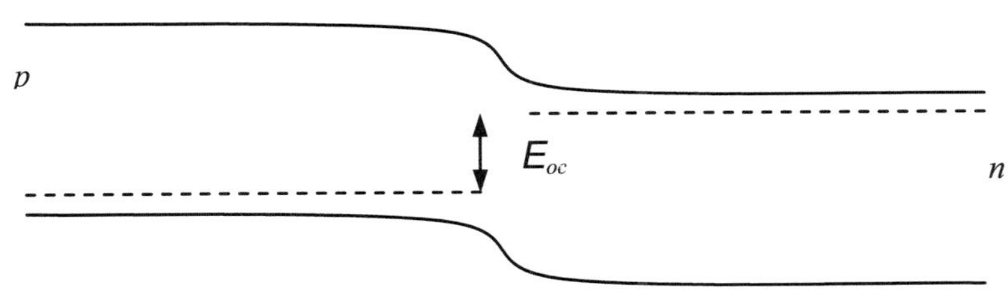

Fig. 8. Schematic band diagram of an illuminated p-n junction of a cell in open circuit conditions

$$V_{oc} \cong \frac{E_g - kT\ln(\frac{1}{CB})}{q} \tag{10}$$

So, the temperature coefficient becomes:

$$\frac{dV_{oc}}{dT} \cong -\frac{k\ln(\frac{1}{CB})}{q} \tag{11}$$

One of the main differences in the technology fabrication between concentrator solar cells and standard solar cell is the requirement for the CPV cells, producing high current density, to have low series resistance.
A simplified formula describing the I-V characteristics of a solar cell taking into account the resistance effect is eq.(12); two electrical resistances can be considered: a series resistance, R_s, and a parallel resistance, R_{shunt}. In a simple one dimensional model they are represented using the solar cell equivalent electrical circuit of fig.(9). It's a rough electrical schematization of the SC, because of the resistances are lumped; a more precise equivalent circuit should require distributed parameters in 3-D (Galiana et al., 2005).

$$J = J_{ph} - J_0 exp\left(\frac{q(V+JR_s)}{AkT}\right) - \frac{V+JR_s}{R_{shunt}} \tag{12}$$

Where J_0 is the dark current of the diode and J_{ph} represents the photo-generated current.

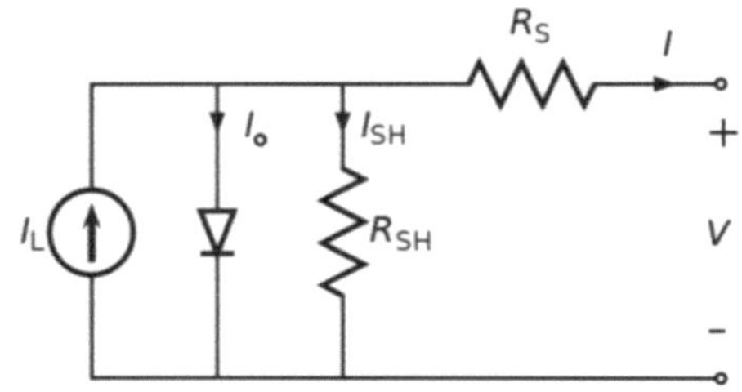

Fig. 9. Simplified 1D equivalent electric circuit of a solar cell

The simplified electrical equivalent circuit of fig.(9) is enough to explain the importance of attaining R_s as low as possible, especially in the case of concentrator solar cells. Indeed, the higher the current, the higher the voltage drop across the series resistance; in this way, the diode senses a voltage higher than that one on the external load, so its exponential behaviour reduces the current in the external circuit when the voltage on the diode is closed to its threshold voltage. The discrepancy between the voltage on the diode and the voltage on the external load gives a shortage in the current delivered from the cell in the region of the I-V characteristic with higher V.

3.1 Silicon solar cells

High efficiency silicon solar cells have been manufactured since the 80's (Green, 1987). These cells were manufactured in labs with microelectronic technology steps and with ultrapure crystals, in order to allow for the maximal performances; efficiency in the order of 27% have been achieved for back contact solar cells under around 100x and in the order of 25% under around 250x for cells produced by Amonix Inc. (Yoon et al., 1994). However, the fabrication processes required for these cells is expensive, and the ultimate device cost is comparable to that for multijunction solar cells on III-V semiconductors. Mainly for this reason the back contact technology is no longer used for CPV under the mentioned value of concentration; Sunpower Corp. commercialized this kind of solar cells until the beginning of 2000th but moved forward and transferred the technology on low cost processes for one Sun module production. The silicon cells are currently used in CPV systems with concentration up to around 100 Suns; the technology used in this range of concentration must not differ so much from that of standard solar cells, in order to allow for an economical convenience of the CPV solutions. One established technology is the laser grooved buried contact (LGBC), in which the metallic contacts of the frontal grid are buried into the bulk of the wafer, as in fig.(10); the high aspect ratio of the fingers allows for low resistance of the contacts, while the large area of metal-semiconductor interface permits to strongly reduce the electrical resistance at the interface of the Shottky energy barrier, keeping a low shadowing of the photo-active material.

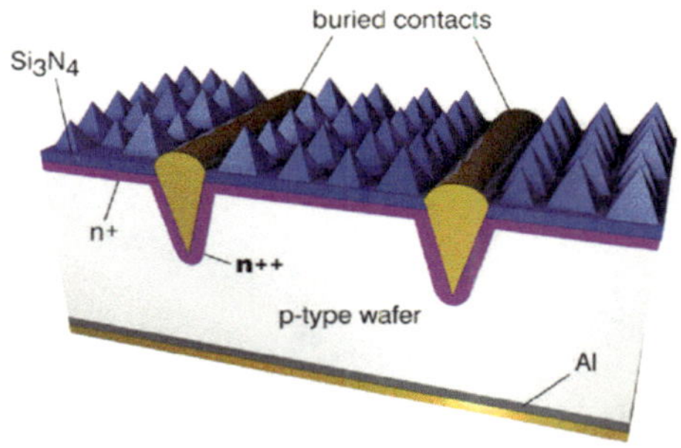

Fig. 10. Cross section of the LGBC silicon solar cell (Cole et al., 2009)

This LGBC concept is employed for the Saturn cells commercialized by BP Solar in flat plate PV modules (Bruton et al., 1994). For concentrated light BP Solar produced cells with this technology for the Euclides concentrators (40x) (Sala et al., 1998); at the Narec PV technology centre, these cells are manufactured and developed for different concentrating solutions, with efficiency approaching the 20% (Cole et al., 2009).

Standard solar cells obtained with screen printing technology and designed for one sun application strongly reduce their efficiency even at 2-3 suns because of ohmic losses due to series resistance; however, some improvements can be achieved through slight design modifications, varying doping concentrations, electroplating parameters, line pitches and other fabrication steps.

3.2 Solar cells of III-V materials

The highest conversion efficiency for solar cells has been obtained with the multijunctions approach. Through epitaxial growth the deposition of crystalline layers of compound semiconductors is possible whenever specific requirements on the lattice parameter are satisfied (Yamaguchi, 2002). Many layers of different semiconductors are stacked in order to create a structure where the first layers appear transparent at the light absorbed by the semiconductors at their bottom. This is obtained decreasing the band gaps of the compound semiconductors, from the frontal surface to the rear. The Germanium is often used as substrate material, both for its lattice parameter as well as for its band gap adapt for the bottom cell function. Unfortunately, some semiconductor compounds with suitable band gaps haven't a lattice matching with the other materials useful for the stack; however, cells growth with lattice matched (LM) technique have achieved the 40% of efficiency under concentration. To further improve the performances of the cells, the metamorphic (MM) approach has been developed (King et al., 2007), delivering record cells efficiency higher than 41% under concentration; with this technique, consisting in the introduction of step-graded buffer layers allowing for stress/strain relief to avoid the formation of dislocations in the layers growth, the flexibility in band gap selection is greatly improved, providing freedom from the constrain of same crystal lattice constant for all the stacked material in the monolithic structure as for LM. In fig.(11) a semplified MM multijunction cell structure from (King et al., 2007) and the distribution of irradiance absorbed for photovoltaic conversion by the three active materials are reported. To electrically connect the integrated sub-cells of different materials, tunnel junctions are formed.

These complex structures represent 3 solar cells series connected. So, the active element producing the lower current limits the current generation. The current produced by each layer depends on the light spectrum too, so spectral variations, as happen with different weathering conditions, can affect the performances of the cells (Muller, 2010).

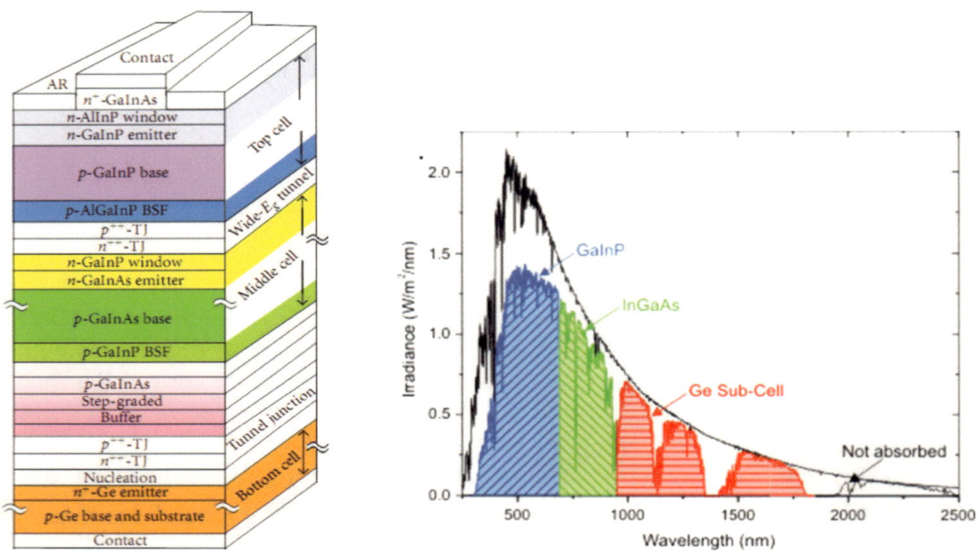

a) b)

Fig. 11. Triple-junctions solar cells; a) stacks of layers of different semiconductor compounds from (King et al., 2007); b) absorbed portions of the solar spectrum (AM1.5) for the three photo-active semiconductors

Theoretically, a cell with 4 junctions can achieve an efficiency of 58% under an AM1.5 spectrum; with a combination of real and known materials, a terrestrial concentration cell with efficiency of 47% is possible. Until now, however, the most performing cells are 3-J solar cells; at the end, for energy production installations, a trade off between costs and performances in field must be carried out. Because of the detrimental effect of the spectral changes becomes more influent increasing the number of monolithically stacked junction, the convenience to use, in the future, 4-J solar cells instead of 3-junctions solar cells for in Sun installation must be demonstrated.

The cost of these devices is decreasing, but it is still in the order of 4€/cm^2. To evaluate the cost contribution of the cells on the global system, let's suppose a collected area of the concentrator of 400 cm^2 and of a cell of 1cm^2 (physical area of the cell, usually higher than the irradiated zone, because of, at least, the area for the pads for contact leads is necessary); with a nominal irradiation level of 850W/m^2 and a module efficiency of 25% the cell generates 8.5W, so the €/W_p contribution of the cell on the overall CPV system cost is of 4/8.5 = 0.47 €/W. It's a significant voice of cost, but it can be reduced increasing the concentration level and with the specific cost reduction of the devices obtained with their volume production, as well as with their efficiency improvement.

New products based on III-V semiconductors are doing their first steps into the CPV market, moving from labs to pilot production lines. The approach of the strain balanced quantum well solar cells (SB-QWSC) (Barnham et al., 2002), appears of great technical interest for the efficiency improvement of multi-junctions solar cells as well as for the possibility to tail the cells on particular optical designs acting on the spectral properties of the light, like as dichroic concentrators (Martinelli et al., 2005).

In order to reduces the cost of these cells high research efforts have been invested, following different routes. From the manufacturing point of view, molecular organic chemical vapour deposition (MOCVD) equipments, industrially used for the epitaxial growth of the compound layers have been developed for high productivity. On the other side, different ways to reduce the cell cost replacing the Germanium or GaAs substrate with cheaper Silicon wafers (Archer et al., 2008) or using peeling-off techniques (Bauhuis, 2010) in order to use the same substrate for different growth have been investigated.

3.3 Solar cells assemblies

In general, the cells for concentration are assembled on supporting substrates, treated similarly to bare dies in electronic technology. So, the process is completely different to that for standard PV assembling, but can take advantages by the huge progresses, standardizations and experiences collected during the last decades by the electronic devices industry.

Depending on the cells nature (materials, sizes and manufacturing technologies) and on the operative working conditions, different mounting technologies are used. Generally, the surface mounting technologies (SMT) directly derived from power electronics are applied. Even in this particular subset of components there's plenty of different solutions. A good assembling is fundamental for the performances of the systems; thermal properties, reliability and optical matching are strongly dependent on the assembling solutions. Generally, thermal substrates are used, in order to drain out the high heat flux generated by the concentrated beam on the small cells; as every PV devices, the cells for concentration decrease their performances, as previously described, with the temperature. A substrate able to efficiently drain the heat out from the cells and spreads it onto a large area for heat exchange with the external air or with other cooling means is required. For this purpose, ceramic materials like alumina (Al_2O_3) or aluminium nitride (AlN) are often use, as in hybrid electronics, when the thermal flux are very high, because of their properties of thermal conductivity; when the thermal budget is lower, cheaper material can be employed as, for example, insulated metal substrate (IMS), i.e. an electronic support fabricated laminating an insulator between a massive mechanical substrate of aluminium and a foil of copper used as electrically conductive layer. Depending on the material and thickness adopted for the insulator layer, the circuit will have consequent thermal properties as well as dielectric capabilities. These insulating materials have usually a thermal conductivity in the range of 0.8 - 3 W/mK. In table (1) a summary of thermal conductivity of useful materials employed in CPV receivers assembling is reported.

The cells are electrically connected at the circuitry on the substrate; the rear contacts are attached using electrically conductive adhesives or soldering, while the frontal contact is connected with soldered ribbons or bonded wires; in fig.(12) two different solutions using soldered leads and wire bonding, with chip on board technology (CoB), are shown.

Material	Thermal conductivity W/mK
Aluminium	204
Copper	390
Tin	67
Silicon	150
Germanium	60
Alumina	25
Aluminium Nitride	160
Silicones	0.1 - 0.2
Electrically conductive adhesives	4 - 5
Thermal conductive adhesives	1 - 4

Table 1. Thermal conductivity of materials usually considered for the assembly of CPV receivers

a)

b)

Fig. 12. CPV solar cells assembled on substrates: a) soldered silicon solar cell (Courtesy of CPower Srl); b) solar cell of 1mm^2 assembled with chip on board technology (Courtesy of CRP - Centro Ricerche Plast-ottica)

Because of the technology used is derived from the electronic industry, the reliability issue related to the assembling with these approaches have been evaluated for long time; the CPV receivers in working condition can suffer different stresses respect to many electronic applications; however, many standards are already defined to verify the level of quality of the assembling processes and some possible defects leading to probable reliability problems can be identified even prior to carry out accelerated aging tests. In fig.(13a) a X-ray picture of a solar cell soldered onto a substrate using a correct surface mounting technology is shown, while in fig.(13b) a cell with an excess of voids in the soldering of the rear cell's surface is sketched. The voids can produce cracking and failures during thermal cycling, as known in electronic technology. (Yunus et al., 2003).

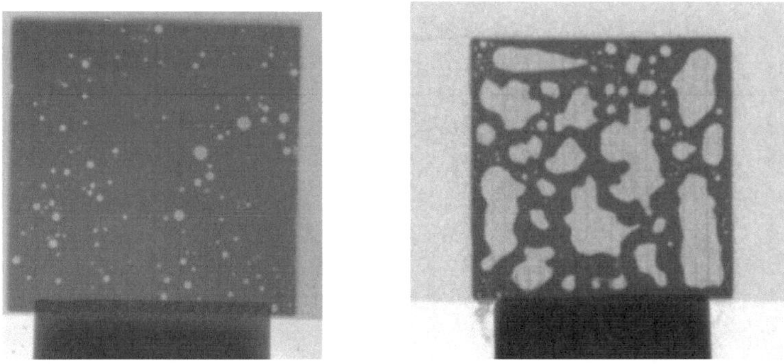

a) b)

Fig. 13. X-ray image of soldered solar cells - a) acceptable soldering with <5% of voids area; b) unacceptable soldering, with high fraction of voids under the cell

Bypass diodes are often mounted on the same substrate of the cells; for multijunctions solar cells these component assumes great importance due to the high sensitivity to reverse bias of these cells, protecting the devices against destructive reverse loads. Currently, each individual cell has its own bypass diode, which can be an integral diode or an external, more standard, Si-diode. Basically, the integral concept consists in separating small area of the multijunction cell via mesa etching, and using the p-n junctions of the cell as protective diode.

Secondary optics, wherever used, are components of the receiver. These components require a high level of precision for their assembling in the module; indeed, the higher is the concentration level to ménage, the higher is the precision of positioning, in order to avoid magnified losses; these secondary optics usually have to work under beams already highly concentrated. In high concentration photovoltaic modules, positioning errors higher than 100 microns can produce not negligible power losses (Diaz et al. 2005); however, this level of precision is usually achieved by high speed pick & place equipments for SMT in electronics, which are employed for the receivers assembly (Jaus et al., 2009).

4. Systems

The CPV system is composed of many parts which must cooperate efficiently; generally, the modules or assemblies must follow the Sun in its apparent motion, to ensure the collection of the direct irradiation from the cells, through the optics. The possibility of the concentrators to catch only the direct portion of the sunlight, with an additional circumsolar light dependent on the acceptance angle of the optics, is an important limitation for the CPV respect to standard photovoltaics. Diversely, the necessity to follow the Sun is not generally a limitation; indeed tracking installations are already in fields for standard, flat plate modules too. The tracking of the Sun gives a significant improvement in the energy collection, because of it allows for a constant maximal intercepted area of the modules for the sunrays. This fact permits to improve the energy production of 30-40% respect to fixed installations of same peak power, with this percentage depending on the latitudes of the installation. So, for an economical point of view, the additional costs introduced by the Sun-tracker have to be balanced by the gain in the energy production; this is the straightforward evaluation in the case of standard modules; for CPV the trackers are fundamental parts of the systems, so, it's an integral element and must be considered as an essential component as well as the inverters or the modules.

For these reasons, high efforts in the designing and production of cheap and reliable trackers are fundamental for the CPV establishment.

As previously described, CPVs, depending on the technology employed for the modules and cells, can use single axis trackers and two-axis trackers. While for the HCPV the 2-axis tracker is compulsory, the low concentration systems can be found, depending on the technology, on 1-axis or 2-axis trackers. In fig.(14) a 2-axis system mounting 25x concentrating modules with high angular acceptance is shown; in this case, the high optical acceptance permits to use standard trackers generally used for flat plate modules (Antonini et al., 2009a).

The most common kind of CPV systems are constituted with panels of many modules. These CPV modules are treated similarly to standard flat plate modules on a tracking structure; in the CPV panels, the rigidity of the structures and the precision of mounting on the frames are more critical than for standard modules, as well as the pointing precision in the Sun tracking. These modules are made of many cell-optics units, electrically connected internally into a closed, water proofed box. Each cell-optics unit play the role of a single cell in a standard flat plate module.

Fig. 14. CPV tracking system in Sun; installation of Rondine™ CPV modules on standard sun-tracker for flat plate modules in Sicily (South Italy). (Courtesy of CPower Srl - www.cpower.it)

An alternative approach uses a large concentrating optics collecting the light onto a dense array of cells. The most classic designs consist of big reflective dishes with paraboloid or similarly curved shapes and dense arrays positioned in the focuses of the concentrators or at the end of a secondary optical elements (Stefancich et al., 2007). In fig. (15) a dense array of silicon solar cells is shown. The main advantage of this approach is that there is a high technology core of small area, which can be assembled with standard equipments for electronics, while in the CPV modules the cells are distributed on all the module surface with consequent high area to be considered for the CPV receivers. However, the dense arrays have some important limitations too; first, an even light irradiation is required on the series connected string of cells. This is because the less illuminated limits the current of all the string. Second, it is necessary to reduce at the minimum the spaces between the cells and to reduces the bus-bars and interconnections areas; indeed, all these zones give optical losses for the photovoltaic concentrator. These two points are not in common with CPV modules, because the light irradiance on the optics-cell units is equal for all, and the connectors and bus bars of the cells are usually kept out of the illuminated region, using for these purposes the large area between the cells in the module receiver.

Fig. 15. 30cm x 30cm dense array of silicon solar cells (Courtesy of CPower srl)

The CPV systems have the advantage of a lower energy payback time (EPBT) respect to standard c-Si modules. The EPBT, an indicator for the energetic sustainability of a system, is the time a system for energy production needs to generated the input energy required during its whole life-cycle. The shorter EPBT for CPVs is because the material used for the concentrators are usually produced with low energy consumption. The high level of purification required for the silicon to achieve the electrical properties essential for the photovoltaic use needs a high energy utilization. To understand the order of magnitude, to produce about 100W of silicon for standard photovoltaic cells with efficiency of about 15%, about 300 kWh are necessary; considering an average annual production of 1400 kWh/kWp, more than 2 years are required to pay back the energy for the solar grade silicon alone. Adding to this energy consumption needed for the silicon purification the other fabrication steps to get a compete standard PV modules, the EPBT usually reported for the modules is in the order of 3-4 years (Stoppato, 2008). The CPVs technologies have only a small fraction of very purified materials, being mainly composed of plastics, glass and metallic frames. This fact leads to shorter energy payback time, in the order of 1 year (Peharz & Dimroth, 2005).

The localization for CPV installation is strongly dependent on the weather conditions; diversely than for standard flat plate modules, the fundamental irradiation data is not just the global irradiation, but it is the direct normal irradiation (DNI), i.e. the component of light collected by the concentrators. The humidity, the clouds, the dust and the pollution scatter the light coming from the Sun deflecting the rays; usually, the best conditions for CPV are in dry and highly sunny climates. The higher DNI/GNI ratios are typical of desert areas or elevated terrains. The evaluation of this parameter is fundamental, and the knowledge of the global irradiation is not sufficient to estimate the energy production of a CPV system; indeed, the yearly average DNI/GNI ratio can vary from 50% to 80% (NREL, 1994). Reliable solar maps for direct irradiance are not yet available for everywhere as for the GNI.

Sometimes, even the DNI is not enough to evaluate the energy production of a system; indeed, the light impinging the cells in a concentrator systems is not necessary the same read from the pyroheliometer, i.e. the instrument used to measure the direct irradiation; this instrument, basically a sensor of irradiation with a tube limiting the angle of incidence for the incoming rays, usually has a view angle of ±2.5° and a limit angle of about ±4°. Depending on the optical solution adopted for the photovoltaic concentrator, the acceptance angle of a CPV system can be higher or lower respect to the pyroheliometer, so the light seen by the cells can be higher or lower respect to the reference instrument. The effect of the soiling on the modules is similar to the scattering effect due to the atmospheric conditions; indeed, the particles deflect the sun rays and can contribute to significant losses. Generally, the higher the acceptance angle of the optics, the lower is the effect of the soiling on the performances; for low concentrator systems with high acceptance angle the losses seem to be comparable with that of standard modules (Antonini et al., 2009b).

The peak power for the CPV modules and assemblies is usually defined under a DNI of 850 W/m^2. Although the conditions for the performances testing of CPVs are not yet defined in international standards, the main producers and research institutions recently refer to the 850 W/m^2 of DNI and module temperature of 25°C; performances tests with the cell's temperature of 60°C are often found too (Hakenjos et al., 2007). The temperature is a more thorny issue for testing respect to the irradiation, because of the temperature in field are usually significantly higher than in lab. The outdoor characterizations are fundamental to evaluate the performances losses due to the heating up of the cells.

The irradiance condition of 850 W/m^2 of DNI has been selected because of a DNI/GNI ratio in the order of 85% is frequently observed in many locations around the world when the GNI is of 1000 W/m^2, i.e. the standard irradiance condition for the test of flat plate modules.

The energy productivity of a CPV installation can be evaluated, similarly than standard installations for flat plate modules, using the energy yield (Y_f) and the Performance Ratio (PR) (Marion et al., 2005). The energy yield represents the energy production for installed peak power of a system; it is measured in kWh/kWp and strongly depends on the location because of it doesn't take care of the incident radiation. It's the first parameter for the comparison of different installations in the same site. Diversely, the Performance Ratio (PR), dimensionless and defined as in (14), normalizes the energy production to the incident irradiation, delivering a useful parameter for the comparison of installation under different irradiation conditions; it quantifies the losses due to temperature, AC/DC conversion, soiling, down-times, failures and mismatching.

$$Y_f = \frac{generated\ Energy}{Peak\ Power\ installed\ (DC)} \tag{13}$$

$$Y_r = \frac{incident\ radiative\ energy}{Reference\ Irradiance} \tag{13}$$

$$PR = \frac{Y_f}{Y_r} \tag{14}$$

The PR can be read as the equivalent time the system has delivered it's nominal peak power (Y_f) respect to the time of equivalent nominal irradiance conditions on the panel.
The Reference Irradiance (the irradiance for the DC peak power estimation) for CPVs is 850 W/m^2 of DNI, instead of 1000 W/m^2 of GNI as for the standard modules. This difference must be taken into account during the comparison of the CPV with other different PV technologies.

5. Market & prospective

Large installations of CPV are not yet common. Until the end of 2009, about 21MW of CPV systems have been reported as set on Sun (Kurtz, 2009; Extance & Marquez, 2010); large part of them (13 MW) are from one HCPV technology of modules based on Fresnel lenses concentrator developed by Amonix Inc. in the last twenty years; the fraction of operational systems with low concentration level (LCPV) is less then 1MW, mainly of Entech Solar products installed in the 90's. Although at the end of 2009 around 70 vendors of photovoltaic concentrator systems have been found (EPRI, 2009), the CPV is a small niche of the photovoltaic market; indeed, at the end of 2009 already 7 GWp of PV modules have been installed and grid connected around the world. The market of CPV is mainly oriented toward solar farms and large installations, because of the necessity to use Sun-trackers. For this kind of large installations, big investments are required; this is one of the first hurdles for the CPV entry into the market. Indeed, to gather large investments for solar energy production, the demonstration of high reliability and durability of the systems is a fundamental issues, which need time and systems in Sun. Moreover, the high competiveness of the other PV technologies and their levels in industrialization and economies of scale is another high obstacle to face for any new PV product, which must have a price lower than the established technologies.
With the aim to demonstrate the reliability and durability of different CPV technologies the ISFOC project has been set in Spain, for the testing in field of some MWp of photovoltaic concentrator systems. In this experimental solar plant hundreds of kWp of different CPV technologies are continuously monitored and the performances are evaluated.
An important step forward for the commercial feasibility of CPV has been done with the publication of the international standard for the design qualification and type approval for CPV modules and assemblies (IEC62108, 2007). The tests defined in this IEC standard are mainly oriented to demonstrate the durability and reliability of the CPV modules. This recently published text, milestone for the CPV deployment, presents some tests more severe than for standard flat plate modules and takes longer time to be concluded (approximately one year).

Fig. 16. Installation of CPV systems of Amonix Inc. in Nevada (USA) (www.amonix.com)

Some pioneers companies, the US based Amonix Inc. and the Australian Solar Systems, have set in Sun large installations since the '90s. However, because of the huge difference between any CPV solution, each technology must proof its reliability independently.

Based also on the durability and reliability of the systems, the Levelized Cost of Energy (LCOE) is a parameter expressed in cents/kWh frequently used to evaluate the economical convenience of a PV solution; this parameter takes into account not just the energy production, but the cost complexities associated with the entire lifetime of a solar plant, from financing through to end of life (Short et al., 2005). The LCOE takes into account installation and commissioning costs, operations and maintenance (O&M), degradation and lifetime, and the output. It calculates the average value of the total energy produced, revalued at the time of calculation based on forward assessments of inflation and costs of financing. Starting considering this parameter, some CPV companies have claimed to can achieve the lowest LCOE in the market, in the order of 10 dollar cents/kWh for the next years (Nishikawa & Horne, 2008).

An important advantage of the CPV approach is the reduced necessity of capital investments (scalability). Both the thin film industry as well as the Silicon standard module production require high capital investments. By reducing the amount of semiconductor material, the initial investment is also reduced. Although no CPV companies have yet demonstrated it, the relative easiness of scale-up of CPV is logical and could be a significant advantage in a rapidly growing market. Some companies have already declared production capacities of many tens of MWp per year in 2009, with large announced growth for the 2010 (Extance & Marquez, 2010).

Because of any CPV systems is composed of many parts, the economical advantage improves with the cost reduction of any components; it can be achieved with the economy of scale consequence of the rising of the number of installed systems. In order to achieve affordable product prices, some CPV companies are moving toward a sort of vertical integration on the value chain, being often producer of trackers, inverters, components, taking care of the field installations and even approaching the processes for the cells manufacturing. In this path CPV companies are rapidly following the way of the largest standard PV groups; indeed, in standard flat plate industry, some companies specialized in the module production, cells manufacturing, tracking fabrication and system integration are still working without integration. But for the largest groups this process is in progress, from the row material purification up to the final installation, in order to achieve the lowest costs.

Aside the CPV module and system producers, there are many companies working in the development of the components required for the installations; in particular, many companies are focused on the production of solar cells for concentration, on the production of specialized optics and for the fabrication of dedicated trackers (Extance & Marquez, 2010).

The CPV is an emerging technology in the photovoltaic sector. The cost of installed kWp is continuously decreasing to try to compete with standard c-Si modules and thin films. The wide range of solutions will lead to the accomplishment of some leaders and to the disappearing of other companies or technical solutions; because of the large amount of investments required for the establishment of a competitive technology, some of these inventions could not found a commercial deployment for financial reasons rather than technological imperfections.

The future of the CPV technologies will be probably defined in the next few years, with the direct comparison of the energy production of the first large solar farms of different photovoltaic technologies in different sites around the world.

6. References

Smestad, G.; Ries, H.; Winston, R.; Yablonovitch, E.; (1990) The thermodynamic limits of light concentrators , *Solar Energy Materials* 21 99-111

ASTM D 1003-97, (1997). Standard Test Method for Haze and Luminous Transmittance of Transparent Plastics, The *American Society for Testing and Materials,* West Conshohocken, PA

Gleckman, P.; O'Gallager, J. and Winston, R.; (1989) *Nature* 339 198

Winston, R.; Minano, J.C.; Benitez, P., (2005) *Non Imaging Optics*, Elsevier

Ries, H.; Gordon, JM.; Lasken, M.; (1997) High-flux photovoltaic solar concentrators with kaleidoscope-based optical designs, *Solar Energy*, 60(1):11-16

Chen, MM.; Berkowitz-Mattuck, JB.; Glaser, PE.; (1963) The use of a kaleidoscope to obtain uniform flux over a large area in a solar or arc imaging furnace, *Applied Optics*, 2(3):265-272

Marta Victoria, César Domínguez, Ignacio Antón, and Gabriel Sala; (2009) Comparative analysis of different secondary optical elements for aspheric primary lenses , *Optics Express*, Vol. 17, Issue 8, pp. 6487-6492

Dutton, S.; Shao, L.; (2007) Raytrace simulation for predicting light pipe transmittance; *International Journal of Low Carbon Technologies*, 2007, 2(4):339-358

Green, A. M. and Ho-Baillie, A.; (2010) Forty three per cent composite split-spectrum concentrator solar cell efficiency, *Progress in Photovoltaics*, Vol.18 Issue 1 pp. 42-47

Ward, J.; Ramanathan, K.; Hasoon, F.; Coutts, T.; Keane, J.; Moriarty, T.; Noufi R.; (2001) Cu(In,Ga)Se_2 Thin-Film Concentrator Solar Cells; NREL/CP-520-31144, presented *at the NCPV Program Review Meeting* Lakewood, Colorado 14-17 October 2001

Andreev, V. M.; Grilikhes, V. A.; Rumyantev, V. D.; (1997) *Photovoltaic Conversion of Concentrated Sunlight*, John Wiley & Sons, Chichester

Yoon, S.; Garboushian, V.; Roubideaux, D. ; (1994) Reduced Temperature Dependence of High-Concentration Photovoltaic Solar Cell Open Circuit Voltage (Voc) at high Concentration Levels; *Proceeding of the 1st World Conference on Photovoltaic Energy Conversion*, pp. 1500-1504, 1994

Siefer, G.; Abbott, P.; Baur, C.; Schleg, T.; Bett, A.W.; (2005) Determination of the temperature coefficients of varions III-V solar cells, *Proceedings of 20th European Photovoltaic Solar Energy Conference*, 6 – 10 June 2005, Barcelona, Spain

Galiana, B.; Algora, C.; Rey-Stolle, I.; and Garcia Vara, I.; (2005) A 3-D model for concentrator solar cells based on distributed circuit units; *IEEE Transactions on Electron Devices*, vol. 52, pp. 2552-2558

Bruton, T.M..; Heasman, K.C.; Nagle, J.P.; Cunningham, D.W.; Mason, N.B.; Russel, R.; Balbuena, M.A.; (1994) Large Area High Efficiency Silicon Solar Cells made by the Laser Grooved Buried Grid Process. *Proc. 12th European PV Solar Energy Conf.* 1994: 761-762

Sala,G.; Arboiro, J.C.; Luque, A.; Antón, I.; Mera, E.; Camblor, E.; Datta, P. ; Gasson, M.P.; Mason, N.B.; Heasman, K.C.; Bruton, T.M.; Cendagorta, M.; Friend, M.P.; Valera, P.; Gonzalez, S.; Dobon, F.; Perez, F.; (1998) *Proceedings 2nd World conference and exhibition on photovoltaic solar energy conversion* 1963-1968, 1998

Cole, A.; Baistow, I.; Brown, L.; Devenport, S.; Heasman, K.C.; Morrison, D.; Whyte, G.; Bruton, T.M.; 2009; ; Technological and financial aspects of laser grooved buried contact silicon solar cell based concentrator systems; *2nd International Workshop on Concentrating Pholtovoltaic Power Plants: Optical Design and Grid Connection*; 9-10 March 2009; Darmstadt; Germany

Yamaguchi, M.; (2002) Multi-junction solar cells and novel structures for solar cells; *Physica E*; Vol.14; 84-90

King, R. R.; Law, D. C.; Edmondson, K. M.; Fetzer, C. M.; Kinsey, G. S.; Yoon, H.; Krut, D.;.Ermer J. H.; Sherif, R. A.; Karam, N. H.; (2007), *Advances in OptoElectronics*, Article ID 29523

Muller, M.; (2010) Spectral Effects in CPV Performances, *NREL Reliability Workshop* Feb 18-19, 2010, Golden, CO

Barnham, K.W.J.; Ballard, I.; Connolly, J.P.; Ekins-Daukes, N.J.; Kluftinger, B.G.; Nelson, J.; Rohr, C..; (2002) Quantum well solar cells; *Physica E* 2002; 14: 27-36

Martinelli, G.; Stefancich, M.; Antonini, A.; Ronzoni, A.; Armani, M.; Zurru, P.; Pancotti, L.; Parretta A.; (2005) Dichroic Flat Faceted Concentrator for PV Use; *Proceedings of the International Conference on Solar Concentrators for the Generation of Electricity or Hydrogen 2005*

Archer, M. J.; Law, D. C.; Mesropian, S.; Haddad, M.; Fetzer, M.; Ackerman, A. C.; Ladous, C.; King, R.R.; Atwater, H.A.; (2008) *Applied Physics Letters* 92, 103503

Bauhuis, G. J.; Mulder, P.; Haverkamp, E. J.; Schermer, J. J.; Bongers, E.; Oomen, G.; Köstler, W.; Strobl G.; (2010) Wafer reuse for repeated growth of III-V solar cells , *Progress in photovoltaics*, (p 155-159) Published Online: Mar 11 2010

Yunus, M.; Srihari, K.; Pitarresi, J.M.; Primavera, A.; (2003) Effect of voids on the reliability of BGA/CSP solder Joints; *Microelectronics Reliability*; 43 (2003) 2077–2086

Díaz, V.; Alonso, J.; Alvarez, J.L.; Mateos, C.; (2005) The Path for Industrial Scale Production of Very High Concentration PV Systems; *Proceedings of the 20th European PV Conference*, Barcellona 2005

Jaus, J.; Peharz, G.; Gombert, A., Ferrer Rodriguez, J. P.; Dimroth, F.; Eltermann, F.; Wolf, O.; Passig, M.; Siefer, G.; Hakenjos, A.; Riesen, S. V.; Bett, A. W.; (2009) Development of Flatcon® modules using secondary optics, *Proceedings of the 34th IEEE Photovoltaic Specialist Conference*, June 7-12, 2009, Philadelphia, PA, USA

Antonini, A.; Butturi, M.A.; Di Benedetto, P.; Uderzo, D.; Zurru, P.; Milan, E.; Stefancich, M.; Armani, M.; Parretta, A.; Baggio, N.; (2009), Rondine® PV Concentrators: field results and developments. *Progress in Photovoltaics: Research and Applications*, 2009; 17:451-459

Stefancich, M.; Milan, E.; Antonini, A.; Butturi, M.A.; Zurru, P.; di Benedetto, P.; Uderzo, D.; Parretta A.; (2007) Experimental results of a tailored dish concnetrator for silicon solar cells; *Proceedings of the 22nd European Solar Energy Conference and Exhibition*, Milano, Italy, 3rd-7th September 2007

Stoppato, A.; (2008) Life cycle assessment of photovoltaic electricity generation, *Energy* 33, 224-232

Peharz, G. and Dimroth F.; (2005) Energy payback time of the high-concentration PV system FLATCON, *Prog. Photovolt: Res. Appl.*, 2005; 13: 627-634

NREL; (1994) Solar Radiation Data Manual for Flat-Plate and Concentrating Collectors

Antonini, A.; Butturi, M.A.; di Benedetto, P.; Uderzo, D.; Zurru, P.; Milan, E.; Parretta, A.; Baggio, N.; (2009) *Proceedings of the 34th IEEE Photovoltaic Specialists Conference*, 8-11 June 2009, Philadelphia, Pennsylvania (USA)

Hakenjos, A.; Wüllner, J.; Lerchenmüller H.; (2007) Field Performance of FLATCON® High Concentration Photovoltaic Systems, *Proceedings of the 22nd European Photovoltaic Solar Energy Conference*, September 3-7, 2007, Milan, Italy

Marion, B.; Adelstein, J.; Boyle, K.; Hayden, H.; Hammond B.; T. Fletcher; Canada, B.; Narang, D.; Shugar, D.; Wenger, H.; Kimber, A.; Mitchell, L.; Rich, G.; Townsend, T.; Performance parameters for grid-connected PV systems, *31st IEEE Photovoltaics Specialists Conference and Exhibition* Lake Buena Vista, Florida, January 3-7, 2005

Kurtz, S.; (2009) Opportunities and Challenges or Development of a Mature Concentrating Photovoltaic Power Industry; *Technical Report NREL/TP-520-43208*

Extance A. & Marquez C.; (2010) The Concentrated Photovoltaic Industry Report 2010; *CPV Today*

EPRI; (2010) 2009 Concentrating Photovoltaic Solar Technology Assessment; 1020895; April 2010

IEC 62108 Ed.1.0; (2007) Design qualification and type approval for CPV modules and assemblies

Short, W.; Packey, D. J.; Holt, T.; (1995) A Manual for the Economic Evaluation of Energy Efficiency and Renewable Energy Technologies, *NREL/TP-462-5173*, March 1995

Nishikawa, W.; Horne, S.; (2008) Key Advantages of Concentrating Photovoltaics (CPV) for Lowering Levelized Cost of Electricity (LCOE), *Proceeding oft he 23rd European Photovoltaic Solar Energy Conference and Exhibition*, 1-5 September 2008, Valencia, Spain, pg. 3765 - 3767

3

Photovoltaics for Rural Development in Latin America: A Quarter Century of Lessons Learned

Alma Cota[1] and Robert Foster[2]
[1]*Universidad Autónoma de Ciudad Juárez*
[2]*New Mexico State University*
[1]*México*
[2]*United States of America*

1. Introduction

Over the past quarter century, Latin America has widely adopted photovoltaic (PV) technologies for social and economic development. Latin America is the world's birthplace for small rural solar electric systems used for residential power, refrigeration, distance education and hybrid systems. The use of PV systems has increased dramatically from an initial concept pioneered by a few visionaries to many thriving businesses throughout the rural regions today.

PV is a viable alternative to conventional large-scale rural grid systems. With the advent of PV as a dependable technology alternative allowing local private enterprise, and made available to the general public, PV systems have become attractive all over Latin America with hundreds of thousands of rural households electrified via solar energy.

During the early 1980s, solar energy pioneers began to disseminate PV technologies in rural Latin America as a solution for providing basic electricity services for non-electrified populations. Some of the first pilot projects in Latin America were undertaken by NGOs, such as Enersol Associates in the Dominican Republic, beginning in 1984. In the late eighties, small solar companies began to form gradually throughout Latin America; the key module manufacturers such as Solarex and Arco sought out distributors for off-grid rural markets.

By the mid-1990s, these activities were followed by large-scale solar electrification activities sponsored by government agencies in Mexico, Brazil, Colombia, Bolivia and Peru. Many of these early governments efforts for large-scale PV electrification faced sustainability issues; planners attempted to force "free" solar electrification projects onto unknowledgeable rural users.

In Mexico, there were large-scale government PV rural electrification projects undertaken under PRONASOL (a Mexican program to better people lifestyle) in the early to mid-1990s with over 40,000 PV systems installed, especially in southern Mexico. In the State of Chiapas more than 12,000 systems were installed. The government also dabbled in village scale PV and wind electrification. Unfortunately, over two thirds of these systems ceased functioning in only a couple of years. The era of large PV electrification projects in Mexico came to a temporary halt in the late 1990s, in large part due to the poor performance and image of these original substandard PV systems. Typical problems on PV systems installations were not related to the PV modules, but rather due to poor quality installations and problems with balance of systems due to inappropriate use of battery technologies and substandard charge controllers.

In response to early system failures, implementing agencies gradually began to adopt more rigid technical specifications that observed international standards that improved the quality and reliability of PV systems. Some examples include the World Bank/Nicaraguan Comission of Energy (Comisión Nacional de Energía de Nicaragua) Program for the electrification of 6,000 homes in the rural regions of Nicaragua, and the World Bank in Bolivia for the PV electrification of 10,000 homes. However, there are still issues of enforcement of standards where they do exist.

To promote a reliable introduction of PV technologies in Latin America, it is of great importance to bring early capacity building that tends to focus on PV specific applications to create a knowledgeable engineering base in country. Sandia National Labs (SNL) and New Mexico State University (NMSU) held many of the early capacity building activities, including the first PV and wind workshop in Central America, in Guatemala in 1992 under the USAID/DOE/US Export Council for Renewable Energy - Latin American Renewable Energy Cooperation Program. Over the next 15 years, hundreds of workshops were held by US government, World Bank, etc. training thousands of engineers and technicians on PV applications such as household lighting, water pumping, refrigeration, communications, clinics, and schools in Brazil, Chile, Ecuador, Honduras, Jamaica, Guatemala, Mexico, Panama, Peru, and the Dominican Republic.

Many of these trained engineers and planners were later responsible for implementing the first PV electrification projects, such as the 1993 EEGSA project in the community of San Buenaventura, Guatemala for 68 homes using 50 W systems. Likewise, the founding of Guatemala's Fundación Solar in 1993 furthered progress by installing over 3,000 PV household-electrification systems, mostly in the Quiché and Verapaz regions.

The Mexico Renewable Energy Program (MREP) was designed to expand the use of renewable energy technologies for Mexico's rural development (Foster *et al.*, 2009, Cota, 2004). MREP was launched in 1992 by the US Department of Energy (DOE) and the US Agency for the International Development (USAID) and was managed by SNL (Richards *et al.*, 1999). Various Mexican program partners have collaborated with MREP, including the Fideicomiso de Riesgo Compartido (FIRCO) for the deployment of PV systems for agriculture. The key application supported by MREP between 1994 and 2000 was PV water pumping systems for livestock and community water supply (Cota *et al.*, 2004), although additional projects included PV lighting (Foster *et al.*, 2004), communication, education (Foster *et al.*, 2003, Ley *et al.*, 2006), ice-making (Foster *et al.*, 2001, Foster 2000, Hoffstatter and Schiff, 2000), and refrigeration systems (Estrada *et al.*, 2003), as well as a few wind-energy projects (Romero Paredes *et al.*, 2003, Foster *et al.*, 1999, Ley and Stoltenberg, 2002). The project continued its work until 2005 and directly installed over 500 solar and wind systems, and spun off with the application of an additional 3,000+ more systems across Mexico. However, the main impact was the capacity building of the Mexican solar energy industry and increasing the quality of installed systems.

2. PV home systems in Mexico

Rural Latin households pay anywhere from US$5-25/month for dry cell batteries and kerosene lighting, the main energy source PV competes against. Rural users mostly use electricity for lighting and entertainment with radio and TV. In 1998, a market study was undertaken in rural Chihuahua by NMSU under the MREP to determine what the average consumer willingness to pay (WTP) was for PV lighting systems (Foster *et al.*, 1998a). Chihuahuans were found to be favorably disposed to the concept of solar PV systems as an alternative source of energy for their homes. At the time, non-electrified households in Chihuahua were already spending about US$25 per month for gas powered lights and small dry cell batteries for radios, and were willing to pay similar amounts of money to displace those services through PV.

In 1999, one hundred forty five innovative high quality PV home lighting systems were installed in the State of Chihuahua as part of the MREP. A total of 120 systems were installed in the Municipality of Moris, as well as an additional 15 systems in the municipality of Nonoava and 10 systems in Bachíniva, totaling 7.3 kW and benefiting about 800 people.

The municipality of Moris is located about 250 km west of Chihuahua City, from which it takes about 8 hours to arrive in vehicle. The terrain consists of steep mountains and 1,000 m deep canyons in the midst of pine forests. The arid climate is hot in the summer (~40°C) and cold in the winter (<0°C). The steep topography makes electric grid access difficult and indeed there is no interconnection with the national electric grid, nor are there paved roads. Over 3/4 of Moris residents do not have access to electricity, and the few that do are mostly on diesel powered mini-grids.

Fig. 1. 50 Wp PV lighting system installed in Talayotes, Moris County, Chihuahua.

2.1 Financing program for household lighting

The State of Chihuahua, working with MREP, designed the first Mexico's first ever pilot renewable energy financing program. The objective was to promote the use of renewable energy technologies in rural areas that lie outside the national electric grid. The financing activities were conducted by the State Trust Fund for Productive Activities in Chihuahua (FIDEAPECH - Fideicomiso Estatal para el Fomento de las Actividades Productivas en el Estado de Chihuahua) (Ojinaga *et al.*, 2000). This state trust fund provides direct loans and guarantees, primarily based on direct lending (e.g., to farmers for tractors). For this project, FIDEAPECH used US$99,000 of MREP seed funding from USAID to support renewable energy projects. FIDEAPECH implemented the revolving fund in which the municipality paid up front 33% of the total cost of PV home lighting systems, end users provided a down payment of 33%, and the remaining 34% was financed for one year by FIDEAPECH. The municipal government provided the loan guarantee and eventual repayment to FIDEAPECH. The total installed cost of each quality code compliant PV home lighting system was about US$1,200. The FIDEAPECH financing program went on to roll over its seed capital four times.

Other financing and leasing programs have been initiated in Nicaragua, Bolivia, Dominican Republic, Honduras, etc. by such organizations as the World Bank and companies like Soluz. These programs have had mixed results and generally PV systems leasing has not been successful in part due to rural seasonal incomes. PV financing programs can be set up in rural Latin America to compete with conventional technologies so long as financing terms are compatible with current rural user expenses and seasonal incomes.

2.2 System design

NMSU worked closely with the Chihuahua Renewable Energy Working Group (GTER) to implement a quality PV lighting system project. NMSU assisted GTER with the development of a technical specification for the PV lighting systems that would comply with the Mexican electrical code (NOM–Norma Oficial Mexicana) (Wiles, 1996). The NOM essentially mirrors the US National Electrical Code (NEC); Article 690 of both directly applies to PV installations. The NOM had not previously been applied in Mexico for the thousands of PV lighting systems installed. Besides meeting legal guidelines, NOM compliance can extend system reliability, lifetime, and safety.

The Solisto PV systems were designed by Sunwize Technologies to meet NMSU specifications based on the Mexican electric code (Wiles, 1996). This is a prepackaged control unit engineered for small-scale rural electrification and long life. The Moris PV systems consist of one 50 W Siemens SR50 module, which was the first deployment of these modules that were specifically developed for the rural lighting market. The PV modules are mounted on top of a 4-meter galvanized steel pole capable of withstanding high winds. The module charges a nominal 12 V sealed gel VRLA battery (Concorde Sun-Xtender, 105 Ah at C/20 rate for 25°C). These are sealed, absorbed glass mat (AGM) and never require watering. The immobilized electrolyte wicks around in the absorbed glass mat, which helps the hydrogen and oxygen that form when the battery is charged to recombine within the sealed cells. The thick calcium plates are compressed within a micro-fibrous silica glass mat envelope which provides good electrolyte absorption and retention with greater contact surface to plates than gelled batteries. The Concorde batteries are in compliance with UL924 and UL1989 standards as a recognized system component. These batteries meet US Navy specification MIL-B-8565J for limited hydrogen production below 3.5% during overcharging (less than 1% in Sun-Xtender's case), which means they are safe for use in living spaces. All batteries were installed inside a spill proof heavy plastic battery case strapped shut and children-proof. Control is maintained through the Solisto power center via a UL listed Stecca charge controller with a 10 A fuse. The system has a dc disconnect and 6 other dc fuses protecting different circuits. The controller uses fuzzy logic to monitor battery charging to avoid under or overcharging the battery and is equipped with an LED lighted display to indicate state of charge. The Solisto power center is still available on the commercial market; Chihuahua marked the first use of these power centers in the world.

The PV system powers three compact fluorescent lamps with electronic ballasts (20 W each). It also has a SOLSUM dc-dc voltage converter (3, 4.5, 6, 7.5, 9 V options) and plug to allow for use of different types of appliances, such as radio and TV. For an extra of US$200, end-users could also elect to install a Tumbler Technologies Genius 200 W inverter, although few chose to do so. Five users immediately decided to install the satellite DirectTV service upon installation, which comfortably allowed them about 3 h of color TV viewing with this service in the evenings. The design of the Solisto SHS assumed that a household using the full set of 3 fluorescent lamps (20 W each) for an average 2 h a day would consume about 120 Wh/day on average.

Given that Chihuahua averages about 6 sun-hours per day annually, and assuming an overall PV system efficiency of 60% for this fairly well designed system (*i.e.*, including battery efficiency losses, module temperature derate, line losses, etc.), the user could expect on average to have about 180 Wh/d of available power. There are seasonal variations and double or more power could be extracted from the battery on any single day, but could not be sustained long-term. As is typical for solar energy users, they quickly learned to live within finite energy system bounds and learned to ration energy use during extended cloudy periods, which are fortunately relatively rare in Chihuahua. As part of the project specifications, the installer was required to provide end-user training on how to properly maintain and operate the PV system, as well as a simple user instruction booklet.

2.3 System evaluation

From 1999 until 2008, the performance of a Solisto PV lighting system was continuously monitored at NMSU's Southwest Region Solar Experiment Station in Las Cruces, New Mexico, simulating usage of about 171 Wh/day. Climate and irradiance conditions in Las Cruces are very similar to those found in Moris, Chihuahua (less than 500 km distant), and the system is housed in an unconditioned house that performs similarly to unconditioned homes in Moris (*i.e.*, no HVAC system). The long-term monitoring provides a reasonable base case with which to compare fielded systems.

Measured parameters include solar irradiance (at 32° tilt), ambient temperature, battery temperature, PV current, battery voltage, and load current. Each parameter is sampled every ten seconds and averaged each hour and recorded. Lights are operated automatically by the data acquisition system with a timing circuit that turns on all 3 lights for two hours at 7:00 a.m., and then again for another two hours at 7:00 p.m., for a total daily usage of four hours for three lights. Note that several different types of fluorescent lights are tested, including the original Moris lights, for a total nameplate rating of 43 W. In Moris loads will vary, but the NMSU monitored system base load provides a meaningful average that utilizes the average daily PV power production. The charge controller has successfully protected the battery from severe abuse from overcharging and deep discharging during prolonged cloudy periods. The nominally regulated voltage on the battery averaged 12.9 Vdc each day, with the lowest battery voltages observed as 11.9 Vdc after cloudy periods. Discharge to charge ratio for the battery indicated a battery roundtrip efficiency of about 83%, with an average daily depth-of-discharge (DOD) of about 13.5%.

2.4 Field surveys

The intent of the Chihuahua pilot project was to demonstrate that simple PV lighting systems could be designed to provide reliable, essentially maintenance free electrical service for many years with full cost recovery. After nearly five years of operation, random field surveys were conducted of 35 homes in Moris and found that over 90% of the Solisto PV home lighting systems have performed exceptionally well without any significant problems (Foster *et al.*, 2004).

Performance was assessed through electrical measurements, visual inspection, and an end-user survey to determine user satisfaction. The 2003 survey results showed that over 80% of the installed systems were operating correctly and as designed, 11% were in fair condition (e.g., most commonly one of three lamps was no longer working), 6% were non-operational, and 3% of systems had been dismantled (e.g., user moved). The high percentage of working PV lighting systems after nearly five years demonstrates a new degree of reliability for PV home lighting systems rarely seen in Mexico before.

In the household survey, 94% of users expressed complete satisfaction with their PV lighting systems, 86% thought that PV was better than their previous gas lighting source, and 62% believed that the PV systems were reasonably priced for the service provided. New and expanded evening activities were also reported such as sewing, TV, reading, and studying. After five years, the PV systems have saved about US$300 in lieu of previous gas and dry cell battery options, while providing superior light and entertainment capabilities. The average rural family income in Moris is about US$3,000 per year (Ojinaga *et al.*, 2000), which represents a monetary savings for these rural families of about 10% per year. There will be additional future replacement expenses as the batteries and lamps come to the end of their useful lives; however, a number of system components like the PV modules are already an investment that will continue to pay off for years to come.

Among the few component failures experienced within the first four years of operation were individual lamps and ballasts in 9 systems. Some of the failed lamps had been since replaced by the users with conventional incandescent bulbs. Blown fuses were found in 6 systems, but the systems were still functional. The few blown fuses were due to users placing large loads above the fuse rating (2.5, 5, 7, and 10 A fuses used) along with users tampering with the system wiring in an attempt to bypass blown fuses rather than replace them. Batteries had been dismantled or swapped out in 4 cases (they had not actually failed), and charge controllers bypassed in 2 systems.

The sealed battery lifetimes have been very good and much better than most similar PV lighting systems used in Mexico, where batteries rarely last more than two years. Of the original Moris sealed maintenance-free 105 Ah batteries installed, only four had been replaced (they had been sold for cash) and typically replaced with a larger battery bank consisting of truck batteries ranging from 65 to 100 Ah. The four original sealed batteries dismantled or sold apparently had not actually failed; the users simply wanted a larger battery bank. In two cases, the owners had disconnected the charge controllers to override the low voltage disconnect. These users did mention that the shallow cycle replacement car/truck batteries did not last as long as the original deep-cycle batteries, but they had not attempted to make the effort to obtain more expensive deep-cycle batteries to expand their battery bank. PV modules proved to be one of the most reliable components, all modules were functional and no module problems had been reported.

The average daily electricity consumption was estimated by asking users their perceived time schedule for hourly use of appliances on an average day. Users were asked in the month of May, thus usage was more reflective of that month than winter months. This survey reflects their opinion and is not measured load data. The mean value was found to be 248 Wh/day (~20 Ah/day). This implies a daily cycling of about 20 % DOD of the battery at 25°C, which implies these batteries should last about 3,000 cycles (~8 years). Given this level of usage, the batteries in Moris eventually lasted from 7 to 9 years before the first battery replacement was needed. With today's LED technologies, even longer lifetime are possible. There was an increase in the electricity consumed in some households from the purchase of additional appliances such as radios and TV, but the PV systems handled the increased loads.

	Daily Average	Max	Min
Insolation	6.1 kWh/m²		
PV Power	32.1 W		
PV Current	2.1 A		
PV Energy	258 Wh/d		
Battery Charge	205 Wh/d		
Battery Discharge	171 Wh/d		
Net Energy	34 Wh/d		
Avg. DOD	13.5%		
Avg. Voltage	12.9 V	14.0 V	11.9 V
Avg. Current	1.2 A	2.8 A	-3.4 A
Temperature	30.5°C	35.1° C	28.2° C
Load Actual Power	39.4 W		
Load Avg. Current	3.1 A		
Load Avg. Voltage	12.4 V		
Load	171 Wh/d		

Table 1. Summary of the performance of the sunwize solisto PV system.

3. PV water pumping

PV water pumping is a small-scale application of great importance all over the world, has particular impact in rural communities where electrical network has not been extended. These systems are characterized by high reliability, long life and minimum maintenance, which translate to lower long-term cost when compared with other alternatives. Also does not require an operator, and its operation does not pollute the environment and produces no noise. Another advantage is that the systems are modular, so it can be adapted to meet the specific needs of the user at any time.

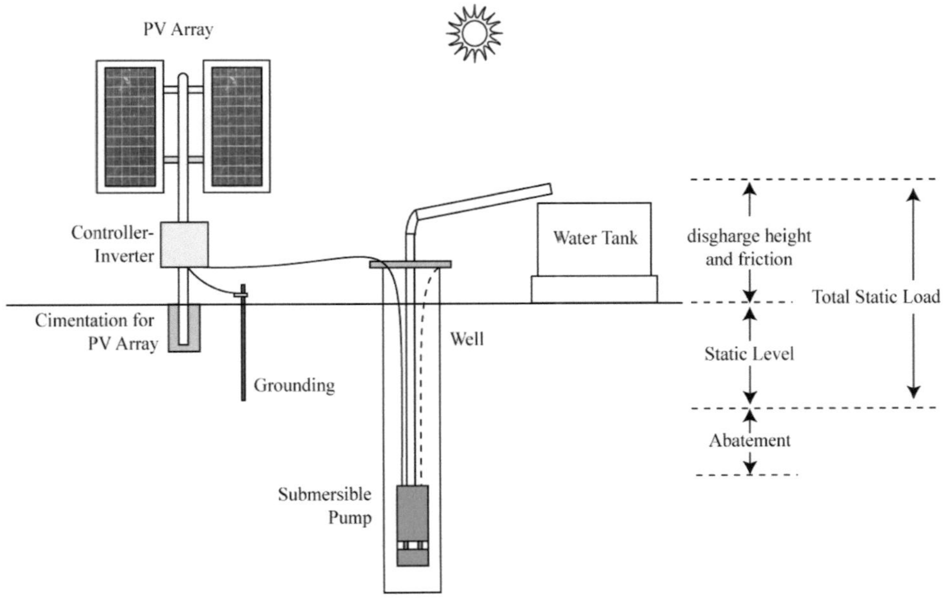

Fig. 2. Diagram of a PV system for water pumping.

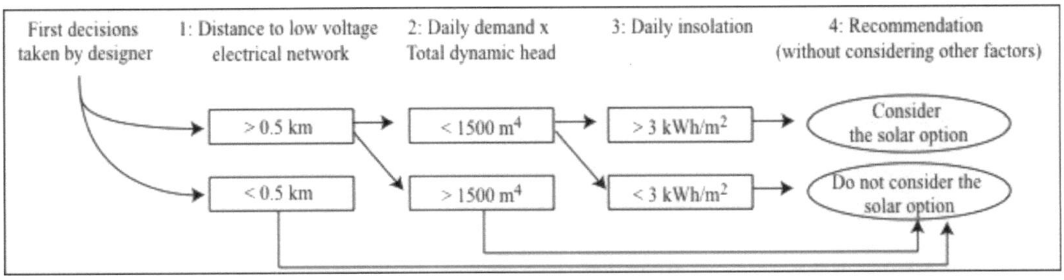

Fig. 3. Diagram for making a decision to use a PV system for pumping water.

PV systems have proven to be an excellent option in meeting water pumping where electrical grid service does not exist. Between 1994 and 2005, over 1,700 PV water pumping systems were installed throughout Mexico, initially as part of a MREP, and later with GEF/World Bank renewables for agriculture program. PV water pumping was largely unknown in Mexico prior to 1994, and MREP paved the way for widespread adoption in Mexico, which leads Latin America in this application.

Given that PV water pumping was largely unknown in Mexico and had a relatively poor reputation prior to 1994, US$2.2 million of USAID pilot hardware funds were used to buy down the PV system risk from the users perspective and were leveraged by additional user cost-share buy-in (~US$1.8 million) and additional Mexican agency implementation and administrative support (~US$0.5 million). DOE funds supported MREP technical assistance to Mexican partners from SNL, NMSU, Ecoturismo y Nuevas Tecnologías, Winrock International, and Enersol Associates. MREP worked with established Mexican agencies for project implementation, in particular FIRCO and the State of Chihuahua (Richards *et al.*, 1999).

Between 1994 and 2000, 206 PV water pumping pilot systems were installed in Mexico as part of the MREP. Most MREP PV water pumping systems were installed in the northern deserts of Mexico in rural areas that suffer from severe water shortages. Underground water is indispensable in these areas to meet daily water needs for domestic, crop, and livestock uses. Traditional water pumping systems powered by diesel or gasoline engines have been used for decades. However, the cost and transportation of fuel, and also engine maintenance, make conventional water pumping technologies expensive for people living in rural areas. One solution to reduce total system and operational cost of conventional water pumping systems is to replace them with PV systems. These may offer a less expensive life-cycle-cost option in many cases. Line extension of the utility grid is prohibitively costly at over US$9,000/km, depending on terrain. Distance to the grid ranges from a few to dozens of kilometers in many cases.

Typical installed system configurations included a PV array (~500 Wp on average), pump, controller, inverter (only for ac powered pumps), and overcurrent protection devices, generally installed in compliance with the Mexican National Electric Code (NOM-Norma Oficial Mexicana), which parallels the US National Electrical Code (NEC).

Table 2 presents a summary of the 206 PV water pumping pilot systems that were installed under MREP in Mexico. A total of 101 kW of PV were installed benefiting 9,389 people. For the first three years, MREP was cost-sharing about 80% of total system costs. After 1996, Mexican counterparts were convinced of the effectiveness of PV technology for water pumping; thus, their willingness to pay gradually increased from about 20% up to 85%, dropping MREP cost-sharing to only 15% by 2000. After 2000, FIRCO has installed over 600 additional PV water pumping systems to date under a World Bank/GEF Renewables for Agriculture Program in Mexico.

	1994	1995	1996	1997	1998	1999	2000	1994-2000
Total kW installed	1.8	2.5	16.9	34.4	26.4	16.6	2.6	101.1
Number of Systems	6	5	24	66	59	41	5	206
Direct Beneficiaries	482	242	1,511	2,705	3,009	1,400	37	9,389
Avg. System Size, Wp	300	507	704	521	446	404	514	491
Avg. $/Watt	$22.01	$22.87	$18.96	$19.06	$19.81	$22.49	$14.77	$19.98
MREP Cost-Share %	78.10	86.50	82.90	63.10	41.90	36.40	15.00	57.60
Mexican Cost-Share %	21.90	13.50	17.10	36.90	58.10	63.60	85.00	42.50

Table 2. Summary of the 206 PV water pumping pilot systems installed under MREP.

After ten years of MREP PV system implementation, FIRCO, NMSU, and SNL conducted a review in 2004 on over 1/5 of the first installed PV pumping systems. The objective of the review was to determine technical status, reliability, and user acceptance of systems after several years of owning and operating such systems. After performing the technical evaluations, it was found that over 3/5 of the surveyed systems were operating appropriately after as much as 10 years of operation. A total of 85% of users thought that PV systems had excellent to good reliability.

Fig. 4. PV water pumping systems in Chihuahua and Baja California Sur, and FIRCO engineer conducting performance evaluation.

3.1 Review of PV water pumping systems

Field surveys began in July of 2003 and continued until March 2004. During these visits, either the owner or the responsible person operating the PV water pumping system was surveyed. A total of 44 questions were included and classified into eight sections, which were: (1) general demographic information and system specifications; (2) information of traditional pumping systems used prior to PV system installation (if any); (3) user perception of vendor and installers; (4) productive and commercial impacts as a result of the use of PV pumping systems; (5) environmental impacts as a result of the use of PV pumping systems (if any); (6) replication of additional systems; (7) user lessons learned, and; (8) other renewable energy applications.

The PV water pumping systems were visually and electrically inspected for electrical performance and pumping productivity. Electrical measurements on the PV array and the controller/inverter were made at the same time to determine water volumetric rate and solar radiation. Wiring, connectors, insulation, junction boxes, breakers, and water pipe were also inspected. Technical field inspections were carried out by engineers from FIRCO, NMSU, and EcoTursimo y Nuevas Tecnologías.

Before installing PV systems, 72% of the visited ranches had conventional pumping systems using gasoline, diesel, car engines, and one used an animal traction system. The typical consumption of gasoline for pumping water ranged from 5 to 10 liters per day for the states of Baja California Sur, Chihuahua, and Sonora. In the state of Quintana Roo, the consumption ranged from less than one liter per day up to 2.5 liters. Northern Mexico is an arid and hot region; livestock and crop production requires more water. Gasoline systems also required about 3 liters of lubricating oil per month. According to user's responses, a conventional gasoline or diesel system only lasts from 4 to 5 years. Solar pumps already exceeded this lifetime in many cases. Once the fossil fuel powered systems started to fail, they had to be repaired 2 or 3 times per year. People who were satisfied with the operation and productivity of PV water pumping systems mentioned that PV systems saved them

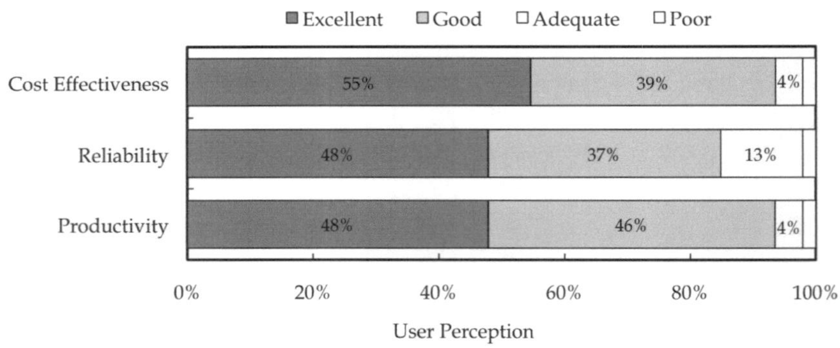

Fig. 5. User perception about cost effectiveness, reliability, and productivity of PV water pumping systems.

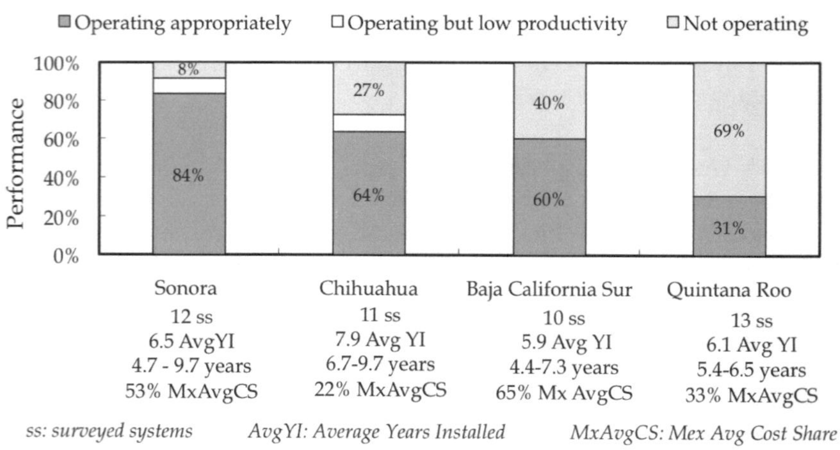

Fig. 6. Performance of surveyed systems by state.

money and time because there is no need to buy and transport fuel, less maintenance is required, and no time is invested in operating the systems on-site as was required before. The survey results found that over 4/5 of the rural Mexican users were satisfied with the reliability and performance of their PV water pumping systems.

The majority of surveyed users in Baja California Sur, Chihuahua and Sonora responded that the work done by vendors and installers ranged from good to excellent regarding installation, training, post-sales service, and the operation and maintenance manual. On the contrary, in the state of Quintana Roo, these answers ranged from bad to adequate on vendor performance (with only two exceptions).

Due to a severe decade long drought in Northern Mexico, the desert ranches in Baja California Sur, Chihuahua and Sonora identify water as a larger issue than in tropical Quintana Roo. Regarding the productive uses of the water, from the 46 surveys, it was found that 100% used the water for livestock watering, 13% also used it for irrigation and 19% for domestic uses.

Figure 7 presents the average cost in dollars per watt of the PV water pumping pilot systems by state and installation year of MREP systems. The continuous line corresponds to the average cost for the installed systems in the State of Chihuahua. During the introduction of PV technology for water pumping, the cost was 22 and 25 dollars per installed watt in 1994 and 1995, respectively. After 1995, a decrease in cost reflecting PV market maturity was observed. By the end of 1999, the average cost was US$12/Wp. Over 40 systems were installed in Chihuahua. Similar results were also seen in Baja California Sur with 40 installations. In other states, the program implemented only a few projects and the PV market had not sufficiently matured and there was less vendor competition. MREP experience shows that key factors for achieving a mature market include training, program size, multiple vendors, quality workmanship, code compliance, and technologies deployed.

A total of 46 of the original 206 installed PV systems (22%) were surveyed to determine reliability and user acceptance of PV technology after owning and operating them for as much as 10 years. The survey was conducted in the states of Baja California Sur, Chihuahua, Quintana Roo, and Sonora.

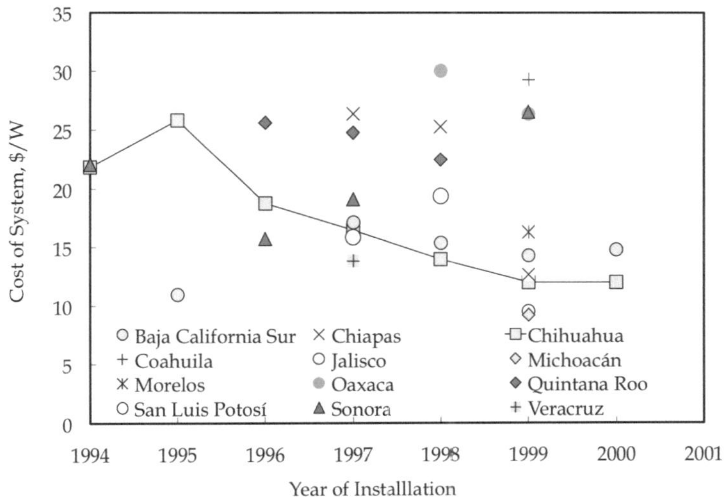

Fig. 7. Average cost of PV water systems by year and by state.

A representative example of successful PV water pumping pilot systems is at Rancho El Jeromín in Chihuahua. The system at Rancho El Jeromín was installed in 1995 utilizing an ASE Americas 848 Wp array to pump 12.5 m^3/day of water daily via a Grundfos pump operating at 40 m total head. The system has not had a single component replaced and has pumped water daily as designed for the past eight years. Full PV system payback was realized in only 2.5 years. Figure 8 presents the life cycle cost analysis for the PV system installed at Rancho Jeromín compared to the conventional diesel system previously used. Since the solar system installation, the owner has saved over US$15,000 in fuel and maintenance, and the PV system should still provide many years of service to come. This was based on initial fuel costs of the mid-1990's of US$1.00 per gallon. Payback would decrease proportionately as fuel prices increase (e.g., at US$2.00 per gallon, payback is half the time.)

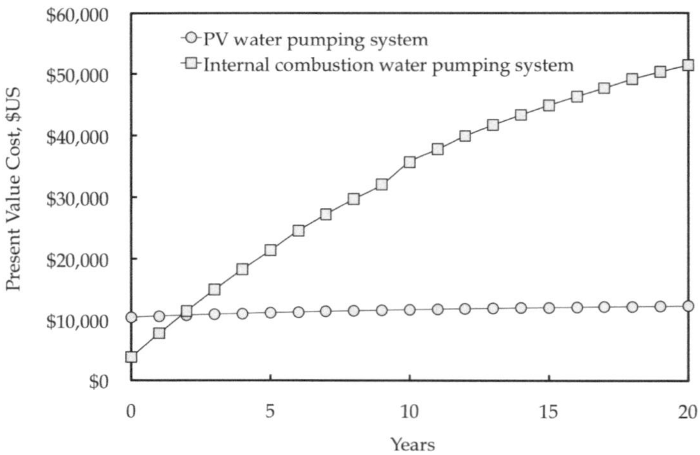

Fig. 8. PV system payback realized in 2.5 years for the Rancho El Jeromín solar vs diesel powered pump. Since system installation in 1997, the rancher has saved over US$30,000 in fuel costs.

The average installed time for all the systems surveyed was 6.5 years. The oldest systems were installed ten years before the review and included the very first system installation in Estación Torres, Sonora utilizing a Grundfos SP3A-10 solar pumping system installed by Applied Power. This system has been operating daily since 1994 with no parts replaced or maintenance of any kind.

4. PV ice-making and refrigeration

In the middle of the Chihuahuan desert lies the Luis Leon Reservoir formed from the waters of the Río Conchos as seen in Figure 9. For over a quarter century, fishermen from the nearby community of Chorreras have fished this man-made lake for bass, catfish, tilapia, sunfish, and carp. Today, there are about 70 fishermen who make a reasonable living from the lake. The community is not serviced by the conventional electric grid, and it is nearly a four-hour drive from the lake to Chihuahua City to get the fish to market. Thus, the fishermen have had to rely on Chihuahuan wholesale merchants to come and purchase fish from them. The fishermen of Chorreras sometimes have lost fish to spoilage due to lack of ice. The fishermen also end up paying relatively high rates for the trucked-in block ice from Chihuahua when it does show up. The fishing cooperative annually harvests about 80,000 kg of fish. However, with no local ice source, they have had to put off fishing or take their chances that ice will arrive on time. The lack of ice also limited their ability to independently sell their fish, particularly during the high demand season of Lent in the spring.

Recognizing this problem, the State of Chihuahua and SNL joined forces to install a renewable energy powered ice-making system. The goal was to install an on-site ice-maker that could adequately meet ice needs. In response, NMSU set up a solar and wind resource monitoring system in 1995 to verify the renewable resources. While the initial concept was to install a wind-powered ice-making system, the wind resource was deemed inadequate with an annual average windspeed measured of only 3.5 m/s (SWTDI, 1999). Thus, the concept for powering the ice-maker from PV came to the forefront with an average annual solar resource of about 6 kWh/m^2/d.

Fig. 9. World's first PV ice-maker developed by SunWize in the heart of the Chihuahuan desert for the fishermen of Chorreras.

The world's first automatic commercial PV ice-making system was installed in March 1999 to serve the inland fishing community. The Chorreras ice-maker system was designed and installed by SunWize Technologies of Kingston, New York, with the assistance of Energía Solar de Ciudad Juárez (ENSO) from Chihuahua. This project was possible due to the support of developing high-value renewable energy applications provided by the New York State Energy Research and Development Authority (NYSERDA), which had teamed with SNL, the State of Chihuahua, and the NMSU to develop, install, maintain, and monitor a PV hybrid ice-maker. The project was done in coordination and with cost-shared funding assistance from the USAID/DOE MREP.

The US$38,000 hybrid system was operated from 1999 to 2002 and produced an average of 8.9 kWh/d at 240 V to the icemaker. The system Coefficient of Performance (COP) was 0.65 and a total of 97% of the energy was supplied by the PV array, while the backup propane generator supplied only 3%. Production of ice varied each month due to changes in insolation and ambient temperatures and averaged about 75 kg of ice/d. About every 9 months, the icemaker water lines would need to be cleaned to remove calcium deposits. With a fixed timer setting, the icemaker operated daily for 3 hours with a dozen 15 minute cycles at night to make ice, except on Sundays (no fishing).

4.1 Ice-making system design

The concept of solar-powered ice production in the remote desert is not a trivial one. High solar insolation certainly maximizes ice-making potential, but likewise high ambient air temperatures of over 40°C in summer also reduce that potential. The northern Chihuahuan desert has high summer ambient temperatures, as well as winter temperatures well below freezing, which is an abusive environment for batteries. These considerations led to some interesting design and operational challenges. Finding an acceptable freshwater source was another challenge since the fishermen wanted to be able to use the ice for personal use as well. This desire eventually resulted in the community building a 7 km gravity flow aqueduct across the rocky desert ground from a clean spring water source.
The PV hybrid system is built on a galvanized steel frame bolted on a concrete platform and consists of the following major components: 2.4 kW PV array (fixed 30° array tilt) with 32 Siemens SP75 solar modules, Ananda Power Technologies (APT) power center, 24 Vdc 2200 Ah battery bank with 2 V cells, two Trace Engineering 3.6 kW modified sinewave inverters provide 240 Vac electricity and one Kohler 6.3 kW propane fueled generator.
Figure 10 shows a one-line diagram of the complete PV hybrid ice-making system. The propane-fueled generator was included in the design to provide backup battery charging and boost ice production when needed for larger fish hauls and/or cloudy weather. Operation is controlled automatically through inverter set points. The batteries absorb high current transients and allow for load shifting to nightime for more efficient summertime ice production when cool ambient temperatures are favorable for maximum ice production. The battery bank is thermally insulated and uses dc fans for cooling and hydrogen venting. Two Trace modified sinewave inverters are stacked together to deliver 60 Hz, 240 Vac single-phase power for the icemaker.

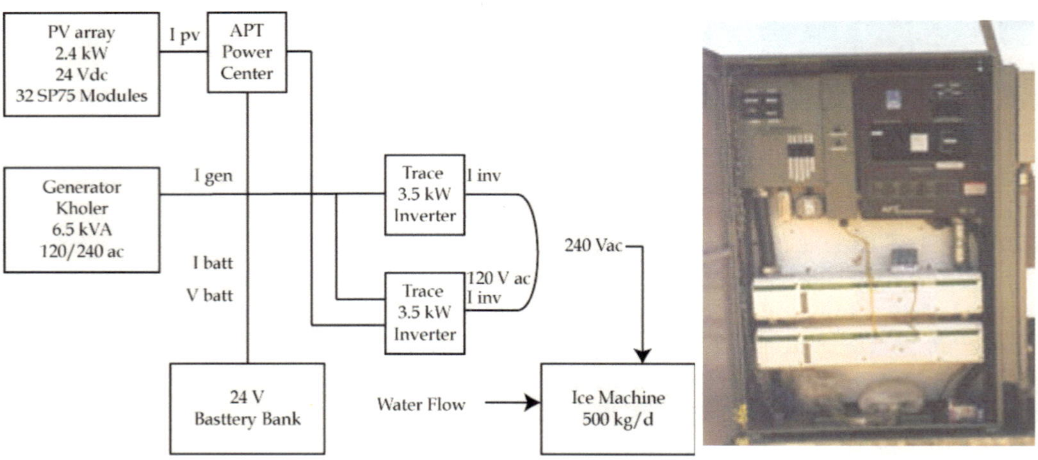

Fig. 10. Icemaker diagram and system Trace DR series inverters and APT power center with disconnects.

The icemaker is a vertical-evaporator compression-cycle unit installed on the roof of the fish storage building. It was designed for low maintenance and high reliability. For this specific application, modifications were made to the icemaker to reduce power consumption. A smaller compressor and condenser heater was modified resulting in a reduced current from approximately 22 to 11.5 Amp at 240 Vac, thus reducing power requirements by about 40%. A 7 km aqueduct was installed by the Chorreras community to the fish storage facility to

provide high-quality water. The polypropylene pipeline was buried 0.3 m in the hard desert rock soil to help provide lower water supply temperatures. Approximately 1.5 km from the icemaker stands a 10 m^3 storage tank on top of a hill that provides consistent gravity water flow with sufficient pressure for the icemaker (Hoffstatter, 2000).
The icemaker is set to run a dozen or so 15-min automatic ice-making cycles each day (about 3 hours a day). The system freezes the water, a crusher breaks the ice into convenient flakes, and the ice falls via gravity into a cold storage room. The PV system typically produces about 90 kg of flake ice per day. However, production of ice can be increased by manual operation of the generator allowing a maximum ice production capacity of more than 400 kg/day. The timer is set to provide no ice on Sundays (a day with minimal fishing) to allow the PV array to fully charge the battery bank and help equalize the batteries each week.

4.2 System reliability

A data acquisition system (DAS) was designed, built, and installed by NMSU for SNL and SunWize to monitor system performance. The DAS was installed in March, 1999 and uses a GOES-based satellite communication system for the remote site. The DAS consists of a Campbell Scientific CR-10X Datalogger, electronic transducers, and an assortment of other sensors. The DAS is used to measure several environmental and system parameters including: PV current, generator run-time, generator current, battery voltage, current and temperature, load voltage and current, ice-room and ambient temperatures, insolation at the array inclination and water flow. The stored data is hourly averaged and transmitted every four hours via the GOES satellite. Monthly data reports allow the project team to monitor system performance and identify any potential problems.
The icemaker is set to operate during the cool, late-night hours during the summer since the high ambient and water temperatures reduce the system ice-making efficiency during the daylight hours. The load is driven solely by the batteries at night while the PV array replaces the consumed energy during the day. This nighttime operation results in deep-battery discharge cycles but increases ice production. In the winter, the system is used to produce ice during the daylight hours, allowing the PV array to provide some energy, extending battery lifetime.
Adjustments to the compressor and modifications on the control timer improved the ice production from a daily average of 80 kg of ice during the first three months of operation to 90 kg.
Daily, weekly, and seasonal weather differences results in variations in the generator run time. During the longer summer days, generator operation is more infrequent. During the first 14 months of operation, the generator provided only 3% of the total energy used.
The batteries are enclosed in a thick-walled, insulated industrial plastic enclosure filled with water and baking soda; however, temperatures in excess of 45°C (hourly average) were recorded while the batteries were being charged. The original passive cooling vents and a small hydrogen vent fan were not cooling the batteries sufficiently after installation; a dc cooling fan was added in July, 1999 to the battery container which remediated high battery temperatures and kept the battery bank below 40°C.

4.3 System performance

Figure 11 summarizes the energy performance of the system from April 1999 - May 2000. The PV array supplied a total of 3,542 kWh (253 kWh monthly average) of energy; the generator delivered a total of 115 kWh of energy. The total energy input to the system (PV plus generator) was 3,657 kWh over the 14 months.

The ac load (ice-maker) consumed a total of 2,075 kWh allowing for an overall system efficiency (energy-out/energy-in) of 57.4%. The battery performance for the first 14 months of operation was found to provide a 50% round-trip efficiency (discharging to charging ratio). Both inverters run continuously, while supplying power to the ac load and the DAS. The roundtrip battery bank efficiency (discharging to charging ratio) was steady throughout the first year indicating little to no change for overall battery bank capacity.

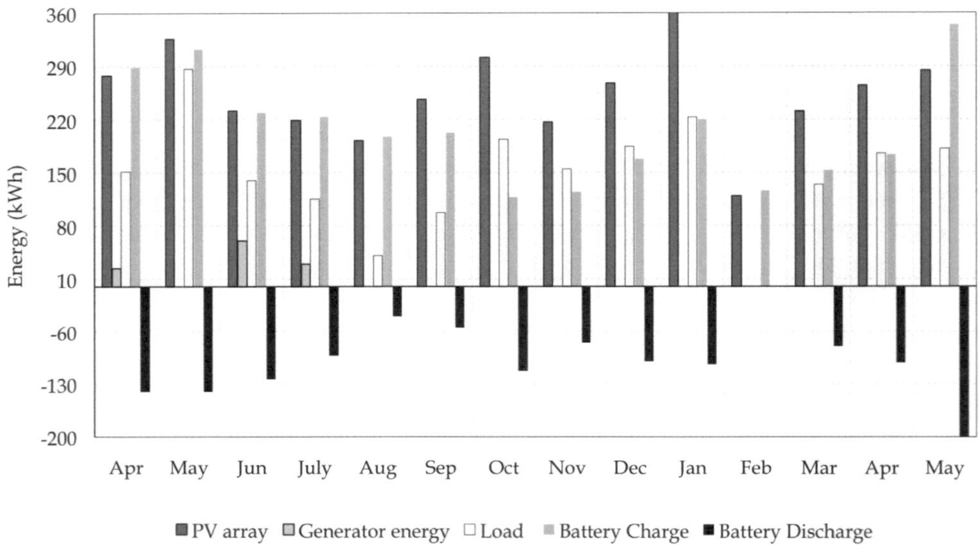

Fig. 11. First year system energy performance.

4.4 Economics of the system

The State Government of Chihuahua purchased the ice-making system for US$38,000 with cost-share assistance from Sandia and USAID. In addition, the State of Chihuahua and the community of Chorreras pitched in additional funds to build the 7 km aqueduct and to rehabilitate the ice-room. NYSERDA funded engineering design and development for this novel system, and Sandia funded the DAS and follow-up system monitoring. Thus, the final cost for this project was about US$150,000. However, the value of the now commercialized PV ice-maker unit is about US$50,000.

Ice production has been found to be about 11.5 kg per sun hour with an overall COP of about 0.65. The system can produce over 25,000 kg of ice per year from the solar alone. Assuming a value of US$0.30 per kg of ice (for this remote site where it must be hauled in), this implies that a simple payback for the ice-making system is under 7 years. Taking into account the value of reduced fish spoilage, actual payback is actually well under 5 years for the PV icemaker. Overall, it is anticipated that ice production over the system lifetime, with future battery replacements and system maintenance, should be about US$0.15 per kg. Of course, having a reliable source of ice in the desert for a cold drink has an intrinsic value that is difficult to express simply in terms of dollars and cents.

It is important to include thermal consideration in the design of battery racks or containers. Even small thermal differences among batteries can contribute to battery decay in the long-run. A strict maintenance schedule and procedure is required for batteries that pays special attention to safety. This schedule includes adding water and monitoring battery temperature and voltage. In periods of low insolation (winter with short days, or cloudy seasons) consideration might be given to adjusting the inverter set points to allow longer generator run times. Manually initiated frequent (monthly) equalization periods are also recommended.

For any type of relatively complicated hybrid system, it is important to not only consider the technical side of the equation, but the institutional side as well. The system has proven that a properly designed, operated, and maintained system can indeed produce a significant and valuable resource, such as ice, even in the middle of the desert. However, such a successful system requires local buy-in and follow-up, and a complicated system such as this if it was simply "parachuted in" would soon not be functioning due to relatively minor problems that require an experienced technician to solve. Long-term commitment and follow-up by the project partners is required for project success. This project is a good example of using renewable energy as a tool to contribute to local economic development in a remote area.

The icemaker performed adequately for the first three years of operation. The project showed that a properly designed, operated, and maintained PV system can indeed produce a significant and valuable resource, such as ice, even in the middle of the desert. Long-term commitment and follow-up by the Mexican project partners was necessary for continued project success. Unfortunately, there were State political changes and the area faced a severe drought. The lake receded over 2 km from the ice house by 2003 and the fisherman moved their catch out to the other end of the resevoir. The ice-making system was shut down and unfortunately has not been operated since. Other Mexican coastal communities attempted to purchase the unused ice-making system, but the Chorreras community refused to sell it in the belief they may one day again reactivate it.

5. PV refrigerators

A significant development for PV refrigeration technology came from SunDanzer in support of NASA. The SunDanzer refrigerator uses thermal storage, and a direct connection is made between the cooling system and the PV panel. This is accomplished by integrating a water-glycol mixture as a phase-change material into a well-insulated refrigerator cabinet and by developing a microprocessor-based control system that allows direct connection of a PV panel to a variable-speed dc compressor. The refrigerator uses a more efficient variable-speed dc compressor. The unit is designed to run on 90 to 150 watts of PV power (needed for compressor start-up), but only draws about 55 W when cycling. During cloudy weather, internal thermal storage keeps products cold for a week, even in a tropical climates. The battery-free unit is designed to work optimally in locations with at least 4 sun-hours per day using a variable speed compressor and peak power tracking.

NMSU began testing solar refrigerators in July 2000 at its facilities and later in the field in 2002. Units were field tested on the Navajo Indian Reservation in New Mexico; in Chihuahua and Quintana Roo in Mexico; and at the highlands of Guatemala through Fundación Solar. The unit offers the most economical method for on-site refrigeration for rural people. Based on these results and lessons learned, only in 2010, did SunDanzer finally launch a commercial battery free solar refrigeration unit that can be purchased today.

Fig. 12. SunDanzer PV direct drive refrigerator piloted in the indigenous Mayan village of Santa Clara, Quiché, Guatemala by NMSU, NASA, and Fundación Solar in 2002.

6. PV for schools

Thousands of rural schools in Latin America do not have grid power. Solar power offers a practical way to meet their power needs. Many early school PV systems often failed and gave the technology a poor image. Around 2000, PV school installations in many parts of Latin America began to show great improvements as the industry matured. Large-scale rural school electrification programs have been implemented in Mexico, Guatemala, Cuba, Honduras, Peru, and Brazil. For instance, the Fundación Solar and the Fundación para el Desarrollo Rural de Guatemala began using PV to bring distance education programs to remote areas that were devastated by Hurricane Mitch in 2000. The PV system is used to power televisions, videocassette recorders, and computers to modernize the educational

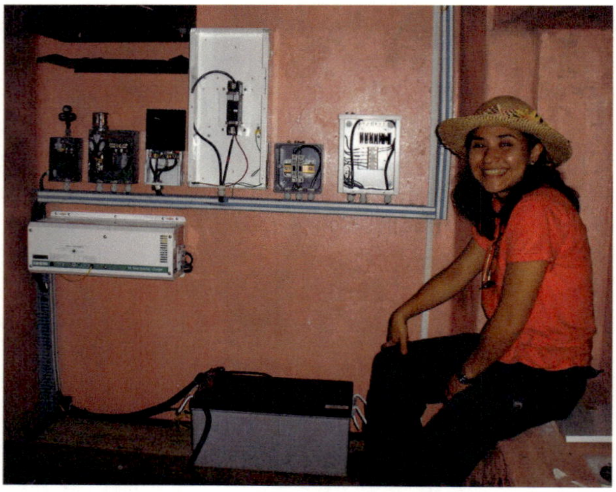

Fig. 13. COHCIT Sosoal PV satellite telecenter with internet connectivity using quality BOS components with SOLARIS installer Ethel Enamorado in Lempira, Honduras.

experience of rural school children. Mexico has over 500 PV powered schools, with some of the best examples being the 54 PV telesecundaria schools in Chihuahua installed in November, 2002 by EDUSAT/State of Chihuahua for satellite education. MREP provided technical advice to avoid common errors.

The Consejo Hondureño de Ciencia y Tecnología (COHCIT) has installed a half dozen quality PV telecenters/schools in rural Honduras with assistance from NMSU. COHCIT with the World Bank set up a first pilot PV powered telecenter in the community of Montaña Grande near Tegucigalpa in 2003. As a result of this, COHCIT installed 5 more telecenters in 2004 with the Inter-American Development Bank and is planning more.

7. PV for protected areas

Renewable energy technologies have been widely applied to support protected area throughout Latin America, especially in Guatemala and Ecuador (Galapagos) (Ley & Stoltenberg, 2002). Mexico with MREP has installed over 70 solar systems in protected areas such as Isla Contoy, El Eden, Montes Azules, and Sian Ka'an Reserves with the Mexican Secretariat of Environment and Natural Resources (SEMARNAT), the Nature Conservancy, World Wildlife Fund, and Conservation International.

Use of solar in protected areas benefits the living conditions of researchers, technicians, and rangers, as well as providing energy for environmental training centers. The solar systems also have the advantage of providing power without the noise or pollution associated with conventional fossil-fueled generators, while reducing the risk of fuel spills in these sensitive biosphere reserves. As always, up front design decisions, user operation, and long-term maintenance issues play an important role for overall system reliability.

Solar energy is an environmentally appropriate example to neighboring buffer communities (often without electricity) surrounding biosphere reserves which can likewise benefit by replicating the protected areas example. Solar systems also provide a useful example for visitors and tourists to take back home.

In addition, the remote protected area facilities benefit economically from solar installations through reduced operation and maintenance costs associated with fossil fuel generators. Actual system life-cycle costs for any particular solar or wind energy system varies and is a function of design, usage, application, and maintenance. With proper system operation and maintenance, the expected solar system lifetimes should exceed 25 or more years (with appropriate battery replacements, etc.).

8. Hybrid systems

The road for hybrid system application in Latin America has been difficult. While the various solar technologies are proven, the institutional and organizational issues for these more complicated systems have proved to be the most difficult to overcome. Some of the key hybrid projects implemented in Latin America include the Campinhas project in Brazil, and the Xcalak and San Juanico systems in Mexico.

In 1992, the State Government of Quintana Roo funded the installation of the world's largest (at that time) wind/solar village hybrid system in Xcalak. The idea was to provide additional hours of power for the community beyond the 3-4 hours per day that the diesel was operated. The combined wind/PV hybrid system hardware cost was approximately US$450,000 and installed by Condumex. The generation system consisted of six Bergey Windpower nominally rated 10 kW Excel wind turbines and 11.2 kW of Siemens PV modules. Energy was stored in two battery strings using 216 GNB Resource Commander batteries for a combined total of 1738 Ah at 220 V. The stored energy was provided to the town's electric grid via an Advanced Energy Systems 40 kW sinewave inverter.

Originally the wind and PV system output was adequate to nearly meet the entire village's electric power demand for 24-h power. However, the village loads rapidly grew after system installation (53% in the first year alone) and there were no electric meters. By 1997 the Xcalak renewables system provided less than 30% of total community power due to significantly increased loads and lack of system maintenance.

After five years, the system ceased to function altogether, in particular due to the failure of the 40 kW inverter, which faced a difficult job in Xcalak with highly unbalanced system loads and corrosion exacerbated by drawing humid air from below ground concrete raceways.

Fig. 14. Xcalak, Mexico wind turbines and PV array (1993).

The early years of the Xcalak hybrid system showed that wind and PV technologies can provide abundant and reliable electric service. However, the lack of institutional planning led to inadequate system maintenance, excessive load growth, and eventual system failure. For hybrid systems to be a viable, an adequate and manageable institutional structure must accompany the technology. To avoid failure, village hybrid systems must include realistic system sizing and proper institutional controls from the onset.

9. Lessons learned

When developing solar projects in Latin America, there is a tendency for some organizations to focus on the technology, while other focus largely on institutional issues. The happy medium takes into account both and promotes partnerships, local capacity building, quality technical design, and monitoring and evaluation.

Some key considerations for any solar project include: Develop solid partnerships, conduct strategic planning, use grass-roots development approach, foster reasonable end-user expectations, create sustainable markets, promote capacity building, size appropriately, obtain user input, develop a professional design, insist on quality, conduct preventive and regular maintenance, anticipate future growth, maintain parts supply inventory, consider safety and security, demand guarantees and warranties, conduct follow-up and evaluate results and think sustainability.

When developing solar projects in Latin America, there is a tendency for some organizations to focus on the technology, while other focus largely on institutional issues. The happy medium takes into account both and promotes partnerships, local capacity building, quality technical design, and monitoring and evaluation. Some key considerations for any solar project include the following:

- Develop solid partnerships: The most sustainable and viable projects are formed when in-country agencies partner with industry. It is important to choose partners carefully.
- Conduct strategic planning: Strategic planning with collaborating partners helps to create realistic goals that makes PV as a useful tool for established programs. Planning should include sufficient promotional activities to accelerate acceptance, including training.
- Use grass-roots development approach: An integrated and grass-roots development approach across a critical mass of different agency types provides a strong base for dissemination and replication. A local and capable champion greatly facilitates local solar development.
- Foster reasonable end-user expectations: Do not oversell PV technologies and capabilities that might disappoint users. End-users want quality systems that work and supplies them the power they need.
- Create sustainable markets: Financing is a major barrier to market growth. Renewables must be cost accessible to rural people and often require smart cost-sharing or financing. Reinforce commitment to sustainability and perceived system value from systems that are donated to ones that users find affordable through micro-credit lending.
- Promote capacity building: In-depth training is critical. It is important not only to train project developers, but also users and local industry (supply side). Success depends largely on the technical capacity of local technicians, users, and administrators while considering gender issues. Adopt participatory techniques in community projects.
- Size appropriately: System sizing and design needs to be focused and realistic as to user needs and loads to avoid unnecessary expenditures on larger systems than required. The system needs to meet the loads now and be expandable for the future. Choose energy saving devices to reduce PV system size and save money.
- Obtain user input: Clearly identify user needs and develop appropriate technical specifications for a system to meet those needs. Consider technical, gender, and cultural issues as well as economic constraints.
- Develop a professional design: Design parameters should be developed by experienced engineers and include realistic system usage, climatic conditions, component selection, O&M considerations, safety, and reliability considerations.
- Insist on quality: Installations should be made by experienced technicians that exhibit good workmanship and meet electrical code requirements. For larger programs, acceptance testing of installed systems should be conducted to verify that contractual obligations have been met.
- Conduct preventive and regular maintenance: O&M is required for long-term successful system operation. There are diverse maintenance levels. Some actions can be undertaken by the end-user, while more complex tasks requiring a skilled technician. Proper tools must be provided. An O&M actions journal is recommended.

- Anticipate future growth: Design a system accordingly for relatively seamless expansion.
- Maintain parts supply inventory: Required for components that are likely to be replaced (e.g., fuses). Build a strong supplier network. Try to use appropriate local components as much as possible to avoid delays in replacement parts. Facilitate links between the end-users (men and women) and equipment suppliers.
- Consider safety and security: Design with safety in mind, meet all applicable codes and standards. Be vigilant as to potential theft, vandalism, etc., and plan accordingly.
- Demand guarantees and warranties: Use reputable vendors who offer guarantees and know what these are. Consider long-term preventive maintenance contracts for system support with the equipment vendor.
- Conduct follow-up and evaluate results: Monitoring and follow-up are key to understanding the true results for any program. End-user surveys can provide valuable feedback in regard to customer expectations, usage patterns, and overall satisfaction. This information helps with future planning.
- Think sustainability: All paths should lead to this and institutions applying solar systems must have a true commitment for long term sustainability. Government agencies face particularly difficult challenges with often changing parties in power. The ultimate goal is to have a well designed and installed solar system that will provide many years of reliable and satisfactory service. The past twenty years have set the stage for future solar development in Latin America, which is growing exponentially.

10. References

Cota, A. (2004a). Azteca Solar: The Mexico Renewable Energy Program. US Department of Energy, Sandia National Laboratories y Southwest Technology Development Institute. New Mexico State University- SWTDI on line. (http://solar.nmsu.edu/publications/index.htm#mex).

Cota, A.; Foster, R.; Gómez, L.M.; Ross, M.P.; Hanley, C.H.; Gupta, V.P.; Montufar, O & Paredes, A.R. (2004). Ten year reliability assessment of photovoltaic water pumping in México. *Proceedings of the American Solar Energy Society: SOLAR 2004*, Portland, Oregon, USA.

Foster, R.; Ghassemi, M. & Cota, A. (2009). *Solar Energy: Renewable Energy and the Environment. In the Series: Energy and the environment.* Series Editor de series: Ghassemi, A. First Ed. CRC Press. ISBN: 352 978-1-4200-7566-3, Boca Raton, FL, USA.

Foster, R.; Ley, D.; Martinez, H.; Estrada, L. & Lara, E. (2006). Nicaraguan renewable energy for rural zones program initiative. *Proceeding from Solar 2006 of the American Solar Energy Society, ASME International Solar Energy Conference – Solar Engineering,* Denver, CO, USA.

Foster, R.; Cota, A. & Estrada. L. (2004). Five-year reliability assessment for SoListo photovoltaic home lighting system in Chihuahua. *Proceedings of American Solar Energy Association Annual Conference*. Portland, Oregon.

Foster, R. E.; Estrada, L.; Gomez, M. & Cota, A. (2004). Evaluación de la confiabilidad de los sistemas FV SOLISTO en Chihuahua, *Proceedings of the 12th Intnl. Symposium on Solar*

Power and Chemical Energy Systems, SolarPACES, 28th Semana de Energía Solar - ANES, Oaxaca, Mexico.

Foster, R.; Estrada, L.; Ojinaga-Santana, L.; Colmenero, J.; & Ross, M. (2003). Utilizing photovoltaics to support distance education in the state of Chihuahua, Mexico. American Solar Energy Society, Austin, Texas.

Foster, R.E.; Estrada, L.; Stoll, S.; Ross, M. & Hanley, C. (2001). Performance and reliability of a PV hybrid icemaking system, *Proceeding of the 2001 Solar World Congress. International Solar Energy Society,* Adelaide, Australia.

Foster, R.E.; Orozco, R. C.& Romero, A. (1999). Lessons Learned from the Xcalak Village Hybrid System: A Seven Year Retrospective, *Proceedings of the 1999 Solar World Congress. International Solar Energy Society,* Vol. I, Israel Ministry of Science, Jerusalem, Israel.

Foster R. E.; Ghosh, S.; Carrillo, O.; Molina, D. & Panico, D. (1998a). Willingness to pay for solar photovoltaic energy lighting systems: The case of rural Chihuahua," *Proceedings of the 1998 Annual Conference,* American Solar Energy Society, 21-25, Albuquerque, NM.

Foster, R. E.; Cisneros, G. & Hanley, C. J. (1998b). Life-cycle cost analysis for photovoltaic water pumping systems in Mexico, *Proceedings of the 2nd World Conference on Photovoltaic Energy Conversion, 15th European PV Solar Energy Conference, 27th US IEEE Photovoltaics Specialists Conference, 10th Asia/Pacific PV Science and Engineering Conference,* Vol. III, ISBN 92-828-5420, 3021-3025,Vienna, Austria.

Hanley, C.; Ross, M.; Montufar, o.; Rovero, C.; Foster, R.E. & Ellis, A. (2001). Introducing photovoltaics to new markets through government development programs: The FIRCO Example in Mexico, *Photovoltaic Systems Symposium*, Sandia National Labs, US Department of Energy, Albuquerque, New Mexico, July 20, 2001.

Hoffstatter, L. & Meraz, V. (2000). One-year maintenance report, Sunwize Technologies & Energía Solar de Ciudad Juárez, October 1999, March 2000.

Hoffstatter, L. and Schiff, M. (2000). Commercial solar electric ice plant - A first year report," *Proceedings of the American Solar Energy Society, ASES 2000 Annual Conference,* Madison, Wisconsin.

Ley D.; Hanley, C.; Foster R. & Mazariegos G. (2006). Rural Honduran PV powered schools and community centers. *Proceeding from Solar 2006 of the American Solar Energy Society, ASME International Solar Energy Conference - Solar Engineering*, Denver, CO, USA.

Ley, D. & Stoltenberg, B. (2002). Hybrid electric systems for the Galapagos. *Proceedings from the 26a Semana Nacional de Energía Solar of the Asociación Nacional de Energía Solar,* Chetumal, Quintana Roo, México.

Ojinaga, L.M.; Rovero, C.; Foster, R. E. & Trespalacios, A. (2000). Programa de financiamiento para energía renovable en Chihuahua, Mexico. *Proceedings of the Millennium Solar Forum 2000, International Solar Energy Society, Asociación Nacional de Energía Solar*, 737-740. Mexico City, Mexico.

Richards, E.H.; Hanley, C.; Foster, R.E.; Cisneros, G.; Rovero, C.J.; Büttner, L.; Ojinaga Santana, L.; Graham, S.; Estrada Gasca, C. A. & Montufar, O. (1999). Photovoltaics in Mexico: A model for increasing the use of renewable energy systems, *Advances in Solar Energy: An Annual Review of Research and Development*, Vol. 13, Energy, American Solar Energy Association, Boulder, Colorado.

Romero-Paredes, A.; Foster, R. E.; Hanley, C. & Ross, M. (2003). Renewable energy for protected areas of the Yucatán Península, *Proceedings of the American Solar Energy Society, SOLAR 2003*, Austin, Texas.

SWTDI (2000). Chorreras system monthly performance summaries, for Sandia National Laboratories, Las Cruces, New Mexico, April 1999 through May 2000.

Wiles, J. (1996). Photovoltaic power systems and the National Electrical Code: Suggested practices (Sistemas de energía fotovoltaica y el Código Eléctrico Nacional: Prácticas recomendadas), Southwest Technology Development Institute, New Mexico State University, Las Cruces, NM USA. Supersedes SAND96-2797.

4

Hybrid Solar Vehicles

Gianfranco Rizzo, Ivan Arsie and Marco Sorrentino
Department of Mechanical Engineering, University of Salerno
Italy

1. Introduction

In the last years, increasing attention is being spent towards the applications of solar energy to electric and also to hybrid cars. But, while cars only fed by sun do not represent a practical alternative to cars for normal use, the concept of a hybrid electric car assisted by solar panels appears more realistic (Letendre et al., 2003; Fisher, 2009). The reasons for studying and developing a Hybrid Solar Vehicle can be summarized as follows:

- fossil fuels, largely used for car propulsion, are doomed to depletion; their price tends to increase, and is subject to large and unpredictable fluctuations;
- the CO_2 generated by the combustion processes occurring in conventional thermal engines contributes to the greenhouse effects, with dangerous and maybe dramatic effects on global warming and climatic changes;
- the worldwide demand for personal mobility is rapidly growing, especially in China and India; as a consequence, energy consumption and CO_2 emissions related to cars and transportation are increasing;
- solar energy is renewable, free and largely diffused, and Photovoltaic Panels are subject to continuous technological advances in terms of cell efficiency; their diffusion is rapidly growing, while their cost, after a continuous decrease and an inversion of the trend occurred in 2004, is continuing to decrease (Fig. 1);

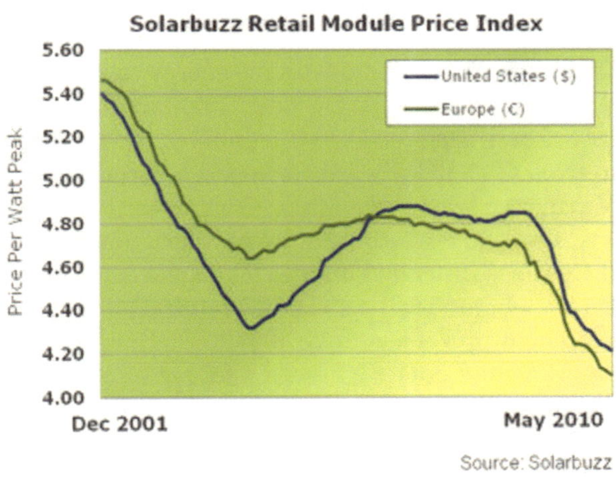

Fig. 1. Trends for cost of photovoltaic modules.

- solar cars, powered only by the sun, in spite of some spectacular outcomes in competitions as World Solar Challenge, do not represent a practical alternative to conventional cars, due to limitations on maximum power, range, dimensions and costs;
- Hybrid Electric Vehicles (HEV) have evolved to industrial maturity, and represent now a realistic solution to important issues, such as the reduction of gaseous pollution in urban drive as well as the energy saving requirements (Guzzella and Amstutz, 1999); the degree of electrification of the fleet is expected to grow significantly in next years (Fisher 2009, Fig. 2).

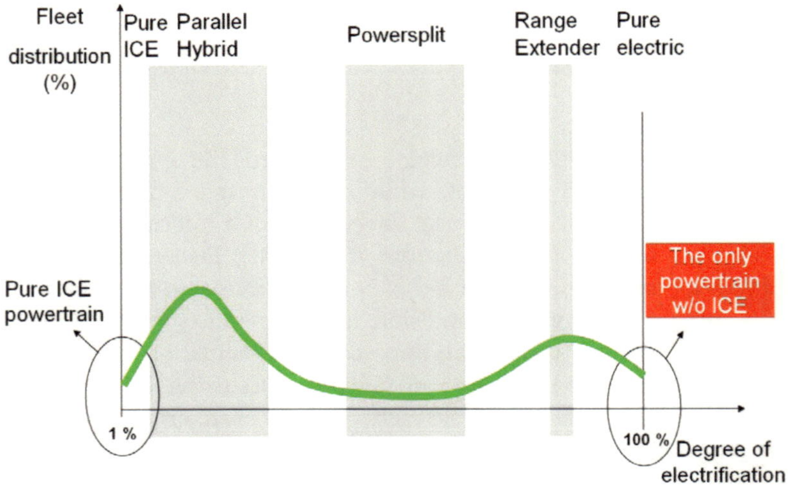

Fig. 2. Degree of electrification. Vision 2025 (Fisher, 2009).

Despite their potential interest, Hybrid Solar Vehicles (HSV) have received relatively little attention in the open literature until a few years ago, particularly if compared with the great effort spent in the last years on other solutions, as fuel cell vehicles, which strongly suffer from the critical issues related to the production and distribution of hydrogen. The scepticism about the direct use of solar energy in cars may be explained by the misleading habit to analyze the automotive systems in terms of power, instead of energy, as discussed in next paragraphs. A proper design of the vehicle-powertrain system may allow meeting a significant share of the total energy required with the energy captured by the panels, during both driving and parking phases, as shown in next paragraphs and evidenced in previous papers (Arsie et al., 2006, 2007). Their economic feasibility appears encouraging: according to some recent studies (Neil C., 2006), PV panels added to hybrid cars could be even more cost effective than PV panels added to buildings. This result has been also confirmed by some recent evaluations, aimed to the estimation of pay-back time of moving and fixed solar roofs for a PV assisted vehicle at different latitudes (Coraggio et al., 2010 II).
Moreover, the presence of a photovoltaic panel on a Plug-In Hybrid Electric Vehicle (PHEV) can enhance the development of Vehicle to Grid (V2G) technology: in this approach, the plug-in vehicles, besides receiving power when parked, can also provide power to the grid. Use of PHEV for V2G can provide benefits to both vehicle owner and the power utility company, apart from the reduced tailpipe emissions and increased mileage, particularly when the number of vehicle connected to the grid is large (Kempton et al., 2001). This technology is now spreading: on September 2009, Delaware's Governor signed a law on

V2G, requiring electric utilities to compensate owners of electric cars for electricity sent back to the grid at the same rate they pay for electricity to charge the battery (www.udel.edu/V2G/). In this context, it is clear that a solar powered vehicles can contribute to power the grid also using solar energy, that is free and renewable. This opportunity prevents also to waste solar energy provided by PV panels on the car when car batteries are fully charged.

In principle, Hybrid Solar Vehicles (HSV) could therefore sum up the advantages of HEV and solar power, by the integration of Photovoltaic Panels in a Hybrid Electric Vehicle. But it would be simplistic to consider the development of a HSV as a straightforward addition of photovoltaic panels to an existing Hybrid Electric Vehicle, that could be considered just as a first step. In fact, the development of HEV's, despite it was based on well-established technologies, showed how considerable research efforts were required for both optimizing the power-train design and defining the most suitable control and energy-management strategies. Analogously, to maximize the benefits coming from the integration of photovoltaic with HEV technology, it is required performing accurate re-design and optimization of the whole vehicle-powertrain system. In these vehicles, in fact, there are many mutual interactions between energy flows, propulsion system component sizing, vehicle dimension, performance, weight and costs, whose connections are much more critical than in conventional and also in hybrid cars (Arsie et al., 2006).

Fig. 3. Astrolab, a Hybrid Solar Vehicle developed by the French company Venturi.

Particularly, the presence of solar panels requires to study and develop specific solutions, since instead of the usual "charge sustaining" strategies adopted in HEV, proper "charge depletion" strategies have to be adopted, to account for the battery recharging during parking (Arsie et al., 2007, 2008). Moreover, advanced look-ahead capabilities are required for such vehicles. In fact, at the end of driving the final state of charge (SOC) is required to be low enough to allow full storage of solar energy captured in the next parking phase, whereas the adoption of an unnecessary constantly-low value of final SOC would give additional energy losses and compromise battery lifetime. The optimal management of battery would therefore require a previous knowledge of the solar energy to be captured in next parking phase, that can be achieved through the real-time access to weather forecast (Coraggio et al., 2010, I).

The impact of solar panels contribution can be significantly improved by adopting suitable Maximum Power Point Tracking (MPPT) techniques, which role is more critical than in fixed plants. The recourse to an automatic sun-tracking roof to maximize captured energy in parking phases has also been studied (Coraggio et al., 2010, II).
Moreover, as it happens for other hybrid vehicles working in start-stop operation, the optimal power split between the internal combustion engine and battery pack must be pursued also taking into account the effect of engine thermal transients. Previous studies conducted by the research group on series hybrid solar vehicles demonstrated that the combined effects of engine, generator and battery losses, along with cranking energy and thermal transients, produce non trivial solutions for the engine/generator group, which should not necessarily operate at its maximum efficiency. The strategy has been assessed via optimization done with Genetic Algorithms, and implemented in a real-time rule-based control strategy (Arsie et al., 2008, 2009, 2010).
In the following, all these topics will be discussed, with reference to the computational and experimental results presented in published papers and achieved during the on-going research.

2. Automotive applications of solar energy

2.1 Photovoltaic panels: efficiency and cost

The conversion from light into direct current electricity is based on the researches performed at the Bell Laboratories in the 50's, where the principle discovered by the French physicist Alexandre-Edmond Becquerel (1820-1891) was applied for the first time. The photovoltaic panels, working thanks to the semiconductive properties of silicon and other materials, were first used for space applications. The diffusion of this technology has been growing exponentially in recent years (Fig. 4), due to the pressing need for a renewable and carbon-free energy (REN21, 2009).

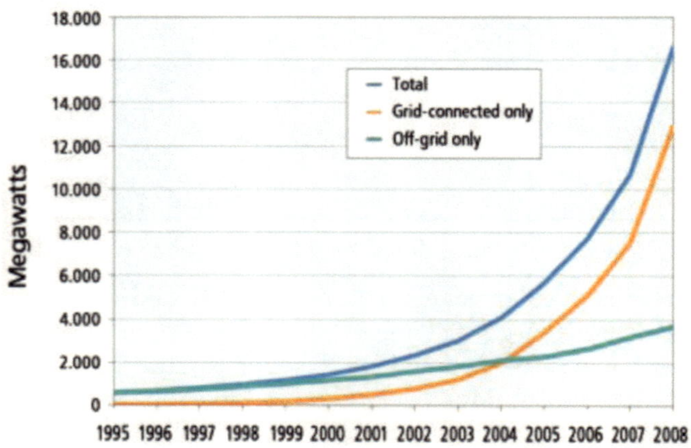

Fig. 4. Solar PV, world capacity 1995-2008

The amount of solar energy is impressive: the 89 petawatts of sunlight reaching the Earth's surface is almost 6,000 times more than the 15 terawatts of average electrical power consumed by humans (Smil, 2006). A pictorial view of the potentialities of photovoltaics is given in Fig. 5, where the areas defined by the dark disks could provide more than the world's total primary energy demand (assuming a conversion efficiency of 8%). The applications range from power station, satellites, rural electrification, buildings to solar roadways and, of course, transport.

In Fig. 6 the trends for the efficiency of photovoltaic cells are shown. Most of the today PV panels, with multicrystalline silicon technology, have efficiencies between 11% and 18%, while the use of mono-crystalline silicon allows to increase the conversion efficiency of about 4%. The recourse to multi-junction cells, with use of materials as Gallium Arsenide (Thilagam et al, 1998), and to concentrating technologies (Segal et al., 2004), has allowed to reach 40% of cell efficiency. Anyway, the cost of these latter solutions is still too high for a mass application on cars.

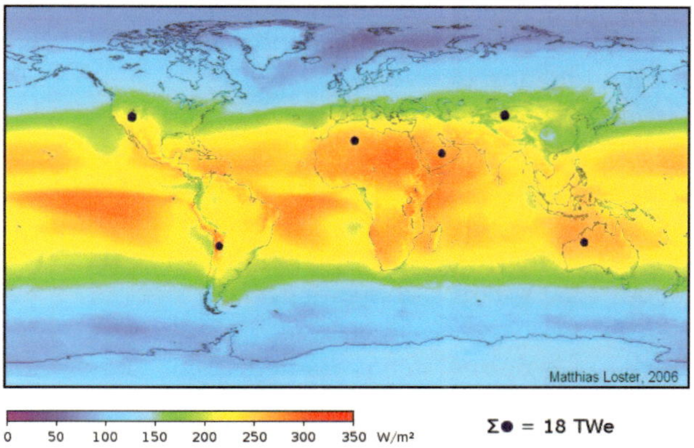

Fig. 5. Average solar irradiance (W/m2) for a horizontal surface (Wikipedia).

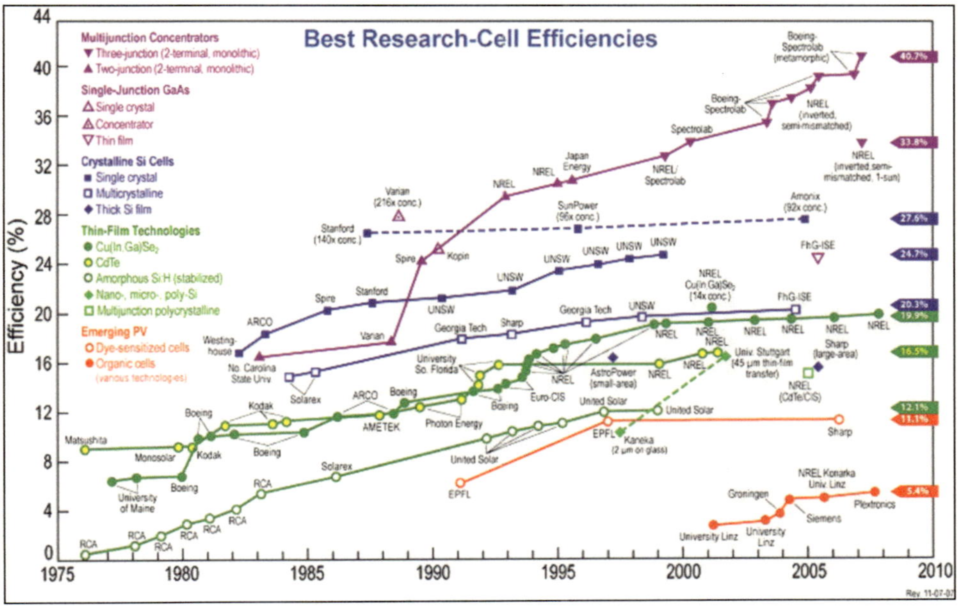

Fig. 6. Trends for efficiency of photovoltaic cells.

About price of solar modules, the market has experienced a long period of falling down of the prices since January 2002 up to May 2004. Afterwards, prices began rising again, until 2006-2007. This inversion has been attributed to the outstripping of global demand with respect to the supply, so that the manufacturers of the silicon needed for photovoltaic production cannot provide enough raw materials to fill the needs of manufacturing plants capable of increased production (Arsie et al., 2006; see also www.backwoodssolar.com). After 2008, the prices began to fall down again, both in USA and in Europe (Fig. 1).

2.2 Solar energy for cars: pros and cons

The potential advantages of solar energy are clear: it is free, abundant and rather evenly distributed (Fig. 5), more that other energy sources as fossil fuels, uranium, wind and hydro. It has been considered that the solar energy incident on USA in one single day is equivalent to energy consumption of such country for one and half year, and this figure could reach embarrassingly high values in most developing countries.

At the same time, also the limitations of such energy source seem clear: it is intermittent, due to the effects of relative motion between Earth and Sun, and variable in time, due to weather conditions (while the former effect can be predicted precisely, the latter can be foreseen only partially and for short term). But the most serious limitation for direct automotive use concerns its energy density: the amount of radiation theoretically incident on Earth surface is about 1360 W/m^2 (Quaschning, 2003) and only a fraction of this energy can be converted as electrical energy to be used for propulsion. Considering that the space available for PV panels on a normal car is limited (from about 1 m^2 in case of panels outfitting 'normal' cars to about 6 m^2 for some solar cars), it emerges that the net power achievable by a solar panel is about two order of magnitude less that the power of most of today cars.

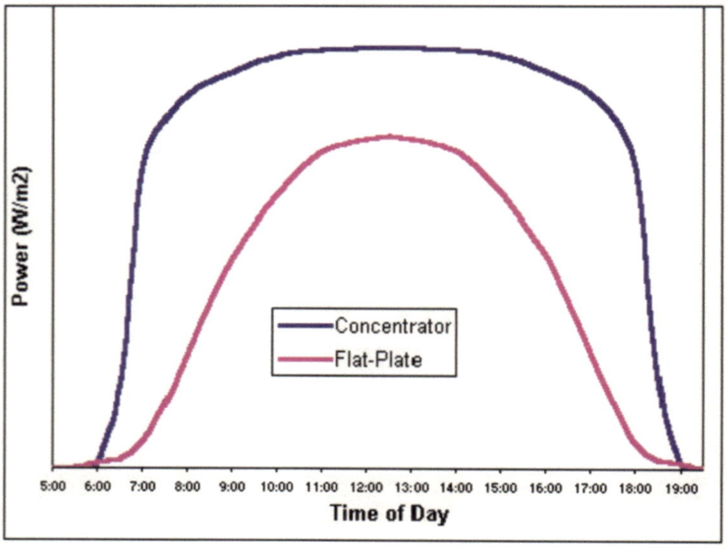

Fig. 7. Solar panel power during a day, for different technologies.

But this simple observation, that explains the scepticism about solar energy in most of the automotive community, is based on the misleading habit to think in terms of power, instead of energy. In fact, for a typical use in urban driving (no more than one hour per day, according to recent Statistics for Road Transport, with an average power between 7 and 10 kW, considering a partial recovery of braking energy), the net energy required for traction can be about 8 kWh per day. On the other hand, a PV panel of 300 W of peak power can operate not far from its maximum power for many hours, especially if advanced tracking techniques would be adopted (Fig. 7). In these conditions, the solar contribution can represent a rather significant fraction, up to 20-30%, of the required energy (Table 1).

	Maximum Power (kW)	Average Power (kW)	Time (h/day)	Energy (kwh/day)
A - Car	70	8	1	8
B - PV	0.30	0.2	10	2
B/A %	0.4 %	2.5 %	1000 %	25 %

Table 1. Incidence of solar contribution in terms of power and energy

It therefore emerges that benefits of solar energy can be maximized when cars are used mostly in urban environment and in intermittent way, spending most of their time parked outdoor, and of course in countries where there is a "sufficient" solar radiation. But, as it will be shown in next sections, feasible locations are not necessarily limited to "tropical" countries.

3. Research issues related to hybrid solar vehicles

There are several research issues related to the application of PV panels on cars. PV panels can be added to a car just to power some accessories, as ventilation or air conditioner, as in Toyota Prius Solar (Fig. 8), or to contribute to car propulsion. Particularly in this latter case, it would be simplistic to consider their integration as the sole addition of photovoltaic panels to an existing vehicle. In fact, the development of HEV's, despite it was based on well-established technologies, has shown how considerable research efforts were required

Fig. 8. Toyota Prius Solar

for both optimizing the power-train design and defining the most suitable control and energy-management strategies. Analogously, to maximize the benefits coming from the integration of photovoltaic with HEV technology, it is required performing accurate redesign and optimization of the whole vehicle-powertrain system, considering the interactions between energy flows, propulsion system component sizing, vehicle dimension, performance, weight and costs. In the following, some of these aspects are described, also based on the author's direct experience on Hybrid Solar Vehicles.

3.1 Solar panel control

The surface of solar panels on a car is limited, with respect to most stationary applications. It is therefore important to maximize their power extraction, by analyzing and solving the problems that could reduce their efficiency. Part of these aspects are common to the stationary plants also, but some of them are quite specific of automotive applications. For example, the need of connecting cells of different types (technology as well as electrical and manufacturing characteristics) within the same array usually leads to mismatching conditions. This may be the case of using standard photovoltaic cells for the roof and transparent ones, in place of glasses, connected in series. Again, even small differences among the angles of incidence of the solar radiation concerning different cells/panels that compose the panel/string may cause a mismatching effect that greatly affects the resulting photovoltaic generator overall efficiency. Such reduction may become more significant at high cell temperatures, with a de-rating of about 0.5%/°C for crystalline cells and about 0.2%/°C for amorphous silicon cells (Gregg, 2005).
These effects are more likely in a car, due to the exigency to cover a curved surface, where differences in solar radiation and temperature can be higher than in a stationary plant. All these aspects are of course enhanced and complicated during driving, due to orientation changes and shadows. In the photovoltaic plants it is mandatory to match the PV source with the load/battery/grid in order to draw the maximum power at the current solar irradiance level.

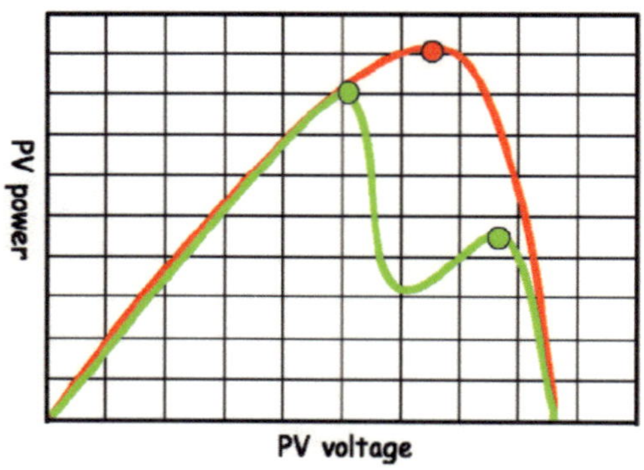

Fig. 9. Power vs. voltage characteristic of a PV field under uniform conditions (red) and with mismatching (green).

To this regard, a switching dc-dc converter controlled by means of a Maximum Power Point Tracking (MPPT) strategy is used (Hohm, 2000) to ensure the source-load matching by properly changing the operating voltage at the PV array terminals in function of the actual conditions. Usually, MPPT strategies derived by the basic Perturb and Observe (P&O)

approach are able to detect the unique peak of the power vs. voltage characteristic of the PV array, in presence of uniform irradiance (Fig. 9, red curve). But, due to mismatching and non uniform irradiation, temperature distribution and manufacturing features, the shape of the PV characteristic may exhibit more than one peak (Fig. 9, green curve). In these cases, the standard MTTP techniques tend to fail, so causing a reduction in power extraction (Egiziano et al., 2007; Femia et al., 2008). More advanced approaches, based on a detailed modelling of the PV field and on numerical techniques, have been developed to face with this problem (Jain, 2006; Liu, 2002).

3.2 Power electronics issues

In a solar assisted electric or hybrid vehicle, particular attention must be spent on power electronics, to enable better utilization of energy sources. To this purpose, high efficiency converter topologies, with different system configurations and particular control algorithms, are needed (Kassakian, 2000; Cacciato et al., 2004).
The use of multi-converters configurations could be advisable to solve the problems of solar generators such as PV modules mismatching and partial shadowing. A comparative study of three different configurations for a hybrid solar vehicle has been recently presented (Arsie et al., 2006, Cacciato et al., 2007). In order to reduce power devices losses, the increase of converter switching frequencies by adoption of soft-switching topologies is also considered. The advantages consist in reducing the size of the passive components and, consequently, the converter weight and volume while decrease the overall Electro Magnetic Interference (EMI), a critical point in automotive applications. Moreover, the converters can be designed by adopting recent technologies such as planar magnetic structures and SMD components, in order to allow the converters to be located inside the photovoltaic modules.

3.3 Optimal design of hybrid solar vehicles

A study on the optimal design of a Hybrid Solar Vehicle has been performed at the University of Salerno, considering performance, fuel consumption, weight and costs of the components (Arsie et al., 2007, 2008). The study, that has determined optimal vehicle dimensions and powertrain sizing for various scenarios, has shown that economic feasibility (pay-back between 2 and 3 years) could be achieved in a medium term scenario, with mild assumptions in terms of fuel price increase, PV efficiency improvement and PV cost reduction.
A prototype of HSV with series structure (Fig. 10) has also been developed (Adinolfi et al., 2008), within the framework on an educational project funded by EU (Leonardo project I05/B/P/PP-154181 "Energy Conversion Systems and Their Environmental Impact, www.dimec.unisa.it/Leonardo). The specifications of the prototype are presented in Table 2.
Vehicle lay-out is organized according to a series hybrid architecture, as shown on Fig. 11. With this approach, the photovoltaic panels PV assist the Electric Generator EG, powered by an Internal Combustion Engine (ICE), in recharging the Battery pack (B) in both parking mode and driving conditions, through the Electric Node (EN). The Electric Motor (EM) can either provide the mechanical power for the propulsion or restore part of the braking power during regenerative braking. In this structure, the thermal engine can work mostly at constant power, corresponding to its optimal efficiency, while the electric motor EM is designed to assure the attainment of the vehicle peak power.

Fig. 10. A prototype of Hybrid Solar Vehicle with series structure developed at the University of Salerno.

Vehicle Piaggio Porter	
Length 3.370 m	**Photovoltaic Panels Polycrystalline**
Width 1.395 m	Surface APV 1.44 m2
Height 1.870 m	Weight 60 kg
Drive ratio 1:4.875	Efficiency 0.125
Electric Motor BRUSA MV 200 - 84 V	**Electric Generator Yanmar S 6000**
Continuous Power 9 KW	Power COP/LTP 5.67/6.92 kVA
Peak Power 15 KW	Weight 120 kg
Batteries 16 6V Modules Pb-Gel	**Overall weight (w driver)**
Mass 520 Kg	MHSV 1950 kg
Capacity 180 Ah	

Table 2. Specifications of the HSV prototype

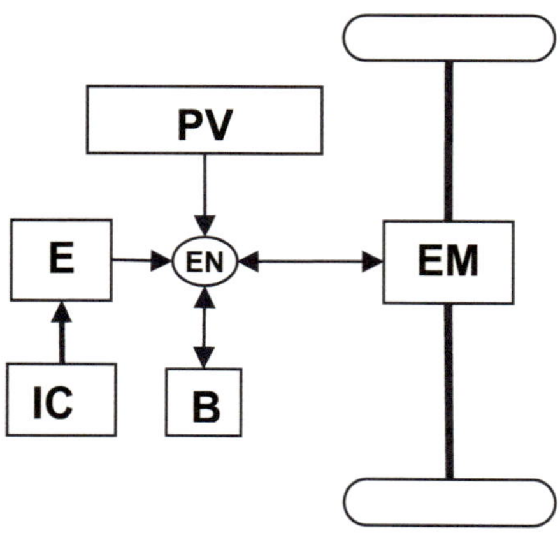

Fig. 11. Scheme of a series Hybrid Solar Vehicle

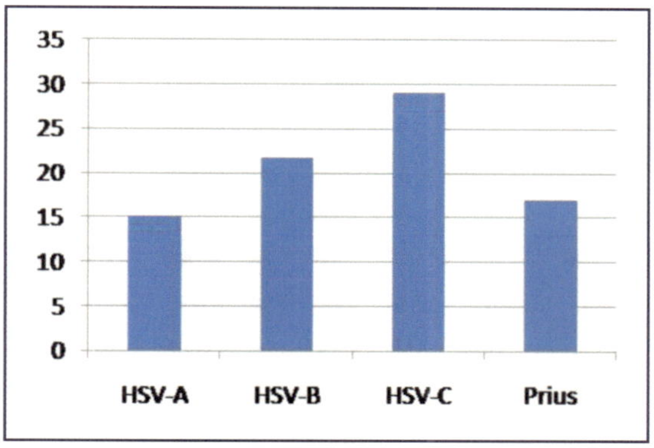

Fig. 12. Fuel Economy (km/l) on ECE Cycle - HSV vs. Toyota Prius. A – actual prototype. B – PV eff.=18% - Batt.=75 Ah. C – B+ 20% weight off – Lithium-Ion Batt.

Experimental and numerical activities have been conducted to develop and validate a comprehensive HSV model (Adinolfi et al., 2008). The model accounts for vehicle longitudinal dynamics along with the accurate evaluation of energy conversion efficiency for each powertrain component. While the actual prototype (HSV-A, Fig. 12) is penalized by a non optimal choice of their components, also due to budget limitations, the simulation model validated over the prototype data shows that very interesting values of fuel economy could be reached by improving the efficiency of solar panels (from 12% to 18%) and optimizing battery capacity and weight (HSV-B), and further reducing vehicle weight by adoption of Lithium-Ion batteries instead of original Lead-Acid (HSV-C).

3.4 Management and control of energy flows

The energy management of Hybrid Solar Vehicles, in spite of many similarities with HEV's, could not simply borrowed from the solutions developed for HEV's: in fact, while in these latter a charge sustaining strategy is usually adopted, in HSV's the battery can be recharged also during parking time by solar energy, and therefore a charge depletion strategy has to be followed during driving, as it happens for Plug-In Hybrid Electric Vehicles (PHEV) (Marano et al., 2009). Anyway, there are again some differences between PHEV and HSV: while for PHEV the recharge is mainly finalized to extend the vehicle range, for HSV's the input energy is free, and solar recharge should be maximized not only to extend the range, but mainly to minimize fuel consumption and CO2 emissions. Therefore, at the end of driving cycle the final state of charge (SOC) should be sufficiently low to leave room for the solar energy to be stored in the battery in the next parking phase. On the other hand, the adoption of an unnecessary low value of final SOC could produce additional energy losses associated to battery operation, so increasing fuel consumption.

In a recent paper (Rizzo & Sorrentino, 2010), the effects of different strategies of selection of final SOC are studied by simulation over hourly solar data at different months and locations, and the benefits achievable by estimating the energy expected in next parking phase are assessed. The simulations are carried out with a dynamic model of a HSV previously developed (Arsie et al., 2007), including a rule-based (RB) energy management strategy. The results have shown that the estimation of the incoming solar energy in next parking phase produces a more efficient energy management, with reduction in fuel consumption, particularly at higher insolation (Fig. 13).

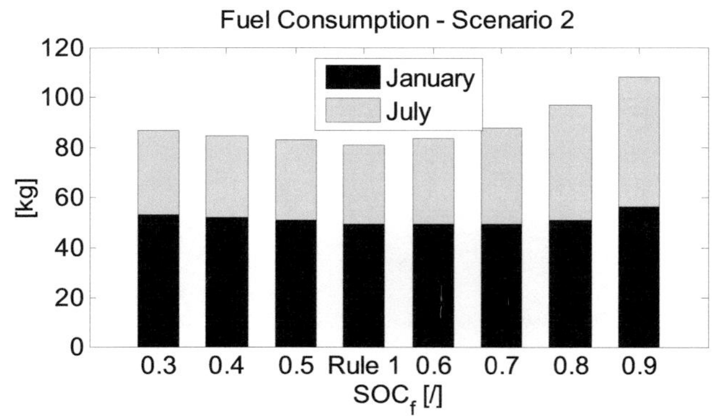

Fig. 13. Effects of optimized (Rule 1) and parametric choice of SOC on Fuel Consumption for a Hybrid Solar Vehicles (Los Angeles, January and July, 1988). ηPV =0.19

The RB control architecture consists of two loops: i) an external loop, defining the desired final state of charge to be reached at the end of the driving cycle; ii) an internal loop, estimating the average power delivered by the internal combustion engine and SOC deviation. The scheme of rule-based control strategy operation is shown in Fig. 14.

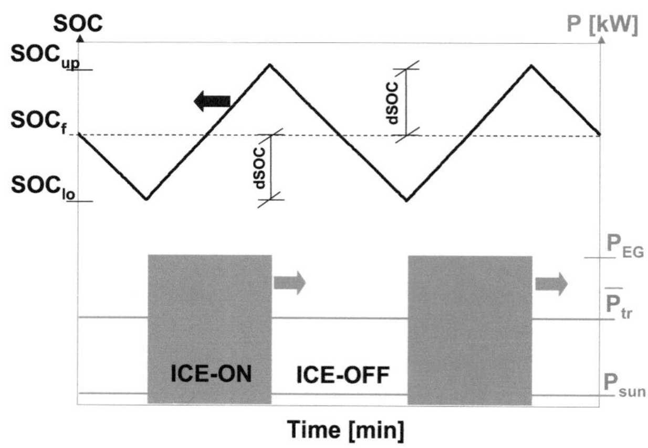

Fig. 14. Schematic representation of the rule-based control strategy for quasi-optimal energy management of a series HSV powertrain.

The results of RB strategy have been successfully compared with a benchmark (non implementable) strategy, obtained by means of a Genetic Algorithm (Sorrentino et al., 2009). In the study, a vehicle dynamic model considering also the effects of engine thermal transients on fuel consumption and power, related to start-stop operation (Fig. 15), has been adopted.

Fig. 16 compares the optimal power of the engine-generator group, operating in start-stop mode, at various vehicle average power (Rizzo et al., 2010). The red line indicates the most efficient ICE-EG operating point (PEG,opt), corresponding to about half nominal power. Such comparison indicates that at high road loads the optimal power values exhibit a load following behavior, whereas at low power demand they always undergoes PEG,opt. These results show that, due to the combined effects of engine losses, of thermal transients and of

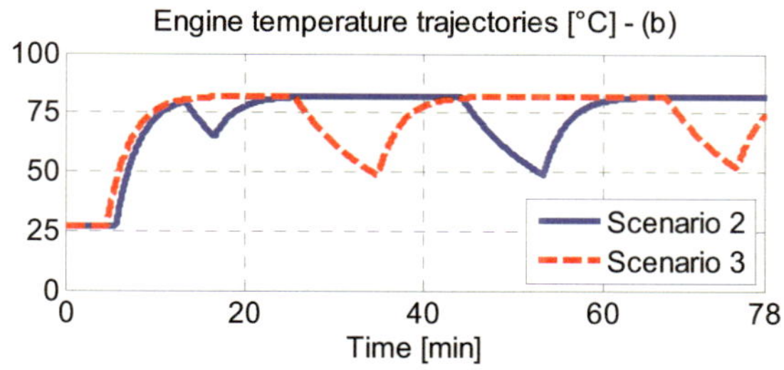

Fig. 15. Simulated engine temperature profiles in a series hybrid electric vehicle with start-stop operation.

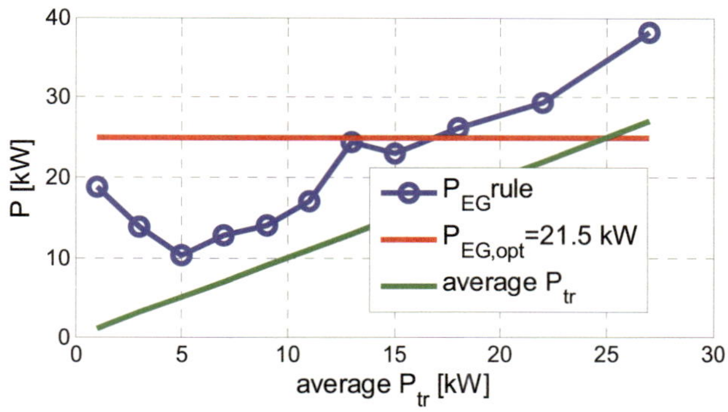

Fig. 16. Optimal generator power vs. average vehicle power for a hybrid electric vehicles with series structure.

electric losses, the optimal choice of generator power in a series hybrid depends in complex way from vehicle power, and that optimal engine power corresponds to the maximum engine efficiency conditions only in a limited power range. A more detailed analysis is reported in the cited paper (Rizzo et al., 2010).

The importance of thermal transients in start-stop operation over fuel consumption and emissions, neglected in most models used for energy management in hybrid vehicles, has been also demonstrated by recent experimental studies (Ohn et al., 2008).

A method for fuel consumption minimization in a Hybrid Solar Vehicle based on application of Model Predictive Control has also been recently proposed (Preitl et al., 2007).

3.5 Effects of panel position and use of moving roofs

In most of solar cars, solar panels are fixed and located at almost horizontal position. This solution, although the most practical by several points of view, does not allow to maximize the net power from the sun. In next figure the mean yearly incident energy corresponding to different position of solar panels is presented, for different latitudes. The data have been obtained by PVWatts (http://www.pvwatts.org/), based on a database of real data covering about 30 years, for different locations in USA.

It can be observed that, with the adoption of a self-orienting solar roof (2 axis tracking), there is an increase of incident energy, varying from about 800 to 600 kWh/m²/year, from low to high latitudes. In terms of relative gain, a moving panel would increase the solar contribution from about 46%, at low latitudes, up to 78%, at high latitudes. Of course, the

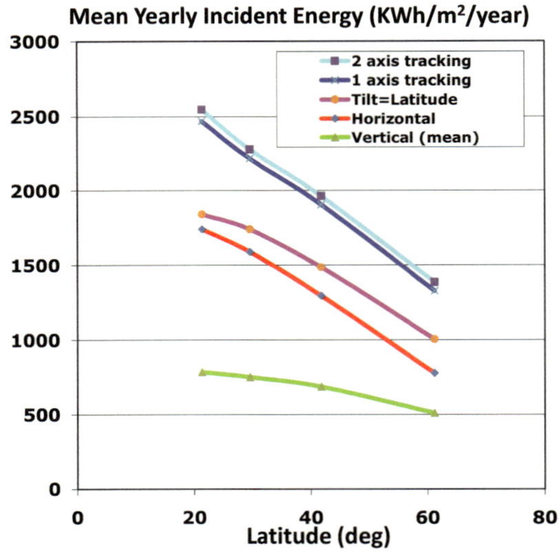

Fig. 17. Effects of panel position and latitude on incident energy

adoption of a moving panel could be feasible only for parking phases, where on the other hand many cars in urban environment spend most of their time. The real benefits would be lower than the ones indicated in the graph, due to the energy spent to move the panel and to possible kinematic constraints preventing perfect orientation. Also, in order to maximize the solar contribution, transparent panel could be incorporated in the windows, and the lateral surface of a car could be also covered by solar panels, as for instance in FIAT Phylla. An estimation of the increase in incident energy can be obtained by considering the mean incident energy on a vertical surface, with random orientation: with respect to the energy incident at horizontal position, their contribution is about 45%, at low latitudes, but up to 65% at higher latitudes.

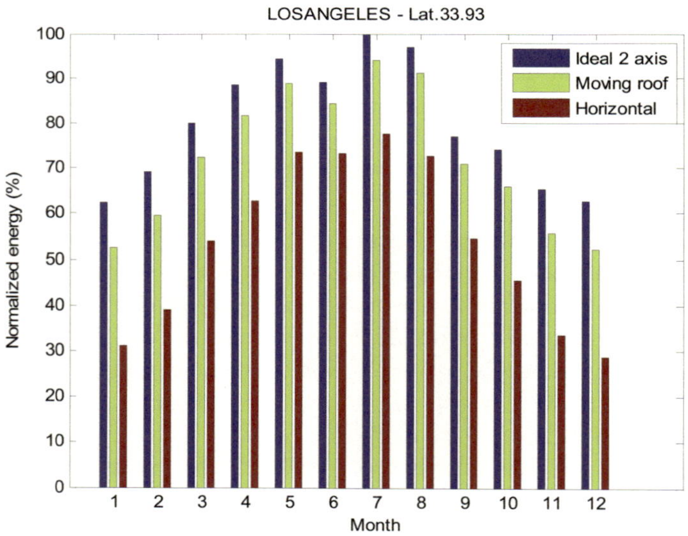

Fig. 18. Energy collected with various options of solar roof (Los Angeles, 1988)

It therefore emerges that the adoption of a moving roof for parking phases, and the utilization of windows and lateral surfaces too, would allow a significant increase of incident energy with respect to the sole utilization of the car roof. Moreover, this increment is particularly significant at high latitudes, so contributing to enlarge the potential market of solar assisted vehicles.

A study on the benefits of a moving solar roof for parking phases in a Hybrid Solar Vehicle has been recently presented (Coraggio et al., 2010). A kinematic model of a parallel robot with three degrees of freedom has been developed and validated over the experimental data obtained by a small scale real prototype. The effects of roof design variables are analyzed, and the benefits in terms of net available energy assessed by simulation over hourly solar data at various months and latitudes (Fig. 18).

3.6 Upgrade of conventional vehicles

A possible remark is that, considering the current economic crisis, it is unlikely that, in next few years, PV assisted EV's and HEV's will substitute for a substantial number of conventional vehicles, since relevant investments on production plants would be needed. This fact would of course impair the global impact of this innovation on fuel consumption and CO_2 emissions, at least in a short term scenario. Therefore, one may wonder if there is any possibility to upgrade conventional vehicles to PV assisted hybrid. A proposal of a kit to be distributed in after-market has been recently formulated and patented by the author (www.hysolarkit.com). Mild-solar-hybridization will be performed by installing in-wheel electric motors on the rear wheels (in case of front wheel drive) and by the integration of photovoltaic panels on the roof. The original architecture will be upgraded with the an additional battery pack and a control unit to be faced with the engine management system by the OBD port. The Vehicle Management Unit (VMU), which would implement control logics compatible with typical drive styles of conventional-car users, receives the data from OBD gate and battery (SOC estimation) and drives in-wheel motors by properly acting on the electric node EN (Fig. 19). A display on the dashboard may advice the driver about the actual operation of the system. The project has been recently financed by the Italian ministry of research (www.dimec.unisa.it/PRIN/PRIN_2008.htm). The results will be published shortly, and presented on the cited websites.

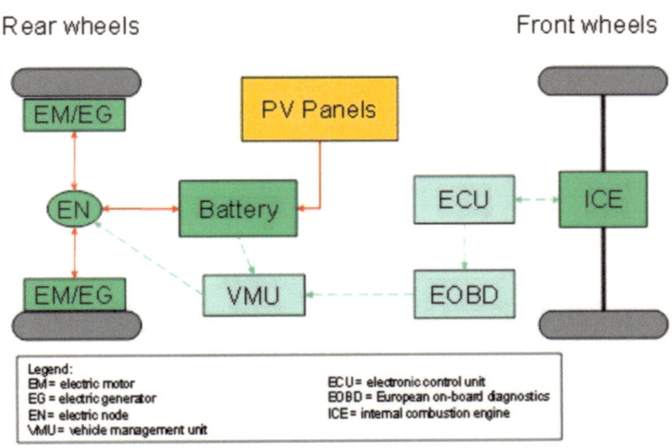

Fig. 19. Scheme of a system to upgrade a conventional car to Mild Hybrid Solar Vehicle.

4. Conclusion

The integration of photovoltaic panels in hybrid vehicles is becoming more feasible, due to the increasing fleet electrification, to the increase in fuel costs, to the advances in terms of PV panel technology, and to the reduction in their cost. Hybrid Solar Vehicles may therefore represent a valuable solution to face both energy saving and environmental issues. Of course, these vehicles cannot represent a universal solution, since the best balance between benefits and costs would depend on mission profile: in particular, significant reductions in fuel consumption and emissions can be obtained during typical use in urban conditions during working days. Moreover, the integration with solar energy would also contribute to reduce battery recharging time, a critical issue for Plug-in vehicles, and to add value for Vehicle to Grid applications.

Putting a solar panel on an existing hybrid vehicle may be just the first step: in order to maximize their benefits, re-design and optimization of the whole vehicle-powertrain system would be required. Particular attention has to be paid in maximizing the net power from solar panels, and in adopting advanced solutions for power electronics. Moreover, these vehicle would require specific solutions for energy management and control, whit more advanced look-ahead capabilities.

The adoption of moving roofs for parking phases and the use of solar panels on windows and lateral sides would enhance solar contribution, beyond the classical fixed panel on the car roof. Moreover, these solutions would reduce the gap between solar contribution at low and high latitudes, so extending the potential market of these vehicles. Interesting opportunities are also related to possible reconversion of conventional vehicles to Mild Hybrid Solar Vehicles, by means of kits to be distributed in after-market.

The perspectives about cost issues of hybrid solar vehicles are encouraging. Anyway, as it happens for many innovations, full economic feasibility could not be immediate, and a financial support from governments would certainly be appropriate. But the recent and somewhat unexpected commercial success of some electrical hybrid cars indicates that there are grounds for hope that a significant number of users is already willing to spend some more money to contribute to save the planet from pollution, climate changes and resource depletion.

5. References

Adinolfi G., Arsie I., Di Martino R., Giustiniani A., Petrone G., Rizzo G., Sorrentino M., (2008), "A Prototype of Hybrid Solar Vehicle: Simulations and On-Board Measurements", Proc.of Advanced Vehicle Control Symposium AVEC 2008, October 6-9, 2008, Kobe (Japan) 917-922 Society of Automotive Engineers of Japan - ISBN: 978-4-904056-21-9

Arsie, I., Rizzo, G., Sorrentino, M., (2006) "Optimal Design and Dynamic Simulation of a Hybrid Solar Vehicle", SAE paper 2006-01-2997, SAE 2006 Transactions - Journal of Engines, vol. 115-3, pp. 805-811.

Arsie, I., Rizzo, G., Sorrentino, M., (2010) "Effects of engine thermal transients on the energy management of series hybrid solar vehicles", Control Engineering Practice (2010), DOI:10.1016/j.conengprac.2010.01.015.

Arsie I, Rizzo G, Sorrentino M (2009) Genetic Algorithms Based Optimization of Intermittent ICE scheduling on a Hybrid Solar Vehicle In: European Control Conference 2009, ECC09, Budapest, August 23-26, 2009.

Arsie I., Rizzo G., Sorrentino M. (2008) A Model for the Optimal Design of a Hybrid Solar Vehicle Review of Automotive Engineering, Society of Automotive Engineers of Japan (JSAE), 2008, ISSN 1349-4724. 29-3: 439-447

Arsie I., Rizzo G., Sorrentino M. (2007) Optimal Design and Dynamic Simulation of a Hybrid Solar Vehicle, SAE TRANSACTIONS- JOURNAL OF ENGINES 115-3: 805-811

Cacciato M., Consoli A., Scarcella G., Testa A. (2004), "A Multhi-Phase DC/DC Converter for Automotive Dual-voltage Power Systems" IEEE Industry Applications Magazine, November/December 2004, pp. 2-9.

Cacciato M., Consoli A, Scarcella G, Scelba G. (2007), Accurate Implementation of a State of Charge Estimator for Hybrid and Elecric Vehicle Battery Packs. 2nd International Workshop on Hybrid Vehicles. 14 September, 2007. (pp. 1-6). Salerno, Italy.

Coraggio G., Pisanti C., Rizzo G., Sorrentino M. (2010, I), Assessment of benefits obtainable in a Hybrid Solar Vehicle using look-ahead capabilities for incoming solar energy, 10th Intnl. Symp. On Advanced Vehicle Control, August 22-26, 2010, Loughborough (UK).

Coraggio G., Pisanti C., Rizzo G., Senatore A. (2010, II), A Moving Solar Roof for a Hybrid Solar Vehicle, 6th IFAC Symposium on Advances in Automotive Control, July 11-14, 2010, Munich (D).

Egiziano L., Giustiniani A., Lisi G., Petrone G., Spagnuolo G., Vitelli M.(2007): "Experimental characterization of the photovoltaic generator for hybrid solar vehicle". Proc of 2007 IEEE International Symposium on Industrial Electronics, June 4-7 2007 Vigo (Spain), pp 329-334.

ESA, Electricity Storage Association, www.electricitystorage.org

Femia N., Lisi G., Petrone G., Spagnuolo G., Vitelli M. (2008), "Analysis of Photovoltaic Systems with Distributed Maximum Power Point Tracking", Proc. of IEEE International Symposium on Industrial Electronics ISIE08, June 30-July 2 2008 pp. 2408 - 2413.

Fischer R. (2009), AVL List GmbH, The Electrification of the Powertrain – from Turbohybrid to Range Extender, 30. Internationales Wiener Motorensymposium 2009

Guzzella L. and Amstutz A. (1999), CAE Tools for Quasi-Static Modeling and Optimization of Hybrid Poweretrains. IEEE Transactions on Vehicular Technology, vol. 48, no. 6, November 1999.

Hohm, D.P.; Ropp, M.E., (2000), "Comparative study of maximum power point tracking algorithms using an experimental, programmable, maximum power point tracking test bed", Conference Record of the Twenty-Eighth IEEE Photovoltaic Specialists Conference, 2000. 15-22 Sept. 2000, pp:1699 - 1702

Jain, A.; Sharma, S.; Kapoor, A., (2006), "Solar cell array parameters using Lambert W-function", Solar Energy Materials & Solar Cells 90 (2006) 25–31

Kassakian, J.G. (2000), Automotive Electronics Power Up IEEE Spectrum, Volume: 37, Issue: 5 May 2000, Pages:34 - 39

Kempton W., Tomić J., Letendre S., Brooks A., Lipman T. (2001), "Vehicle-to-Grid Power: Battery, Hybrid, and Fuel Cell Vehicles as Resources for Distributed Electric Power in California", Report prepared for California Air Resources Board and the California Environmental Protection Agency, 2001.

Letendre S., Perez R., Herig C. (2003), Vehicle Integrated PV: A Clean and Secure Fuel for Hybrid Electric Vehicles, Proc. of the American Solar Energy Society Solar 2003 Conference, June 21-23, 2003, Austin,TX.

Liu, S., Dougal, R.A. (2002), "Dynamic multiphysics model for solar array", IEEE Trans. On Energy Conversion, Vol. 17, No. 2, June 2002, pp. 285-294.

Neil C. (2006), Solar Hybrid Vehicles, http://www.energypulse.net/centers/article/article_display.cfm?a_id=1267

Preitl Z., Bauer P., Kulcsar B., Rizzo G., Bokor J. (2007), Control Solutions for Hybrid Solar Vehicle Fuel Consumption Minimization In: Proceedings of the 2007 IEEE Intelligent Vehicles Symposium, Istanbul, Turkey, June 13-15, 2007.

Quaschning V. (2003), "Technology fundamentals - The sun as an energy resource". Renewable Energy World 6 (5): 90–93.

REN21, Renewables - Global Status Report - 2009 Update, http://www.ren21.net/pdf/RE_GSR_2009_update.pdf

Rizzo G., (2010), Automotive Applications of Solar Energy, 6th IFAC Symposium on Advances in Automotive Control, July 11-14, 2010, Munich (D).

Rizzo G., Sorrentino M., (2010), Introducing Sunshine Forecast to Improve On-Board Energy Management of Hybrid Solar Vehicles, 6th IFAC Symposium on Advances in Automotive Control, July 11-14, 2010, Munich (D).

Segal A., Epstein M., Yogev A., (2004), Hybrid concentrated photovoltaic and thermal power conversion at different spectral bands, Solar Energy 76 (2004) 591–601

Smil V., (2006), Energy at the Crossroads, Global Science Forum Conference on Scientific Challenges for Energy Research, Paris, May 17-18, 2006, http://www.oecd.org/dataoecd/52/25/36760950.pdf

Statistics for Road Transport, UK Government, http://www.statistics.gov.uk/CCI/nscl.asp?ID=8100

Thilagam, A., Singh, J., Stulik, P., (1998), Optimizing Gallium Arsenide multiple quantum wells as high-performance photovoltaic devices, Solar Energy Materials and Solar Cells, Vol: 50, 1-4, January, 1998 pp. 243-249, Elsevier

5

Degradation of Space Exposed Surfaces by Hypervelocity Dust Bombardment – Example: Solar Cell Samples

H. M. Ortner[1,2]
[1]*Darmstadt University of Technology, Dept.of Materials Science,*
[2]*Present address: Osterbichl 16, A 6600 Breitenwang,*
[1]*Germany*
[2]*Austria*

1. Introduction

The analysis of cosmic particles by secondary ion mass spectrometry (SIMS) has developed into an essential tool of cosmophysics and -chemistry as well as of applied space-research. This way it is feasible to gain important information about the origin, the evolution and the structure of our solar system (Brownlee, 1978; Grün et al., 2001). In addition, the discrimination between terrestrial and cosmic particles is critical for an estimate of damage of space exposed surfaces by the impact of such particles. This is especially important for the multitude of satellites in near-earth space, i.e. in low earth orbits, fig.1.

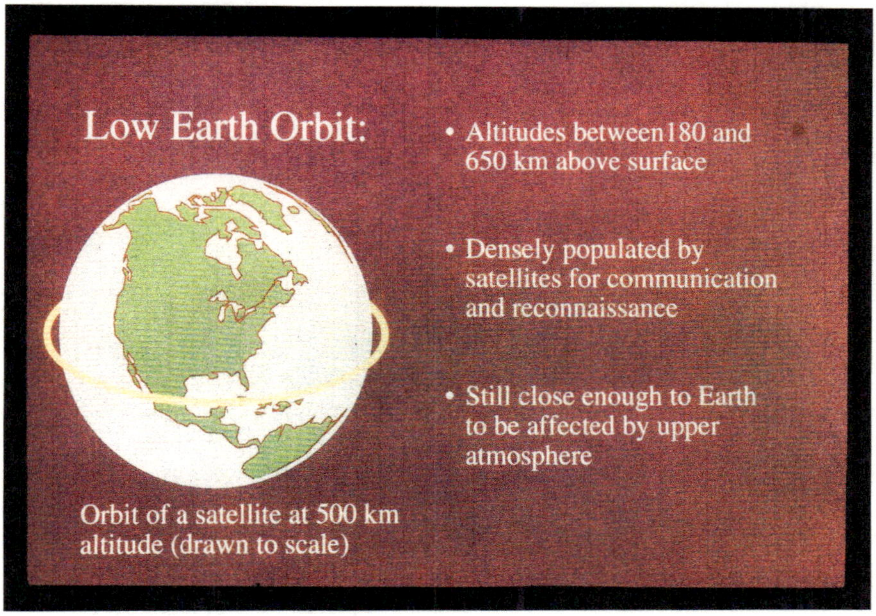

Fig. 1. Draft of a satellite orbit in 500 km altitude

Low earth orbits (LEO), i. e. the altitude between 180 and 650 kilometers above the earth's surface, is one of the busiest traffic zones in space. Nevertheless, the conditions in LEO are harsh. It is a region of intensive hard UV-radiation and the little oxygen still present from the earth's atmosphere is highly-reactive atomic oxygen. It is also a region of high temperature variations between -100°C to +100°C and, as will be discussed in more detail later, a region full of manmade space debris –in addition to cosmic dust micrometeorites (Murr & Kinard, 1993)
Particles are travelling there with velocities of around 10 km/s. If they hit material surfaces they almost completely evaporate due to their high impact velocity and cause the formation of a crater, which is up to one order of magnitude larger than the impacting particle, fig.2.

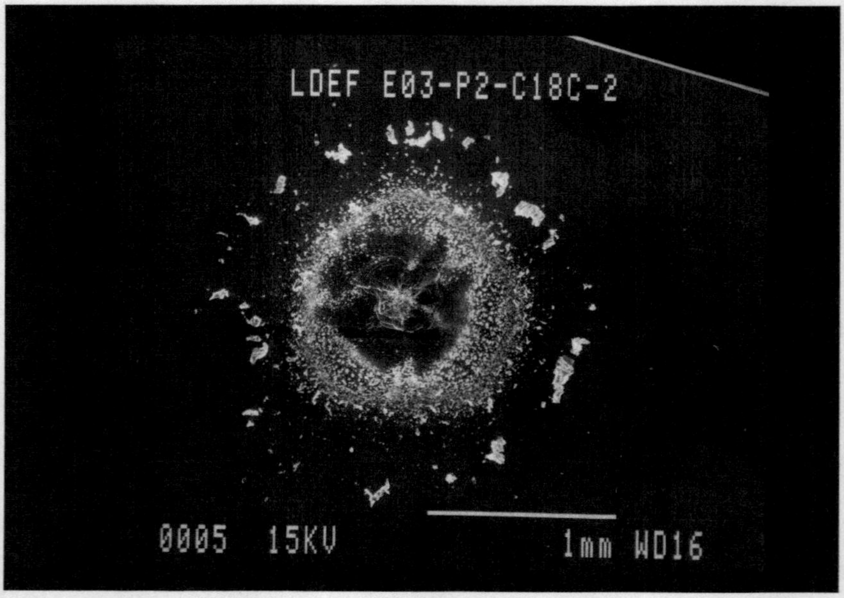

Fig. 2. SEM-micrograph of an impact crater on a germanium surface caused by a cosmic dust particle. In order to clearly differentiate between ions generated from material and such of the impacted particle in SIMS-analysis, it is advantageous to use rather exotic and highly pure substrates such as gold or germanium. The particles scattered around the impact crater are Ge-particles and not remnants of the impacted cosmic particle which evaporated completely. Only extreme traces of its matter are detectable by SIMS.

This turns out to be a serious problem for space technology because the impact of a multitude of such particles will quickly deteriorate space exposed surfaces. The mean life time e.g. of solar panels for the generation of energy for satellites is thus seriously reduced.

2. Cosmic dust: An essential part of matter in the universe

The investigation of cosmic dust particles has thus developed to an interesting and fascinating area of cosmophysics and –chemistry (Stadermann, 1992). Cosmic dust constitutes an essential part of matter in the universe. The earliest hint of the existence of dust in our solar system came from the observation of the zodiacal light. This can be observed with bare eye shortly before sunrise or shortly after sunset, over the Eastern or Western horizon, respectively. Already in the 18th century, Cassini interpreted this Zodiacal light as light-reflection and - scatter caused by a giant cloud of dust particles in the ecliptic. Today, it is known from spectroscopic investigations of the reflected sun light that these dust particles have diameters between 0.1 and 100 µm. The zodiacal dust cloud exhibits the form of a flat disk and extends over the whole inner range of the solar system.
From theoretical considerations it is known that the dust particles of this cloud do not move on Kepler-orbits around the sun but instead move on spiral orbits into the sun (Stadermann, 1992). This "Poynting-Robertson -Effect" is caused by a retardation of orbiting particles by an interaction with the solar radiation. For a 10 µm particle the life time is limited to about 100,000 years before it is burned up in the sun. Some are also trapped by the earth's gravity and may enter its atmosphere. Cosmic particles up to about 50 µm can efficiently radiate away the heat which is generated by their slowing down in the earth's atmosphere due to friction. Greater particles cannot do this effectively enough and hence, burn up in the upper layers of the atmosphere. This leads to the apparent paradox that microscopic dust particles as well as meteorites as big as one's fist survive the entrance into the earth's atmosphere while particles of the size of a grain of sand burn as shooting stars. The macroscopic meteorites survive their travel through the atmosphere because of a totally different reason: They fall so quickly that their inner part does not heat up while only their outer layers evaporate. Once decelerated from cosmic velocities, the cosmic dust particles which are of prime interest to us take a long time for their trip from the earth's outer atmosphere to the earth's surface: depending on atmospheric conditions (wind, weather) this part of their trip can last several months. They usually endure this travel relatively sound and this is the reason why our planet is daily gaining several tons due to the trapping of extraterrestrial material (Stadermann, 1992). This gain in part is counterbalanced by a loss of hydrogen, helium, atomic oxygen and possibly carbon (mainly as methane) in the exosphere as a result of non-thermal escape mechanisms (Shizgal & Arkos, 1996)

3. Problems with sampling of interplanetary dust

The seemingly simplest way - the direct collection of cosmic dust in space with a dedicated space exposed device is in practice rather problematic. The problem is the high velocity of several km/s with which these particles travel. If they hit a collecting device without deceleration they almost completely evaporate in fractions of a second. A part of the evaporated material will condense around the crater which is formed upon the particle impact while only a minor fraction of the original projectile will survive the impact as debris inside the crater, fig.2.
An ideal collector for cosmic dust particles would gently decelerate the often fragile particles. And this is exactly what happens in the outer realms of the earth's atmosphere. Eventually, the particles are sedimenting down with quite low velocities. Interestingly this also causes a density of cosmic particles in the earth's atmosphere that is many orders of

magnitude higher than in space. In order to prevent a mixing of cosmic particles with terrestrial aerosols the sampling has to be carried out in the stratosphere. In the 1960s it was tried to collect cosmic dust with high flying balloons. However, the yield was very modest. Therefore, NASA initiated a program in the 1970s in which cosmic dust was collected with U2-planes flying in the stratosphere (Stadermann 1992). For this purpose, palm sized collecting surfaces have been prepared which were coated with silicon oil. These collecting surfaces were exposed to the air stream of planes travelling at an altitude of 20 km (twice as high as most commercial traffic) beneath a wing of the plane for several hours. Nevertheless, only a single particle greater than 5 μm is caught per hour. Of these very few collected particles in the clean surrounding every second particle is still of terrestrial origin. Often ash particles from volcanic eruptions are found which had been injected into the stratosphere. Hence, after greater volcanic eruptions (as, e. g. of the Pinatubo in 1991) the collection of cosmic dust in the stratosphere has to be discontinued for several months because the volcanic dust cloud is dispersed quickly and thoroughly around the earth.

It goes without mentioning that during sample preparation and investigation no additional contamination can be tolerated, work has to be performed under strict cleanroom conditions and, due to the dust grain size, mostly under the microscope. Hence, particles are removed one by one from the collector surface and subsequently cleaned from the silicon oil. They are thereby viewed in the light microscope. Afterwards they are characterized closer in the scanning electron microscope (SEM). Fig. 3 shows some typical particle morphologies of extraterrestrial particles.

Modern new detection systems for hypervelocity microparticles using piezoelectric material have rather recently been developed (Miyachi et al., 2004). Furthermore, a dust cloud of Ganymede has also been detected by in situ measurements with the dust detector onboard the Galileo spacecraft (Krüger et al., 2000).

4. Secondary Ion Mass Spectrometry (SIMS) – the key instrumentation for cosmic dust analysis

It is difficult to gain information on the nature of impacting particles due to the fact that most of the particle matter is evaporating during the impact. The minute amounts of particle matter which remain on the material surface in and around the impact crater can only be detected by a very sensitive method of topochemical analysis. SIMS is the topochemical method with the highest detection sensitivity and, hence, it is the method of choice for such investigations. In addition, the ability of SIMS to distinguish between various isotopes of an element is the key to differentiate between terrestrial and cosmic particles (Stadermann, 1990). It has been observed in LEO that the most serious degradation is caused by terrestrial aluminium oxide particles (Corso, 1985). The origin of such particles was a solid rocket fuel (Al-powder) which was used by one of the nation's leading in space technology. It was finally feasible to ban this technology in favour of liquid fuels for rocket propulsion which do not generate Al_2O_3-particles. The outstanding significance of SIMS for such investigations consequently led to the development of the NanoSIMS (Schuhmacher et al., 1999) which exhibits a dramatically improved lateral resolution in the ten-nanometer domain (as compared to a lateral resolution in the single μm-range for a conventional SIMS instrument). It also has a multi-detection system which is important since the amount of material to be sputtered is very limited in this special application, fig.4.

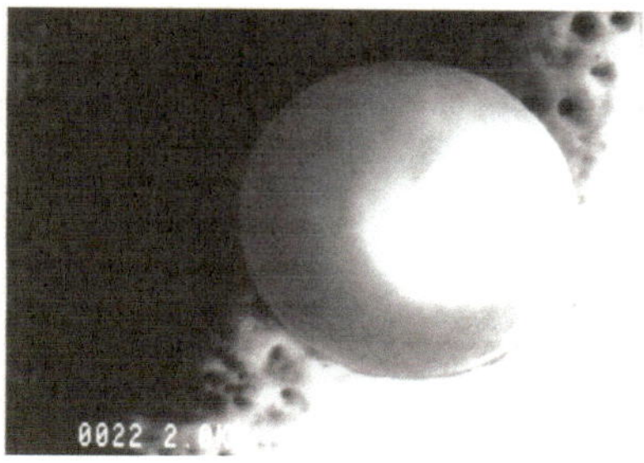

3a. Spherical particle. Main elemental composition: Mg, Si, O (traces Ca, Fe).The morphology of the particle indicates that it once was in a realm where the temperature was higher than its melting temperature. Another possibility would be the emission from a melt.

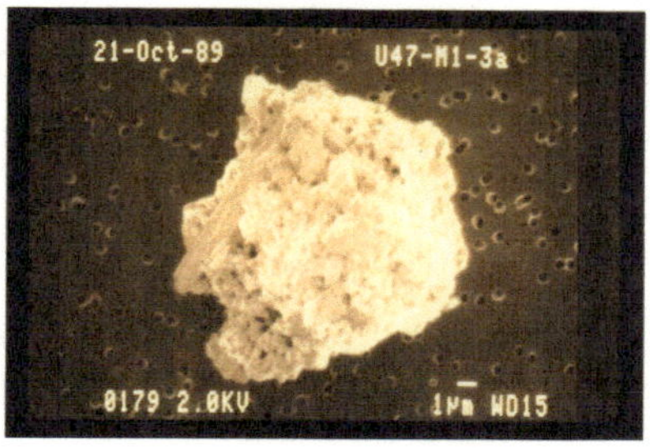

3b. This particle seems to be a conglomerate of smaller particles. Main elemental components: Mg, O (N, C, H).

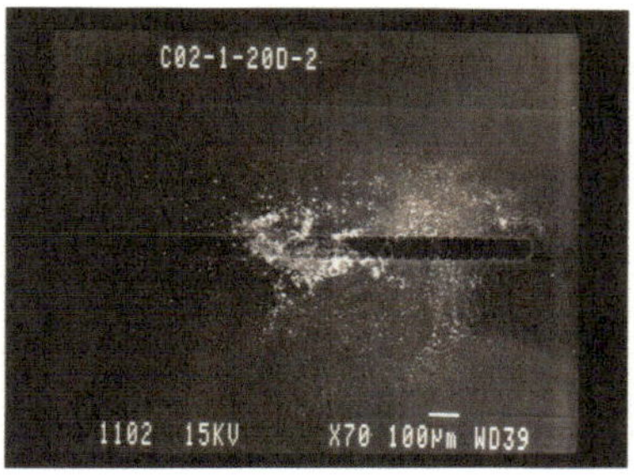

3c. Precipitate of an LDEF impact on germanium. The broad dark stripe is the trace of the ion beam with which the analysis was carried out.

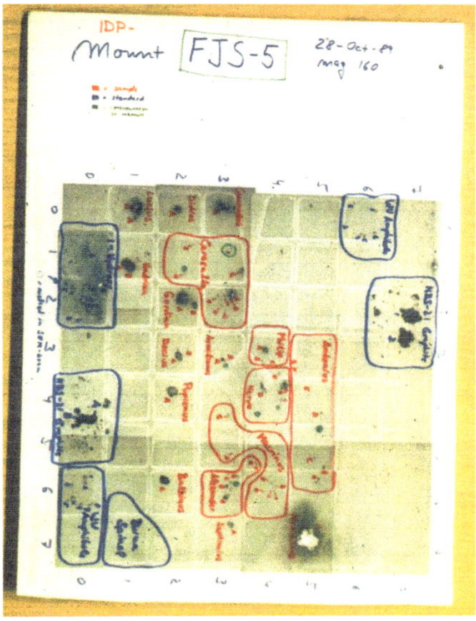

3d. Particle storage sheet of Stadermann

Fig. 3. SEM-micrographs of some typical particle morphologies of extraterrestrial particles (Stadermann, 1990)

Fig. 4a shows the ion optical system of the NanoSIMS of CAMECA (Courtesy of CAMECA, Paris). Fig. 4b shows the NanoSIMS 50 installed in the laboratory of the Physics Dept. at Washington University in St. Louis

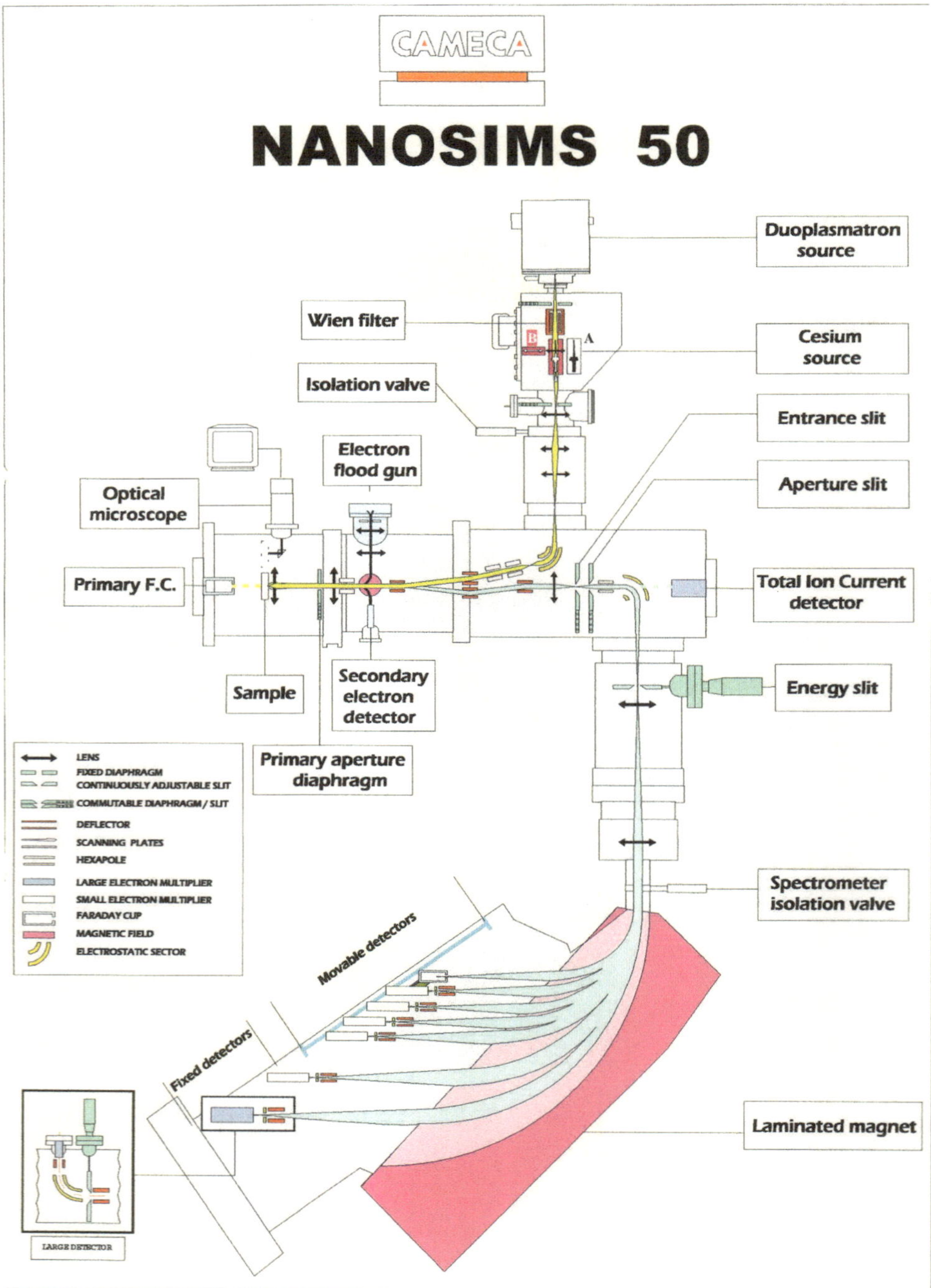

Fig. 4a. The NanoSIMS 50 of CAMECA (Courtesy of CAMECA, Paris)

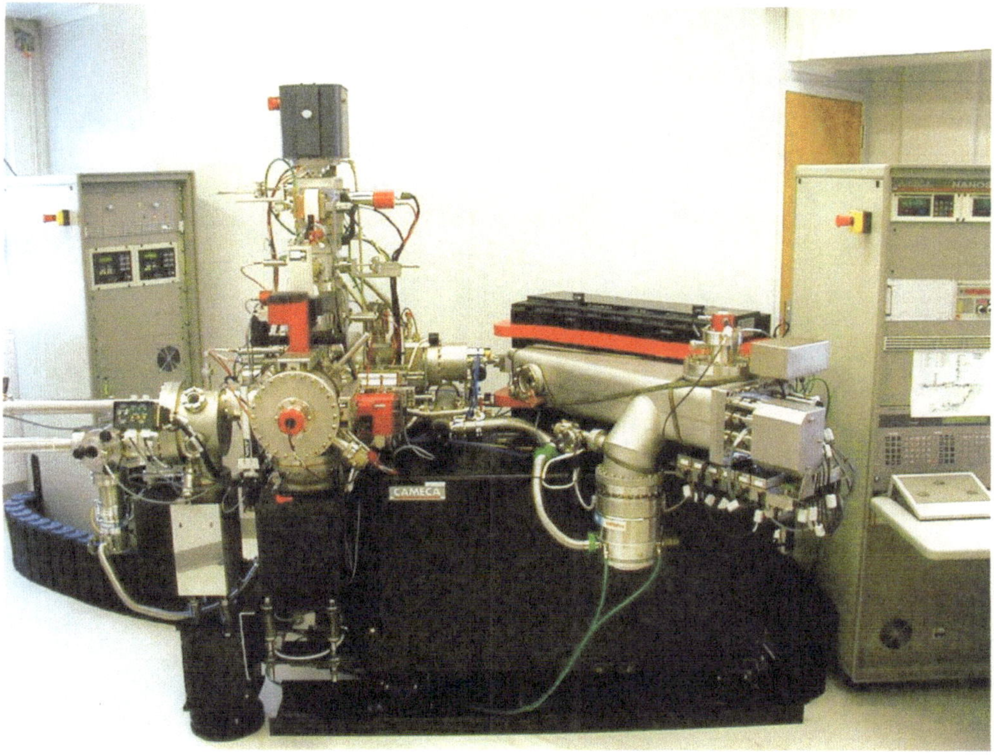

Fig. 4b. The NanoSIMS 50 in the laboratory of the Washington Univ., Physics Dept., St. Louis, MO, USA

The impact crater of fig. 2 demonstrates impressively how space exposed surfaces eventually deteriorate by impact of many such particles.

5. The significance of material degradation of space exposed surfaces – the LDEF experiment

This has alarmed the American National Aeronautics and Space Administration (NASA) to an extent that a respective materials degradation experiment was organized, the LDEF-experiment (Long Duration Exposure Facility). The heart of this action was a large cylindrical satellite with a length of 9 m which is shown in fig. 5.
This satellite was of the size of a bus and was brought into a Low Earth Orbit in an altitude of 476 km in 1984 (Murr & Kinard, 1993). It contained more than 10,000 test material plates which were exposed to the rather unfriendly environment of the LEO for degradation studies. These surfaces were exposed to bombardment by micrometeorites and near-earth space debris of man-made origin which led to a deterioration of the plates' surfaces. The LDEF day was only 90 min long as well as its night. With this frequency, the temperature varied from +100°C to -100°C! In addition during sunshine a most intensive UV-radiation was also hitting the surface. This effect combined with atomic oxygen (Atox) which is also present due to the last traces of the earth's atmosphere in this altitude. The combined action of these influences resulted in interesting corrosion and erosion phenomena (Murr &

Fig. 5. View of the LDEF-experiment exposed in LEO (Courtesy of NASA Langley Research Center)

Kinard, 1993). The satellite was not retrieved after the planned exposure time due to the Challenger disaster. Only in 1990 after 34,000 earth orbits in 2105 days the LDEF-experiment was retrieved in the last possible moment by the Space Shuttle, fig. 6. It was taken into the shuttle in an altitude of only 333 km shortly before the satellite would have burned down in the upper atmosphere.

However, due to its very long exposure time, corrosion and erosion phenomena were very pronounced and a lot of interesting and alarming observations were made (Mandeville, 1991). One of the most alarming finds was that more than 80% of all investigated particle impact craters by SIMS turned out to be caused by terrestrial (man-made) and not by cosmic particles. The highest percentage of these particles was Al_2O_3-particles stemming from solid state rocket fuels. This was the reason why Russia finally changed over to liquid fuel systems. However, not only Al_2O_3-particles of terrestrial origin had been detected. Titanium- and cadmium-rich particles were also registered. They originated from paints with which rocket surfaces had been painted. Particles of stainless steel, mineral particles and such of silver-solder had also been detected (Murr α Kinard, 1993). The geometry of the impact craters of particles allowed calculations of the velocity of impacting particles. SIMS-results on the composition of extraterrestrial particles yielded another interesting detail: Many analyzed cosmic particles exhibited nearly the same composition as so called chondritic (C1) meteorites (main constituents: Si, Al, Mg, Fe, Ca, O) (Stadermann, 1990). It is believed that the solar nebula from which our solar system developed 4.5 billion years ago had the same chondritic composition. Eventually the planets and other bodies developed, the composition of which varies considerably and deviates from this original composition because of diverse chemical processes (so called fractionations). However, material with chondritic composition is still found in some meteorites and many cosmic dust particles. This is an indication that these objects are of "primitive" nature, i. e. very old and unchanged material (Stadermann, 1992).

Fig. 6. Recovery of the LDEF-experiment from LEO.

6. Rocket and other space debris: mortal danger in near Earth space

It must be mentioned that surface erosion by cosmic dust is not the only danger of material degradation in space. Especially near the earth, there is eminent danger of collision with much greater "particles" of space debris. The reason is a rising number of debris items with more than 10 cm diameter, mainly rocket parts and abandoned satellites which all circle around the earth with about 36 000 km/h. Another 100 000 parts with diameters between 1 and 10 centimeters and another billion of parts with diameters below 1 cm complete this symphony of danger for space vehicles near the earth (Spiegel, 1995). Among the very small parts are also minispheres of human debris which were ejected from space vehicles. It goes without saying that a collision with such parts can cause heavy damage of a satellite or a space vehicle. In November 1995 the US space Shuttle "Columbia" was hit by a small part - presumably an electronic structural part. After return to earth an impact crater of six millimeters in depth and two centimeters in diameter was detected in the hatchway of the shuttle. If this part would have hit the oxygen tank of the shuttle an explosion would have been inevitable. Since it is to be expected that the number of such parts will rise in near earth space it could be that in a couple of years a safe travel of space vehicles in this region will not be possible any more (Spiegel, 1995, Schmundt, 2003). This would cause a throwback of mankind into a technological "stone age". If used up satellites can no longer be replaced, satellite television, GPS, wireless global phone calls, and many other services of today will cease to operate. Hence LEO has become something like an international waste disposal. Well over 150 000 scrap parts of earlier space missions race around the earth: Old and inoperable satellites, rocket parts, diverse metal parts, astronauts gloves, metal tools etc. (Schmundt, 2003). They have become the primary danger for space flights in LEO. No wonder that the NASA has installed a watch center which has registered larger pieces of space-scrap in something like a space-emergency map in order to save space shuttles and rockets from collisions with scrap (Schmundt, 2003). The European Space Operations Centre (ESOC) in Darmstadt (Germany) has also installed a Space Debris-working group with the same responsibility. Furthermore, a series of Space Debris Conferences has been installed by the European Space Agency (ESA) which took and take place in Darmstadt (Germany). The Fifth European Conference on Space Debris just took place from March 30 to April 02, 2009 at the ESA Space Operations Center (ESOC) in Darmstadt (Congrex 2009). It was the largest dedicated event on space debris issues. It was co-sponsored by the British, French, German and Italian space agencies, the committee on Space Research and the International Academy of Astronautics.

Most astonishingly there exists no regulation which would forbid the generation of space scrap in spite of the fact that an international Office for Outer Space Affairs (OOSA) is established in Vienna (Austria). Nevertheless, the prevention of generation of space scrap seems to be the most meaningful and urgent requirement for the prevention of a breakdown of space flight. There are several proposals of how to clean LEO from scrap but none of them has materialized due to financial problems. Without massive political pressure in all nations active in space flight a cleaning campaign in LEO is a hopeless case (Schmundt, 2003). The LEO is to a limited extent 'self-cleaning' because all debris will eventually enter the atmosphere and burn up, but the time scale of this process is highly altitude-dependent and will take tens to hundreds of years, even if no additional space debris was produced any more.

7. Degradation of solar cell samples

Solar cell arrangements and thermal MLI-blankets represent typical surfaces of satellite bodies. Hypervelocity impacts of microscopic projectiles can perforate these materials and may ultimately lead to system performance degradation. In order to quantify the particle impact history of samples returned from space it is necessary to know the relationship between impact parameters and resulting crater morphology. For this purpose particles were shot at solar cells and thermal MLI-blankets under controlled conditions at the Ernst-Mach-Institute as part of the ESA/ESTEC project (Schäfer & Schneider, 1994). The objective in that collaboration was to determine to which extent it is possible to deduce impact parameters such as angle, velocity, projectile mass and, possibly, composition from the study of those impact craters. The analytical methods we used included optical and scanning electron microscopy, electron microprobe, and - to a limited extent - secondary ion mass spectrometry (Stadermann et al., 1997; Heiss & Stadermann, 1997). The investigations demonstrated that it is possible to determine some of the parameters of a hypervelocity impact onto solar cell arrangements and thermal Multi-Layer-Insulation blankets by studying crater characteristics like its dimensions or inclination. For several relationships, correlation curves have been found which can be used for calibration. Thus, it is possible to determine important characteristics of the particle environment in space by the study of impact features on satellite components retrieved from low earth orbit. The chemical composition of particle residues can only partly be determined on the retrieved satellite components that consist of complex materials. Essentially, an identification of projectile residues is only possible if the elemental contents of projectile and target are very different.

Fig. 7. shows the famous Hubble Space Telescope (HST) during the first servicing mission in December 1993 after the installation of the solar panels.

It should be mentioned that astronauts gave the HST a probably last overhaul in May 2009 (Carroll, 2010). It will give Hubble several more years of life riding high above Earth's atmosphere haze. "The best times for this telescope are ahead of it" says Hubble Project Scientist Ken Sembach of the Space Telescope Science Institute. With the telescope's now greater imaging sensitivity and resolution, its new images will be spectacular.

8. Conclusion

The probably most interesting conclusion of this survey is the rarely so obvious observation how closely related an initially purely basic research can be to applied materials research. What started out as a study of the elemental properties of cosmic dust evolved into an investigation of the degradation of space exposed surfaces and the safe differentiation between cosmic and terrestrial particles on materials surfaces. Thus the considerable danger of a certain rocket fuel technology for the life time of satellites in LEO could clearly be demonstrated.

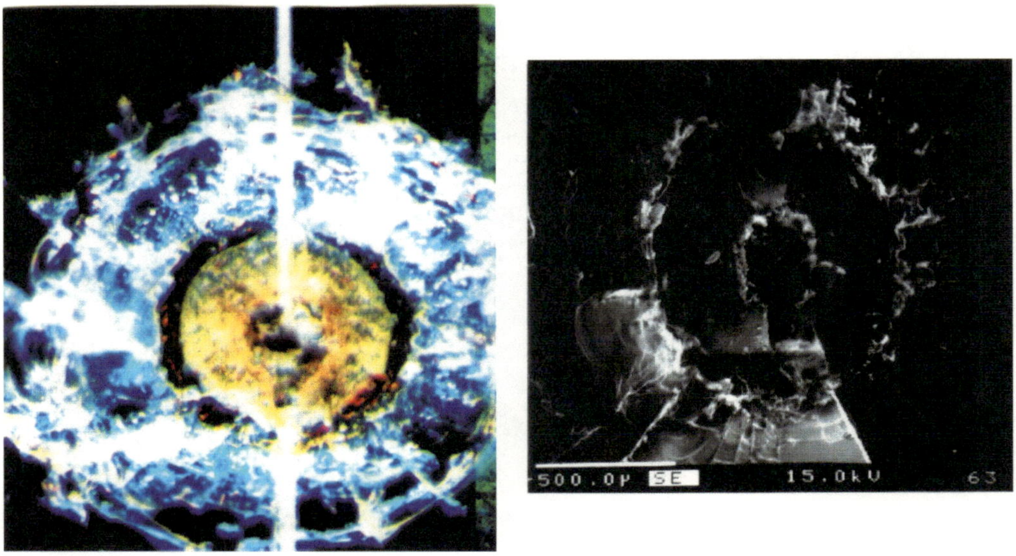

Fig. 8. shows a photo and a secondary electron image (of a scanning electron microscope) of a particle impact on a solar cell sample of the Hubble Space Telescope (Heiss & Stadermann, 1997)

This consequently caused the elimination of this technology in spite of previous great political tensions caused by this discussion. A further respective consequence was the introduction of small protective shields in flight direction especially for satellites in LEO which should operate for a long time. Basic research has thus definitely influenced applied space technology. The positioning of satellites in LEO for long duration is of great importance not only for military reconnaissance but also for modern communication and positioning systems, for global catastrophe survey and many other geopolitical surveys. Without the availability of proper topochemical and analytical technology a respective life time evaluation would not have been feasible. It seems also worth mentioning that SIMS is the key topochemical method for these investigations. The need to analyze particles in the single micrometer range triggered the development of a new SIMS-instrument generation with a much advanced lateral resolution in the upper nanometer range - the NanoSIMS 50 of CAMECA (Schuhmacher et al., 1999). Respective mass spectrometric results are not presented here in order not to unduly lengthen this article. The most important results are given in (Stadermann, 1990 and 1992).

The Hubble Space Telescope is powered by solar cells which are mounted on two flexible solar array wings. These solar array wings have been replaced after almost four years in space. One of these solar array wings was brought back to earth, while the other one was jettisoned. The retrieving of this solar array was a unique possibility for the investigation of the conditions in the low earth orbit. This wing was exposed to a permanent flux of micro particles. Since it was of interest whether these particles were man-made debris or micro-meteoroides, the European Research and Technology Centre (ESTEC) decided to disassemble two of the ten solar panel assemblies for further investigation in which our group participated.

It is not surprising that the investigation of Interplanetary Dust has developed to a most important scientific discipline in the Space Sciences. Dust is an essential component throughout space. It constitutes a considerable part of the total matter of our universe (Brownlee, 1978, Grün et al., 2001). Consequently, material degradation by micrometeorites is one of the common phenomena space flights have to cope with. It is also a fascinating result of this young scientific discipline that a part of the particles which come into the vicinity of our earth will eventually be trapped by the earth's gravity pull. Hence, our planet is collecting cosmic dust daily in a quantity of several tons. This leads to the conclusion that cosmic dust is not as exotic a material as we usually think. And every time we dust our window sills or book shelves we can be sure to clean particles from Deep Space, too.

9. References

Brownlee, D.E. (1978). In: Cosmic Dust, McDonnel J.A.M. (Editor), Wiley, Chichester, 1978.

Carroll, Ch. (2010). Hubble renewed. *Natl. Geographic,* February: 122-129

Corso, G. J., (1985). Potential effects of cosmic dust and rocket exhaust particles on spacecraft charging. *Acta Astronaut 12*: 265-267

Dambeck, Th., (2110). Cosmic navigators (in German), *Bild der Wissenschaft* 2/2010, 48-55

Grün, E. Gustafson BAS, Dermott S. & Fechtig H (Editors)(2001). *Interplanetary Dust,* Springer Berlin, Heidelberg

Heiss, C. H. & Stadermann, F. J. (1997). Chemical Analysis of Hypervelocity Impacts on the Solar Cells of the Hubble Space Telescope with EPMA-EDX and SIMS. *Adv. Space Res.* 19/257-260

Krueger, H, Krivov, A.V. & Gruen, E. (2000) A dust cloud of Ganymede maintained by hypervelocity impacts of interplanetary micrometeorids. *Planet a Space Sci.* 48: 1457-1471.

Mandeville, J.C. (1991). Study of cosmic dust particles on board LDEF: The Frecopa experiment. *Adv. Space Res.* 12, (12) 101-(12)107.

Miyachi, T., Hasebe, N., Ito H., Masumura T., Okada H., Yoshioka H. et al., (2004) Real time detector for hypervelocity microparticles using piezoelectric material. *Adv. Space Res. 34: 935-938.*

Murr L.E. & Kinard, W.H. (1993). Effects of Low Earth Orbit; *Am. Sci.* 81: 152-165

Ortner, H.M. & Stadermann, F.J. (2009). Degradation of space exposed surfaces by hypervelocity dust bombardment, and refractory materials for space. *Int. J. Refract. Metals &Hard Mater.* 27/6, 949-956.

Schäfer, F. & Schneider, E. (1994). Impact experiments on solar cell samples and thermal blankets, *ESTEC Purchase Order* # 142359

Schmundt, H. (2003). Spring Cleaning in Space (in German). *Der Spiegel* 12: 184-186

Schuhmacher, M., Rasser, B., De Chambost, E., Hillion, F., Mortz, Th, & Migeon H,N. (1999). Recent instrumental developments in magnetic sector SIMS.*Fresenius J. Anal. Chem.* 365: 12-18

Shizgal, B.D. & Arkos G.G. (1996) Nonthermal escape of the atmospheres of Venus, Earth and Mars. *Rev. Geophys.* 34, 4: 483-505

Space Flight: Trundling Parts (in German) (1995). *Der Spiegel* 46: 215

Stadermann, F.J. (1990). *PhD-Thesis*: Measurement of isotope- and element- frequencies in single interplanetary dust particles by secondary ion mass spectrometry (in German), *University of Heidelberg*

Stadermann, F.J. (1992). Cosmic dust particles –samples from the original solar nebula (in German) *Physik in unserer Zeit* 23: 197-203

Stadermann, F.J., Heiss C.H. & Reichling M. (1997). Evaluation of impact craters on solar cell samples and thermal MLI-blankets, *Adv. Space Res.* Vol. 20, No. 8: 1517-1521

6

Solar Energy Absorbers

Himanshu Dehra
Noise Behaviour Institute, McGill University
Department of Public Health Medicine, McGill University
Energy Management Research Centre, McGill University
Department of Mining Engineering, McGill University
Department of Building Engineering, Concordia University
Centre for Building Studies, Concordia University
Concordia Institute of Information Systems Engineering, Concordia University
Département de Communication Sociale et Publique, UQÀM
Akal Takht (1-140 Avenue Windsor, Lachine, Québec)
Concordia

1. Introduction

The most of the solar energy is absorbed by moving planets, their satellites and their surrounding environment viz., planet surface, planet atmosphere, forests, farms, rivers, ponds, lakes & seas, living beings and civil structures (e.g. buildings, green houses, thermal power plants, collectors, panels, roads, bridges, ports, canals). The life and its activities are reliant upon the sun's radiant energy which apart from the earth is also stocked up by green plants. In addition to the primary role of light in living economy, a continual environment of mixed radiations from various sources of radiations produce other effects, reactions and adaptations, which have susceptibility to influence the life activities of the living organisms living in a continual environment. The solar radiation is passed through the earth's atmosphere and while passing, the solar radiation is reflected, scattered, and absorbed by gas molecules, ozone, water vapour, clouds and dust. The length of atmospheric path travel by sun rays is determined by the air mass m, the ratio of the mass of atmosphere in the actual earth-sun path to the mass which would exist if the sun were directly overhead at sea level (m=1.0).

The sunlight is the major source of radiations on the earth. The spectrum of sunlight includes ultraviolet radiation, visible light, infrared rays and radio waves. The x-rays are generated by solar flares and their ionization due to absorption occurs high in the earth's atmosphere. X-rays also reach the earth's atmosphere from various celestial sources. About 60 per cent of the energy of sunlight is in the invisible infrared region's indefinite limit in radiation spectrum of sunlight. The sunlight radiations of shorter wavelengths are absorbed in the earth's atmosphere before such radiations reach the surface of earth. The ozone layer is formed high above the atmosphere through absorption of ultraviolet radiation by oxygen. The reversible reaction, again turn the ozone to absorbs longer ultraviolet rays, re-forming oxygen. The radioactive emanations consist of three components: i) gamma rays, which are penetrating radiations of very short wavelength but otherwise like x-rays; ii) alpha particles,

positively charged helium nuclei; and iii) beta particles, rapidly moving electrons. The artificial radioactive elements are formed by bombardment with high energy particles such as helium nuclei. The most of the radiation in ultraviolet region of radiation spectrum is absorbed by the ozone in the upper atmosphere, whilst part of the radiation in the shortwave region of the radiation spectrum is scattered by air molecules, for communication of blue colour appearance of sky to our eyes. The strength of the absorption of solar energy varies with wavelength and absorption bands are formed at regions of strong absorption. The important atmospheric gases forming part of absorption bands are ozone (O_3), water vapour (H_2O), carbon dioxide (CO_2), oxygen (O_2), methane (CH_4), chlorofluorocarbons (CFC) and nitrogen dioxide (NO_2).
The scope of the chapter is to present detailed theoretical aspects of solar energy absorbers, their radiation properties, radiation sources, diffraction and measurement of radiation sources. The importance of selection of roughness factors based on fluid flow is pointed out. The human environmental health is presented for metabolism of your body to intense solar radiation and heat. Mathematical analysis of a solar thermosyphon and experimental results for applications of solar collectors to the environment, human health and buildings are elaborated later in the chapter.

2. Theory

The rate of electromagnetic radiation emitted at a rate E_x from the surface of a solar energy absorber is given by the Stefan-Boltzmann equation as follows:

$$E_x = \varepsilon\sigma T^4 \qquad (1)$$

Where, E_x is exitance of a solar energy absorber, T is temperature in K, σ is Stefan-Boltzmann constant, 5.67 x 10^8 W/(m^2.K^4) and ε is hemispherical emittance for a surface of solar energy absorber. The theoretical maximum value of hemispherical emittance possible from the surface of a solar energy absorber is 1.0. The radiation emitted from the surface of a solar energy absorber for ε=1.0, at normal emittance is called blackbody radiation.
Measurement of Radiation: The intensity of all radiation is measured in terms of amounts of energy per unit time per unit area. When radiation is measured in terms of its heating power, it is only necessary to absorb all the incident radiation on a black surface and convert the radiation to heat which may be taken up in water and measured by a thermometer as in heliometers used for measuring the energy of sunlight. The small amount of radiation is measured by placement of thermocouples in water or on the black receiving surface.

2.1 Radiation properties

Source and Sink: A line normal to the plane, from which energy is imagined to flow uniformly in all directions at right angles to it, is a source. It appears as a point in the customary two-dimensional energy flow diagram. The total energy flow per unit time and unit length of line is called the strength of the source. As the flow is in radial lines from the source, the current of energy flow is at a distance r from the source, which is determined by the strength divided by the energy flow area.
The radiation of the sun, direct rays from the sun and diffuse rays from the sky, clouds, and surrounding objects incident on a transparent surface of a solar energy absorber is partly transmitted and partly reflected. In addition to this some part of the radiation is absorbed by

the selective coating on the surface of a solar energy absorber. The part of the incident flux that is reflected is called the reflectance ρ, the part absorbed is called the absorptance α, and the part transmitted is called the transmittance τ. The sum of reflectance, absorptance and transmittance is unity, or

$$\rho + \alpha + \tau = 1 \tag{2}$$

The radiation incident on the surface of a solar energy absorber has non-constant distributions over the directions of incidence and over the wavelength (or frequency) scale. The radiation properties transmittance, reflectance and absorptance are properties of a specific thickness for a sample of selective material of a solar energy absorber. The emittance ε of the surface of a solar energy absorber is the ratio of the emission of thermal radiant flux from a surface to the flux that would be emitted by a blackbody emitter at the same temperature. The angular dependence for radiation properties is explained through a solid angle formed by all rays joining a point to a closed curve. For a sphere of radius R, the solid angle is the ratio of the projected area A on the sphere to the square of length R. A sphere has a solid angle of 4 π steradians. The solar radiation incident on a point at a surface of a solar energy absorber comes from many directions in a conical solid angle. For a cone of half angle θ, the solid angle defined by the circular top and point bottom of that cone is given by

$$\Omega = 2\,\pi\,(1\text{-}\cos\theta) \tag{3}$$

In measurement of the transmittance or reflectance, a sample is illuminated over a specified solid angle. The flux is then collected for a given solid angle to measure reflectance or transmittance. A conical solid angle is bound by right circular cone. The source of solar radiation is sunlight. The radiation properties of sunlight necessary for performance analysis of daylighting and lighting are defined as follows:

The luminous flux is the time rate of flow of light. A receiver surface of a solar energy absorber receives watts of sunlight and it emits luminous flux. The measure of the rate of success in converting watts of sunlight to lumens is called efficacy.

The illuminance on a surface of a solar energy absorber is the density of luminous flux incident on that surface. The luminous flux travels outward from a source, it ultimately impinges on many surfaces, where it is reflected, transmitted and absorbed.

Luminous intensity is the force generating the luminous flux. A source of sunlight is described as having a luminous intensity in a particular direction. The inverse square law of illumination states that the illuminance on a surface perpendicular to the line from the point source of sunlight to the surface of a solar energy absorber varies directly with the intensity of the source and inversely with the square of the distance from the source of sunlight to the surface of a solar energy absorber.

The luminance of a source or a sink is defined as the intensity of the source or the sink in the direction of an observer divided by the projected area of the source or sink as viewed by an observer. The luminance of the source or sink in the direction of the observer is the intensity in that direction divided by the projected area.

The luminance exitance is the density of luminous flux leaving a surface of a solar energy absorber. The reflectance is the ratio of the luminous flux reflected from a surface to the luminous flux incident on that surface. The transmittance is the ratio of the luminous flux transmitted through a surface to that incident on the same surface.

Quantity of Sources: Quantity of sources is luminous energy and is related to luminous flux, which is luminous power per unit time.

2.2 Radiation sources

The sources of radiation are classified according to the type of wave of interference (Dehra, 2007c, Dehra 2006):

Light: The light is a visual sensation evaluated by an eye with seeing of a radiant energy in the wavelength band of electromagnetic radiation from approximately between 380 to 765 nm (nm = nanometer = $(10^9 + 1)^{-1}$ meter). The units of light are based on the physiological response of a standard (average) eye. The human eye does not have the same sensitivity to all wavelengths or colors. The solar energy spectrum in the visible region contributes in adding daylight as a visual sensation to the human body.

Sound: The sound is a hearing sensation evaluated by ear due to fluid pressure energy in the frequency band approximately between 20 Hz and 20,000 Hz. The units of sound are based on the physiological response of the standard (average) ear. The human ear does not have the same sensitivity to the whole frequency band.

Heat: The heat is a sensation of temperature evaluated by a radiant energy in the wavelength band of electromagnetic radiation from approximately between 0.1 μm to 100 μm (μm = micrometer = $(10^6 + 1)^{-1}$ meter). The units of heat are function of sensation of temperature. The sensation of temperature is a measure of hotness and coldness. Thermal comfort is an evaluation of comfort zone of temperature on the basis of physiological response of a standard (average) human body. The solar energy spectrum in the ultra violet radiation region contributes to sensation of discomfort of the human body.

Electricity: The electricity is a sensation of shock evaluated by skin of an observer due to an electromagnetic energy stored in a conductor short-circuited by a human body either due to pass of direct current or an alternating current.

Fluid: The fluid is a combined sensation of ventilation and breathing evaluated by the amount of fluid passed either externally or internally through a standard (average) human body.

Fire: The fire is a sensation of burning caused due to combined exposure of skin to radiation energy and fluid acting on a standard (average) human body.

2.3 Diffraction of radiation sources

The diffraction of radiation sources is termed as interference of noise. The interference of radiation sources are based on areas of energy stored in a wave due to interference, speed of wave and difference of power between two intensities of wave (Dehra, 2008b).

Noise of Sol: The noise of sol (S) is noise occurring due to difference of intensities of power between two solar systems. The amplitude of a solar energy wave is defined as the power storage per unit area per unit time. The solar power is stored in a packet of solar energy wave of unit cross sectional area and of length s, the speed of light.

Noise of Therm: The noise of therm is noise due to difference of intensities of power between two heat power systems. The amplitude of a heat wave is defined as the power storage per unit area per unit time. The heat power is stored in a packet of heat wave of unit cross sectional area and of length s, the speed of light.

Noise of Photons: The noise of photons is noise due to difference of intensities of power between two lighting systems. The amplitude of a light beam is defined as the power storage per unit area per unit time. The light power is stored in a packet of light beam of unit cross sectional area and of length s, the speed of light.

Noise of Electrons: The noise of electrons is noise due to difference of intensities of power between two electrical power systems. The amplitude of an electricity wave is defined as the power storage per unit area per unit time. The electrical power is stored in a packet of an electricity wave of unit cross sectional area and of length s, the speed of light.

Noise of Scattering: The noise of scattering is noise due to difference of intensities of power between two fluid power systems. The amplitude of a fluid wave is defined as the power storage per unit area per unit time. The fluid power is stored in a packet of fluid energy wave of unit cross sectional area and of length s, the speed of fluid.

Noise of Scattering and Lightning: The noise of scattering and lightning is a noise due to difference of intensities of power between two fire power systems. The amplitude of a flash of fire is defined as the power storage per unit area per unit time. The fire power of light is stored in a packet of flash of fire of unit cross sectional area and of length s, the speed of light. The fire power of fluid is stored in a packet of flash of fire of unit cross sectional area and of length s, the speed of fluid.

Noise of Elasticity: The noise of elasticity is a noise due to difference of intensities of power between two sound power systems. The amplitude of a sound wave is defined as the power storage per unit area per unit time. The sound power is stored in a packet of sound energy wave of unit cross sectional area and of length s, the speed of sound.

2.4 Measurement of interference of radiation sources

The measurement equations for measuring interference of radiation sources are presented herewith (Dehra, 2008b).

Noise of Sol: The solar power intensity I is the product of total power storage capacity for a packet of solar energy wave and the speed of light. The logarithm of two solar power intensities, I_1 and I_2, gives power difference for two solar power intensities. It is mathematically expressed as:

$$\text{Sol} = \log \left(I_1\right) \left(I_2\right)^{-1} \tag{4}$$

Where, Sol is a dimensionless logarithmic unit for noise of sol. The decisol (dS) is more convenient for solar power systems. Since a decisol (dS) is 1/11th unit of a Sol, it is mathematically expressed by the equation:

$$dS = 11 \log \left(I_1\right) \left(I_2\right)^{-1} \tag{5}$$

Noise of Therm: The heat power intensity I is the product of total power storage capacity for a packet of heat energy wave and the speed of light. The packet of solar energy wave and heat energy wave, have same energy areas, therefore their units of noise are same as Sol.

Noise of Photons: The light power intensity I is the product of total power storage capacity for a packet of light energy wave and the speed of light. The packet of solar energy wave and light energy wave, have same energy areas, therefore their units of noise are same as Sol.

Noise of Electrons: The electrical power intensity I is the product of total electrical storage capacity for a packet of electricity wave and the speed of light. The packet of solar energy wave and an electricity wave, have same energy areas, therefore their units of noise are same as Sol.

Noise of Scattering: The fluid power intensity I is the product of total power storage capacity for a packet of fluid energy wave and the speed of fluid. The logarithm of two fluid power intensities, I_1 and I_2, gives power difference for two fluid power intensities. It is mathematically expressed as:

$$Sip = \log (I_1)(I_2)^{-1} \tag{6}$$

Where, Sip is a dimensionless logarithmic unit for noise of scattering. The decisip (dS) is more convenient for fluid power systems. Since a decisip (dS) is 1/11th unit of a Sip, it is mathematically expressed by the equation:

$$dS = 11 \log (I_1)(I_2)^{-1} \tag{7}$$

The water is a standard fluid used with a specific gravity of 1.0 for determining the energy area for a fluid wave.

Noise of Scattering and Lightning: The intensity, I, of flash of fire with power of light, is the product of total power storage capacity for a packet of fire wave and the speed of light. The intensity, I, of flash of fire with power of fluid, is the product of total power storage capacity for a packet of fire wave and speed of fluid.

The combined effect of scattering and lightning for a noise due to flash of fire is to determined by superimposition principle.

- The packet of solar energy wave and a flash of fire with power of light, have same energy areas, therefore their units of noise are same as Sol. The flash of fire with power of light may also include power of therm.
- The packet of fluid energy wave and a flash of fire with power of fluid, have same energy areas, therefore their units of noise are same as Sip. A multiplication factor of a specific gravity of fluid is used in determining the areas of energy for the case of fluids other than water.

Noise of Elasticity: The sound power intensity I is the product of total power storage capacity for a packet of sound energy wave and the speed of sound.

The logarithm of two sound power intensities, I_1 and I_2, gives power difference for two sound power intensities. It is mathematically expressed as:

$$Bel = \log (I_1)(I_2)^{-1} \tag{8}$$

Where, Bel is a dimensionless logarithmic unit for noise of elasticity. The decibel (dB) is more convenient for sound power systems. Since a decibel (dB) is 1/11th unit of a Bel, it is mathematically expressed by the equation:

$$dB = 11 \log (I_1)(I_2)^{-1} \tag{9}$$

3. The roughness factors

The utilisation of solar energy is based on selective design of solar energy absorbers. The minimal flow resistance is required for critical design so that there is maximum absorptance of solar energy at the optimum roughness of the surface. The solar collectors and ducts used for heating, ventilation and air conditioning (HVAC) and hot water have fluid resistance due to friction losses and dynamic losses. For fluid flow in conduits, the friction loss is calculated by Darcy equation:

$$\Delta p_f = \frac{f\,L}{D_h}\,\frac{\rho\,V^2}{2} \tag{10}$$

Where, Δp_f is friction loss in terms of total pressure (Pa); f is friction factor, dimensionless; L is duct length, m; D_h is equivalent hydraulic diameter, m; V is velocity of fluid, m/s and ρ is density of fluid, kg/m^3. For a region of laminar flow (Reynolds number less than 2000), the friction factor is a function of Reynolds number only.

For turbulent fluid flow, the friction factor depends on Reynolds number, duct surface roughness, and internal protuberances such as joints. The region of transitional roughness zone lies in between the bounding limits of hydraulically smooth behaviour and fully rough behaviour and for this region of transitional roughness, the friction factor depends on both roughness and Reynolds number. For this transitionally rough , turbulent zone the friction factor, f is calculated by Colebrook's equation. Colebrook's transition curve merges asymptotically into the curves representing laminar and completely turbulent flow.

$$\frac{1}{\sqrt{f}} = -2\log\left(\frac{\varepsilon}{3.7\,D_h} + \frac{2.51}{Re\sqrt{f}}\right) \tag{11}$$

Where, ε is absolute roughness factor (in mm) for material of a solar energy absorber and Re is Reynolds number. Reynolds number is calculated by using the following equation:

$$Re = \frac{D_h \cdot V}{\upsilon} \tag{12}$$

Where, υ is kinematic viscosity, m^2/s. For standard air, Reynolds number is calculated by:

$$Re = 0.0664\,D_h\,V \tag{13}$$

The roughness factors, ε are listed in Table 1.

4. Human environmental health

Your body acts as a solar energy absorber, which enable your senses for interpretation of our surrounding environment. Your body when exposed to solar radiation releases heat by radiation and conduction. The amount of heat you loose is a function of the difference in temperature between the surface of your body and the environment. The greater is the difference in temperature, the greater the heat loss would be. The heat would be released from your body, if the surface temperature of your body is higher than that of the environment. If due to excessive solar radiation, the environmental temperature rises above your body temperature, you will gain heat from the environment.

Another important method of loosing heat is through evaporation. After swimming, when you come out of the water, there is evaporation of water from your skin and you feel cool.

Duct Material	Roughness Category	Absolute Roughness, ε, mm
Uncoated carbon steel, clean (0.05 mm) PVC plastic pipe (0.01 to 0.05 mm) Aluminium (0.04 to 0.06 mm)	Smooth	0.03
Galvanized steel, longitudinal seams, 1200 mm joints (0.05 to 0.10 mm) Galvanised steel, continuously rolled, spiral seams, 3000 mm joints (0.06 to 0.12 mm) Galvanised steel, spiral seam with 1, 2 and 3 ribs, 3600 mm joints (0.09 to 0.12 mm)	Medium smooth	0.09
Galvanised steel, longitudinal seams, 760 mm joints (0.15 mm)	Average	0.15
Fibrous glass duct, rigid Fibrous glass duct liner, air side with facing material (1.5 mm)	Medium rough	0.9
Fibrous glass duct liner, air side spray coated (4.5 mm) Flexible duct, metallic (1.2 to 2.1 mm when fully extended) Flexible duct, all types of fabric and wire (1.0 to 4.6 mm when fully extended) Concrete (1.3 to 3 mm)	Rough	3.0

Table 1. Roughness factors for some common duct materials.

The water molecules on your body surface must have minimum amount of energy for evaporation. The faster moving water molecules can overcome the forces holding them in the liquid state and bound off into the air as water vapour molecules. The slower and therefore cooler molecules are left behind. Heat then flows from the warmer surface of your skin to the cooler water molecules. This flow of heat transfers energy to the water, speeding the water molecules up so that more of them escape. This cooling of your skin surface also cools any blood which tends to flow through that part of your body. Sweating is a noticeable way to lose heat by evaporation. During the process of sweating, water continuously evaporates from your skin. There is also a small loss of water from the surface of the lungs when you breathe. The amount of water that evaporates, when you breathe or sweat, depends on the humidity of the air. When the humidity of the surrounding air is high, water evaporates much more slowly and therefore contributes less to the cooling process.

4.1 Effects of intense heat

Your presence in a room with high air temperature, radiation and conduction do not work in your favour for loss of body heat. Instead of loosing heat from the surface of your body to the surroundings, you gain heat. You can survive, but now sweating is the only mechanism you have for losing heat. The normal response of your body is intense heat strains of the circulatory system. This follows because the hypothalamus responds to the increased heat by causing the blood vessels in your skin to expand. This leads to a decreased resistance to blood flow and your blood pressure tends to fall. Reflexes which prevent large changes in

blood pressure then begin to operate and the decreased resistance to blood flow is compensated for by the heart working harder. The expanded blood vessels make it possible for large amounts of blood to pool in the vessels of your skin at the expense of other organs. If as a result, the blood supply to your brain becomes sufficiently low, you will faint.
Sweating may also create a circulatory problem because of the salt and water loss. Excessive fluid loss causes a decreased plasma volume. This may slow down the output of blood from the heart, which could lead to decreased blood flow to the skin, which in turn could reduce sweating. If this happened, your main avenue for heat loss would be closed. In that event heat production would continue and your body temperature would rise until your whole system is collapsed. The body's ability to control heat loss is limited. When heat can not be lost rapidly enough to prevent a rise in body temperature, a vicious circle may occur. When heat regulation fails, the positive feedback loop (Heat production - metabolism - temperature control) goes into operation; if unchecked it ends in heat stroke and death.
In order to support the case of heat loss from your body, a mathematical analysis of a solar thermosyphon is illustrated. This is followed by presenting some experiments conducted on photovoltaic duct wall. Your body follows the thermosyphon principle for loss of heat. The example of photovoltaic duct wall illustrates the production of heat, metabolism for heat production rate and temperature control in your body.

5. Mathematical analysis of a solar thermosyphon

The mathematical analysis has been performed for steady heat conduction and heat transport analysis of a solar thermosyphon (Dehra, 2007d). The analysis has been conducted on system geometry of a solar thermosyphon with discretisation of its total covered volume into surface and air nodes located by formulation of the control volumes. As illustrated in Fig. 1, thermosyphon is placed along the y-axis with y = 0 near the bottom end of the system boundary and y = H near the top end of the system boundary. The solar thermosyphon is rectangular in cross-section with width W in z-direction and air-gap length, L in x-direction. The thermal conductivities of outer wall and inner wall are assumed to be constant along their dimensions-L, W and H. The inner wall is well-insulated with thermal conductance u_i. The outer wall is of good thermal conductance (u_o) for conducting heat flux of solar irradiation. The heat transfer between building space and well-insulated inner wall is nil. The heat transfer between side walls of length L, and height H and surrounding zone is nil. The air passage of thermosyphon system is connected with the building space through a damper operating system. The physical domain of the thermosyphon is analysed as a parallel-plate channel. The climatic and thermal design data has been kept constant in the steady heat flow analysis of a solar thermosyphon. Single climatic variable of ambient air temperature, solar irradiation and building zone air temperatures are known constants in the analysis. The unique characteristics of the improved numerical solution method are: i) inclusion of conduction heat flow along height of outer and inner walls of thermosyphon; and (ii) inclusion of radiation exchange calculations using radiosity-irradiation method by assuming enclosure between outer and inner walls of thermosyphon. The resultant affect of conjugate heat exchange and heat transport on temperature distribution in thermosyphon has improved the accuracy of the numerical method over analytical method.
The key assumptions and initial conditions used in mathematical analysis are: (i) outer wall is thin, light weight and good conductor of heat; (ii) the net solar heat flux, q_o on the outer wall is quasi steady-state and distributed uniformly over the surface; (iii) inner wall is light

weight and good insulator for heat; (iv) temperature variation only along y-ordinate, being taken as lumped in x and z-coordinates; (v) heat conduction (diffusion) equation term with negligible value for air is not included in the energy balance; (vi) heat transfer between the side walls/inner wall of the thermosyphon and the surrounding environment is negligible; (vii) temperatures of ambient air (T_a) and single building air zone (T_s) are specified. As illustrated in Fig. 2, nodal or lattice points are created in the rectangular mesh at which temperatures are to be approximated. The nodal points are created after dividing the thermosyphon system into control volumes. The distance between control volume nodes on x-y plane is $\Delta x_o=(t_o+L)/2$, $\Delta x_i=(t_i+L)/2$ for outer wall and inner wall in x-ordinate and Δy in y-ordinate. The control volumes are lumped sub system, in which temperature represented at the node represent the average temperature of the volume. The computational grid is developed by drawing five vertical construction lines at distance x = 0, t_o, (t_o + L/2), (t_o + L), and (t_o + t_i + L) apart and ten horizontal construction lines at Δy distance apart starting from $y=\Delta y/2$. Nodes are located at all the intersections of the construction lines. The control volumes are formed by drawing horizontal and vertical lines that exist midway between adjoining construction lines. The control volumes formulated are solid up to width of the outer or the inner wall and continued with made up of air of width (L/2). Surface nodes are located midway and air nodes are located on the edges of the control volume. Air-nodes are common to the two adjoining solid-air and air-solid control volumes.

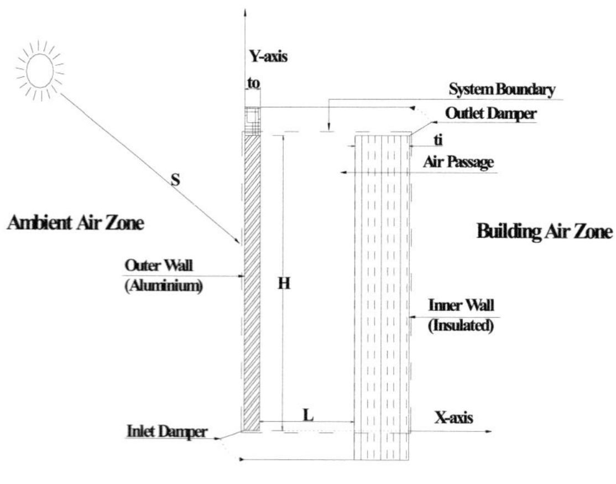

Fig. 1. Schematic of a solar thermosyphon integrated to building air zone

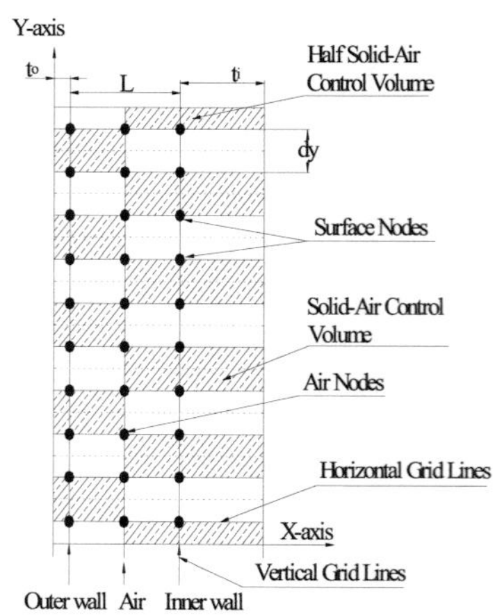

Fig. 2. Discretisation of a solar thermosyphon into control volumes, cell faces and nodes

5.1 Initial Boundary Value Problem (IBVP)

Initial boundary value problem is formulated as per initial conditions and boundary conditions. For the outer wall with uniform heat flux, heat conduction equation is written with boundary conditions as (Dehra 2007d):

$$\frac{\partial^2}{\partial x^2}T_o + \frac{\partial^2}{\partial y^2}T_o + \frac{q_o}{k_o} = 0 \qquad \text{in} \quad 0 < x < t_o \qquad 0 < y < H \qquad (14)$$

$$-\left(\frac{\partial}{\partial x}T_o\right) = \frac{\alpha\,S - h_a\left(T_o - T_a\right)}{k_o} \qquad \text{at} \quad x = 0 \tag{15}$$

$$\frac{\partial}{\partial x}T_o = \frac{h_o\left(T_o - T_f\right) + hr\left(T_o - T_i\right)}{k_o} \qquad \text{at} \quad x = t_o \tag{16}$$

For the inner wall with insulation, heat conduction equation with boundary conditions is:

$$\frac{\partial^2}{\partial x^2}T_i + \frac{\partial^2}{\partial y^2}T_i = 0 \qquad \text{in} \quad L + t_o < x < L + t_o + t_i \qquad 0 < y < H \tag{17}$$

$$\frac{\partial}{\partial x}T_i = \frac{h_s\left(T_i - T_s\right)}{k_i} = 0 \qquad \text{at} \quad x = L + t_o + t_i \tag{18}$$

$$-\left(\frac{\partial}{\partial x}T_i\right) = \frac{h_i\left(T_i - T_f\right) + hr\left(T_i - T_o\right)}{k_i} \qquad \text{at} \quad x = L + t_o \tag{19}$$

Heat transport equation for air with its boundary value as:

$$\theta(y)\left[\frac{\partial}{\partial y}T_f(y)\right] + T_f(y) - \left[\frac{T_o(y) + T_i(y)}{2}\right] = 0 \qquad \text{in} \quad 0 < y < H \tag{20}$$

$$T_f(y) = T_f(k) \qquad \text{at} \quad y = \left(k + \frac{1}{2}\right)\cdot\frac{H}{n} \qquad k = 0..(n-1) \tag{21}$$

In Eq. 21, *k* varies from 0 to (n-1), where n is number of nodes in y-ordinate. $\theta(y) = \theta$ is constant within the control volume at steady flow conditions, defined by following expression:

$$\theta = \frac{v\,\rho\,L\,c_p}{h_c\,W} \tag{22}$$

5.2 Semi-analytical method

The partial differential equations are solved by applying initial conditions and lumped parameter assumption to get the analytical solution. The temperatures of outer wall, inner wall and air are obtained as:

$$T_o = \frac{h_a\,T_a + h_o\,T_f + hr\,T_i + \alpha\,S}{h_o + hr + h_a} \qquad \text{for} \quad \left(\frac{\partial}{\partial x}T_o\right)_{x=0} = \left(\frac{\partial}{\partial x}T_o\right)_{x=t_o} \tag{23}$$

$$T_i = \frac{h_s\cdot T_s + h_i\cdot T_f + hr\cdot T_o}{h_i + hr + h_s} \qquad \text{for} \quad \left(\frac{\partial}{\partial x}T_i\right)_{x=t_o+L} = \left(\frac{\partial}{\partial x}T_i\right)_{x=t_o+L+t_i} \tag{24}$$

$$T_f = \frac{T_o + T_i}{2} + \left[T_f \cdot (k) - \frac{T_o + T_i}{2} \right] \cdot e^{-2 \cdot \frac{\Delta y}{\theta}} \tag{25}$$

Where, $\Delta y = H/n$ is discretisation height of the control volume. Equation (6) is applicable with in the control volume and it predicts particular solution for each Δy from the values of $T_f(k)$ at previous air node. The exponential solution of Equation (6) is semi-analytical in nature because of its applicability for the nodes with in the physical domain of Fig. 1(b).

The numerical solutions are obtained by creation of additional heat exchange paths in the computational grid. The additional heat exchange paths are created by incorporating conduction heat flow along height of walls of thermosyphon and integrated radiation heat exchange between composite surface nodes of outer and inner walls of thermosyphon (Dehra, 2004). The numerical analysis involves (i) construction of nodal networks; (ii) energy balance on the surface nodes located at solid-air edges of the walls; (iii) energy balance on control volume for air passage; and (iv) computer solution of system of algebraic equations. The energy balance equations for the N nodes involves formulation of ($U_{N,N}$)-matrix with conductance terms and heat source elements ($Q_{1,N}$). Conductance terms describe entropy flux over the discretised area (in W/K units) at the node. Inverse of U-matrix is multiplied with heat source matrix to give temperature solution of the thermal network. In writing nodal equations in matrix form, sign notation is adopted for automatic formulation of U-matrix with unknown temperatures and heat source elements. Sum of all incoming heat source elements and U-matrix conductance terms multiplied with temperature difference with respect to the unknown temperatures at other nodes are equal to zero. The energy balance is written in equation form for any general node (m,n) as per sign notation:

$$\sum_{n=1}^{N} \left(U_{m,n} \times \Delta T_{m,n} \right) + \sum_{n=1}^{N} Q_{m,n} = 0 \tag{26}$$

Where $U_{m,n}$ is the conductance at node (m,n), $\Delta T_{m,n}$ is the difference between unknown temperature at the node (m,n) and unknown temperature at surrounding heat exchange node. Q_N is heat source term at the node (m,n). The detail of numerical method is provided in (Dehra, 2004) and is omitted here by presenting its numerical solution procedure (Dehra, 2008a):

Step 1. Thermal properties are used to initialise the numerical solution. The conductance values are calculated as per constitutive relations for conduction, convection, radiation and heat transport.

Step 2. The corrected iterated value of the mass flow rate as depicted in Table 2 is obtained from the numerical solution and is used for obtaining thermal capacity conductance values.

Step 3. The heat transfer coefficients are calculated using temperatures obtained from the analytical solution. The values of convective heat transfer coefficients obtained from semi-analytical solution are also used in obtaining the numerical solution.

Step 4. The effect of integrated radiation heat exchange between surface nodes of outer and inner wall is considered with radiosity-irradiation method assuming enclosure analysis (Dehra, 2004). The radiation heat exchange factors are calculated for each node using script factor matrix of size (20 X 20). Using radiation heat exchange factors radiation conductance values are calculated, which also form matrix (20 X 20).

Step 5. Once conjugate heat exchange conductance values for 30 nodes are calculated, U-matrix of size (30 X 30) is formulated. The U-matrix is formulated by obtaining off-diagonal and diagonal entries as per constitutive relations and sign convention. The inverse of U-matrix (30 X 30) is multiplied with heat source element matrix (1 X 30), to obtain temperatures at 30 nodes (30 X 1) as per Equation (26).

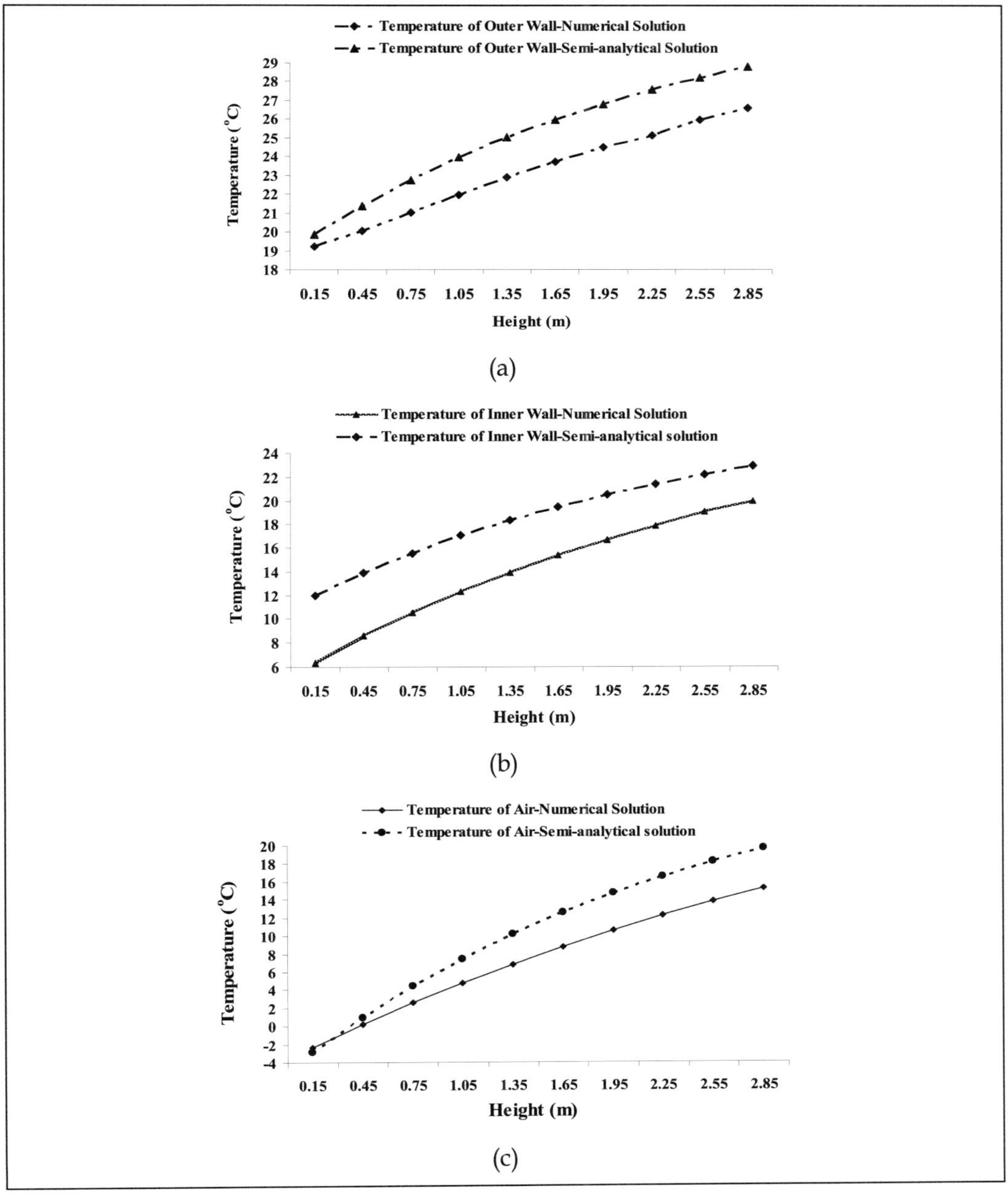

Fig. 3. Comparison of temperature profiles from semi-analytical and numerical solutions with height of solar thermosyphon for (a) outer wall; (b) inner wall; and (c) air

Figure 3 has compared the results obtained from traditional analytial model and numerical model. A matrix solution procedure is adopted for solving energy balance nodal equations at surface and air nodes. The improved numerical method has considered the effect of thermal storage by incorporating conduction heat flow factors in y-direction for outer and inner walls of thermosyphon. The heat conduction and radiation heat exchange between surface nodes has improved the accuracy of a traditional analytical solution for predicting buoyancy-induced mass flow rate through a solar thermosyphon. The conduction and convection conductance terms are based on discretisation height Δy, thermal capacity conductance (mc_p) is based on air-gap length Δx, whilst integrated radiation conductance terms are based on both height Δy and width Δx of the grid. The constitutive relations for obtaining conductance terms for conductance (U's) matrix are calculated over discretised control areas in y-z plane for conduction heat flow, radiation and convection heat exchange. Whilst, heat transport conductance terms are calculated from mass flow rate crossing the control volume in x-z plane assuming no leakage or infiltration sources in the thermosyphon.

6. Photovoltaic duct wall

In an effort to enhance overall efficiency of PV module power generation system, a novel solar energy utilization technique for co-generation of electric and thermal power is analyzed with a photovoltaic duct wall system. A full scale experimental facility for a photovoltaic duct wall was installed at Concordia University, Montréal, Concordia (Dehra, 2004). The photovoltaic duct wall was comprised of a pair of glass coated PV modules, ventilated air passage and polystyrene filled plywood board. In this case duct wall with air ventilation acts as cooling channel for PV modules by reducing surface temperature of solar cells in PV modules and slightly increases its efficiency for electric power generation. With air as fluid medium, assessment of the potential use of photovoltaic duct wall to be used as a source of co-generation of electric and thermal power can be performed by thermal analysis of material properties of photovoltaic duct wall system (PV module, air and plywood board). The thermal analysis of a photovoltaic duct wall has been performed through experimental and numerical investigations. The measurement data collected from the experimental setup was for solar intensities, currents, voltages, air velocities and temperatures of air and composite surfaces. The measured temperatures were obtained as a function of height of photovoltaic duct wall. The heat transfer rate from a photovoltaic duct wall is a measure of heat storage and thermal storage capacities of its various components. The steady state heat transfer rate has been predicted by performing two dimensional energy and mass balances on discretised section of photovoltaic duct wall, to get solutions of one dimensional heat conduction and heat transport equations. The assumptions of steady state heat transfer and lumped heat capacity are validated by comparing heat losses along all major dimensions. The non-consideration of transient analysis has been justified by comparing thermal losses along all major dimensions.

6.1 Experimental setup

The photovoltaic duct wall was installed on south facing façade of prefabricated outdoor room. The outdoor room was setup at Concordia University, Montréal, Québec, Concordia for conducting practical investigations (Dehra 2004, Dehra 2007a, Dehra 2009). The photovoltaic duct wall was vertically inclined at 10° East of South on the horizontal plane. The test section of photovoltaic duct wall was assembled in components with two commercially available PV modules, air passage with air-gap width of 90 mm, plywood board filled with polystyrene as insulation panel, side walls made up of Plexiglas and all parts connected with wooden frames. The photovoltaic duct wall section was constructed with two glass coated PV modules each of dimensions: (989 mm X 453 mm). The PV modules were having glass coating of 3 mm attached on their exterior and interior sides. The plywood board was assembled with 7 mm thick plywood board enclosure filled with 26 mm polystyrene. The overall thickness of plywood board with polystyrene was 40 mm. The exterior dampers were made of wood covered with an aluminium sheet. The heating, ventilating and air-conditioning (HVAC) requirements were met in the outdoor room by a baseboard heater, an induced-draft type exhaust fan and a split window air conditioner (Dehra, 2004). The heating was supplemented by conditioning from the fresh air entering from the inlet damper through photovoltaic duct wall. However, during the mild season of autumn for the duration of conducting experimental runs, neither baseboard heater was used nor air-conditioning unit was used for auxiliary heating or cooling inside the pre-fabricated outdoor room.

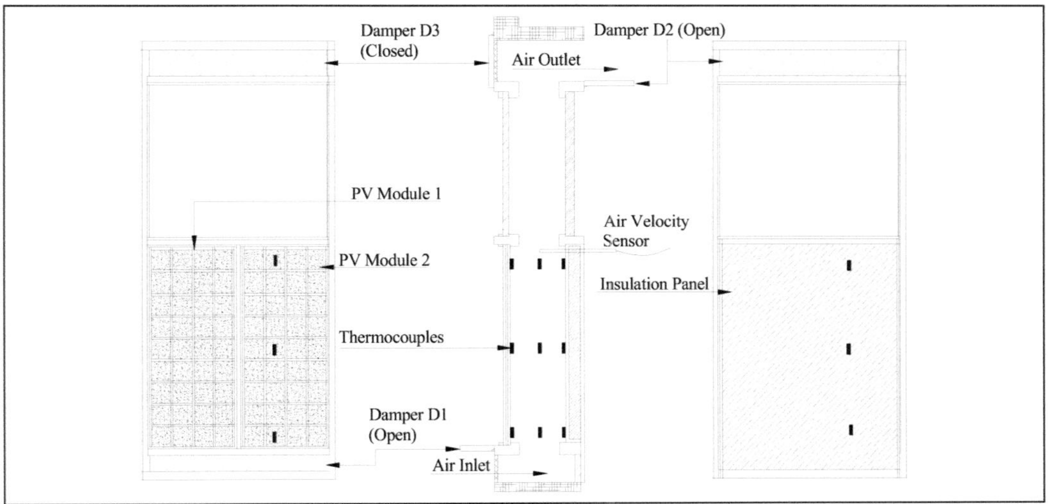

Fig. 4. Schematic of the Experimental Setup

The pair of PV modules used for conducting experimental investigations was connected in series for generation of electric power with a rheostat of maximum varying resistance up to 50 Ω. T-type thermocouples were used for obtaining thermal measurements from the test section of photovoltaic module. As is illustrated in Fig. 4, three thermocouple sensors were placed at the top, middle and bottom locations in the PV module, air-passage and insulation panel of plywood board filled with polystyrene were used to measure local temperatures. Two thermocouples were used to measure the inside test room air temperature and ambient air temperature. The hybrid air ventilation created for the PV module test section was by natural wind, or through buoyancy effect in the absence of wind (Dehra, 2004). The fan pressure was used to achieve higher air velocities by operation of the exhaust fan fixed on opposite wall with respect to wall of the test section (Dehra, 2004). The slight negative pressure was induced for drawing low air velocities in absence of wind-induced pressure from the inlet damper into the test section through the test room (Dehra, 2004). Air velocity sensor was placed perpendicular to the walls of the PV module test section to record axial air velocities near its outlet. The thermocouple outputs, currents, voltages, solar irradiation and air velocity signals were connected to a data logger and a computer for data storage. The measurements collected from the sensors were recorded as a function of air velocities or mass flow rate from the test section with use of fan pressure. The experimental data from the data acquisition system was collected and stored every two minutes in the computer (Dehra, 2004).

6.2 Temperature plots

The measurements collected from the sensors were recorded as a function of air velocities through the PV module test section (Dehra, 2004): (i) Hybrid ventilation without use of fan pressure; and (ii) Hybrid ventilation with use of fan pressure. The temperature measurements were obtained from the PV module test section as a function of air velocities. The temperatures were obtained for glass coated PV module, air passage and insulating panel in the wooden frame. The set of sample measurements obtained from outdoor experiments is presented in Table 2. The temperature plots for PV module, insulating panel and air for different hybrid ventilation conditions are illustrated in Figs. 5 and 6. The temperature plots are obtained against the height of PV module test section for the data provided in Tables 2 to 4. The variation of mean temperatures of PV module, insulating panel and air with approximately steady solar noon irradiation with varying mass flow rates for the case of fan-induced hybrid ventilation and buoyancy-induced hybrid ventilation are plotted in Figs. 7 to 9.

Run No.	S (W m^{-2})	E_p (W)	T_o (° C)	T_s (° C)	V (m s^{-1})
	Fan-induced hybrid ventilation				
1	716.1	30.7	15.2	22.4	0.68
2	716.1	30.7	13.4	22.4	0.53
	Buoyancy-induced hybrid ventilation				
3	697.5	28.9	13.2	25.1	0.13
4	697.5	28.8	13.3	24.9	0.17

Run No.	$T_p(b)$ (° C)	$T_p(m)$ (° C)	$T_p(t)$ (° C)	$T_b(b)$ (° C)	$T_b(m)$ (° C)	$T_b(t)$ (° C)	$T_a(b)$ (° C)	$T_a(m)$ (° C)	$T_a(t)$ (° C)
	Fan-induced hybrid ventilation								
1	35.4	33.8	36.8	20.6	24.7	29.1	18.8	21.7	19.4
2	35.9	34.6	37.9	20.9	25.0	29.5	19.3	22.5	19.9
	Buoyancy-induced hybrid ventilation								
3	40.8	44.9	46.8	27.9	34.8	38.0	21.3	29.5	29.8
4	39.9	45.0	46.8	28.4	35.0	38.3	21.7	28.3	29.8

Distance as per locations shown in Fig. 4	$T_p(b)$ (° C)	$T_p(m)$ (° C)	$T_p(t)$ (° C)	$T_b(b)$ (° C)	$T_b(m)$ (° C)
y (cm)	15	55	94	15	55
z (cm)	60	60	60	60	60
x (mm)	6.2	6.2	6.2	96.2	96.2

Distance as per locations shown in Fig. 4	$T_b(t)$ (° C)	$T_a(b)$ (° C)	$T_a(m)$ (° C)	$T_a(t)$ (° C)	Air velocity sensor
y (cm)	94	15	55	94	99
z (cm)	60	60	60	60	60
x (mm)	96.2	51.2	51.2	51.2	51.2

Note: x is horizontal; y is vertical; z is adjacent 3rd axis of x-y plane

Table 2. Outdoor measurements obtained from the experimental setup

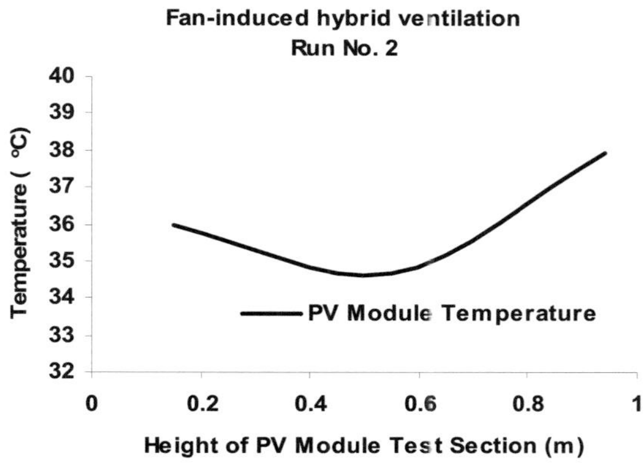

Fig. 5.(a) Temperature plot of PV module for fan-induced hybrid ventilation with height of PV module test section

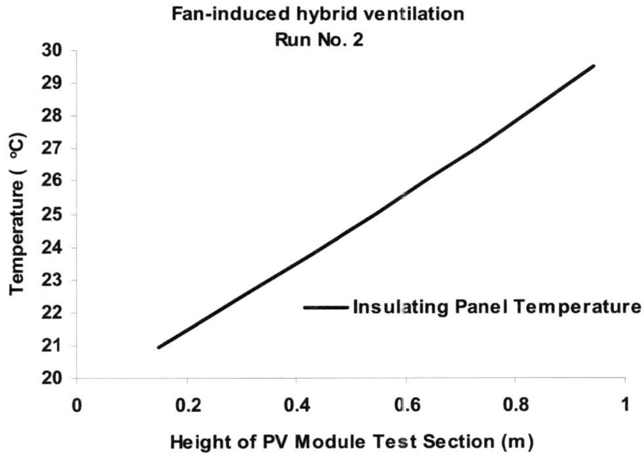

Fig. 5.(b) Temperature plot of insulating panel for fan-induced hybrid ventilation with height of PV module test section

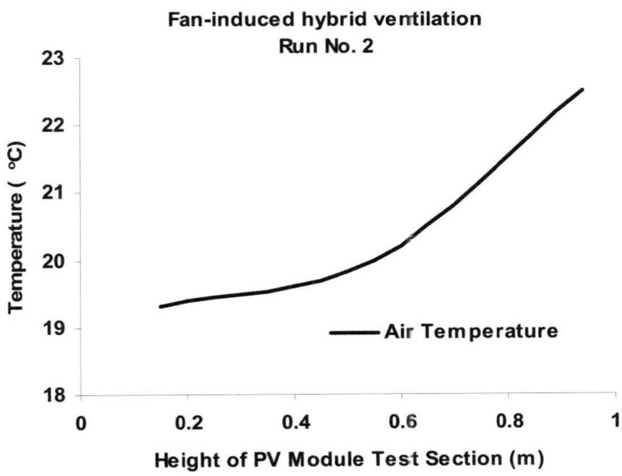

Fig. 5.(c) Temperature plot of air for fan-induced hybrid ventilation with height of PV module test section

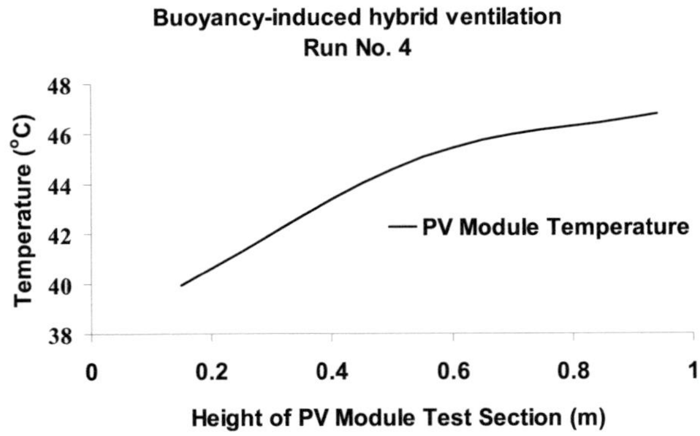

Fig. 6.(a) Temperature plot of PV module for buoyancy-induced hybrid ventilation with height of PV module test section

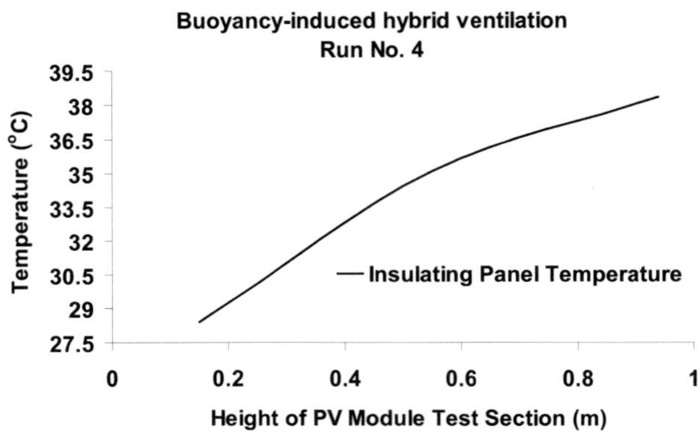

Fig. 6.(b) Temperature plot of insulating panel for buoyancy-induced hybrid ventilation with height of PV module test section

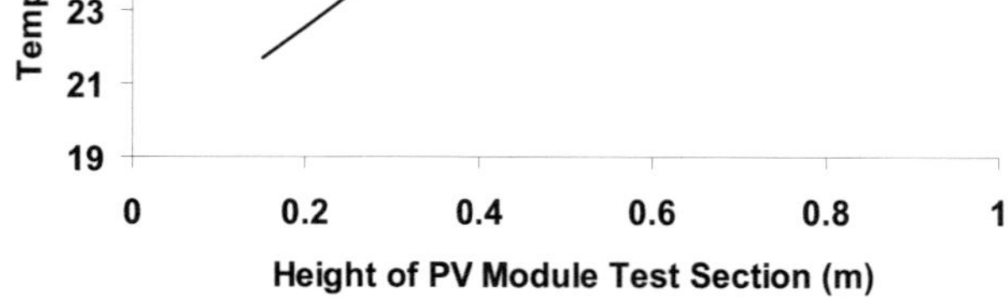

Fig. 6.(c) Temperature plot of air for buoyancy-induced hybrid ventilation with height of PV module test section

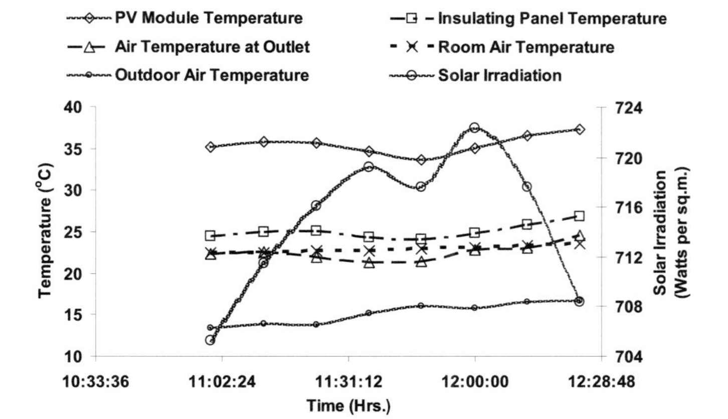

Fig. 7. Variation of mean temperatures of PV module, air and insulating panel with solar irradiation under fan-induced hybrid ventilation

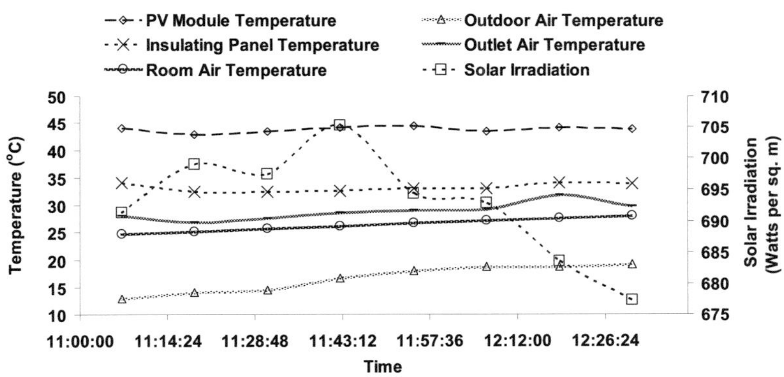

Fig. 8.(a) Variation of mean temperatures of PV module, air and insulating panel with solar irradiation under buoyancy-induced hybrid ventilation

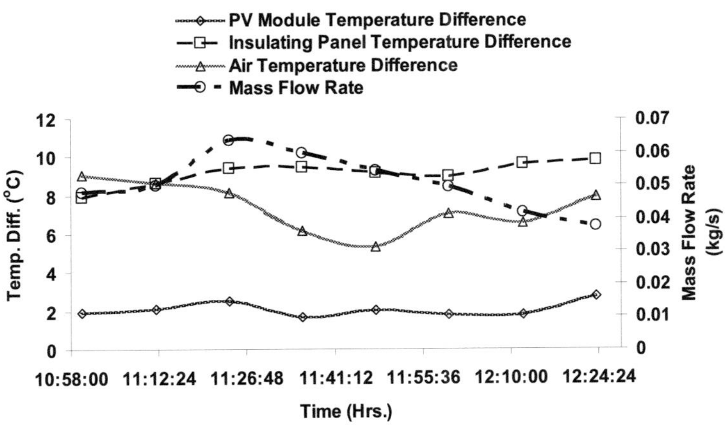

Fig. 8.(b) Temperature difference for PV module, insulating panel and air with height of PV module test section for under fan-induced hybrid ventilation

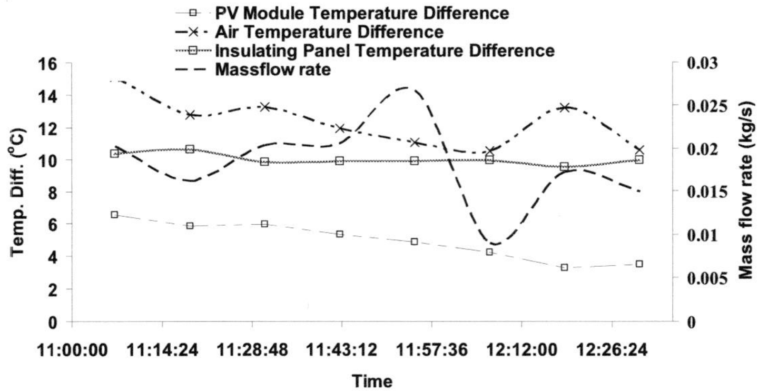

Fig. 9. Temperature difference for PV module, insulating panel and air with height of PV module test section under buoyancy-induced hybrid ventilation

6.3 Sensible heat storage capacity

The glass coated PV module test section with wooden frame was composed of non-homogeneous materials having different densities, specific heats and thicknesses (2007 a). The pair of glass coated PV module was having three layers of material viz., a flat sheet of solar cells, with glass face sheets on its exterior and interior sides. The surface temperature of PV module was assumed to be uniformly distributed in the three layers. The heat capacity of the wooden frame and sealing material was having negligible effect on the temperature of PV module, air or insulating panel because wood was used as construction material and moreover the magnitude of the heat capacity of wood framing material was not proportional to the face area of glass coated PV modules. Table 3 has presented sensible heat capacities of glass coating, solar cells, air and polystyrene filled plywood board. For the critical case of buoyancy-induced hybrid ventilation of Run no. 4 in Table 2, it was observed that the difference of temperatures recorded by the top and bottom sensors for PV module, air and insulating panel were 6.9 °C, 8.1 °C and 9.9 °C respectively. The temperature differences were used for obtaining sensible heat storage capacities of various components in y-ordinate. The heat storage capacities calculated were 59.6 kJ, 0.755 kJ and 510.7 kJ for

Component	ρ_n (kg m^{-3})	C_n (J Kg^{-1} K^{-1})	d_n (m)X10^{-3}	$d_n\rho_nC_n$ (J m^{-2}K^{-1})	$H_{pv\text{-}T}$ (J K^{-1})
Glass coating	3000	500	3	4500	4171.5
PV module	2330	677	0.2	315.48	292.45
Glass coating	3000	500	3	4500	4171.5
Sub-total	-	-	-	-	8635.5
Air	1.1174	1000	90	100.56	93.22
Plywood	550	1750	7	6737.5	6245.66
Polystyrene	1050	1200	26	32760	30368.5
Plywood	550	1750	7	6737.5	6245.66
Sub-total	-	-	-	-	42953.0
Total	-	-	-	-	51588.5

Note: Heat capacities were calculated for face area of PV module test section of 0.927 m^2.

Table 3. Sensible Heat Storage Capacities

PV module, air and insulating panel respectively. The values of heat capacities predicted were negligible in comparison with the total daily solar irradiation on PV modules on the day of conducting outdoor experiments. The values were estimated by assuming the constant surface properties and ideal still air at the instance of collection of measurements. Similar value of heat storage capacity in x-ordinate was obtained by assuming same proportionate temperature difference along thicknesses in x-ordinate. It was found to be nil in comparison with the value of heat capacity obtained for y-ordinate.

6.4 Thermal time constant

Thermal time constant is the time required for the outlet air temperature from the PV module test section to attain 63.2 percent of the total difference in value attained in air temperature following a step change in temperature of outdoor air crossing the inlet opening (Dehra 2007 a). The data was selected observing a step change in the ambient air temperature. The selected data was in steady state before and after the time-interval during the unsteady state response of the outlet air temperature with the step change in ambient air temperature. Thermal time constant under buoyancy-induced hybrid air ventilation was estimated between 8-10 minutes in comparison to 2 minutes estimated under fan-induced hybrid air ventilation. Therefore duration of time interval for obtaining measurements from the data logger was selected for a minimum of two minutes to record any subtle temperature changes. The graphs of outdoor and outlet air temperatures were plotted against the time-interval of measured data for the cases of buoyancy and fan-induced hybrid ventilation are illustrated in Fig. 10(a) and 10(b).

Thermal time constants of the PV module test section were function of ambient air temperatures and air velocities and were therefore approximately calculated under conditions of buoyancy-induced and fan-induced hybrid air ventilation.

6.5 Thermal storage capacity

Thermal storage capacities of various components of PV module test section are obtained from their thermal conductivities. Time constant ($T=\rho_d C_{pd} d_d / h_d$) for each component is

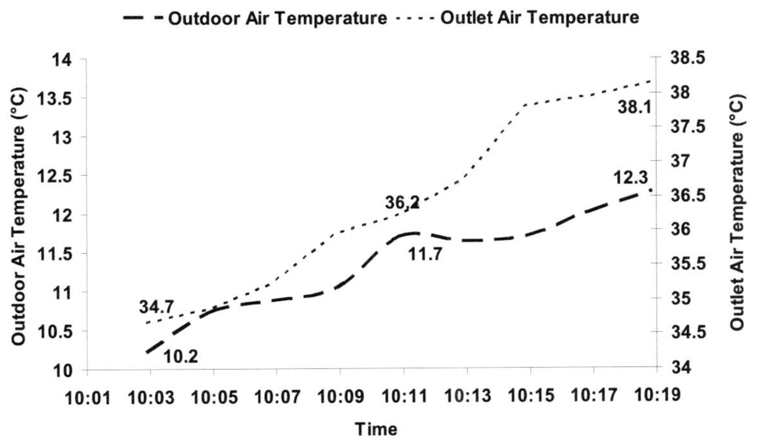

Fig. 10.(a) Changes in outlet air temperature from PV module test section with a step change in outdoor air temperature under buoyancy-induced hybrid ventilation

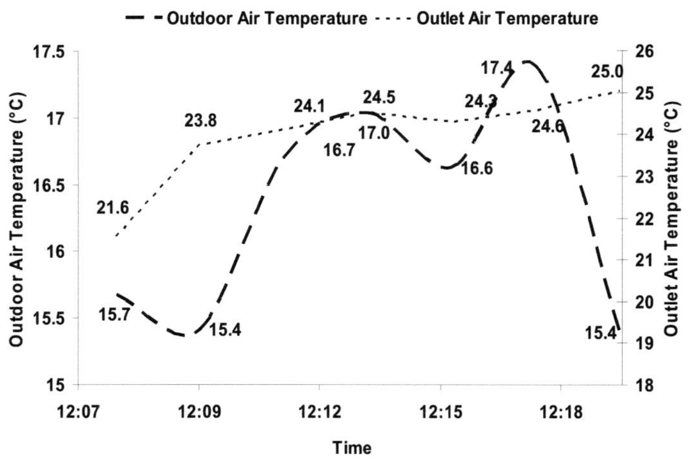

Fig. 10.(b) Changes in outlet air temperature from PV module test section with a step change in outdoor air temperature under fan-induced hybrid ventilation

calculated in units of time from their heat capacities and film coefficients ($Wm^{-2}K^{-1}$). Thermal storage capacity of PV module test section along its height is 15.9 KJ. Thermal storage capacity in x-direction is negligible in comparison with the thermal storage capacity in y-direction. Therefore temperature measurements were also felt necessary along the height of PV module system to consider pattern of heat flow and heat transport. The thermal storage capacities of components of PV module test section are presented in Table 4.

Component	k_d ($W\ m^{-1}\ K^{-1}$)	$d_n\rho_n C_n$ ($J\ m^{-2}K^{-1}$)	H_d ($Wm^{-2}K^{-1}$)	T (sec)
PV module	0.91	9315.48	10	932
Air	0.02624	100.56	10.0	10
Plywood	0.0835	6737.5	10.0	674
Polystyrene	0.02821	32760	1.0	32760
Plywood	0.0835	6737.5	10.0	674
Component	ΔT_V (K)	ΔT_H (K)	Q_V (KJ)	Q_H (J)
PV module	6.9	0.04	5.8	0.2
Air	8.1	0.75	0.0	0.0
Plywood	9.9	0.40	0.55	0.16
Polystyrene	9.9	0.40	9.0	9.6
Plywood	9.9	0.40	0.55	0.16
Total	-	-	15.9	10.12

Table 4. Thermal Storage Capacities

Notes to the Table 4:

i) Equivalent thermal conductivity of glass coated PV module was calculated to be 0.91 $Wm^{-1}\ K^{-1}$; ii) temperature differences along y-direction i.e. along height of PV module test section (0.993 m) were obtained from Table 2 for Run No. 4 in the case of buoyancy-induced hybrid ventilation.

7. Conclusion

Solar energy absorbers use their selective properties for utilisation of solar energy. The theory, analysis and methodology for selective design of solar energy absorbers is presented. The radiation properties, sources of radiation, diffraction and measurement of radiation are presented. The environmental health is discussed by presenting physiology of solar radiation effects on the skin surface and mechanism of entropy generation by solar energy absorbers. The supporting examples of solar thermosyphon and photovoltaic duct wall are presented. The heat and body temperature control for human environmental health is involved by presenting the physiology and metabolism of heat loss from the human body surface. The modeling and experimental results for solar thermosyphon and photovoltaic duct wall are elaborated for illustrating the cases of solar energy absorbers as solar collectors and panels.

8. Acknowledgements

The Author (Guru of the Sikhs Dr. Himanshu Dehra) is a Doctor of Public Health Medicine in *Noise Behaviour*. The sovereign King *Guru Shri Shri Shri 1008* Dr. Himanshu Dehra (Emperor of Earth, Wind, Sky and Lord of Moon) is grateful to his thirty two Queens for their worships and allegiances for his everlasting Kingdom and is a capital industrialist of enterprises:

I. Quality Tools & Measurement Systems (QTMS)
II. Quality Guard & Protection Systems (QGPS)
III. Quality Detection & Prevention Systems (QDPS)
IV. Quality Defence & Security Systems (QDSS)
V. Quality Arms & Ammunition Systems (QAAS)
VI. *Gati* Transportation Systems (GTS)
VII. *Mati* Logistics Systems (MLS)
VIII. *Indrani* Corporation (IC)
IX. *Indrani* Securities & Holdings (ISH)
X. *Indrani* Investments & Finances (IIF)
XI. *Indrani* Projects & Controls (IPC)
XII. *Indrani* Laws & Books (ILB)
XIII. *Indrani* Routes & Travels (IRT)
XIV. *Indrani* Herbs & Medicines (IHM)
XV. *Indrani* Designs & Furnishings (IDF)
XVI. *Indrani* Flags & Decorations (IFD)
XVII. *Indrani* Languages & Communications (ILC)
XVIII. *Indrani* Resources & Employees (IRE)
XIX. *Indrani* Commands & Forces (ICF)
XX. *Indrani* Events & Plans (IEP)
XXI. *Indrani* Agencies & Societies (IAS)
XXII. *Indrani* Religions & Worships (IRW)
XXIII. *Indrani* Forests & Timbers (IFT)
XXIV. *Indrani* Productions & Films (IPF)
XXV. *Indrani* Maps & Atlases (IMA)
XXVI. *Indrani* Reports & Presses (IRP)

XXVII. *Indrani* Crops & Animals (ICA)
XXVIII. *Indrani* Palaces & Monuments (IPM)
XXIX. *Indrani* Foods & Dairies (IFD)
XXX. *Indrani* Farms & Lands (IFL)
XXXI. *Indrani* Empires & States (IES)
XXXII. *Indrani* Marks & Trades (IMT)

9. References

Dehra, H. (2004). A Numerical and Experimental Study for Generation of Electric and Thermal Power with Photovoltaic Modules Embedded in Building Façade, submitted/unpublished Ph.D. thesis, Department of Building, Civil and Environmental Engineering, Concordia University, Montréal, Québec, August 2004.

Dehra, H. (2006). A 1-D/2-D Model for an Exterior HVAC Rectangular Duct with a Steady Solar Heat Flux Generation, orally presented at the proceedings of the Second International Green Energy Conference, Oshawa, Ontario, June 25-29 2006, 1240-1251, 0978123603.

Dehra, H. (2007a). The Effect of Heat and Thermal Storage Capacities of Photovoltaic Duct Wall on Co-Generation of Electric and Thermal Power, in the proceedings of American Institute of Chemical Engineers - AIChE 2007 Spring Meeting, Houston, Texas, USA, April 22-26, 2007, Session 36a.

Dehra, H. (2007b). On Solar Building Energy Devices, 18th IASTED International Conference on Modelling and Simulation, Montréal, Québec, May 30-June 1, 2007, 96-101, 9780889866645.

Dehra, H. (2007c). A Unified Theory for Stresses and Oscillations, Canadian Acoustics, (September 2007), Vol. 35, 3, 132-133, 07116659.

Dehra, H. (2007d). Mathematical Analysis of a Solar Thermosyphon, International conference on Advances in Energy Research, Department of Energy Science and Engineering, Indian Institute of Technology, Bombay, December 2007, Macmillan, USA, 023063432X.

Dehra, H. (2008a). The Entropy Matrix Generated Exergy Model for a Photovoltaic Heat Exchanger under Critical Operating Conditions, International Journal of Exergy, Vol. 5, Issue 2, 2008, 132-149, 17428297.

Dehra, H. (2008b). The Noise Scales and their Units, Canadian Acoustics, 36, 3, (September 2008) 78-79, 07116659.

Dehra, H. (2009). A Two Dimensional Thermal Network Model for a Photovoltaic Solar Wall, Solar. Energy, Vol. 83, Issue 11, (November 2009) 1933-1942, 0038092X.

7

Space Power System – Motivation, Review and Vision

Harijono Djojodihardjo
Aerospace Engineering Department
Faculty of Engineering, Universiti Putra Malaysia
43400 UPM SERDANG, Selangor, Malaysia
Department of Industrial Engineering
Faculty of Engineering, Universitas Al-Azhar Indonesia
Jalan Sisingamangaraja, Jakarta 121010, Indonesia

1. Introduction

The world's population growth is exhausting the world's limited supply of non-renewable energy sources, and along with it, introducing significant anthropogenic environment and climate change. Although the economic and business inertia trend to cling to the lure of non-renewable energy resources, it is imperative that in the foreseeable future extensive renewable or green energy sources should be progressively utilized to replace the non-renewable ones and to sustain reasonable living standard for the entire world's population.
The world's dream of world's socio-economic and technological equity and networking is still far from being a reality, and many of the pioneering technology breakthrough for the benefits of humanities, to some extend, contribute to their widening gap. Then it will be timely and appropriate that a new vision for world's "green" energy be shared by and contributable to a fair distribution of world's population, presently still grouped into countries with unbalanced capacity distribution. In particular, judging from the large population distribution and growth in developing countries compared to the developed ones, the need and growth for energy resources will also be more or less similar.
Mankind success in space exploration has opened up their vision of the uniqueness of the world we live in and the need for conserving our environment (Djojodihardjo, 2009; Djojodihardjo & Varatharajoo, 2009), as illustrated in Figure 1. Such vision which has inspired mankind to develop technological capabilities in atmospheric and space flight as well as exploring new fromtiers beyond the earth's atmosphere has been profoundly articulated as far back as in 400 BC by Socrates in the well known verse: "Man must rise above the Earth -- to the top of the atmosphere and beyond -- for only thus will he fully understand the world in which he lives."
Mankind has acquired further wisdom and intelligence to observe, identify and respond to the challenges posed by the observed global climate change from advances in space science, technology and exploration, and its close relationship to sustainable life on earth and mankind dramatically rising demand of energy. Such state of affairs is summarized in Figure 1. With the appreciable climate change that has been observed and of great concern

not only by scientists but world population at large, the need for global energy and its production demands new approaches and paradigm.
Through advances in space science, technology and exploration, mankind also acquired awareness of the presence of our sun as an inexhaustible source of energy, which may then offer a host of additional solutions to meet the need of world expanding population and increasing demand for energy.
The globalization process that have been taking place at impressive pace due to progress and wide accessibility of information technology, products and network has made each individual is figuratively speaking a world citizen, with more or less common objectives, interest and role in the world problems. Therefore, for a synergetic effort in Space Solar Power imperatives, open participation should be encouraged, and barriers should be identified and resolved.

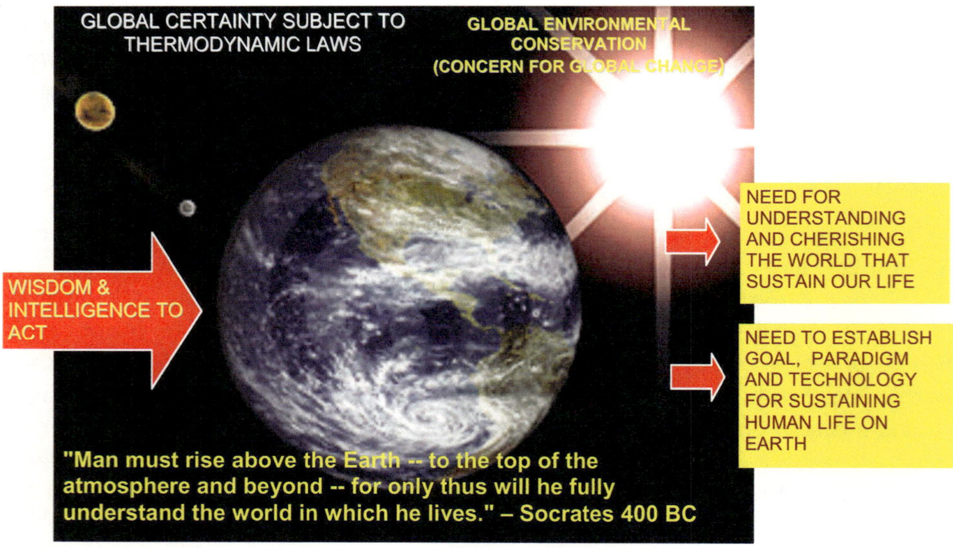

These words due to Socrates in 400 BC has inspired mankind to explore new fromtiers beyond the earth's atmosphere which eventually convince mankind for the preciousness of our unique world and the the need to for sustainable llife on earth and the significance of space exploration to that end.

Fig. 1. Mankind Vision for Sustainable World

Hence, as implicitly stipulated in Figure 1, in this article attention will be focused on three issues: 1. the need and efforts to meet basic and mandatory mankind quality of life, which at present will be directed to achieving high or acceptable Human Development Index (such as described in the United Nations Framework Convention on Climate Change), with a perspective on mankind dramatic increase in energy demand[1].; 2. understanding the global climate change process in order to identify practical conceptual approaches and policies in meeting mankind energy demand; 3. status, progress, projection and development of global energy utilization and technologies.

---

[1] The efforts to meet the basic and mandatory mankind demand for quality of life will be discussed after the discussions on environment and energy, since these may shape up the former, although in general these issues are interconnected.

Through a meticulous review of these issues to direct our wisdom and intelligence to establish vision for future solutions of man kind needs and problems, the present article will explore the prospects of space power system as one of the feasible, viable, and equitable solutions for world population as a whole and entity.

It is with such overview that the following aspects associated with Space Solar Power or Solar Power Satellite is discussed:

1. Review of Global Environment Imperatives which necessitates proactive initiatives for Sustainable Power System leading to Space Power System; Environment and Climate Change Imperatives
2. Energy Demand and Space Power System adressing the whole world, with particular reference to the developing world
3. Significance of Space Power System to the developing world, including the devloping countries
4. Space power system: review of selected architecture and technologies
5. Space Power System as a Unifying Agent for Global Networking
6. Stimulating Positive Attitude in Developing Countries: The Microsatellite Tool
7. Universal SPS Program Initiatives - arms reaching but novel paradigm - establish productive and resourceful partners-in-arms by expanding opportunities at creative circles
8. Technological options considerations in view of overall strength and weakness, gains and losses

2. Environment and climate change imperatives

The Global Climate Change and the Green-House Effect are terminologies that have been widely discussed and of global concern and global efforts in the last decades, and brought to the attention of world top policy-and-decision-makers through a series of World Summits (also known as the Earth Summits), which have been organized under the auspices of the United Nations Environment Programme (UNEP) and the World Meteorological Organization (WMO).

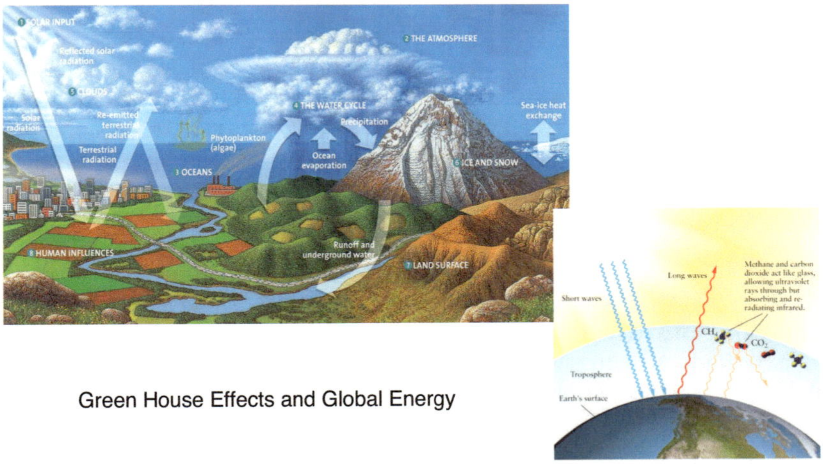

Fig. 2. Global Warming due to Unsustainable Use of Fossil Fuels is of Mankind Concern

To this end, UNEP and WMO established the Intergovernmental Panel on Climate Change (IPCC) to provide policy makers with authoritative scientific information in 1988. Related to this, the United Nations Framework Convention on Climate Change (UNFCCC or FCCC), which is an international environmental treaty, was produced at the United Nations Conference on Environment and Development (UNCED), held in Rio de Janeiro from 3 to 14 June 1992.

The objective of the treaty is to stabilize greenhouse gas (GHG) concentrations in the atmosphere at a level that would prevent dangerous anthropogenic interference with the climate system (United Nations Framework Convention on Climate Change). Actually the greenhouse effect is important, since without it, the Earth would not be warm enough for humans to live. But if the greenhouse effect becomes stronger, it could make the Earth warmer than usual that may impose serious problems for humans, plants, and animals, and thus, the environment and what is now recognized as Global Warming.

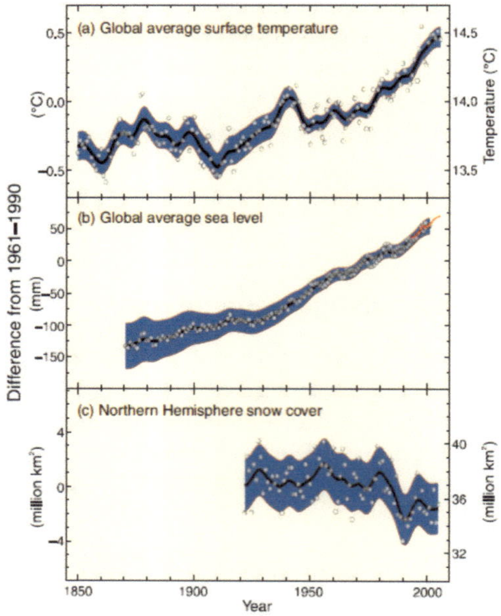

Fig. 3. Observed changes in (a) global average surface temperature; (b) global average sea level from tide gauge (blue) and satellite (red) data and (c) Northern Hemisphere snow cover for March-April. All differences are relative to corresponding averages for the period 1961- 1990. Smoothed curves represent decadal averaged values while circles show yearly values. The shaded areas are the uncertainty intervals estimated from a comprehensive analysis of known uncertainties (a and b) and from the time series (c) (Bernstein et al, 2007).

As illustrated in Figure2, energy from the sun enters the Earth's and filtered by the atmosphere in the form of electromagnetic radiation in the visible spectrum, with small amounts of Infra-Red and Ultra-Violet radiation. The incoming solar energy has a very short wavelength and passes through the atmospheric gases unaffected to reach the Earth's surface. The Earth's surface absorbs the solar energy and releases it back to the atmosphere as infrared (IR) radiation, some of which goes back into space. Some of the IR radiation emitted by the Earth is absorbed by the gases in the atmosphere that re-emit the energy as heat back towards the Earths surface. The Counter Radiation and Greenhouse Effects consist of

- Long-wave radiation from the earth's surface is absorbed by atmospheric gases: CO_2, H_2O, CH_4, other gases
- The increase in atmospheric gases increases the heat absorption by the lower atmospheric layers
- Cloud layers (water droplets) also absorb radiation.

The unbalanced production of these greenhouse gases and their absorption by the Earth's surface has been identified as the major contributor to what is now identified as Global Warming, which will be further elaborated and discussed subsequently.

Bernstein et al's report (2007), which is based on the assessment carried out by the three Working Groups of the Intergovernmental Panel on Climate Change (IPCC) provides an integrated view of climate change as the final part of the IPCC's Fourth Assessment Report (AR4). Some of their findings are reproduced or adapted below.

Observed changes in climate and their effects

Warming of the climate system is undeniable, as is now evident from observations of increases in global average air and ocean temperatures, widespread melting of snow and ice and rising global average sea level, and illustrated in Figure 3. The report stipulates that eleven of the last twelve years (1995-2006) rank among the twelve warmest years in the instrumental record of global surface temperature (since 1850). The 100-year linear trend (1906-2005) of 0.74 [0.56 to 0.92]°C1 is larger than the corresponding trend of 0.6 [0.4 to 0.8]°C (1901-2000) given in the Third Assessment Report (TAR) of IPCC. The temperature increase is widespread over the globe and is greater at higher northern latitudes. Land regions have warmed faster than the oceans.

Rising sea level is consistent with warming, as indicated in Figure 3(b). Global average sea level has risen since 1961 at an average rate of 1.8 [1.3 to 2.3] mm/yr and since 1993 at 3.1 [2.4 to 3.8] mm/yr, with contributions from thermal expansion, melting glaciers and ice caps, and the polar ice sheets.

Causes of change

Global GHG emissions due to human activities have grown since pre-industrial times, with an increase of 70% between 1970 and 2004, as indicated in Figure 4. Global atmospheric concentrations of CO_2, methane (CH_4) and nitrous oxide (N_2O) have increased markedly as a result of human activities since 1750 and now far exceed pre-industrial values determined from ice cores spanning many thousands of years.

Figure 5 shows that decadal averages of observations for the period 1906-2005 (black line) plotted against the centre of the decade and relative to the corresponding average for the period 1901-1950. Lines are dashed where spatial coverage is less than 50%. Blue shaded bands show the 5 to 95% range for 19 simulations from five climate models using only the natural forcings due to solar activity and volcanoes. Red shaded bands show the 5 to 95% range for 58 simulations from 14 climate models using both natural and anthropogenic forcings. Most of the observed increase in global average temperatures since the mid-20th century is very likely due to the observed increase in anthropogenic GHG concentrations. It is likely that there has been significant anthropogenic warming over the past 50 years averaged over each continent (except Antarctica), as indicated by Figure 5.

Long term CO_2 concentration estimates within 150,000 years are also given by Walker et al (2010) as exhibited in Figure 6. The striking peak during last few decades may indicate the anthropogenic causes of the change in CO_2 concentration.

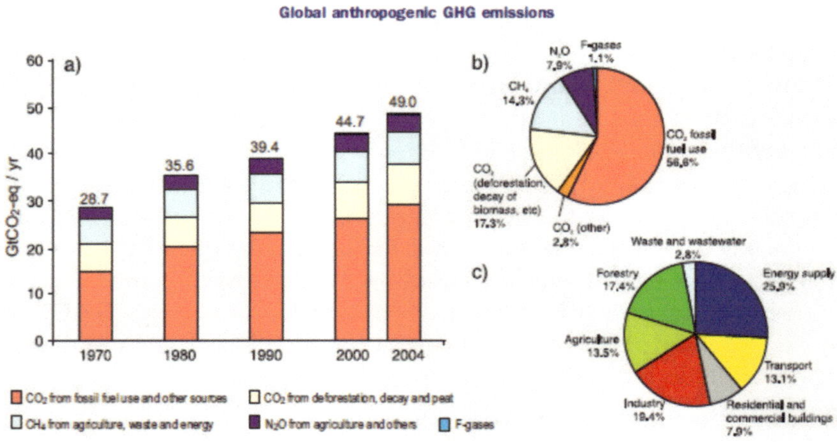

Fig. 4. Global annual emissions of anthropogenic Green House Gases (GHGs) from 1970 to 2004 (Bernstein et al, 2007).

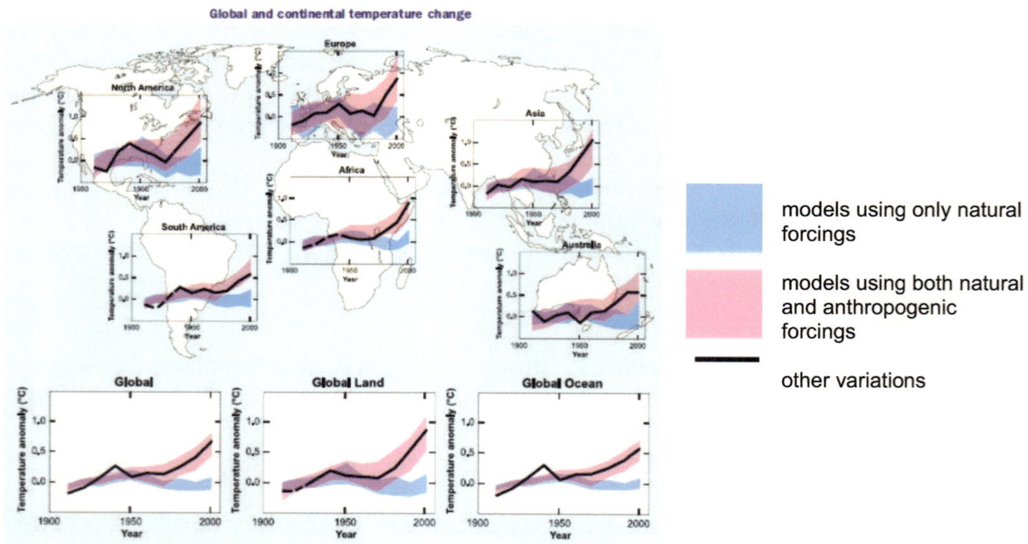

Fig. 5. Comparison of observed continental- and global-scale changes in surface temperature with results simulated by climate models using either natural or both natural and anthropogenic forcings (Bernstein et al, 2007).

Projected climate change and its impacts

There is high agreement and much evidence that with current climate change mitigation policies and related sustainable development practices, global GHG emissions will continue to grow over the next few decades. The IPCC Special Report on Emissions Scenarios (SRES, 2000) projects an increase of global GHG emissions by 25 to 90% (CO2-eq) between 2000 and 2030 (Figure 7), with fossil fuels maintaining their dominant position in the global energy mix to 2030 and beyond. More recent scenarios without additional emissions mitigation are comparable in range. Continued GHG emissions at or above current rates would cause further warming and induce many changes in the global climate system during the 21st century that would very likely be larger than those observed during the 20th century.

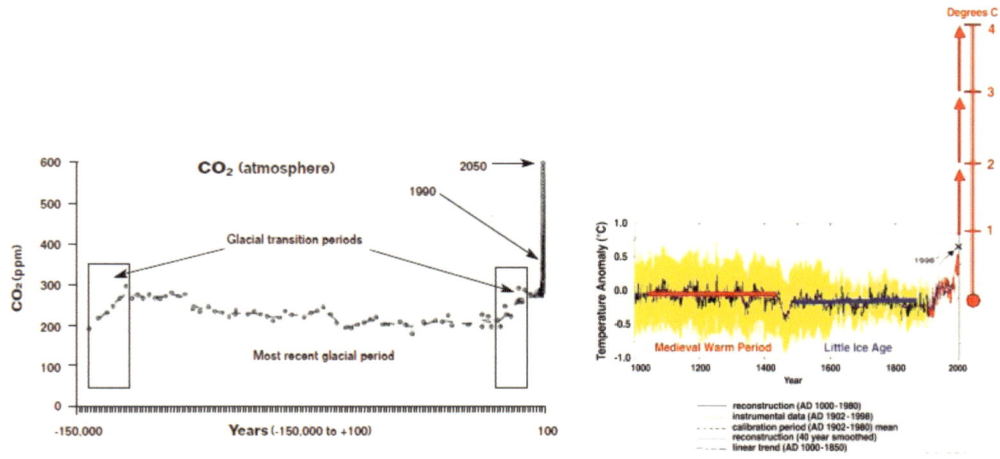

Fig. 6. Estimates of CO2 concentrations for the last 150,000 years with a projection to A.D. 2050 based on present-day rates of emission. The corresponding temperature anomaly is shown on the right hand side (From Walker, Keim and Arndt, 2010).

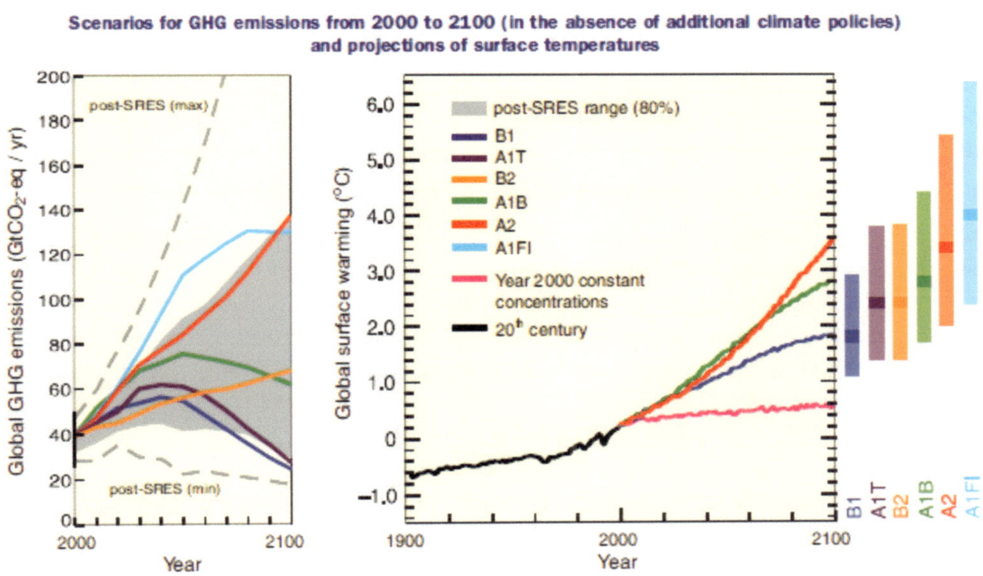

(a) (b)

Fig. 7. (a) Global GHG emissions (in GtCO2-eq) in the absence of climate policies: six illustrative SRES marker scenarios (colored lines) and the 80th percentile range of recent scenarios published since SRES (post-SRES) (gray shaded area) (Bernstein et al, 2007).

In Figure 7, the dashed lines show the full range of post-SRES scenarios. The emissions include CO2, CH4, N2O and F-gases. (b) Solid lines are multi-model global averages of surface warming for scenarios A2, A1B and B1, shown as continuations of the 20th-century simulations. These projections also take into account emissions of short-lived GHGs and aerosols. The pink line is not a scenario, but is for Atmosphere-Ocean General Circulation Model (AOGCM) simulations where atmospheric concentrations are held constant at year 2000 values. The bars at the right of the figure indicate the best estimate (solid line within each bar) and the likely range assessed for the six SRES marker scenarios at 2090-2099. All temperatures are relative to the period 1980-1999 (Bernstein et al, 2007).

Thus these scientific studies and facts have led to the conclusion that human influences have:

1. *very likely* contributed to sea level rise during the latter half of the 20th century
2. *likely* contributed to changes in wind patterns, affecting extra-tropical storm tracks and temperature patterns
3. *likely* increased temperatures of extreme hot nights, cold nights and cold days
4. *more likely than not* increased risk of heat waves, area affected by drought since the 1970s and frequency of heavy precipitation events.

Such situation certainly need fully fledged and visionary mitigation efforts to change the situation drastically, subject to the effectiveness of such measure due to natural causes and elapsed time required for such actions take effects. A wide array of adaptation options is available, but more extensive adaptation than is currently occurring is required to reduce vulnerability to climate change. There are barriers, limits and costs, which are not fully understood. Both bottom-up and top-down studies indicate that there is high agreement and much evidence of substantial economic potential for the mitigation of global GHG emissions over the coming decades that could offset the projected growth of global emissions or reduce emissions below current levels indicated in Figure 6 and Figure 7. While top-down and bottom-up studies are in line at the global level there are considerable differences at the sectoral level.

Natural Forcings to Counteract Assessed Green House Gases effects

Sensitivity experiments indicate that a level of solar variability as reconstructed over the past 1000 years is insufficient to mask the predicted 21st century anthropogenic global warming. Volcanic forcing could counteract the anthropogenic greenhouse warming, but this requires (i) a permanent level of very high volcanic activity, (ii) a volcanic forcing increasing with time, (iii) a huge stratospheric aerosol burden (unlike anything we have seen in the recent past).

Bernstein et al (2007) carried out study of various mitigation scenarios, which results in a range of future emission scenarios is exhibited in Figure 8.

Another projection of policy impact on global climate is exhibited in Figure 10. These scenarios indicate that:

i. There is an urgent need for global mitigation policy and action to mitigate the GSG global warming effect to allow sustainable development for mankind to take favorable effect.
ii. Even with appropriate immediate mitigation action, their favorable effect to the global environment will take more than hundred years to return the situation to previous situation, as Figure 10 illustrates.

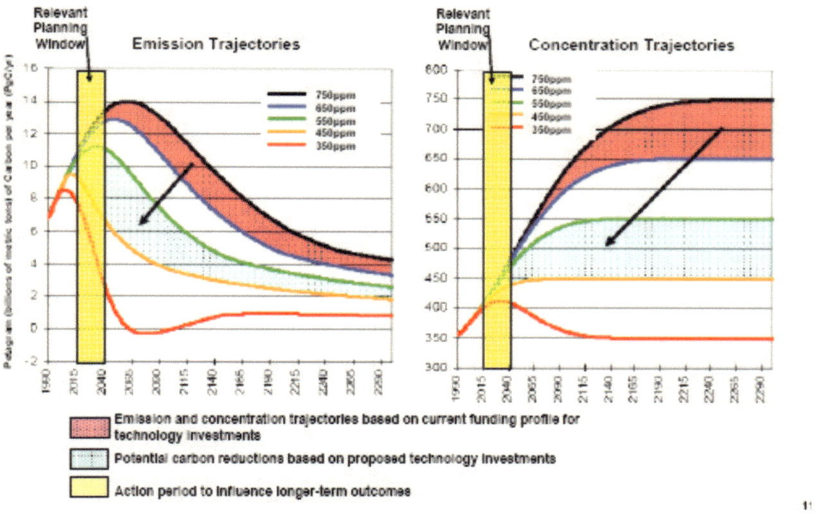

Fig. 8. Policy Impact on Global Climate (adapted from Ghoniem, (2008)

3. Energy demand and space power system adressing the whole world, with particular reference to the developing world

World Energy Demand is very much related to Economic Development, and without the global concern of global environmental sustainability, will probably be ever increasing. Such trend will probably change in a decade or so, as projected by various studies, as illustrated in Figure 9. In the US Energy Information Administration IEO2009 projections (2010), total world consumption of marketed energy is projected to increase by 44 percent from 2006 to 2030. The largest projected increase in energy demand is for the non-OECD economies, as illustrated in Figure 10(a). Grillot (2008) made a forecast, based on UNDP and DOE data that the World energy consumption will increases about 60% from 2004 to 2030~2030. Associated with this, the Carbon emission is projected in Figure 10(b).

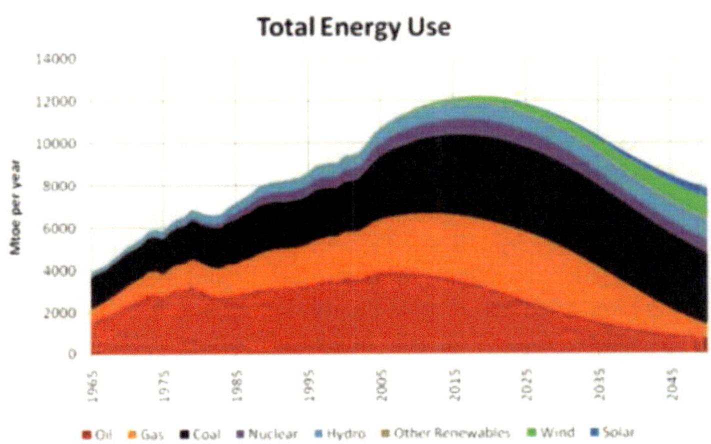

Fig. 9. World total energy utilization projection, as projected from 1965 to 2045. (Source: Chefurka, 2010).

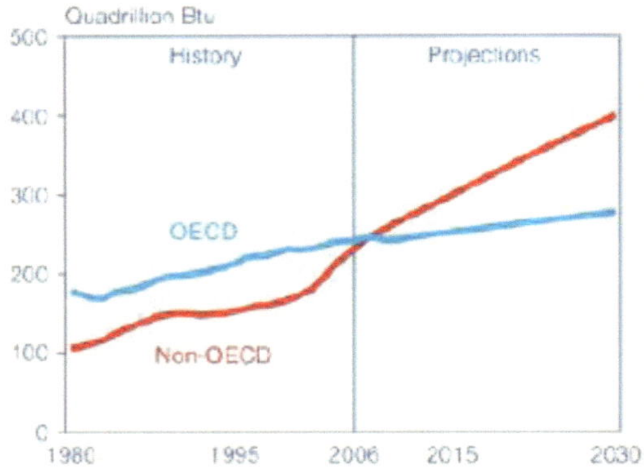

(a)

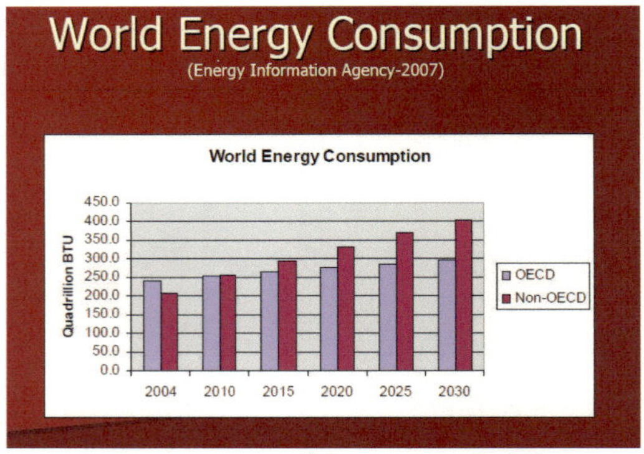

(b)

Fig. 10. (a) World marketed energy consumption, in Quadrillion Btu., in OECD and Non-OECD countries, 1980-2030, indicating higher rate of increase in developing countries (Source: US Energy Information Administration, 2010) (b) World Energy Consumption from 2004 to 2030 (Grillot, 2008). Both (a) and (b) indicate higher rate of increase in developing countries.

Much of the growth in world economic activity between 2006 and 2030 is expected to occur among the nations of non-OECD Asia, where regional GDP growth is projected to average 5.7 percent per year. China, non-OECD Asia's largest economy, is expected to continue playing a major role in both the supply and demand sides of the global economy. *IEO2009* projects an average annual growth rate of approximately 6.4 percent for China's economy from 2006 to 2030—the highest among all the world's economies. Although the difference in world oil prices between the high and low oil price cases is considerable, at $150 per barrel in 2030, the projections for total world energy consumption in 2030 do not vary substantially among the cases. There is, however, a larger impact on the mix of energy fuels consumed. The projections for total world energy use in 2030 in the high and low oil price cases are separated by 48 quadrillion Btu , as compared with the difference of 106 quadrillion Btu between the low and high economic growth cases.

The potential effects of higher and lower oil prices on world GDP can also be seen in the low and high price cases. In the long run, on a worldwide basis, the projections for economic growth are not affected substantially by the price assumptions. There are, however, some relatively large regional impacts. The most significant variations are GDP decreases of around 2.0 percent in the high price case relative to the reference case in 2015 for some regions outside the Middle East and, in the oil-exporting Middle East region, a 5.5-percent increase in GDP in 2015.

The regional differences persist into the long term, with GDP in the Middle East about 6.2 percent higher in 2030 in the high oil price case than in the reference case and GDP in some oil-importing regions (such as OECD Europe and Japan) between 2.0 percent and 3.0 percent lower in the high price case than in the reference case.

Economic viability will play a critical role in determination of the optimal energy option. The current worldwide energy market is dominated by fossil fuels, making any alternative difficult to implement due to lack of existing infrastructure, as well as commercial and practical interest driven, although may not be visionary. Not only will the technical feasibility and cost of both green and space based power sources be well understood and appreciated, but also the necessary technological learning curve and economic pressure.

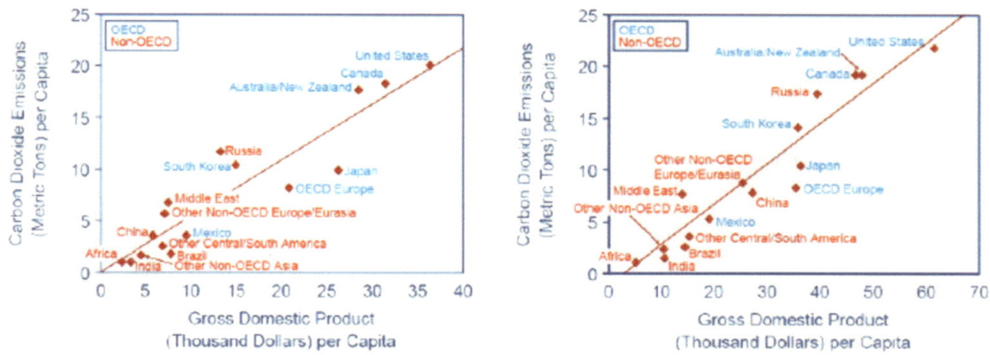

Fig. 11. (a) Carbon Dioxide Emissions and Gross Domestic Product per Capita by Region, 2004 ; (b) Carbon Dioxide Emissions and Gross Domestic Product per Capita by Region, 2030, (Grillot, 2008).

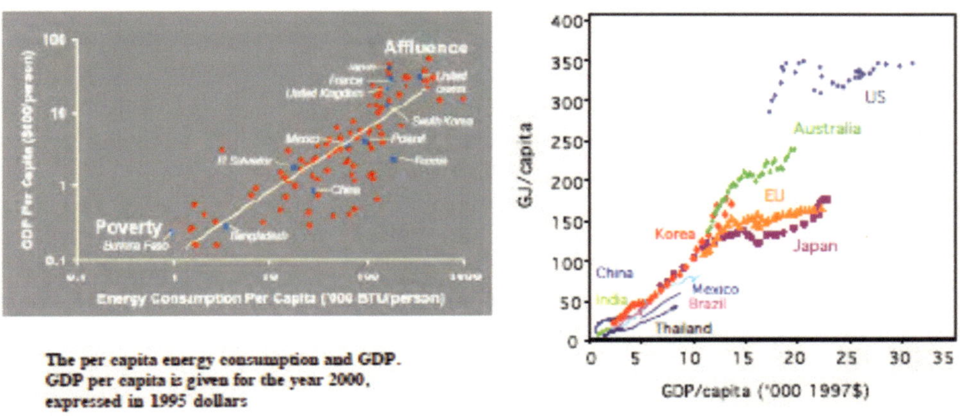

(a) (b)

Fig. 12. The trends in energy utilization is driven by developing economies (Ghoniem (2008), using data from UNDP Human Development report (2003))

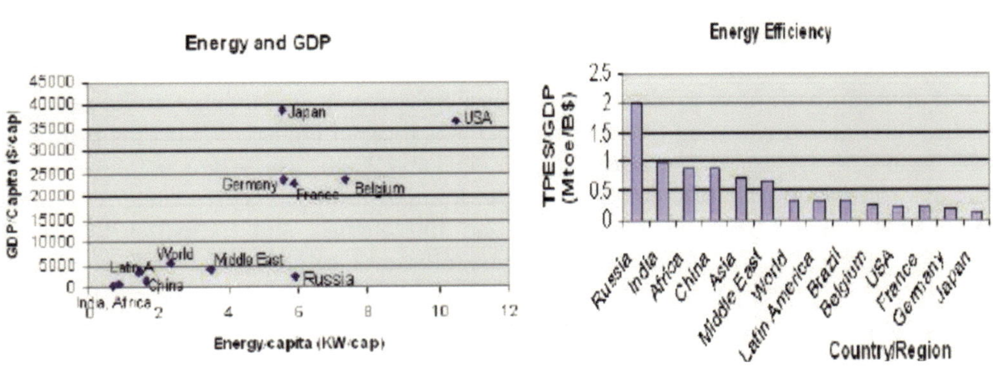

Fig. 13. (a) GDP versus energy consumed per capita in selected countries and the world, which indicates that it is driven by developing economies. (b) Energy efficiency of selected countries and regions (Schmitt, 2007).

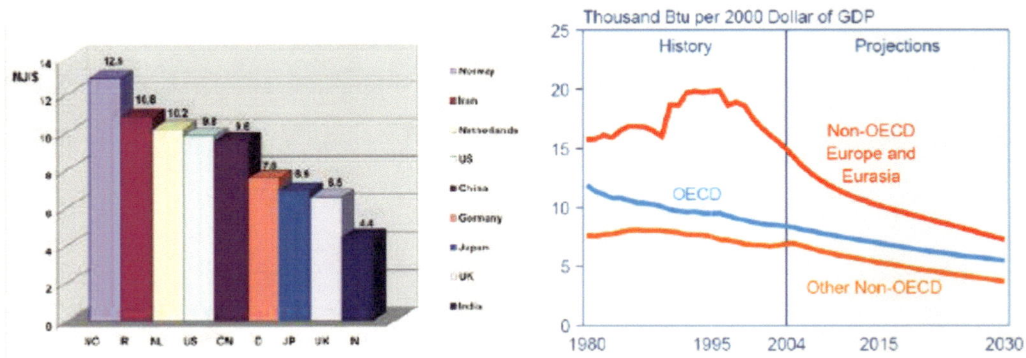

Fig. 14. (a) Energy Intensity of different economies[2] The graph shows the amount of energy it takes to produce a US $ of GNP for selected countries. GNP is based on 2004 purchasing power parity and 2000 dollars adjusted for inflation (US Energy Information Administration 2010). (b) Energy Intensity by Region, 1980-2030 (Grillot, 2008)

The trends reflected from the results of these studies as illustrated in Figures 12 to 14 indicate that the world energy utilization is increasing commensurate with population increase and economic development as indicated by GDP's of individual countries. However, the encouraging information reflected here, as illustrated in Figure 14, is the energy intensity, which tends to decrease in 2030.

It will be imperative how these trends relate to the UN Millennium Goal and Human Development Index. Energy can be considered to be a key factor in promoting peace and alleviating poverty. Solar power from space can help keep the peace on Earth. In September 2000 the world's leaders adopted the UN Millennium Declaration, committing their nations to stronger global efforts to reduce poverty, improve health and promote peace, human rights and environmental sustainability.

The Millennium Development Goals that emerged from the Declaration are specific, measurable targets, including the one for reducing—by 2015—the extreme poverty that still grips more than 1 billion of the world's people. These Goals, and the commitments of rich and poor countries to achieve them, were affirmed in the Monterrey Consensus that emerged from the March 2002 UN Financing for Development conference, the September 2002 World Summit on Sustainable Development and the launch of the Doha Round on international trade (UN Development Report, UNDP, 2008).

As reflected by Figures 9, 10 and 12, the world is facing an energy crisis on two fronts. There are not enough fossil fuels to allow the developing countries to catch up to the developed countries and global warming (Figures 3, 6 and 7) is threatening to cut short the production of the fossil fuels we can access today (UNDP, 2003 and 2008). These two factors necessitate the active role of relevant stake-holders in developing countries as represented by the triple-helix of government, research institutions and universities, and industries to establish integrated policies, action plans and budgetary measures to accelerate local participation and contribution to the global market that address sustainable development issues and green initiatives, with particular reference to energy issues. Active role of government in developing countries taking advantage of the research initiatives by local research and

[2] Energy intensity is energy consumption relative to total output (GDP or GNP)

academic institutions in utilizing locally available and/ or renewable technology will be necessary, as transitional stage towards more sustainable energy mix structure.

Economic viability will play a critical role in determination of the optimal energy option. The current worldwide energy market is dominated by fossil fuels, making any alternative difficult to implement due to lack of existing infrastructure. Not only will the technical feasibility and cost of both green and space based power sources be investigated, but also the necessary technological learning curve and economic pressure.

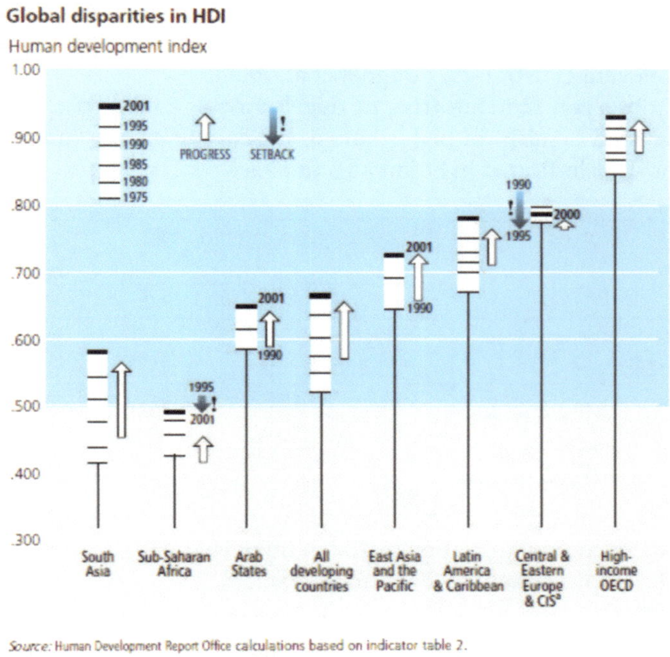

Fig. 15. Human Development Index Assessment on various geographical regions (UNDP, 2008)

In addition, international cooperation and industrial and developing countries economic interactions should also be directed towards these two factors: human resources development and industrial development transactions that is intricately related to environmental policy issues. Such initiatives should be based on long term and global vision rather that short term and local interest if an overall gain is desired, and should be seriously dedicated to overcome local and / or short term hurdles.

With respect to energy model and energy policy, the following which demand real solutions should be given due considerations (Ghoniem, 2008):

i. Energy consumption rates are rising, fast.
ii. Energy consumption rates are rising faster in the developing world.
iii. The developing world can not afford expensive energy.
iv. Oil is becoming more expensive, so is gas.
v. Massive and cheap coal reserves and resources should not distract synergetic efforts for green energy
vi. CO2 will become a dominant factor (as illustrated in Figures 8 and 11).

4. Significance of space power system to the developing world

People all over the world are more or less aware about solar power satellites, although their comprehension, initiative and creativity in addressing related problems do depend to a large extent to the above mentioned differentiations. It is also an observed fact that since the inception of the idea of SPS, the world has experienced tremendous increase in energy utilization.

The Solar Power Satellite (SPS) system is a candidate solution to deliver power to space vehicles or to elements on planetary surfaces and to earth to meet increasing demand of electricity. It relies on RF or laser power transmitting systems, depending on the type of application and relevant constraints (Cougnet et al. 2004).

It has also been observed that the fruit of developments taking place in the developing countries is manifested in terms of higher rate of increase of energy utilization compared to the industrial world, as indicated in Figures 10 and 12.

Region	Consumption, (million toe)			Increase, 1985-95 (%)
	1985	1990	1995	
Asia excluding Japan	924.0	1219.6	1591.1	72.2
Total Europe	1656.2	1739.3	1725.2	4.2
Japan	362.7	428.3	490.2	35.2
USA	1739.0	1930.7	2069.4	19.0
ASEAN (5)	84.4	135.9	201.2	138.4
World	694.1	7855.2	8135.8	17.1

Table 1. The trends in energy utilization is driven by developing economies (adapted from UNDP Human Development report, 2003, and Ghoniem, 2008)

There are not enough fossil fuels to allow the developing countries to catch up to the developed countries and global warming is threatening to cut short the production of the fossil fuels we can access today.

Space solar power is potentially an enormous business. Current world electrical consumption represents a value at the consumer level of nearly a trillion dollars per year; clearly even if only a small fraction of this market can be tapped by space solar power systems, the amount of revenue that could be produced is staggering (Landis, 1990). To tap this potential market, it is necessary that a solar power satellite concept has the potential to be technically and economically practical.

Possibly the most interesting market is third-world "Mega-cities," where a "Mega-city" is defined as a city with population of over ten million, such as São Paolo, Mexico City, Shanghai, or Jakarta. By 2020 there are predicted to be 26 mega-cities in the world, primarily in the third world; the population shift in the third world from rural to urban has been adding one to two more cities to this category every year, with the trend accelerating.

Even though, in general, the third world is not able to pay high prices for energy, the current power cost in mega-cities is very high, since the power sources are inadequate, and the number of consumers is large. Since the required power for such cities is very high-- ten billion watts or higher-- they represent an attractive market for satellite power systems, which scale best at high power levels since the transmitter and receiver array sizes are fixed by geometry. In the future, there will be markets for power systems at enormous scales to feed these mega-city markets. Therefore, it is very attractive to look at the mega-city market as a candidate market for satellite power systems (Landis, 1990).

Therefore, it is imperative that Space Power System should be viewed and analyzed as a challenging but realistic answer to the need to meet electrical energy needs for developing countries, just like satellite communication has proven itself since its visionary projection by Arthur Clark and its utilization in the past five decades.

To be economically viable in a particular location on Earth, ground based solar power must overcome three hurdles. First, it must be daytime. Second, the solar array must be able to see the sun. Finally, the sunlight must pass through the bulk of the atmosphere itself. The sky must be clear. Even on a seemingly clear day, high level clouds in the atmosphere may reduce the amount of sunlight that reaches the ground. Also various local obstacles such as mountains, buildings or trees may block incoming sunlight.

In addition, global concern and interest point toward the need for the world community to progressively but urgently change for environmentally friendly and green energy utilization. Hence one should examine existing power sources as well as near term options for green energy production including cellulosic ethanol and methanol, wind-power, and terrestrial and space solar power(Supple & Danielson, 2006; Andrews & Bloudek, 2006).

The prevailing economic gaps between developing (non-OECD) and industrialized (and space-fairing) countries also introduces significant gaps that place developing countries as by-standers in the global efforts for space technology utilization for appropriate development. It is therefore imperative to carefully examine:

a. Options, resources and policies related to establishing devloping countries vision on the inter-related relevance and promise of space, energy and environment
b. Economic development considerations as viewed from developing country
c. Human capital development considerations as viewed from developing country

These aspects can be discussed in view of two extreme factors: Policy impact on global climate, which is illustrated in Figure 8, and Human Development Index (HDI), Figure 15. HDI measures overall progress in a country in achieving human development,

The utilization of terrestrial solar energy has increased significantly in industrialized countries, and to a lesser extent in many developing countries, due to economic competitiveness and local industrial support. In this conjunction, analogous to the use of domestic communication satellite without waiting for well established terrestrial microwave communication network (which has proved to be very gratifying judged from a multitude of objectives, which was the case of Indonesia), the utilization of Solar Power Satellite services without waiting for well established terrestrial solar power may prove to be appropriate. Therefore, the idea suggested by Landis (1990) to utilize space solar as a "plug and play" replacement for ground solar arrays could be attractive for developing countries. Table 2 shows the advantages of using space solar as a "plug and play" replacement for ground solar arrays. From the point of view of a utility customer, a rectenna to receive space-solar power looks just like a ground solar array-- both of them take energy beamed from outer space (in the form of light for solar power, in the form of microwaves for the space solar power) and turn it into DC electricity.

Such exercise may be beneficial in establishing energy policy which has multiple goals, which addresses economic, national security, as well as environmental issues, as illustrated below, as adapted from Supple & Danielson (2006).

Economic

- limit consumer costs of energy
- limit costs & economic vulnerabilities from imported oil
- help provide energy basis for economic growth elsewhere
- reliably meet fuel & electricity needs of a growing economy

Homeland And National Security

- minimize dangers of conflict over oil & gas resources
- avoid energy blunders that perpetuate or create deprivation

Environmental

- improve urban and regional air quality
- limit greenhouse-gas contribution to climate-change risks
- limit impacts of energy development on fragile ecosystems

A wide variation of different energy production technologies was examined and Monte Carlo analyses were generated to take into account the data variability in the rapidly changing energy field (Mankins, 2008). Initial model results indicate that the shortage of fossil fuels can be overcome within a reasonable time period.

A Natural Synergy: Ground-based solar as the precursor to space solar power

Ground solar precedes space solar

1.Ground solar economically feasible in good locations as soon as solar array price reaches SSP targets
2.Does not need to wait for development of the beaming technology or low-cost space transportation

Upgrade ground-solar facilities to space solar

1. SPS receiver looks and operates just like a solar array
A. Both receive power from space and converts it to electricity
B. Utilities see "plug and play" replacement that operates at night
2. SPS rectennas can be put at the same location as solar facility
- rectenna sites bought and paid for by ground solar
- energy distribution infrastructure already in place
- rectennas can be made transparent to sunlight
- or advanced solar array can be designed with integral rectenna built in

Approach:

- Design a SPS to capitalize on the synergy between ground solar and space solar [4,14].
- Such a satellite concept must use ground-based solar when it is economical to do so, but fill in for ground-based solar when ground based solar is inadequate.

Table 2. A Natural Synergy: Ground-based solar as the precursor to space solar power (Landis, 1990)

The SPS system is characterized by the frequency of the power beam, its overall efficiency and mass. It is driven by user needs and SPS location relative to the user. Several wavelengths can be considered for laser transmission systems. The visible and near infrared spectrum, allowing the use of photovoltaic cells as receiver surface, has been retained. Different frequencies can be used for the RF transmission system. The 35 GHz frequency has been considered as a good compromise between transmission efficiency and component performances.

5. Space power system: review of several architecture and technologies

Through advances in space science, technology and exploration, mankind has also been acquiring awareness of the presence of our sun as an inexhaustible source of energy, as depicted in Figure 16, which may then offer a host of additional solutions to meet the need of world expanding population and increasing demand for energy.

Since its introduction by Dr. Peter Glaser (Glaser, 1968a; Glaser, 1968b; Ledbetter, 2008) for which it was granted US patent in 1973 (Glaser, 1973), Solar Power Satellite as a means to supply inexhaustible power from the Sun for use on the Earth and/ or other space objects of human interest has gained much attention and endeavor, in particular with the global concern on environmental issues and sustainable development. Energy from the Sun is inexhaustible, as clearly underscored in Figure 16.

Solar Power Satellite then reflects mankind vision and scientific and technological progress on the problem solving end but also global concern for energy and environmental sustainability on the problem end. Even it has been strategically recognized that Solar power from space can help keep the peace on Earth (Mankins, 2008), which should be intimately related to mankind observed certainty on human population "Exponential growth", and which has led to a multitude of practical consequences.

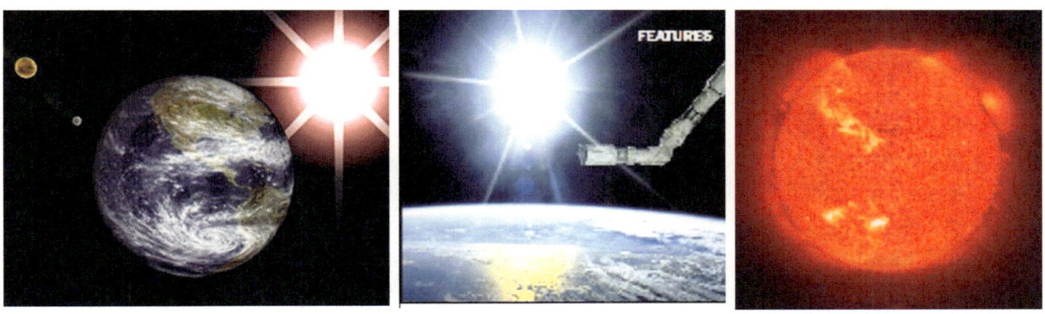

(a) (b) (c)

Fig. 16. Impressive views of the Sun: (a) The Sun, the Earth and the Moon, (b) The Sun as seen from Space Shuttle Endeavour (c) and the Sun observed surface, all of which emphasize its tremendous potential as an inexhaustible source of energy (http://www.google. com.my/imglanding).

Therefore, Solar Power Satellite has become a universal human interest regardless of human earth-based and anthropogenically defined differentiations: geographical origin, color, creed, wealth, intelligence and the like. A space solar power generation system can be designed to work in synergy with ground solar power. In fact, progress in space-based photovoltaic technology has been the driver for their earth-based utilization, as well as making them more economical. Hence in a broader sense, terrestrial-based and extra-terrestrial based Solar Power should provide a feasible answer for meeting mankind energy needs. The principle and operation of Solar Power Satellites is illustrated and summarized in Figure 17 (Harkins et al, 2008).

Elecricity has been produced and used in space from sunlight by hundreds of satellites in operation today. One may say that technological progress in land-based Photo-voltaic Electricity Generator to an affordable techno-economic state is due to a large extent by progress in space-based Photo-Voltaic Cells. As introduced by Dr. Peter Glaser in 1968, the concept of a solar power satellite system with square miles of solar collectors in high geosynchronous orbit is to collect and convert the sun's energy into a microwave beam to transmit energy to large receiving antennas (rectennas) on earth.

In 1999 NASA formed SERT, the Space Solar Power Exploratory Research and Technology program to perform design studies and evaluate feasibility of Solar Power Satellites (SPS). The concept has now evolved into a broader one: Space-Based Solar Power (SSP), which incorporates the concept and design of Sun Tower, as illustrated in Figure 18.

The general benefits of Space-Based Solar Power is that there is no pollution after construction, no GHG during power generation, the source of energy is free and it has a large amount of energy potential.

The advantage of placing the Solar Power Generator in space rather than on the surface of the Earth are, among others: less atmosphere for sunlight to penetrate for more power per unit area, any location on the Earth can receive power, the Satellite can provide power up to 96% of the time, the solar panels do not take up land on Earth while there are figuratively speaking infinite space is available in space and the initiative will promote growth of space, solar, and power transmission technology. On the other hand, there are significant problems to overcome with SSP. These are, among others, very expensive initial cost, power transmission by microwave and/or lasers still has to be developed to counter their possible harmful effects, cosmic rays can deteriorate panels, very large receiving antennas on earth may be required, maintenance problems and to avoid solar winds displace it off course would need a complex propulsion system.

Technological options considerations in view of overall strength and weakness / gains and losses. Several studies have been carried out for various Solar Power Satellites, including those located in Low Earth Orbits. These are the LEO and MEO SPS, Geostationary SPS and Supersynchronous SPS.

Some visionary concepts have been introduced, such as the Solar Power Satellite / Space-Based Solar Power to be located at an orbit around the Lagrange point L2, illustrated in Figure 19. Recent study on Innovative Power Architectures has been carried out by Landis

Fig. 17. The principle and operation of Solar Power Satellites (Harkins et al, 2008)

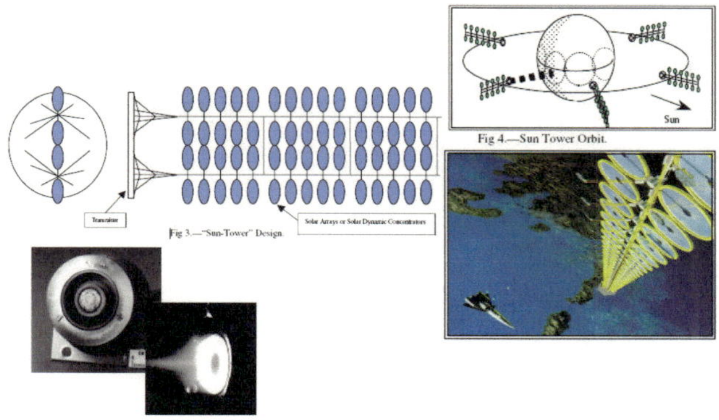

- The space-based antenna needs to be at least 1 km in diameter, making it far larger than any satellite ever proposed.
- Receiving antenna (an array of wires) must cover 20,000 acres.
- Sidebands not worth capturing
- Laser alternative to microwave power transmission.

Fig. 18. Design Ideas of Sun Tower (Harkins et al, 2008)

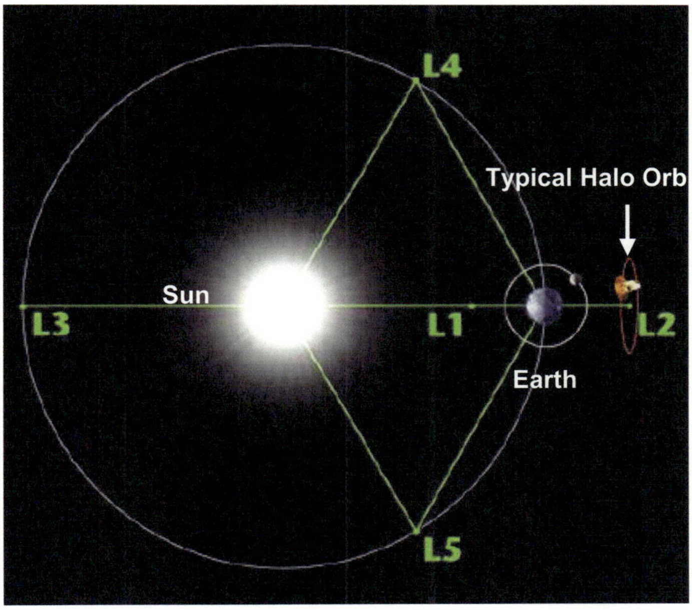

Fig. 19. Lagrange Points of the Earth-Sun system (not to scale). The Earth-sun L2 point distance is 1.5M km from Earth. Also shown an example of a typical halo orbit around L2

(2004), with some concepts of Earth-Sun L2 Design details are exhibited in Table 3. Three new concepts for solar power satellites were invented and analyzed:

i. a solar power satellite in the Earth-Sun L2 point,
ii. a geosynchronous no-moving parts solar power satellite, and
iii. a non-tracking geosynchronous solar power satellite with integral phased array.

The space power system designed to be located at Earth-Sun L2 will be radically different from conventional GEO Space power concept. As illustration, individual concentrator/PV/solid-state-transmitter/parabolic reflector element is exhibited in Figure 20. and An integral-array satellite has been proposed and invented and has several advantages, including an initial investment cost approximately eight times lower than the conventional design.

1	❖ Since the sun and Earth are nearly the same direction, it can feature:	
		• Integrated solar concentrator dish/microwave transmission dish
		• Integrated solar cell/solid state transmitters
		• **No rotating parts or slip-rings**
2	❖ Frequency: 30 GHz: ❖ efficiency is lower than 2.45 GHz, but much tighter beam	
		• transmitter diameter: 3 km
		• receiver diameter: 6 km
		• 3 ground sites, receive 8 hours per day
3	❖ 33,000 16.5 meter integrated PV concentrator/transmitter elements	
		• Concentrator PV efficiency 35%
4	❖ Larger distance from the sun means less solar radiation intensity compared to geosynchronous orbits, MEO and LEO	

Table 3. Earth-Sun L2 Design details (Landis, 2004)

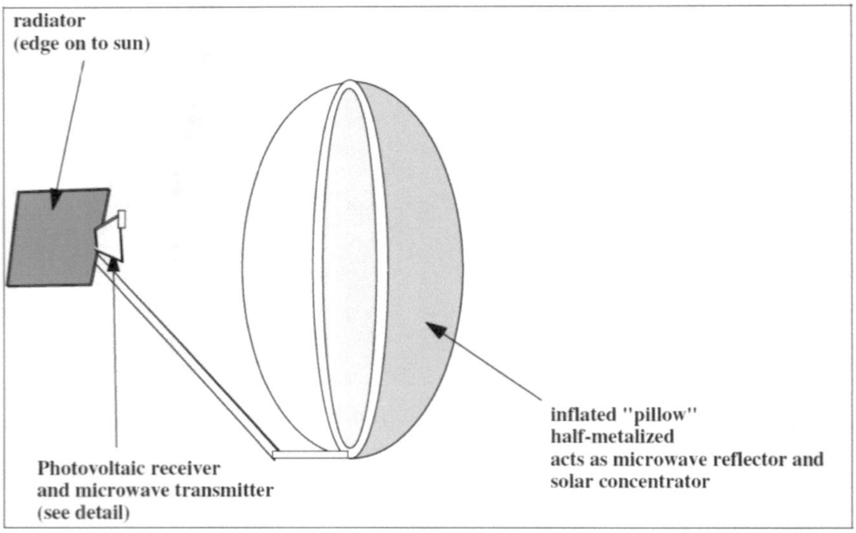

Fig. 20. Individual concentrator/PV/solid-state-transmitter/parabolic reflector element (Landis, 2004)

The following criteria, among others, will have to be used for a credible analysis of solar power satellite economic benefits and rate of return: Satellite power generation should fit electrical demand profile, Satellite power generation should generate power at the maximum selling price, and actual data on electrical demand and price should be used in its concept, design, implementation and operation

A novel scheme to implement Space Solar Power (SSP) to generate abundant, clean, and steady electric power "twenty-four hours a day every in a year" (or "24/365") in Space from solar energy, and conveyed down to Earth has been proposed by Komerath [26]. To overcome the massive cost to build large collector-converter satellites in Geosynchronous Earth Orbit (GEO) or beyond, the Space Power Grid (SPG) approach has been proposed by Komerath [27-28] breaks through this problem by showing an evolutionary, scalable approach to bringing about full SSP within 25 to 30 years from a project start today, with a viable path for private enterprise, and minimal need for taxpayer investment. This paper deals with the interplay of technology, economics, global relations and national public policy involved in making this concept come to fruition.

Given their high retail costs and unsteady nature, terrestrial solar-electric and wind power sources still remain secondary and subsidized. The key feature of Komerath's concept is to use the potential of the space-based infrastructure to boost terrestrial "green" energy production and thus benefit from the concerns about global warming and energy shortage.

In this first paper on the concept, the scope of the project, possible benefits and the obstacles to success are considered. It is seen that the inefficiency of conversion to and from microwave poses the largest obstacle, and prevents favorable comparison with terrestrial high-voltage transmission lines. However, competitive revenue generation can come from the nonlinearity of cost with demand at various places on earth. Point delivery to small portable, mobile receivers during times of emergencies is necessary. The benefits to 'green' energy generation make the concept attractive for public support as a strategic asset. This also sets a market context for concepts to convert solar power directly to beamed energy - a prospect with many applications. The following description is taken from Komerath (2007), with stages illustrated in Figures 21 and 22.

Briefly, the SPG approach is a 3-phase process to bring about full SSP. In Phase 1, no power is generated in Space. Instead, Space is used as the avenue to exchange power generated by renewable-energy plants located around the world. This is a breakthrough because renewable power plants today are unable to compete with local alternatives such as nuclear and fossil thermal power, due to their inherently unsteady, fluctuating nature. The sun only shines during the day, and not very well in cloudy weather, on Earth's surface. Wind power fluctuates wildly. The ideal locations for wind, solar and tidal/wave power plants are typically far from their customers, hence demanding the installation of new high voltage

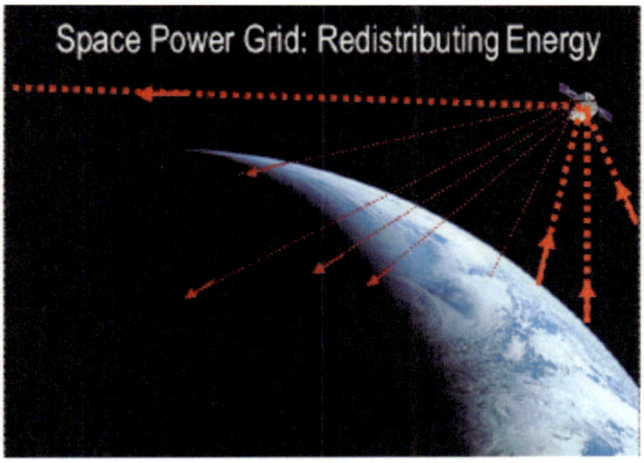

Fig. 21. Space Power Grid satellite receiving and redistributing beamed power (Komerath, 2007).

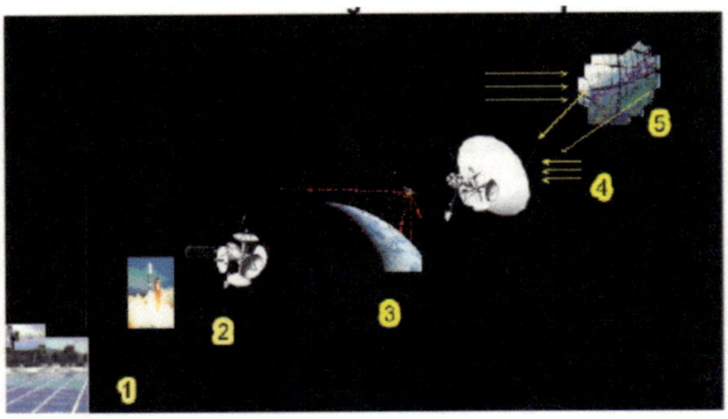

Fig. 22. Evolutionary path to full Space Solar Power (Komerath, 2007).

power grids in an age when land rights and environmental impact policies impose high costs on such infrastructure. In addition, to qualify for "base load" status, renewable power plants must install auxiliary power generators amounting to essentially 100% of their standard capacity, in order to be able to guarantee a steady level of output, and the ability to respond to demand surges. Such auxiliary generators are usually fossil burners, and relatively inefficient.

Once the Phase 1 SPG is in place and essentially self-sustaining by synergy with the renewable power industry, the satellites are gradually replaced, as each reaches about 17 years of age, with new, larger Phase 2 satellites that incorporate collector-converters for solar power, using the technological advances of the 23 years since project start. These put a small amount of space-generated power into the already-functioning grid, at a much lower generation cost.

Phase 3 consists of launching several very large, but ultra-light collector/ reflector satellites to high orbits. These will contain no converters (thereby reducing their mass by 2 orders of magnitude) but simply collect and focus sunlight on to the Phase 2 collector-converters in L/MEO. Phase 3 then allows for expansion until the constellation in L/MEO reaches saturation. To double terrestrial primary energy availability, some 300 square kilometers of ultra-light reflectors will be needed, in high orbits.

Such a system involving global power exchange obviously requires international collaboration on a global scale. Komerath et al (2007) proposes a global public-private Consortium, partially based on the model for the European Space Agency, where member nations and private corporations collaborate to reduce risk, make low-interest funding available, and organize the construction of major Space infrastructure. This set up is also shown to be open a path towards resolving some of the most vexing obstacles in space resource utilization, arising from current Space Law. On a national level, moving towards the Space Power Grid approach requires some fundamental realignments that synergize the Space and Energy enterprises with the environmental / Climate Change control movement.

The scheme proposed by Komerath is a bridging between the present economic and technological capabilities with acquired technologies and global economic capabilities to be acquired in the future, taking into account synergistic and cooperative efforts among countries.

6. Space power system as a unifying agent for global networking

The vision of Space Power System is a world of synergy: synergy of development efforts, bridging the Technological and Economic Gaps, Economic and Business Partnership and unify world communities for common concern and interest. Space Power System may serve to unify global efforts in its conception, design, implementation which may include manufacturing, assembly, commissioning and operation, and so forth. These objectives are quite in line with the objectives stipulated in many UN/ UNDP studies: to promote public policies to improve people's health and education, to ensure environmental and energy sustainability, and to promote truly global partnership, in which the rich countries should not follow the paradigm of charity, but what rich countries can do to cooperate with the less developed ones to serve as partners in reaching the common goals as articulated in the UN Millennium Goals.

In particular, Energy utilized by world population manifests itself in the following forms:

a. Primary forms:

sunlight, biomass, hydropower, wind, ocean currents, waves, tidal energy, fossil fuels: coal, crude oil, gas, oil shale, tar sands, methane, clathrates, geothermal energy, ocean heat, fission fuels (U, Th), and fusion fuels (H2, Li, He3)

a. Secondary forms

gasoline (from oil), diesel and heating oil (from oil, biomass), jet fuel (from oil), electricity (from anything), hydrogen (from anything) , alcohol (from any biomass), charcoal (from wood), "town gas" (from coal), and synfuels (from coal, oil shale, biomass)

The state of affairs of primary energy availability and utilization is reflected in Table 3 and Figure9.

Associated with energy policy and vision, the following key questions could be posed (adapted from Supple et al (2006):

- Could the ever-increasing need for energy be a unifying driver for world community in synergistic and unified effort for inexhaustible source of energy but sustainable for the global environemnt?
- What is the current and future composition of energy utilization in industrial and developing country and will it tend towards acceptable range of energy utilization per capita? Or by source? by sector? by region?
- What scenarios have been forecast for the future? Do they or should they lead to unified and synergistic efforts rather than competitive ones?
- Could Space Power System (and Renewable Energy Initiatives) be a driving force for concerted and unified efforts by and for the world community as a whole? Could this be incorporated and envisioned by individual and mutual energy policy?

These questions need to be resolved by stakeholders in the near future. Indications for such desire and emerging paradigm could well be in the offing, as reflected as a motto in the UNDP 2003 summary report (2003, 2008): **Climate change—together we can win the battle**

In this conjunction, the following factors could be identified as pushing and pulling factors:

Pushers:

- Sustainability
- Investment - and economic transactions
- Government policy and triple helix mutual interaction
- The role of international and United Nations bodies, in particular the UN-COPUOS
- Lessons learned from hard facts derived from other countries' experience

Pullers:

- Common universal goal

- Extending hands - together we can win the battle
- New paradigm
- Cooperative efforts, synergy and networking

Analysis along such frame of thought has been carried out by Sathaye (2006). Table 4 displays indicator analysis carried out which could be used in analyzing how countries of differing economic category develop their energy policy to respond to world's sustainability requirements.

	Buildings	Industry	Transportation
Activities	•Population •No.of households •(eletrification, urban/ rural) •M2 residential •M2 commercial	•GDP •Production: –Economic –Physical	•Personal –Pax-km •Freight –Tonne-km
Structure	By sub-sector Residential commercial By end-use Heating/cooling Refrigeration Appliances Euipment Lighting Cookinng Water-heating	By sub-sector Iron & steel Non-ferrous Cement Pulp & paper Chemicals etc Product mix	•By mode –Road –Rail –Air –Watere •By Vehicle Type –Passenger car –Truck –etc
Energy Intensity	Technology •Energy intensities –Efficiency –Usage –Site features –Energy carrier	Technology •Energy intensities –Efficiency –Usage –Energy carrier	Technology •Energy intensities –Efficiency –Usage –Energy carrier

Table 4. Indicators Analysis: Drivers, Indicators Sector and Technology Structure (Sathaye, 2006)

Considerations of all these issues will lead to the need for narrowing the Technological and Economic Gaps, and in turn unify world communities for common concern and interest and the prospects and potential of Space Power System.

7. Stimulating positive attitude in developing countries: The microsatellite tool

It is well known that the United Nations, in particular through its Committee on Peaceful Uses of Outer Space and its Office of Outer Space Affairs, is dedicated to bring the benefits of Space Science and Technology, in particular, to the outreach of the developing nations. In his keynote address at the First Asian Conference, Abiodun (2004) recognizes that the space enterprise has become one of the fundamental foundations of industrialization, and will be much more so in the foreseeable future. Accordingly, translating the national and global policies into successful national operational programs that can take advantage of progress in space science and technology to address global and national strategic and relevant issues demands a strong political will by the government(s) concerned, as well as public understanding and proactive attitudes. Such a political will demands commitment to human resources development at all required levels, institutional building, and far-sighted funding plan(s) - all needed to ensure the successful implementation of the national space policy. Its success would be further enhanced if the government is able to ensure the full support of all stakeholders, the decision makers at various levels and the society at large, in the different aspects of its space activities resulting in demonstrable tangible and positive impact in the lives of the people, in particular to meet those demands implicitly implied by the objective of defining the Human Development Index.

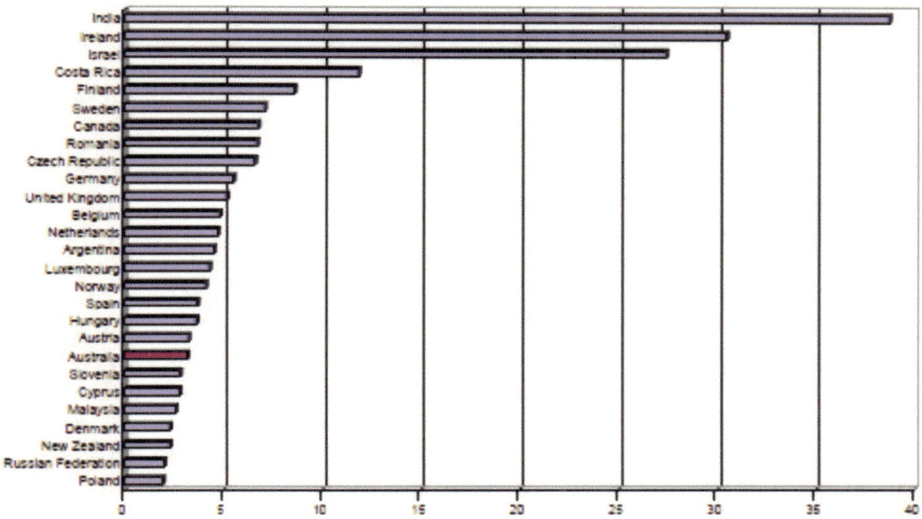

Fig. 23. Share of IT services in total services exports, 2006 (per cent)(Houghton, 2009)

The systematic implementation of the program associated with such space policies within the global spirit and environment for cooperation, has placed each country in the world a specific space-capability categorized as Space-fairing nations, Space-capable nations, Space-aspiring nations and Space-aiming nations (Abiodun, 2004).

Many developing countries have taken steps to take advantage of the advances in space science and technology for their national interest, in pace and promoting their national development goal, and even accelerating their status, using the categorization above, from space aiming to space fairing nations.

In addition to looking into the interest, technological and industrial capabilities in space science and technology for addressing national and global issues, it is an observed fact and necessity that each country to a large extent utilize and develop Information and Communication technologies (ICT) in almost every aspect of life, in particular related to those activities related to productivity and creativity. ICT is very intimately related to Space Technology in a large host of social and economic activities.

Both developed and developing countries face many environmental challenges, including climate change, improving energy efficiency and waste management, addressing air pollution, water quality and scarcity, and loss of natural habitats and biodiversity. Houghton (2009) has looked and explored how the Internet and the ICT and related research communities can help tackle environmental challenges in developing countries

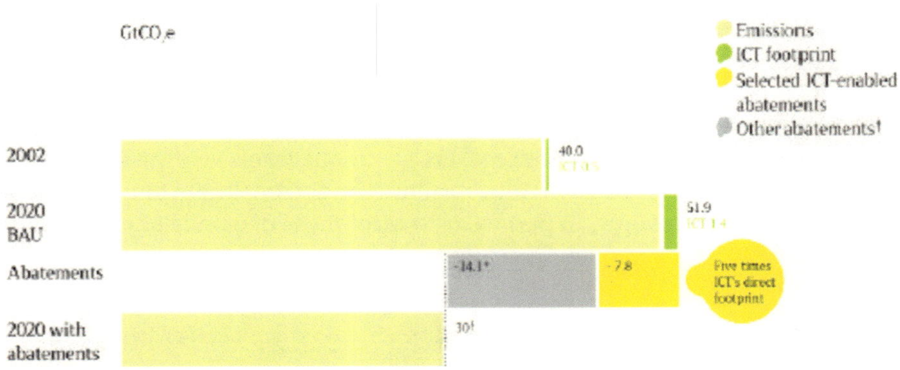

Fig. 24. ICT Impact: The global footprint and the enabling effect (Houghton, 2009).

through more environmentally sustainable models of economic development, and examines the status of current and emerging environmentally friendly technologies, equipment and applications in supporting programs aimed at addressing climate change and improving energy efficiency. Such concern could be extended to addressing energy and space technology, the present focus interest., in particular in addressing and developing positive attitude and enabling technological capbilities for establishing a synergistic global society for space endeavor of mutual concern in mega-scale. Figure 23 exhibits the share of IT services in total services exports for the year 2006 (in per cent), while Figure 24 exhibits ICT Impact and the global footprint and the enabling effect. Such information which may lead to further study relating manpower and technological capabilities for particular nation in carrying out more full-fledged space related activities addressing a broad spectrum of applications, including environmental challenges, improving energy efficiency and exploring novel energy initiatives.

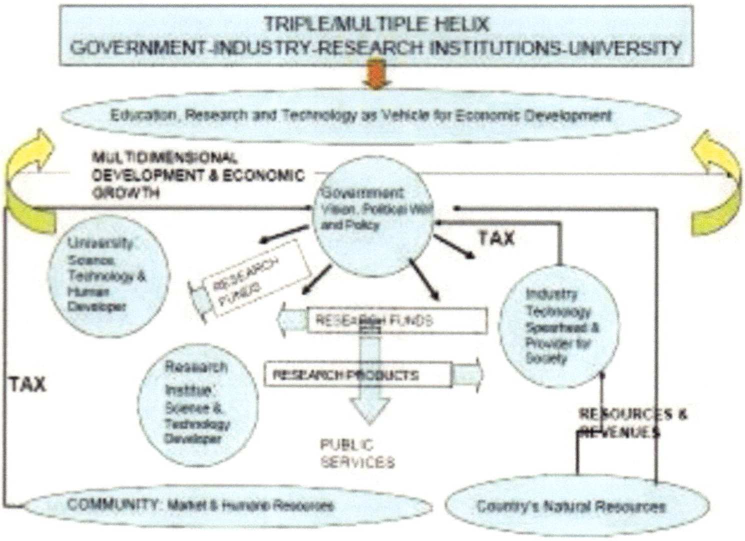

Fig. 25. Strategic (Triple-Helix) Partnership for Capacity building (Djojodihardjo, 2003; Djojodihardjo et al, 2007).

Another vehicle to in-country capacity building and establishing space-related human capital, technology and industry is by following a micro-satellite development oriented towards the needs of the country, which can be initiated at university and research institution level. Recent advances in Commercial Off-The-Shelf (COTS) sensor- and storage technology is enabling a completely new class of micro-satellites. Ground Sampling Distances (GSD) smaller than 5 m that was only possible on larger satellites, are now possible on satellites with a mass of less than 60kg due to smaller pixel sizes, refractive optics and accurate ADCS pointing and viewing control (Mostert, 2007). New paradigms for the application of remote sensing are expected to develop as a result of the possibility of having a personal remote sensing resource to complement more conventional Earth resource large satellites.

Such development are expected to lead to significant repercussions in the national decision circles which could lead to vision sharing and commitments of the four entities representing the responsible players and strategic stakeholders in "space ventures": the government, the academic and research institutions, the private industry and the local and international organizations, which may lead to the following new socio-political paradigm (Salatun et al, 1975; Djojodihardjo et al, 2007): a. Space endeavor is essential for sustainable national development, b. Space endeavor contribute to national capacity building and establishment of infrastructure essential for industrial and knowledge-based economy.

Examples of successful national microsatellite program in non-space-fairing countries using triple-helix partnership (Figure 25) are exhibited in Figure 26.

Fig. 26. Examples of national microsatellite product: (a) LAPAN-TUBSAT as one of the TUBSAT based microsatellite development (shown here: DLR-, MAROC- and LAPAN-TUBSAT); b. TiungSAT-1 as one of the UoSAT based microsatellite development (with UoSat-1, TMSAT and TiungSAT-1 as examples); c. SUNSAT 1 from Stellenbosch University, South Africa (Mostert et al, 1998) which leads to follow on impressive microsatellites development.

SUNSAT micro satellite development demonstrated the potential of local triple-helix synergistic efforts for capacity development, which indicates also the space technology research and industrial capabilities progress in the conducive global cooperation in space-related venture. Most of these locally conceived microsatellite development served at least three purposes: establishing elements of integrated space technology capabilities locally, application of nationally oriented space application mission and establishing conducive national atmosphere with reference to strategic decision-making process and public acceptance for further space technology developments.

Space endeavor contributes to the betterment of human quality of life, economic and employment opportunities. At the conspicuous tip of the iceberg, microsatellite has projected itself as a new tool and paradigm for affordable space venture oriented towards basic human needs and quest for rapid progress. Mostert (2008) also demonstrated how satellite engineering is indeed a catalyst for development. Satellite and Space Engineering has posed a Complex Value Proposition, and successful initiatives in this direction could lead to geo-political Influence, National Pride and Strategic Capability, motivated Science and Technology Community in the country, provides added value to Government Services, stimulates Industrial Development, and provide commercially useful data/information.

As was proved to be the case in the Indonesian Aircraft Technology program in 1976-1998, there is an intricate relationships between capacity development and development of space technology, which could be well reflected in Table 4 (adapted from Mostert, 2008).

Areas of Concern	Contribution of space technology
Shortage of Locally Available Technical Skills since many top local scientists & engineers are not locally employed and are engaged in space activities in the developed world	• High technology attracts & retains high quality engineers & scientists • Attracts young people to science and maths disciplines - fulfulling careers • Stimulates R&D • Independent space capability; contributes to world knowledge

Table 5. Intricate relationships between capacity development and development of space technology.

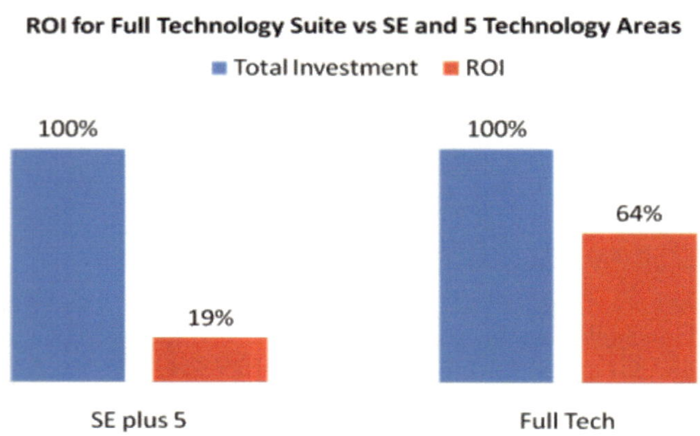

Fig. 27. Return on Investment (ROI) on Satellite Engineering Investment (from Mostert, 2008).

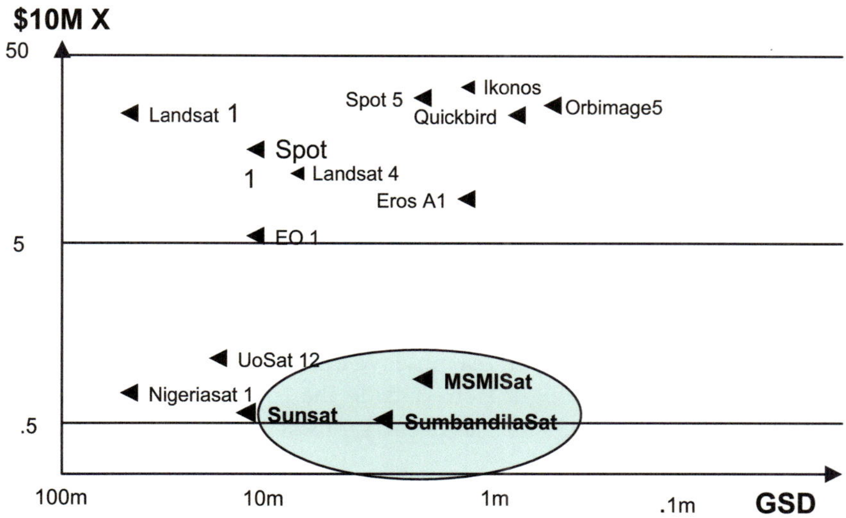

Fig. 28. Trends in Cost of Ground Sampling Distance (Mostert, 2008)

Desired triple-helix partnership (Etzkowitz, 2000) as schematically depicted in Figure 24 should be established in such space venture for effective efforts and capacity building with utmost national benefits. Micro-satellite programs established in Indonesia and Malaysia (Djojodihardjo et al, 2007; Triharjanto et al, 2007, Hardhienata et al, 2007, Md.Said et al, 2008) are a small example of such partnership for space technology development, implementing a combination of paradigms, and taking advantage of the technological progress and global cooperation atmosphere.

With such structure, synergy between private venture and government leadership could still be strengthened. It is heartening that since 2000, effort to develop microsatellite has gained new momentum by the support of key government officials. Such vision and atmosphere is required since budget for realizing developing country program space activities are allocated through an intricate mix of triple helix setting in developing country Cooperation between space communities, involving government space agency and strategic domestic industry with other regional countries as well as space fairing countries has developed favorably in the past decades. Of interest is the Return of Investment of national space venture. An encouraging example is elaborated by Mostert (2008), which is summarized in Figure 26. The viability of such efforts is also illustrated in Figure 27.

8. Universal SPS program initiatives – arms reaching but novel paradigm – establish productive and resourceful partners-in-arms by expanding opportunities at creative circles

Space Solar Power initiatives which give promise to the world energy and environment solution in not too distant futures could best be addressed by global cooperation and global networking, thus establishing less technologically (and thus industrially) endowed nations as partners in synergistic space ventures. The challenge for space program initiatives for non-space fairing developing countries, can be addressed in more positive partnerships capitalizing on international cooperation paradigms now only promoted by the UN system. Decision makers and all stake holders in developing countries are also encouraged to develop National Development and Capacity Building Vision and National Program and sustainable development solutions capitalizing on enhancing the quality and capacity of their human capital through Coordinated Efforts and Broad Based Strategic Partnerships, thus aiming for more equitable distribution of Human Development Indices around the world and within each country.

The development of microsatellite in some developing countries, exemplified by Malaysia and Indonesia reflects the following situation.

1. Microsatellite development has to bear direct relevance to the national development objectives, and can draw strategic synergistic efforts and partnerships among stakeholders in the country, which may involve university community
2. Technology providers are sought from outside the country which appeals both form affordability and promise for in-country technology and human resources development.
3. New paradigm for space technology development that offer time-cost-effective, affordable and strategic participation and role for local human resources and local technology development / local space technology infrastructure are available and progressing.
4. Microsatellite development scheme undertaken should be able to convince national decision makers of its orientation towards basic human needs and overall effectiveness and success in the quest for rapid progress.

It is encouraging that the relevance of space related initiatives to down-to-earth basic human needs to a larger extend has been shared by stakeholders, by significant breakthrough in overcoming social, cultural and technological handicaps. The emergence and growing development of strategic technologies, cooperative spirit, globalization of information and vision for global techno-economic networking is considered to be responsible to the favorable paradigm shift that has enabled and conducive to the translation of Space Imperative for National Development and Capacity Building through Vision, Coordinated Efforts and Strategic Partnerships in relevant, mission-oriented and affordable space initiatives of these countries.

Universities can play significant role in the national and regional efforts to master new sciences and technologies which require high expertise in the relevant basic sciences and along with national research institutions in the region can form indigenous broad-based scientific and technological infrastructure conducive for effective development of advanced technologies.

Progressive regional initiatives that within conceivable future will contribute to the development of regional cooperative space program, close networking and coordination of research institutions, joint governmental funding, etc, as has been carried out in many joint bilateral as well as ASEAN programs, should be enhanced. To this end, efforts should be carried out by relevant stakeholders to establish political will and vision for such initiatives. Effective regional or geographical cooperation as carried out and developed by the European Space Agency that has now carried out significant, strategic and far-sighted space initiatives could serve as an encouraging example. Regional or Asian cooperative initiative which capitalize on Remote Sensing Space System dedicated for tropical equatorial countries as stipulated in [19] and have been followed by national initiatives in the region should be a logical and feasible step. The Asian region, in particular the ASEAN region, is poised to evolve added regional approach in the not too distant future in the ever expanding space frontier, which also implies ever expanding opportunities, including progressive participation in a global program for Space Solar Power.

9. Conclusion

The SPS system appears as a promising solution for meeting future energy (electrical) needs of mankind on earth, notwithstanding other needs associated with mankind venture on space objects. A comprehensive review and analysis have been carried out on global participation on solar power satellite initiatives. The following conclusions can be made:

1. Space Power System initiatives could be extended to the developing world in progressive fashion, to address global urgent need for sustainable energy and equitable development.
2. Space Power System Global initiative has the potential to reduce the Technological and Economic Gaps, and in turn unify world communities for common concern and interest.
3. Space programs already carried out in many developing countries, including microsatellite development, will serve as a vehicle and policy stimulus for global partnership in space projects of interest to global community.
4. A universal SPS Program Initiatves could be envisioned with the assistance of international organization, in particular the UN COPUOS and its Office of the Outer Space Affairs.
5. Comparative analysis of several architecture and technologies could be carried out to look into feasible means of progressive participation of developing countries. Simulation by microsatellite-concept could be of interest.
6. A wide range of technological options for SPS configurations are available and could be given due considerations in view of overall strength and weakness.

10. References

Abiodun, A.A., 2004. The Roles Of Governments, International Organizations *And* The Private Sector In The Promotion Of Space Science And Technology, Keynote Address/ Paper invited for presentation as Keynote Address at the First Asian Spacc Conference, Chiang Mai, THAILAND, November 23-26.

Andrews, D.G. and Bloudek, B. (2006). Space and the Green Energy Options, paper IAC-08-C3.3.1 , 59th International Astronautical Congress/ The World Space Congress-2006, 29 September and 3 October 2008, Glasgow, Scotland.

Bernstein, L. (2007). Climate Change 2007: Synthesis Report Summary for Policymakers, An Assessment of the Intergovernmental Panel on Climate Change, A summary approved in detail at IPCC Plenary XXVII, Valencia, Spain, 12-17 November 2007.

Bertrand, C., Van Ypersele, J-P. and Berger, B., Are Natural Climate Forcings Able To Counteract The Projected Anthropogenic Global warming? (2001) *Climatic Change* 55: 413–427, © 2002 *Kluwer Academic Publishers. Printed in the Netherlands.*

Boechler, N., Hameer, S., Wanis, S. and Komerath, N., 2007, An Evolutionary Model for Space Solar Power, Parameter Selection for a Space Power Grid

Chefurka, P. (2010). Energy and GDP in 2050, The Growth of Destitution, http://www.paulchefurka.ca/WEAP2/Energy_GDP_2050.html

Cougnet, C., Sein, E., Celeste, A. and Summerer, L. (2004). Solar Power Satellites For Space Applications, IAC Paper, IAC-04-R-3-06, 55th International Astronautical Congress 2004 - Vancouver, Canada.

Djojodihardjo, H., 2003. Internal Report, International Cooperative Center for Engineering Education Development (ICEED), Toyohashi University of Technology.

Djojodihardjo, H., Md. Said, M.A., and Parman, S., 2007. Translation Of Space Imperative For National Development And Capacity Building Through Vision, Coordinated Efforts And Strategic Partnerships, IAC-07-E3.1.3, presented at 58th International Astronautical Congress 23-28 September, Hyderabad, India.

Djojodihardjo, H. (2009). Beyond Fossil Fuels, Proceedings of the National Seminar on Alternative Energies to substitute Fossil Fuels, (in Indonesian), Universitas Jenderal Achmad Yani (UNJANI), Bandung.

Djojodihardjo, H. and Varatharajoo, R. (2009).Space Power System Initiatives: Establishing World Vision And Capacity, paper IAC-09-C3.3.3, 60th International Astronautical Congress, 12 – 16 October 2009, Daejeon, Republic Of Korea

Etzkowitz, H., 2000. The Triple Helix of University - Industry – Government: Implications for Policy and Evaluation, www.sister.nu, ISSN 1650- 3821, Institutet för studier av utbildning och forskning, Drottning Kristinas väg 33D, SE-114 28 Stockholm.

GhoniemA.F. (2008). The Energy Challenge and Emerging Opportunities, The International Conference on Mechanical And Manufacturing Engineering, Johor Bahru, May 21-23.

Glaser,P.E., 1968a. *Science* 22: Vol. 162. no. 3856, pp. 857 – 861, November..

Glaser, P.E., 1968b. "Power from the Sun: It's Future," *Science Vol. 162*, 957–961.

Glaser, P.E., 1973. "Method and Apparatus for Converting Solar Radiation to Electrical Power". *United States Patent 3,781,647, December 25.*

Grillot, L.R., (2008). Energy Scenarios, College of Earth and Energy, University of Oklahoma, http://www.sec.ou.edu/mee/pdf/Energy%20Outlook%20Sem%2007/Energy%20 Scenarios_Grillot.pdf.

Hardhienata, S., Nuryanto, A., Triharjanto, R.H. and Renner, U., 2007. www.ilr.tu-berlin.de/RFA/sat/lapan/ paper/ hardhienaSchmitt,J., (2007). *Energy "Consumption" and GDP, in* The Second Law of Life, Energy, Technology, and the Future of Earth As We Know It, an interactive study, http://secondlawoflife. wordpress.com/2007/05/17/energy-consumption

Harkins, J., Livingston, D., Wong, A., and Sanders, A., 2008. Space-Based Solar Power, *sexton.ucdavis.edu/CondMatt/cox/hw16007/spacepvs.ppt*

Houghton, J., 2009. "ICT and the Enivironment in Developing Countries: an Overview of Opportunities and Developments", Centre for Strategic Economic Studies, Victoria University, AUSTRALIA; Communications & Strategies 76(4), pp.39-60.

Komerath, K., 2007. The Space Power Grid: Synergy Between Space, Energy and Security Policies, Email: komerath@gatech.edu

Komerath, N., Venkaty, V. and Butchibabu, B., 2007. Parameter Selection for a Space Power Grid.

Landis, G. (1990). "An Evolutionary Path to SPS," *Space Power, Vol. 9* No. 4, pp. 365–371.

Landis, G.A., 2004. Reinventing the Solar Power Satellite, NASA/TM—2004-212743, NASA Glenn Research Centre, Cleveland, Ohio ,February 2004, also appearing as IAC Papers IAC–02–R.3.06 and IAC–02–R.1.07.

Ledbetter, W., 2008. "An Energy Pioneer Looks Back," *Ad Astra,* Vol. 20, No. 1, p. 26.

Mankins, J.C., 2008. Space-Based Solar Power, Inexhaustible Energy From Orbit, Ad-Astra, The Magazine of Space Society, *www.nss.org/adastra/AdAstra-SBSP-2008.*

Md Said, M.A., Jamil, I., Fadly, M. and Djojodihardjo, H., 2008. Design Philosophy, Features And Development Of Sun-Sensor And Magnetic Field Probe For Nano Satellite, paper IAC-08-B4.1.10, 9th UN/IAA Workshop on Small Satellite Programmes at the Service of Developing Countries, 59th International Astronautical Congress, 29 September and 3 October, Glasgow, Scotland

Mostert, S., Cronje, T. and du Plessis, F., 1998. The SUNSAT Micro Satellite Program: Technical performance and limits of imaging micro satellites, http://www.jamsat.or.jp/oscar/sunsat/mirror/sspapers/cnes.pdf

Mostert, S.,2008, Satellite Engineering as Catalyst for Development, Paper IAC-08.B4.1.2, 9th UN/IAA Workshop on Small Satellite Programmes at the Service of Developing Countries, 59th International Astronautical Congress, 29 September and 3 October, Glasgow, Scotland.

Salatun, J, Djojodihardjo, H. and Alisyahbana, I., 1975. *Satellite Orbital Considerations for Remote Sensing in Indonesia,* UN-FAO-LAPAN Joint Workshop on Satellite Remote Sensing and Space Communication, Jakarta.

Sathaye, J.A., de la Rue du Can, S., Levine, M., Price, L., Sinton, J. and Zhou, N., 2006. Global Energy Demand: Global Developing--Country Indicators, Presented at the Introduction to Energy Indicators Workshop, International Energy Agency, Paris, April 26, 2006, http://ies.lbl.gov/ppt/ gedindicators.pdf

Supple, D. and Danielson, D. (2006). Energy 101: World and U.S. Energy Overview , MIT Energy Club Discussion, September 27, 2006, R&D Pub, Stata Center (32-410).

Triharjanto, R.H., Hasbi, W., Widipaminto, A., Mukhayadi, M. and Renner, U., 2007, www.ilr.tu-berlin.de/RFA/sat/lapan/ paper/triharj1.pdf

UN, The United Nations Framework Convention on Climate Change(2010). http://unfccc.int/ essential_background /convention/background/items/1353.php.

UNDP Human Development report 2003, (2003). *hdr.undp.org/en/reports/global/hdr2003/;* www.unic.un.org.pl/hdr/hdr2003/hdr03_complete.pdf; *also* Human Development Report 2003, Millennium Development Goals: A compact among nations to end human poverty, Published for the United Nations Development Programme (UNDP) New York Oxford, Oxford University Press 2003.

UNDP (2008). Human Development Report 2007/2008, Fighting Climate Change, Human Solidarity in a Divided World, Published for the United Nations Development Programme.

US Energy Information Administration. (2010). International Energy Outlook 2010, Highlight, Report #: DOE/EIA-0484(2010) Release Date: May 25, 2010, http: //www.eia.doe. gov/oiaf/ieo/

Walker, H., Keim, B. and Arndt, M.B. (2010) -Natural and Anthropogenic Factors Affecting Global & Regional Climate, (2010), http://www.necci.sr.unh.edu/necci-report/NERAch3.pdf

8

Shape Measurement of Solar Collectors by Null Screens

Víctor Iván Moreno-Oliva,[1] Rufino Díaz-Uribe[2]
and Manuel Campos-García[2]
[1]*Universidad del Istmo (UNISTMO), Campus Tehuantepec, Oaxaca,*
[2]*Universidad Nacional Autónoma de México, CCADET-UNAM, México, D.F.*
México

1. Introduction

The optical performance of solar concentrating collectors is very sensitive to inaccuracies of components and assembly. Because of the finite size of the sun and errors of the collector system (e.g., tracking, receiver alignment, mirror alignment, mirror shape, and mirror specularity) the intensity of light at the focal receiver is reduced.

Among the principal methods for testing optical surfaces, the Ronchi and Hartmann tests have been popular for many years for testing slow (F/# >> 1) spherical and aspherical surfaces (Cornejo-Rodríguez, 2007; Malacara-Doblado & Ghozeil, 2007). Both methods for testing optical system are useful mainly for surfaces of revolution. Previous works (Shortis & Johnston, 1996; Pottler & Lüpfert, 2005) have described the application of photogrammetry to the characterization of solar collectors. Briefly, close-range photogrammetry involves the use of a network of multiple photographs of a target objet (a solar collector component in this case) taken from a range of viewing positions, to obtain high-accuracy, 3D coordinate data of the object being measured. Furthermore, photogrammetry is self contained and requires little external information if only the shape and size of the object is of interest.

Other works propose a system called Scanning Hartmann Optical Test (SHOT), and Video Scanning Hartmann Optical Test (VSHOT), in these methods a mirror is typically positioned at a distance slightly greater or less than twice its focal length (f). Then a laser beam is steered by a 2-axis scanner to a point on the mirror. After reflecting off the mirror, the laser beam returns to a location near its source and reaches a CCD camera. A computer video board digitizes the CCD camera's image and the centroid location of the reflected spot is calculated. The laser is scanned quickly across the surface of the mirror and this process is repeated many times. The concentrator's slope at each point of reflection and a polynomial surface fit to the measured slopes is then calculated. The accuracy of the VSHOT device depends upon the geometry of the test setup (F#, distance, laser spot size, etc.) and the accuracy of input data (distance, scanner calibration, video calibration, etc.).

On the other hand, the null screen method consists of a screen with an array of points or lines that by reflection on an ideal surface, gives a perfect square array of points or lines at a CCD camera, while any departure of the ideal geometry is indicative of shape errors of the surface. The shape can be obtained through the general and exact formula proposed by (Díaz-Uribe, 2000).

In this Chapter we describe the principles of the null screen method and show the implementation for concave (Campos-García et al., 2008) and convex (Díaz-Uribe & Campos-García, 2000) surfaces and also some new developments in the null screen method (section 2). The application of the null screen principles to the testing of solar collector components (parabolic trough solar collector PTSC (section 3) and parabolic dish solar collector systems (section 4)) are described.

2. Null screen principles (testing fast convex and concave surfaces)

The null screen method has been successfully used in testing optical surfaces of revolution, both concave and convex, of small and medium size. The method is regarded as an extension of the Hartmann test. The Hartmann test (Malacara-Doblado & Ghozeil, 2007) uses a perforated screen for sampling the wave front when the surface is illuminated by a point source; by reflection on the mirror, the light passing through the holes gives a set of bright spots on a plane screen parallel to the perforated screen, close to the point source (or its conjugated plane). The positions of the bright spots give the derivative of the wave front at each point of incidence on the mirror. The essential principle of the null screen method is to design screens, when you know the ideal shape of the surface under test. The null screen contain an array of curves or spots such that, when the screen is observed by reflection on the real surface, the image in the optical system consists of an array of perfectly square straight lines or bright spots if the test surface is perfect; if the surface is of high optical quality the distortion of the lines is null (hence the null term in the name). Otherwise, if the image of the array is not square, the deformations are due to imperfections, defocus or misalignment of the test surface. With this technique the alignment of the optical system is relatively easy.
To test a surface of revolution, the points of the designed null screen are plotted on a sheet of paper with the help of a laser printer (or a plotter, depending on the size of the screen); then, the paper is rolled into a cylindrical shape and inserted into a transparent acrylic cylinder which supports the paper. In the test of concave surfaces the diameter of the cylindrical screen must be many times smaller than the diameter of the surface under test (Fig. 1a), and for convex surface the diameter of the test surface defines the minimum diameter used for the cylindrical screen, in this case the test surface lies inside the cylindrical screen (Fig. 1b). Figure 1 also shows the inverse trace of a ray starting from a point P_1 of the image plane (CCD plane) passing through the small aperture lens stop (point P). This ray reaches the test surface at the point P_2 and, after reflection, the ray hits the cylindrical screen at P_3. The distances a and b are the CCD-pinhole distance and the distance between the pinhole and the vertex of the surface respectively; they are related by

$$b = \frac{aD}{d} \pm \beta, \tag{1}$$

where D is the diameter of the test surface, d is the length of the smallest side of the CCD and β is the sagitta at the rim of the surface. The plus sign refers to concave surface and the minus to convex surface, and which for a conical surface is given by

$$\beta = \begin{cases} \dfrac{D^2}{8r} & \text{for } k = -1 \\[2ex] \dfrac{r}{k+1}\left\{1 - \left[1 - \dfrac{(k+1)D^2}{4r^2}\right]^{1/2}\right\} & \text{for } k \neq -1. \end{cases} \tag{2}$$

where k is the conic constant of the test surface.

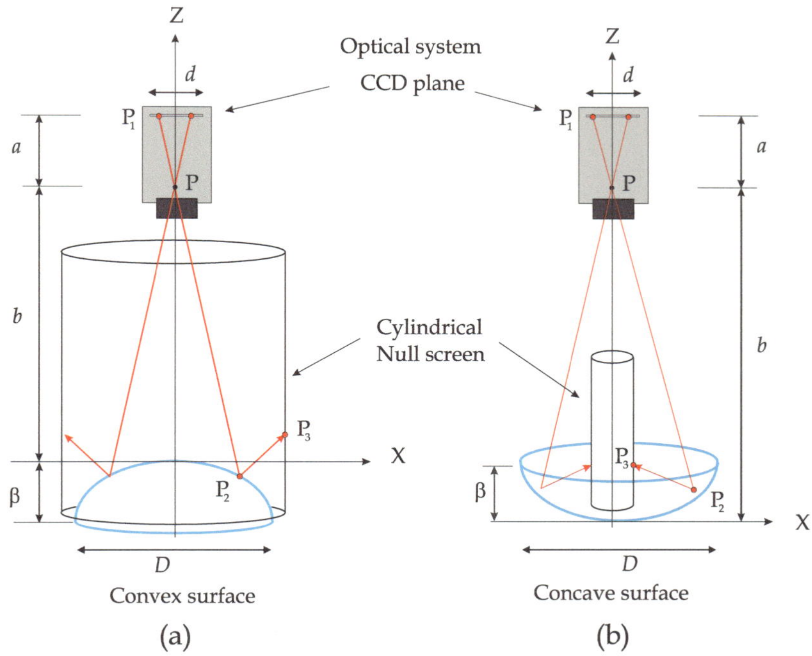

Fig. 1. Setup for testing a) a convex and b) a concave

2.1 Practical implementation of the screen

As a proof of principle, we show a qualitative test on two surfaces. The first is a concave hyperbolic surface, and the second is a convex spherical surface.

2.1.1 Hyperbolic concave surface

The concave hyperbolic surface used was built as a mold for casting the secondary convex mirror of an infrared telescope; the surface was 459 mm in diameter (F/0.5087). The screen was designed with the values given in Table 1. Figure 2b shows the actual null screen before being wrapped around an acrylic cylinder; the units are in millimeters.

Element	Symbol	Size
Surface radii of curvature	r	467 mm
Conic constant	k	-1.345
Surface diameter	D	459 mm
Camera lens focal length	a	12 mm
CCD length	d	6.6 mm
Stop aperture-surface vertex	b	889.81 mm

Table 1. Design parameters for the test of the hyperbolic surface.

Figure 2a shows the null screen wrapped around the acrylic cylinder and the hyperbolic test surface; finally, Figure 2c shows the resultant image of this screen after reflection on the test surface. It is clear that at the center of the surface, the image is almost a perfect square array of grid lines; going to the edge of the surface, the grid lines are deformed depending on the slope of the surface.

(a)

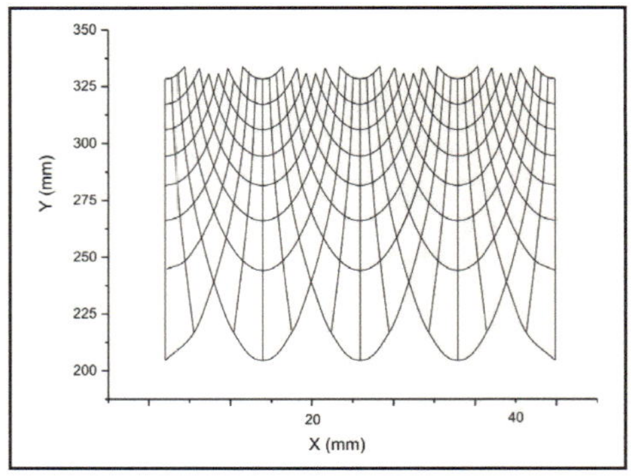

(b)

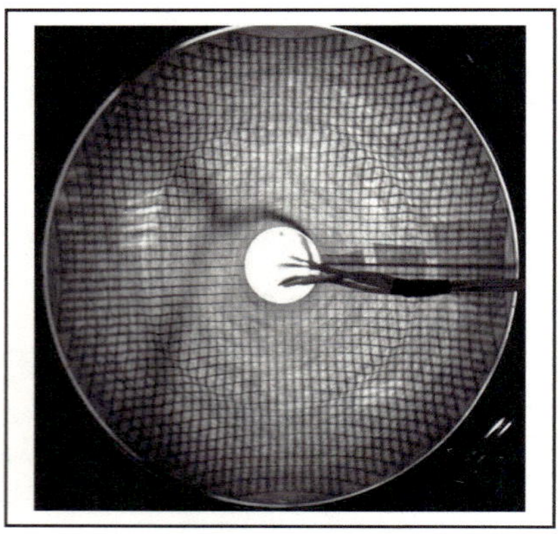

(c)

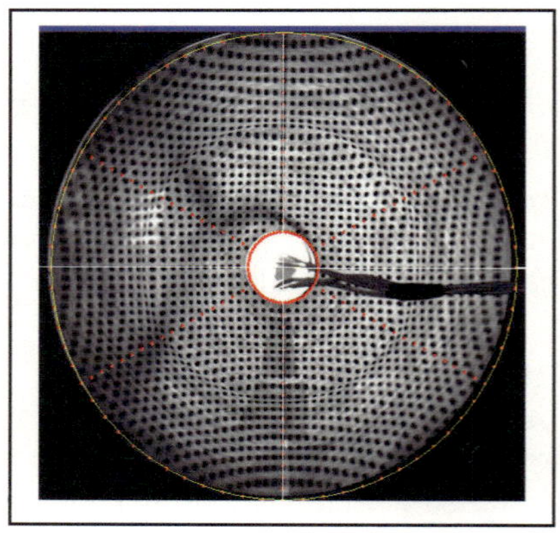

(d)

Fig. 2. a) The hyperbolic surface with the null screen, b) Flat printed null screen with grid lines for qualitative testing, c) resultant image of the screen shown in (b) reflection on the test surface and d) resultant image by a null screen with drop shaped spots for quantitative testing.

For a quantitative testing of the surface, a null screen with drop-shaped spots is used (Fig. 2d) to simplify the measurement of the positions of the spots on the CCD plane, which are estimated by the centroids of the spots on the image of the null screen.

2.1.2 Spherical convex surface

The spherical convex surface used was a steel ball with a diameter of 40 mm; the proposed cylindrical null screen was 60 mm in diameter. For a qualitative evaluation of the shape of the surface, we designed a screen to produce a square array of 19x19 lines on the image plane. Figure 3a shows the spherical surface, in Fig. 3b the flat printed null screen is shown, and the image of the cylindrical screen after reflection on the spherical surface is shown in

Fig. 3c; the image is almost a perfect square grid but, in this case, the departures from a square grid which can be seen are probably due to a defocus of the surface and some printing errors, and not to deformations of the surface.

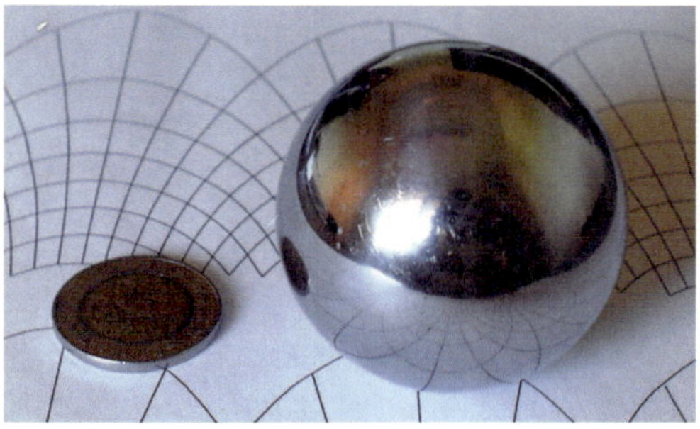

(a)

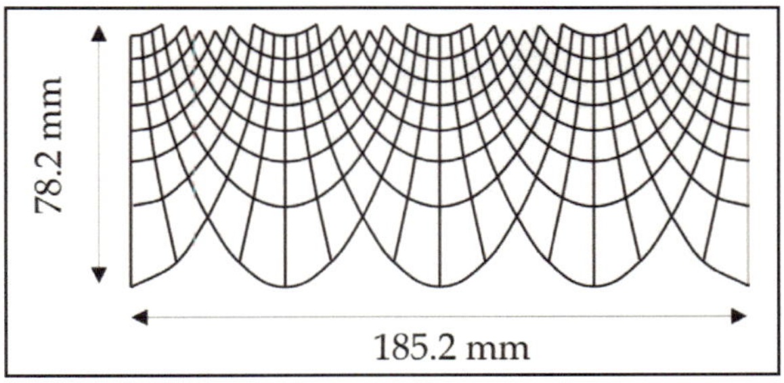

(b) (c)

Fig. 3. a) Spherical surface (steel ball), b) flat printed null screen with grid lines for qualitative testing, and c) the resultant image of the screen after reflection on the test surface.

2.2 Surface shape evaluation

The shape of the test surface can be obtained from measurements of the positions of the centroids of the spot images on the CCD plane through the formula (Díaz-Uribe, 2000)

$$z - z_0 = \int_{p_o}^{p} \left(\frac{n_x}{n_z} dx + \frac{n_y}{n_z} dy \right), \tag{3}$$

where n_x, n_y, and n_z are the Cartesian components of the normal vector **N** on the test surface, and z_0 is the sagitta for one point of the surface. The value of z_0 is not obtained from the test, but it is only a constant value that can be ignored.

The evaluation of the normals to the surface consists of finding the directions of the rays that join the actual positions P_1 of the centroids of the spots on the CCD and the corresponding Cartesian coordinates of the objects of the null screen P_3. According to the reflection law, the normal **N** to the surface can be evaluated as

$$\mathbf{N} = \frac{\mathrm{r}_r - \mathrm{r}_i}{\left|\mathrm{r}_r - \mathrm{r}_i\right|}, \tag{4}$$

Where $\mathbf{r}_i$ and $\mathbf{r}_r$ are the directions of the incident and the reflected rays on the surface, respectively; the reflected ray passes through the pinhole P and arrives at the CCD image plane at P_1 (Fig. 4). For the incident ray $\mathbf{r}_i$ we only know the point P_3 at the null screen, so we have to approximate a second point to obtain the direction of the incident ray by intersecting the reflected ray with a reference surface; the reference surface can be the ideal design surface or a similar surface close to the real one.

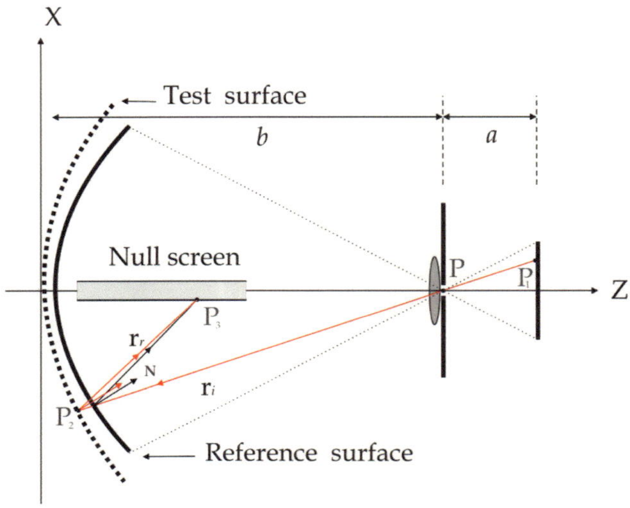

Fig. 4. Approximated normals.

The next step is the numerical evaluation of Eq. (3). The simplest method used for the evaluation of the numerical integration is the trapezoid rule (Malacara-Doblado & Ghozeil, 2007). An important problem in the test with a null screen is that the integration method accumulates important numerical errors along the different selected integration paths. It is well known (Moreno-Oliva et al., 2008a) that a bound to the so called truncation error can be written as

$$|\varepsilon| \leq \frac{h^2}{12}(b-a)M, \tag{5}$$

here h is the maximum separation of two points along the integration path, (b-a) is the total length of the path and M is the maximum value of the second derivative of the integrand along the path. Díaz-Uribe et al. (2009) have shown that for spheres this error is negligible; for other surfaces it can be very significant.

To reduce the numerical error, some authors have proposed the use of parabolic arcs instead of trapeziums (Campos-García et al., 2004), or the fit of a third degree polynomial that describes the shape of the test surface locally(Campos-García & Díaz-Uribe, 2008).

There are other integration methods going from local low order polynomial approximations (Salas-Peimbert et al., 2005) to global high order polynomial fitting to the test surface (Mahajan, 2007) in the latter case, the Least Squares method is commonly used but some

other fitting procedures, such as Genetic Algorithms (Cordero-Dávila, 2010) or Neural Networks, have been also used.
By far the simplest integration method is the trapezoid rule method; however, since the error increases as the second power of the spacing between the spots of the integration path, to minimize the error, it is desirable to reduce the spacing between spots (see eq. (5)). This implies more spots in the design of the null screen; there is, however, a physical limit on the number of spots; if the spot density is too large,the spot images can overlap because of defocus, aberrations or because of diffraction. A method to increase the number of points, thus reducing the average separation between them, is to use the so called point shifting method (Moreno-Oliva et al., 2008a; Moreno-Oliva et al., 2008b). The basic idea is to acquire a total of m pictures, each with different null screen arrangement and containing n spots on the image; the spots will be shifted from their positions in other pictures, making a total of $m \times n$ evaluation points, with an average separation of

$$h_m = \frac{h}{\sqrt{m}} . \tag{6}$$

Then, the bound to the truncation error is reduced as the original bound for only one image (n points), divided by m

$$|\varepsilon_m| \le \frac{h^2}{12m}(b-a)M \le \frac{|\varepsilon|}{m} . \tag{7}$$

In order to implement this method in the lab, small known movements are applied to the cylindical screen along the axis of the surface under test. With this method it was possible to reduce the accumulated numerical error by up to 80%, with respect to the error for a single screen without scrolling. In Fig. 5a the image for the initial position of the screen is show; and figure 5b is the image for the final position of the screen. A total of ten images were captured. Each image was independently captured and processed to obtain the centroids of the spots, Fig. 5c shows the plot of the spot centroids for all the captured images.
Another method to implement the same idea is to design a screen such that its image in the optical system is an array of dots or spots in a spiral arrangement (Moreno-Oliva et al., 2008b). In this case the movement of the screen or surface is made by rotation around the axis of the surface to obtain, a high density of points depending on how the screen or the surface is rotated. Figure 6(a) shows the image of a screen with spots ordered in a spiral arrangement. The plot of the positions of the centroids for the spots from twelve images captured on each rotation step of the test surface is shown in Fig. 6(b). The screen is designed to increase the density of points with respect to the original radial distribution of the image at the initial position. In Fig. 6(b) a set of equally spaced spots along the radial direction is observed.
One of the main disadvantages of the previous methods, where a movement is applied to the cylindrical screen, is the introduction of errors due to mechanical translation or rotation devices. In a more recent work, the use of LCD flat panels was proposed, for the test of convex surfaces (Moreno-Oliva et al., 2008c); the screens are arranged in a square array and the surface under test is placed in the center. The screens display the required geometry in a sequence so that each distribution of points produces an array of equally spaced spots in the image plane, and the sequence causes these points to move. By taking a picture for each step and merging the centroids of the spot images is possible have a greater density of equidistant spots for better evaluation.

(a)

(b)

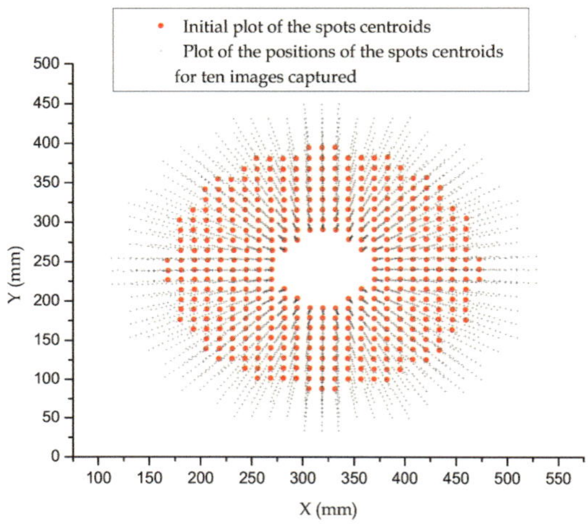

(c)

Fig. 5. a) Image of the screen at the initial position, b) Image of the screen at the final position, c) Plot of the centroid positions of the spots for ten images captured by using the point shifting method.

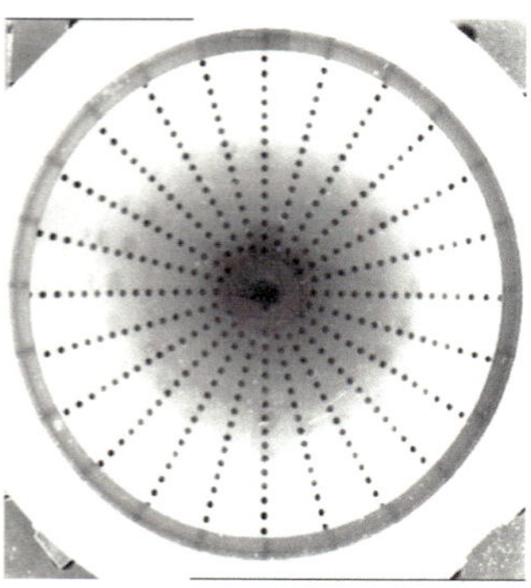

(a)

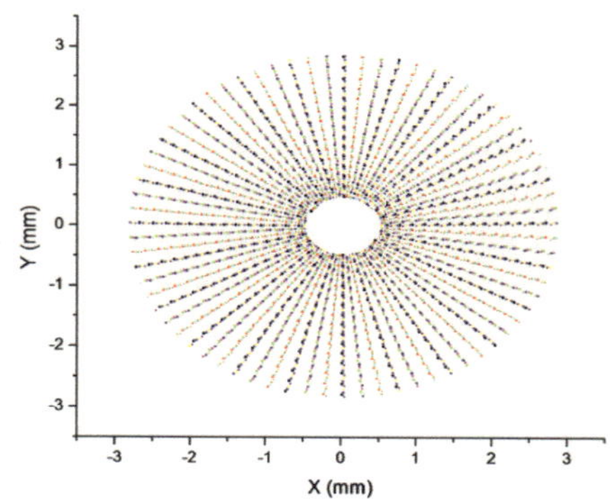

(b)

Fig. 6. a) Image of the screen at the initial position, b) Plot of the position of the centroids for the spots at each rotation step of the test surface.

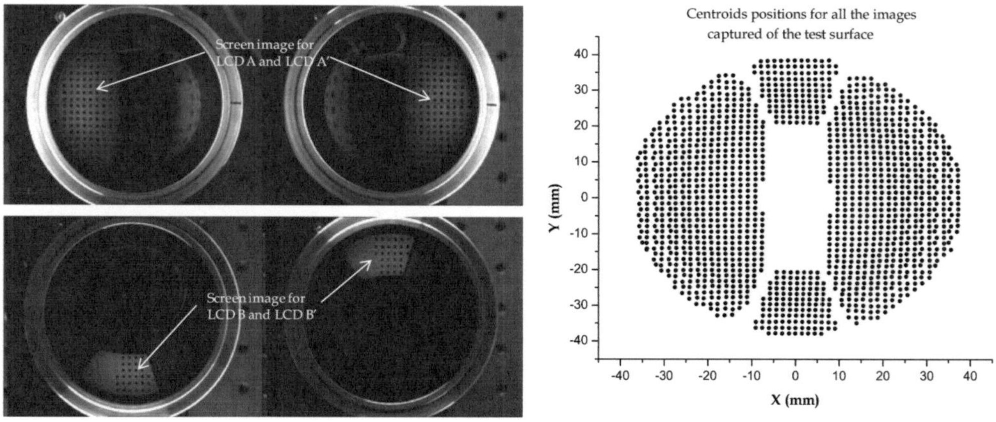

(a) (b)

Fig. 7. (a) Image of each LCD monitor showing a sequence of flat null screens and (b) plot for many sequences of all the LCD monitors.

The screen in this method consisted of four LCD flat panels (LCD A, A′ and LCD B, B′), the distance between LCD A and A′ is smaller than the distance between LCD B and B′, for this reason the image area covered by LCD A and A is greater than that covered by LCD B and B (Fig. 7a). Each LCD displayed a sequence of dynamic flat null screens, and the number of sequences can be increased to the density of equidistant spots. Figure 7b shows the plot of the centroids for all the screens displayed.

3. Testing a parabolic trough solar collector (PTSC)

3.1. Testing a PTSC by area

3.1.1 Screen design

The null screen method can also be used for testing other surfaces without symmetry of revolution such as off-axis parabolic surfaces (Avendaño-Alejo, et al., 2009). This method has also been used in the testing of parabolic trough solar collectors (PTSC). In both cases the use of flat null screens was proposed; the screen is designed in the same way as the cylindrical screens described above, using inverse ray tracing starting on the array of points in the image plane and intercepting the reflected ray on the surface with the flat screen.

The proposal is to use two flat null screens parallel to the collector trough; physically, they are located on each side of a wood or plastic sheet; each side is useful for testing half of the surface of the PTSC. Figure 8 shows the schematic arrangement for the proposed evaluation for a PTSC with flat null screens.

The design of the screen starts on a CCD point P_1, with coordinates $(x,y,a+b)$; the ray passes through the point $P(0,0,b)$ (pinhole of the camera optical system), and arrives at the test surface at $P_2(X,Y,Z)$; after reflection, the ray hits the point $P_3(x_3,y_3,z_3)$ on the null screen (see Fig. 8).

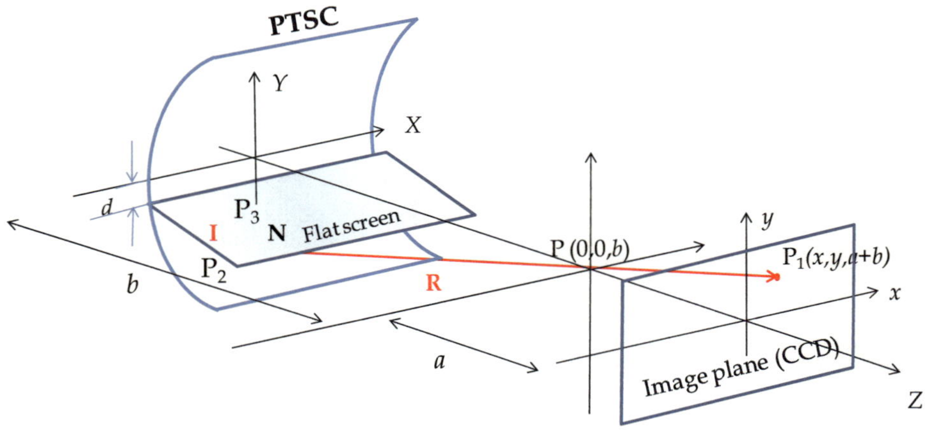

Fig. 8. Setup for the testing for a PTSC with null screens.

The equation for the PTSC is given by.

$$Z = \frac{Y^2}{2r}, \tag{8}$$

where r is the radius of curvature at the vertex. Then, the coordinates of the point P_2 are found by

$$X = tx\,, \tag{9}$$

$$Y = ty\,, \tag{10}$$

$$Z = at + b = \frac{Y^2}{2r}, \tag{11}$$

where

$$t = \frac{1}{y^2}\left(ar \pm \sqrt{a^2r^2 + 2y^2rb}\right). \tag{12}$$

Here, a is the distance from the aperture stop to the CCD plane and b is the distance from the aperture stop to the vertex of the surface. Then, using the Reflection Law written as

$$\mathbf{I} = \mathbf{R} - 2(\mathbf{R} - \mathbf{N})\cdot\mathbf{N}\,, \tag{13}$$

where **I**, **R**, and **N**, are the incident, reflected and normal unit vectors associate with each corresponding ray. As we are performing an inverse ray trace, the real incident ray is the reflected ray of our tracing. Then, as the normal vector (not normalized) is given by

$$\mathbf{N} = \left(0, \frac{Y}{r}, -1\right), \tag{14}$$

the normalized Cartesian components of the vector **I** are given by

$$I_x = \frac{x}{\sqrt{x^2+y^2+a^2}}, \quad I_y = \frac{y(r^2-Y^2)+2arY}{(Y^2+r^2)\sqrt{x^2+y^2+a^2}}, \text{ and } I_z = \frac{2ryY-a(r^2-Y^2)}{(Y^2+r^2)\sqrt{x^2+y^2+a^2}} \quad (15)$$

Finally, the intersection with the flat null screen gives the coordinates of the point P_3

$$x_3 = tx + s\frac{x}{\sqrt{x^2+y^2+a^2}}, \quad (16)$$

$$y_3 = ty + s\frac{y(r^2-Y^2)+2arY}{(Y^2+r^2)\sqrt{x^2+y^2+a^2}}, \quad (17)$$

$$z_3 = at + b + s\frac{2ryY-a(r^2-Y^2)}{(Y^2+r^2)\sqrt{x^2+y^2+a^2}}, \quad (18)$$

where s is a parameter determined by the condition that the point P_3 is on the flat screen. The equation for this condition is

$$y_3 = d, \quad (19)$$

where d is the distance between the XZ plane and the flat null screen. Substituting Eq. (19) in Eq. (17) yields

$$d = ty + s\frac{y(r^2-Y^2)+2arY}{(Y^2+r^2)\sqrt{x^2+y^2+a^2}}, \quad (20)$$

and solving for s, we get

$$s = \left[\frac{(Y^2+r^2)\sqrt{x^2+y^2+a^2}}{y(r^2-Y^2)+2arY}\right](d-ty). \quad (21)$$

To test the whole area of the PTSC with only one image, it is necessary use two flat null screens in the positions d and $-d$ with respect to the Y axis.

3.1.2 Quantitative surface testing

With the aim of testing a PTSC with the parameter data given in table 2, a null screen was designed. The test surface and the screen designed for it are shown in Fig. 9; the resultant image of the screen after reflection on the test surface is also shown.

Parameter	Symbol	Size
Full aperture	ΔY	3.0 m
Length	L	1.2 m
Focal Length	f	1.0 m
Vertex radius of curvature	r	2.0 m
Stop aperture-CCD plane	a	12.5 mm
CCD length	d	8.1 mm
Stop aperture-surface vertex	b	5192.12 mm

Table 2. Design parameters for the test of a PTSC

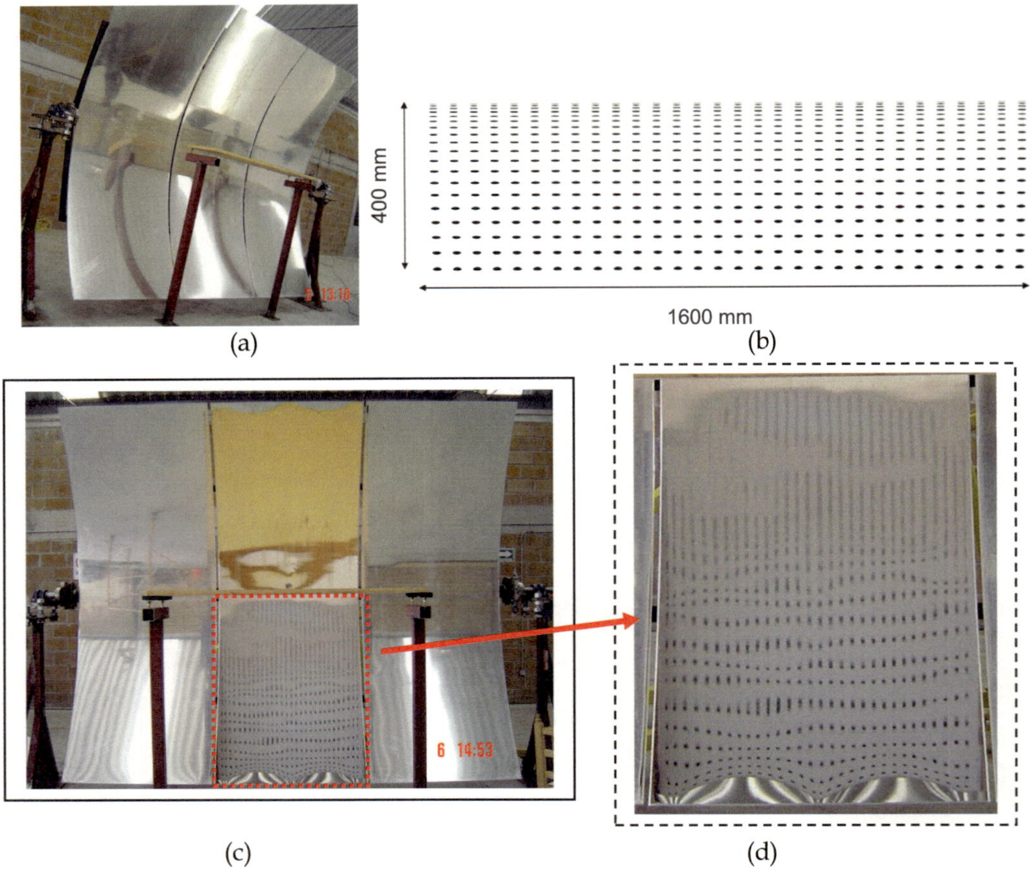

Fig. 9. a) PTSC component, b) flat printed null screen with drop shaped spots for quantitative testing (400x1600 mm), c) image of the screen after reflection on the test area surface, and d) detail of the image.

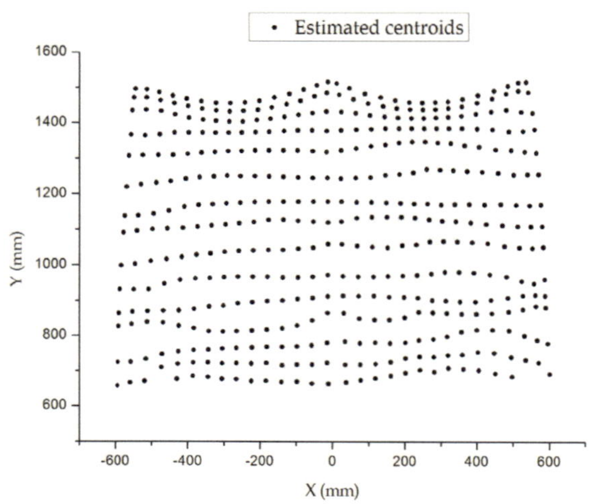

Fig. 10. Plot of the centroid positions for some spots of the flat null screen.

In Fig. 9 the PTSC before assembly is shown, for final assembly it is possible to use a flat null screen for alignment of the PTSC sections. In this example only the result of the test of the lower central panel of the PTSC component is shown. In the qualitative result for the test of a central panel (Fig. 9c) it can clearly be observed that, in general the image shows deformations near the edge of the surface; in the upper part of the image (Fig. 9d) it can be observed that there are doubled or elongated spots. This behavior is due to some small deformations of the test surface. In this case it is not possible to separate the doubled spot images and the surface cannot be tested in this zone, the only spots for which its positions can be determined on the CCD plane (centroids) are show in Fig. 10.
The proposed flat null screen consists of 600 spots, and only 443 were processed for quantitative evaluation.
Having the information of the positions of the centroids on the CCD plane, the normals to the surface are evaluated and the shape of the surface is obtained by using Eq. (3). The method used for the discrete evaluation was the trapezoidal method, which can be written as

$$z_m = -\sum_{i=1}^{m-1}\left\{\left(\frac{n_{x_i}}{n_{z_i}}+\frac{n_{x_{i+1}}}{n_{z_{i+1}}}\right)\frac{(x_{i+1}-x_i)}{2}+\left(\frac{n_{y_i}}{n_{z_i}}+\frac{n_{y_{i+1}}}{n_{z_{i+1}}}\right)\frac{(y_{i+1}-y_i)}{2}\right\}+z_1\,, \tag{22}$$

Here m represents the number of points along some integration path; z_1 is the value for the initial point, which represents only a rigid translation of the surface so it can be approximated by Eq. (11).

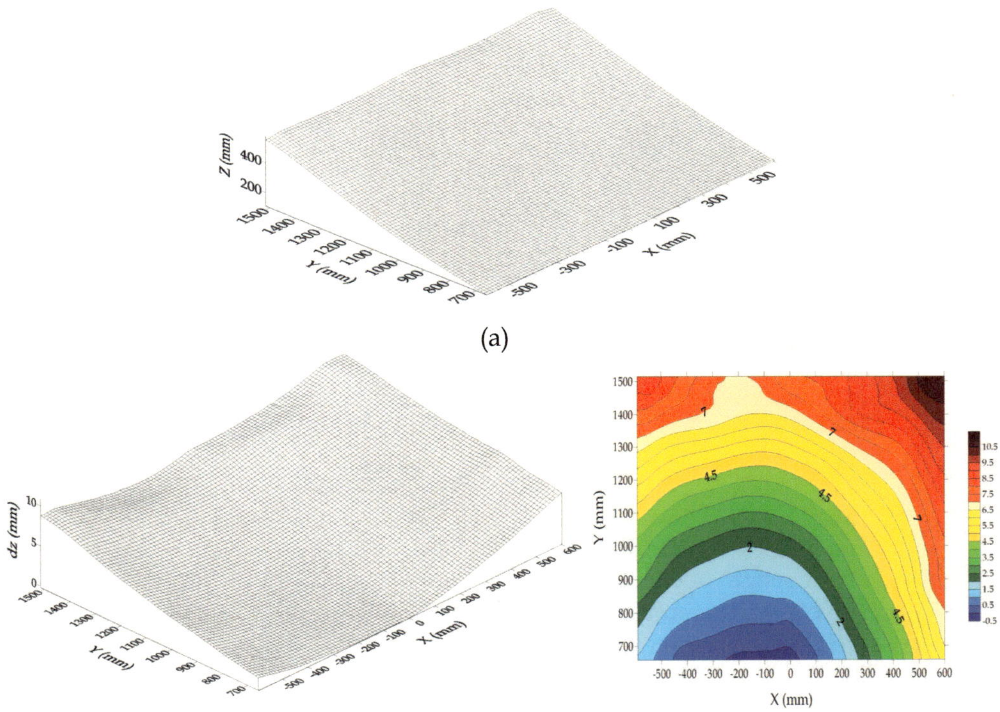

Fig. 11. a) Evaluated surface, b) Differences in sagitta between the measured surface and the best fit, and c) Contour map of differences in sagitta.

Figure 11a shows the evaluated surface (lower central panel of PTSC); Fig. 11b shows the differences in sagitta (z coordinate) between the evaluated surface and the best fit. In this case the P-V difference in sagitta between the evaluated points and the best fit was $\Delta z_{p\text{-}v}$ = 11.08 mm and the rms difference in the sagitta was Δz_{rms} = 4.89 mm.

3.2 Testing a PTSC by profile

An alternative method for testing the PTSC is given by (Moreno-Oliva et al., 2009); here the test is made by testing one profile at a time with two flat null screens and by scanning the PTSC. All the calculations were made in a meridional plane (X, Y), and for simplicity in the calculus we use an approximation using ellipses instead of drop shaped spots (Carmona-Paredes & Díaz-Uribe, 2007).
A ray starting at point P_1 (α, $a+b$) on the image plane passes through the pinhole located on the Y axis at a distance b, P (0, b) (Fig. 12), away from the vertex of the surface; this ray arrives at the test surface at the point P_2 (x_2, y_2). After reflection on the PTSC the ray hits the surface at the point P_3 (x_3, y_3) on the flat null screen.

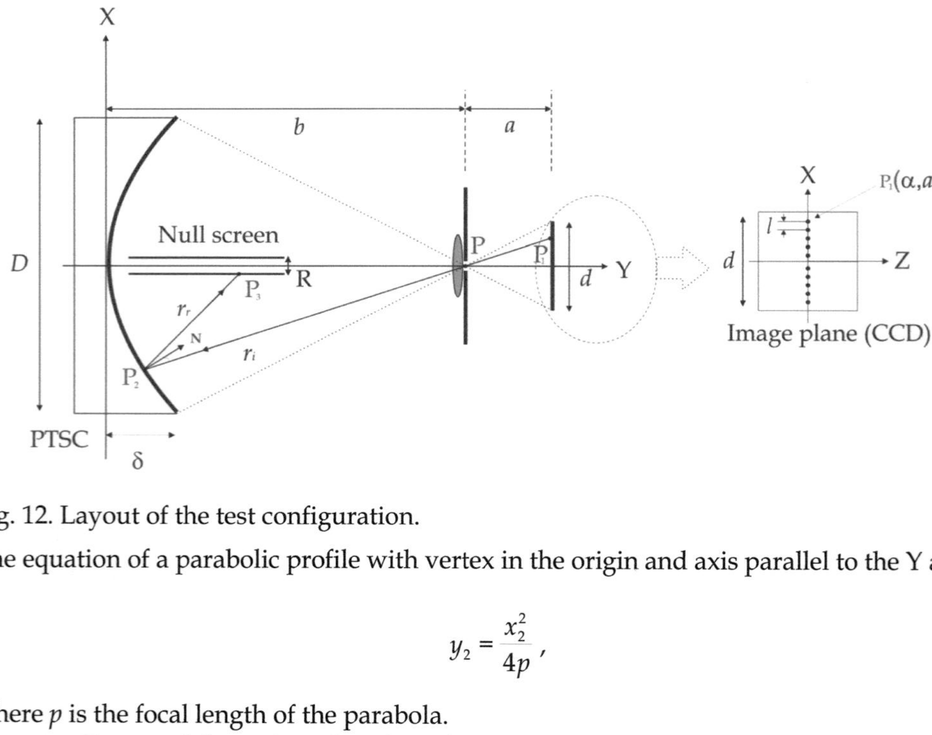

Fig. 12. Layout of the test configuration.

The equation of a parabolic profile with vertex in the origin and axis parallel to the Y axis is

$$y_2 = \frac{x_2^2}{4p}, \tag{23}$$

where p is the focal length of the parabola.
The coordinates of the points that describe the parabolic profile $P_2(x_2, y_2)$, in terms of the parameters of the optical system and the focal length of the parabola p are

$$x_2 = \frac{2pa - 2\sqrt{p^2a^2 + pba^2}}{a}, \tag{24}$$

and the intersection points on the flat null screen $P_3(x_3, y_3)$ are given by

$$y_3 = y - \left[\frac{4ap^2 - 4pxa - x^2 a}{x^2 a - 2apx - 4pa - 2pxa} \right] (x - x_3) , \tag{25}$$

where $x_3 = \mathbf{R}/2$ is constant, **R** is the separation between the flat null screens.
In the meridional plane, with the inverse ray tracing it is only possible to obtain the coordinates of the spots from their center and the vertices along the direction parallel to the Y-axis of each spot in the CCD plane. For each spot on the CCD we obtained three points on the flat null screen (Fig. 13), and according to reference (Carmona-Paredes & Díaz-Uribe, 2007) we can use an approximation using ellipses instead of the drop shape for simplicity in the calculations.

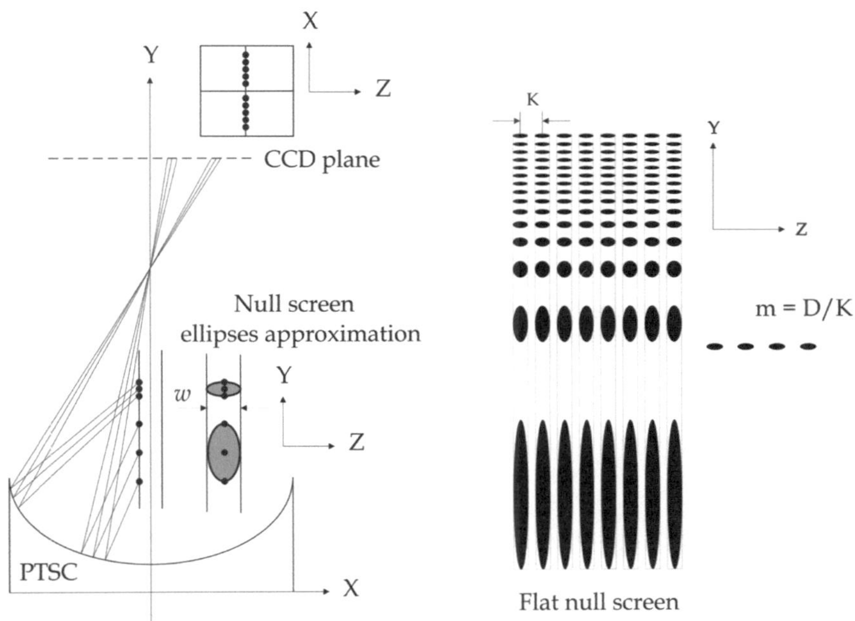

Fig. 13 Inverse ray tracing on the X-Y plane, the elliptical approximation in the Y-Z screen plane, and the flat null screen for testing the PTSC component.

To test the PTSC, the optical system was displaced a distance K and an image for each profile of the PTSC was captured, the PTSC was scanned along the trough (axis Z), m was the number of linear arrangements of spots of the flat null screen, and D the trough length.

4. Testing parabolic dish solar collector systems

In reference (Campos-Garcia et al., 2008) the procedure to obtain the shape of fast concave surfaces is described for a general conic. The same method can be applied to testing of parabolic dish solar collector systems and the equations are simplified if, instead of using a general conic only a parabolic surface is considered. The layout of the test configuration is similar to that of Fig. 1b, starting with one of the points of the proposed arrangement at the

CCD plane $P_1(\rho_1, \phi, a + b)$, where P_1 is given in cylindrical coordinates ($\rho_1 > 0$; $0 \leq \phi \leq 2\pi$; a, $b > 0$), and the calculations are made for a conic with constant $k = -1$; a ray passing through the point $P(0,0,b)$ (the pinhole of the camera optical system) reaches the surface at the point $P_2(\rho_2, \phi + \pi, z_2)$, where

$$\rho_2 = \frac{ar - \left[a^2r^2 - 2r\rho_1^2 b\right]^{1/2}}{\rho_1}, \tag{26}$$

$$z_2 = \frac{\rho_2}{\rho_1} a + b, \tag{27}$$

here $r = 1/c$ is the radius of curvature at the vertex, a is the distance from the aperture stop to the CCD plane, and b is the distance from the aperture to the vertex of the surface.
After reflection on the surface the ray hits the cylindrical screen at P_3 (ρ_3, $\phi + \pi$, z_3), where

$$\rho_3 = R, \tag{28}$$

$$z_3 = \frac{-a\rho_2^2 + ar^2 - 2r\rho_1\rho_2}{\rho_1\rho_2^2 - \rho_1 r^2 - 2r\rho_2 a}(-R - \rho_2) + z_2, \tag{29}$$

R is the radius of the cylindrical screen. Distances a and b are chosen in such a way that the image of the whole surface fits the CCD area; they are related by Eq. (1), where D is the diameter of the test surface and β is the sagitta at the rim of the surface, which for a parabolic surface is given by Eq. (2). The method for the surface shape evaluation is as given in section 3.1

5. Conclusion

This Chapter gives a general view of the latest developments of the null screen method and its application in the measurement of the shape of solar collectors. The null screen principles principle has many advantages when compared to other methods; the method does not require a special optical system and its implementation is not very expensive, it is also possible to apply the method to any collector system geometry. With new developments in null screen methods (section 3) it is possible to increase the precision and sensitivity of the quantitative evaluation.

6. References

Avendaño-Alejo, M., Moreno-Oliva, V.I., Campos-García, M. & Díaz-Uribe, R. (2009), Quantitative evaluation of an off-axis parabolic mirror by using a tilted null screen. *Applied Optics*. 48, 1008-1015.

Campos-García, M., Díaz-Uribe, R. & Granados-Agustín, F. (2004). Testing fast aspheric surfaces with a linear array of sources. *Applied Optics*. 43, 6255-6264.

Campos-García M., Díaz-Uribe R. (2008), Quantitative shape evaluation of fast aspherics with null screens by fitting two local second degree polynomials to the surface normals, AIP Conf. Proc. 992, 904-909.

Campos-García, M., Diaz-Uribe, R., & Bolado-Gómez, R. (2008). Testing fast aspheric concave surfaces with a cylindrical null screen. *Applied Optics*. 47, 6, (February 2008) 849-859.

Cordero-Dávila, A., & González-García, J., Surface evaluation with Ronchi test by using Malacara formula, genetic algorithms and cubic splines, in International Optical Design Conference (IODC)/Optical Fabrication and Testing (OF&T) Technical Digest on CD-ROM (Optical Society of America, Washington, DC, 2010), JMB46.

Cornejo-Rodríguez, A. (2007). Ronchi Test, In: *Optical Shop Testing*, D. Malacara, (Wiley, New York), pp. 317-360.

Diaz-Uribe, R., & Campos-García, M. (2000). Null-screen testing of fast convex aspheric surfaces. *Applied Optics*. 39, 16, (June 2000) 2670-2677.

Díaz-Uribe, R. (2000). Medium precision null screen testing of off-axis parabolic mirrors for segmented primary telescope optics; the case of the Large Millimetric Telescope. *Applied. Optics*. 39, 2790-2804.

Díaz-Uribe, R., Granados-Agustín, F., & Cornejo-Rodríguez, A. (2009) "Classical Hartmann test with scanning", Opt. Express, 17, 13959-13973.

Mahajan, V. N., ZernikePolynomials and Wavefront Fitting, In: *Optical Shop Testing*, D. Malacara, 3rd. Ed. (Wiley, New York), pp. 498-546.

Malacara-Doblado, D., & Ghozeil, I. (2007). Hartmann, Hartmann-Shack, and other screen tests, In: *Optical Shop Testing*, D. Malacara, (Wiley, New York), pp. 361-397.

Moreno-Oliva, V.I., Campos-García, M., Bolado-Gómez, R., & Díaz-Uribe, R. (2008a). Point Shifting in the optical testing of fast aspheric concave surfaces by a cylindrical screen. *Applied Optics*. 47, 5, 644-651.

Moreno-Oliva, V.I., Campos-García, M., & Díaz-Uribe, R. (2008b). Improving the quantitative testing of fast aspherics with two-dimensional point shifting by only rotating a cylindrical null screen. *Journal of optics A: Pure and Applied Optics*. 10, 1-7.

Moreno-Oliva, V.I., Campos-García, M., Avendaño-Alejo, M., & Díaz-Uribe, R. (2008c). Dinamic null screens for testing fast aspheric convex surfaces with LCD´s. *Proceedings of 18th IMEKO TC 2 Symposium on Photonics in Measurement*, M. Jedlicka, M. Klima, E. Kostal, P. Pata, eds., (Czech and Slovak Society for Photonics, Czeck Republic, 2008). (1P5, 6pp).

Moreno-Oliva, V.I., Campos-García, M., Granados-Agustín, F., Arjona-Pérez, M.J., Díaz-Uribe & Avendaño-Alejo (2009). Optical testing of a parabolic trough solar collector by null screen with stitching. *Proceedings of SPIE in Modeling Aspects in Optical Metrology II*. edited by Harald Bosse, Bernd Bodermann, Richard M. Silver, Vol. 7390 (October 2009) 739012.

Pottler, K., & Lüpfert, E. (2005). Photogrammetry: A Powerful Tool for Geometric Analysis of Solar Concentrators and Their Components. *Journal of Solar Energy Engineering*. 127, 94-101.

Salas-Peimbert, Malacara-Doblado, Durán-Ramírez, Trujillo-Schiaffino & Malacara-Hernández (2005). Wave-front retrieval from Hartmann test data. *Applied Optics*. 44, 4228-4238.

Shortis M., & Johnston G. (1996). Photogrammetry: An Available Surface Characterization Tool for Solar Concentrators, Part 1: Measurement of Surfaces. *ASME J. of Solar Energy Engineering*, 118,146-150.

9

Theory, Algorithms and Applications for Solar Panel MPP Tracking

Petru Lucian Milea[1], Adrian Zafiu[2], Orest Oltu[1] and Monica Dascalu[1]
[1]"Politehnica" University of Bucharest,
[2]Romanian Academy Research Institute for Artificial Intelligence
Romania

1. Introduction

The photovoltaic panel is a power source whose parameters depend on some external factors like incident light angle, shading, ambient temperature etc. Some of these factors are unpredictable and, for this reason, so is the evolution of cell parameters. The most known parameters of the photovoltaic panel are the open circuit voltage (V_{oc}) and short circuit current (I_{sc}). These values define the points where the I(V) graph curve of the panel intersects the two axes (I and V), like in Fig. 1.

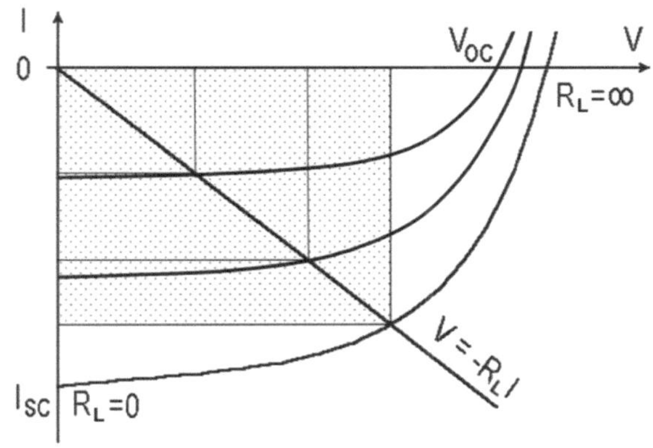

Fig. 1. Some *I(V)* characteristics of a photovoltaic panel under different work conditions

Every point on the I(V) curve has specific values of V_i and I_i, defining the power as $P_i = V_i \cdot I_i$. For a specific I(V) curve there is only one point corresponding to the maximum power. This is named maximum power point or MPP. For any power source it is always good to supply electrical consumers at this value or close to it. For a specific resistive load, R_L, the $I_{RL}(V)$ load characteristic is a line given by the equation $I_{RL}=-I=V/R_L$. This line intersects the panel characteristic in a point which is near or far from MPP (Fig. 1.). According to this position, the power transferred to the load can be only a fraction of the power that panel can supply at MPP. To correct this unbalance and prevent the associated lose of usefully power, some methods, generically named MPPT (Maximum Power Point Tracking) are used.

MPPT methods are designed to tune the electrical current to the value corresponding to MPP. This means, in other words, to adapt the impedance of the consumers to the optimal impedance for the best power transfer. For this reason, we named the resulted circuits as "impedance adapters". For the studied case, characterized by the almost permanent modification of the I(V) curve, the MPP varies almost at every moment, therefore the power maximization process needs a dynamic impedance adaptor.
In the studied scientific literature, we found some algorithms which are purposed to implement this function, but the adaptation speed may be a problem for some of them (Yang et al., 2008). We start this study in order to find solutions for simple, fast and accurate (efficient) MPPT. For this purpose, we propose two computational algorithms and a impedance adjustment method which use DC/DC converters. The tracking method was designed to be able to track the MPP for an unknown type of solar panel (viewed as a black box) and external conditions (irradiance and temperature). The algorithms were simulated with a dedicated application and the results were compared with other algorithm results and also with some experimental data.

2. The mathematical model

We propose a mathematical model to estimate the maximal power point of a panel, starting from the simplest model of a photovoltaic cell.
The photovoltaic cell *I(V)* characteristic, presented in Fig 2, is given by a equation derived from the Shockley diode equation:

$$I(V) = I_0\left(e^{V/(a \cdot V_T)} - 1\right) - I_L, \tag{1}$$

where, I_0 is the reverse saturation current, V_T is the thermal voltage ($V_T=kT/q$, with $q=1.602x10^{-19}C$ the electron charge, T – jonction temperature and $k=1.381x10^{-23}J/K$ is Boltzmann constant), a is known as the diode ideality factor (for silicon diodes a is between 1 and 2) and I_L is cell illumination current.

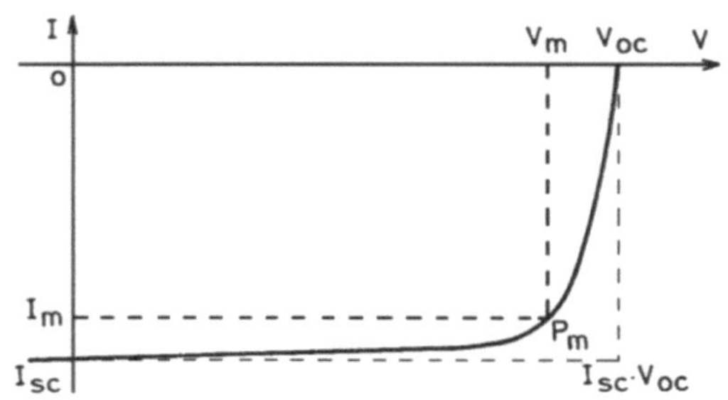

Fig. 2. The parameters of I-V cell illumination characteristic

The maximal power point, P_m, corresponds to the point where the power transferred from the panel/cell to the consumer is maximal. The ratio of P_m and the product $I_{SC} \cdot V_{OC}$ (the dotted areas) define the fill factor, *FF*, which represents a measure of the resistive losses of the device.
For the cell illumination current we choose to use a formula based on datasheet parameters of the cell or panel (Chenni et al., 2007):

$$I_L = \left(I_{Lref} + \alpha\left(T - T_{ref}\right)\right)G/G_{ref}\,, \tag{2}$$

where: $G_{ref}=1kW/m^2$ represents the irradiation at $AM1,5$, $T_{ref}=25^0C$, $I_{Lref}(A)$ has value I_L at G_{ref} and T_{ref}, $a(A/K)$ is the temperature coefficient at short circuit.
The reverse saturation current, I_0, is given by the relation:

$$I_0 = DT^3 e^{-q\varepsilon_g/akT} = DT^3 e^{-\varepsilon_g/aV_T} \tag{3}$$

For the ideal photovoltaic cell (without resistive losses – Fig. 3.a) we have equation:

$$I_1 = I_0\left(e^{\frac{V_1}{aV_T}} - 1\right) - I_L = I_{0ref}\left(\frac{T}{T_{ref}}\right)^3 e^{\frac{\varepsilon_g}{aV_T}\left(\frac{T}{T_{ref}}-1\right)}\left(e^{\frac{V_1}{aV_T}} - 1\right) - \frac{G}{G_{ref}}\left(I_{Lref} + \alpha\left(T - T_{ref}\right)\right) \tag{4}$$

The real photovoltaic cell contains also energy dissipation elements. For the approximate model, presented in figure 3.b, the relation $I(V)$ becomes:

$$I_2 = I_0\left(e^{\frac{V_2 - I_2R_s}{aV_T}} - 1\right) - I_L - \frac{V_2 - I_2R_s}{R_p} \tag{5}$$

In order to adapt the model for a panel, we take into account that this is formed by a matrix of NS x NP cells (Fig. 4.). The cell parameters are scaled as follows: $I_L^P = N_p I$, $I_0^P = N_p I_0$, $V_3 = N_s V_2$, $I_3 = N_p I_2$, $I_3 = N_p I_2$, $R_s^P = R_s \dfrac{N_s}{N_p}$, $R_p^P = R_p \dfrac{N_s}{N_p}$.

We obtain the global $I(V)$ equation of the panel:

$$I_3 = I_0^P\left(e^{\frac{V_3 - I_3R_s^P}{aN_sV_T}} - 1\right) - I_L^P - \frac{V_3 - I_3R_s^P}{R_p^P} \tag{6}$$

The MPP condition for transferred power is defined as follows:

$$\left.\frac{dP}{dV}\right|_{MPP} = \left.\frac{d(I_{RL}\cdot V)}{dV}\right|_{MPP} = \left.\frac{d(-I\cdot V)}{dV}\right|_{MPP} = -V\left.\frac{dI}{dV}\right|_{MPP} - I\left.\frac{dV}{dV}\right|_{MPP} = 0 \tag{7}$$

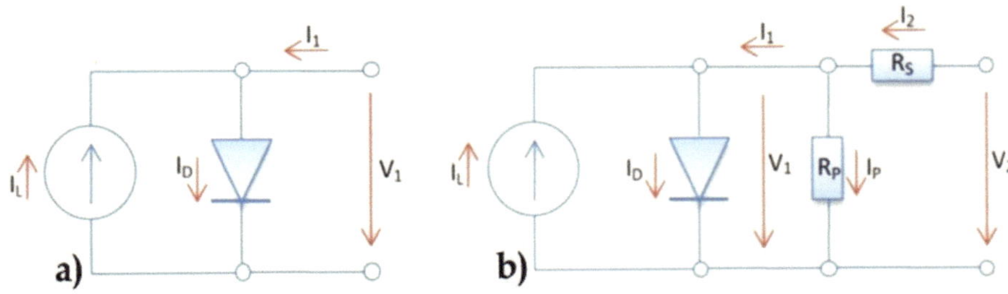

Fig. 3. Two equivalent circuits of PV cell: a) an ideal circuit, b) an equivalent circuit with serial and parallel resistive loses

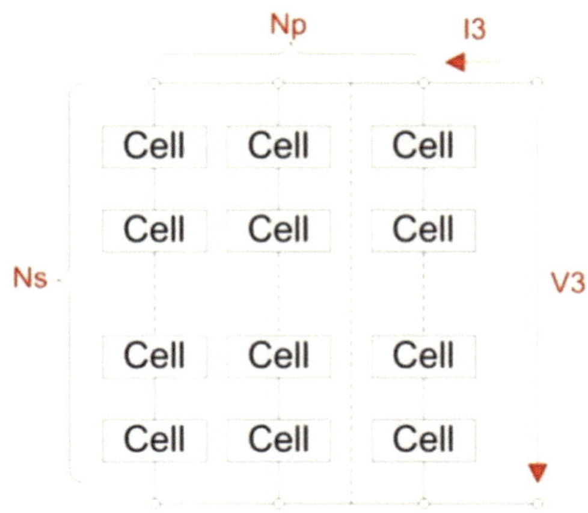

Fig. 4. A photovoltaic panel, as a matrix of N_S x N_P cells

Denoting with R_X the value of resistive load at MPP, R_X=-V_m/I_m, we obtain:

$$\left.\frac{dI}{dV}\right|_{MPP} = -\frac{I}{V} = \frac{1}{R_X} \tag{8}$$

For the studied cases, from the relation (8), we obtain:

$$R_{x1} = aV_T / (I_0 e^{V_1 / aV_T}) \tag{9}$$

$$R_{X2} = R_S + 1/(1/R_p + 1/R_{X1}) \tag{10}$$

$$R_{X3} = R_{X2} N_S / N_P \tag{11}$$

which can be replaced in the following equation, to obtain the PV cell/panel MPP current:

$$I = -\frac{V}{R_X} \tag{12}$$

The equation (12) is a nonlinear implicit equation and has to be solved numerically. For the ideal model of a PV cell, the equation to be solved is:

$$\left(\frac{V_1}{aV_T} + 1\right) e^{\frac{V_1}{aV_T}} = \frac{\frac{G}{G_{ref}}\left(I_{Lref} + \alpha\left(T - T_{ref}\right)\right)}{I_{0ref}\left(\frac{T}{T_{ref}}\right)^3 e^{\frac{\varepsilon_g}{aV_T}\left(\frac{T}{T_{ref}} - 1\right)}} \tag{13}$$

Also we solved this equation graphically (Fig. 5.), at the intersection of the curves *I(V)* (cell/panel characteristic) and $I_m(V)$, where we denoted I_m=-V/R_X, under specific conditions. From the graphics we observed the relatively constant value of *FF*, for relatively high variations of *G* and *T*. A numerical solution can be also obtained following this principle.

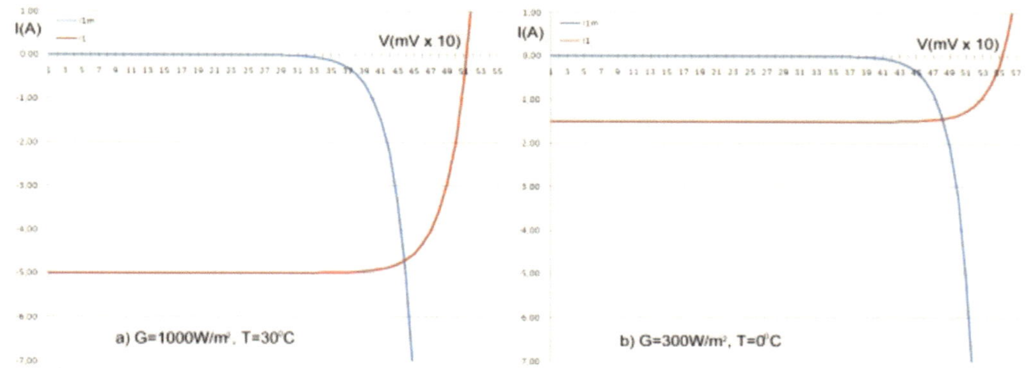

Fig. 5. The graphic solution of the MPP equation for two datasets of G and T

In practical applications, the MPP tracking is made in an iterative manner, similar with numerical solution, trough iterative adjustment of the charge's impedance to the necessary value, with the aid of a DC/DC converter (boost, buck or buck-boost converter). For this type of application we implement two types of algorithms which are presented in section 4.

3. MPPT advantages evaluation, based on proposed model

To assess the benefits of using a MPPT circuit, we consider that it is based on a buck-boost converter. We'll make a comparative analysis of the energetic transfer from photovoltaic panel to a battery by using a direct load circuit and a MPPT circuit respectively.

In this evaluation we will estimate the energetic transfer in a summer day, when the panel temperature varies between $T_0=290K$ at sunrise and $T_{max}=330K$ at middle day, when the Sun is at meridian. To reduce the computational complexity we suppose that the panel tracks the direction to the Sun, so the direct irradiation is always upright on the panel.

In this simplified hypothesis we suppose that global radiation *AM1* has value $G_0=1000W/m^2$ and the global radiation during all day is inverse with the optical path (air mass) throw atmosphere (m_r).

$$G = \frac{G_0}{m_r}, \tag{9}$$

where m_r is given by (Milea, 2010):

$$m_r = 531 \cdot \left(\sqrt{1/265{,}25 + \cos^2(z)} - \cos(z) \right) \tag{10}$$

To estimate the panel temperature, we used a simplified formula, considering that it is directly dependent with the Sun's position. Under this aproximative model, for a cloudless summer day, the diurnal variation of the panel temperature will be (Milea, 2010):

$$T = T_0 + \frac{\Delta T}{m_r}, \tag{11}$$

where $T=290K$ and $\Delta T=40K$.

For a panel with $N_S\ x\ N_P$ cells, based on the ideal model, the open circuit voltage formula is:

$$V_{OC} = a \cdot V_T \cdot N_S \cdot \ln\left(1 + \frac{I_L}{I_0}\right) \quad (12)$$

The short circuit current is:

$$I_{SC} = N_P \cdot I_L = N_P \frac{G}{G_{ref}} \left(I_{Lref} + \alpha\left(T - T_{ref}\right)\right) \quad (13)$$

To model a solar panel for off-grid application with a DC bus of 12V, under reference conditions, we choose N_S=44 taking into account that V_{OCcref}=0,51V, for a single cell. The obtained open voltage of the panel is V_{OCref}=22,64V, which is a common value for such applications.
For the mentioned parameters, we determined the values of V_{OC}, I_{SC} and P_{MAX} for G and T values corresponding to different hours of a cloudless summer day. The results are plotted in Fig. 6.

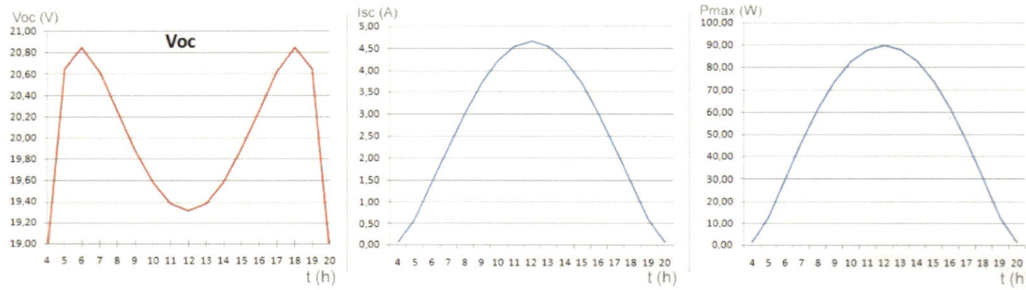

Fig. 6. V_{OC}, I_{SC} and P_{MAX} curves at different hours of a cloudless summer day

Note that the voltage is always greater than 19V, while the current and maximum reference power curves have a gait similar to that of the optical path.
If direct charging of accumulator batteries, since the panel voltage is significantly higher than the battery's one (Milea, 2010), we can assume with a very good approximation that the battery will charge at the panel's short-circuit current:

$$P_{DIR} = V_{bat} \cdot I_{SC} = V_{bat} \cdot N_P \cdot \frac{G}{G_{ref}} \left(I_{Lref} + \alpha\left(T - T_{ref}\right)\right) \quad (13)$$

From (9) şi (13), because the irradiance references, G_0 and G_{ref}, are identical, we obtain:

$$P_{DIR} = \frac{V_{bat} \cdot N_P}{m_r} \left(I_{Lref} + \alpha\left(T_0 + \frac{\Delta T}{m_r} - T_{ref}\right)\right) \quad (14)$$

For the cosine function of the zenith angle of a summer day, in Bucharest, Romania, we used an approximated relation (Milea, 2010):

$$\cos(z) = 0,2813 + 0,6488 \cdot \cos(0,2618 \cdot t + 2,8117), \quad (15)$$

where t is the legal time of Bucharest, expressed in hours.

If the battery is charged using a MPPT charger, approximating *FF=0.8* as constant throughout the day (Fig. 5.), we get:

$$P_{MPP} = V_m \cdot I_m = FF \cdot V_{OC} \cdot N_P \cdot I_{SC} \cdot \eta = FF \cdot V_{OC} \cdot N_P \cdot \eta \cdot \frac{G}{G_{ref}}\left(I_{Lref} + \alpha\left(T - T_{ref}\right)\right), \quad (16)$$

where η =*0,95* is the efficiency of the MPPT charger.
From (9) and (16), we obtain:

$$P_{MPP} = \frac{FF \cdot V_{OC} \cdot N_P \cdot \eta}{m_r}\left(I_{Lref} + \alpha\left(T_0 + \frac{\Delta T}{m_r} - T_{ref}\right)\right) \quad (17)$$

The instant ratio of powers in the two cases will be:

$$\frac{P_{MPP}}{P_{DIR}} = \frac{FF \cdot V_{OC} \cdot \eta}{V_{bat}} = \frac{FF \cdot \eta \cdot a \cdot V_T \cdot N_S \cdot \ln\left(1 + \frac{I_L}{I_0}\right)}{V_{bat}} = \frac{\eta \cdot a \cdot FF \cdot N_S}{V_{bat}} V_T \cdot \ln\left(1 + \frac{I_L}{I_0}\right) \quad (18)$$

Based on these relations we calculate the powers supplied by a solar panel to the battery by direct connection (P_{DIR}), and through a MPPT charger (P_{MPP}). All calculations are made for a cloudless summer day. The two powers and their ratio are plotted in Fig. 7.

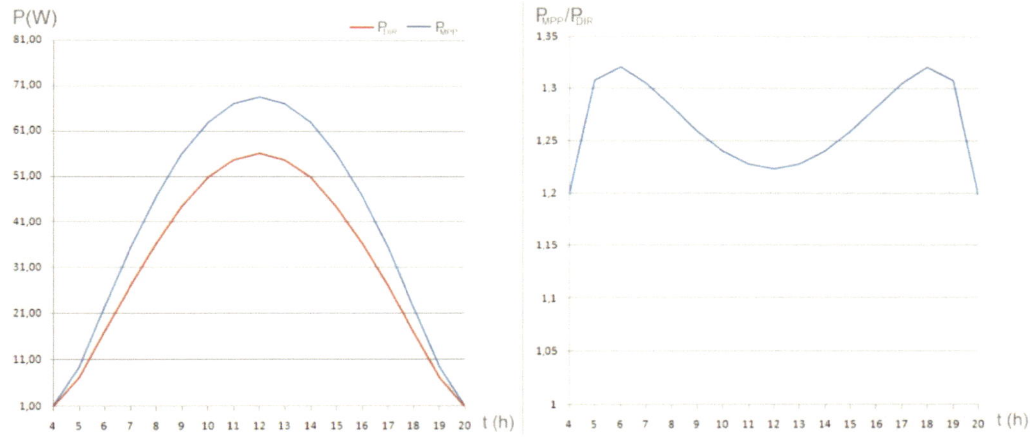

Fig. 7. Direct power and MPPT power (left) and the power ratio (right)

If we evaluate the daily average power gain of using MPPT, starting from average hourly values, we get:

$$\frac{\overline{P_{MPPs}}}{\overline{P_{DIRs}}} = 1,2 + \frac{2 \cdot \left(\frac{0,11}{2} + 0,11 + \frac{0,01}{2} + 6 \cdot 0,02 + \frac{6 \cdot 0,1}{2}\right)}{16} = 1,2 + \frac{0,59}{8} = 1,274 \quad (19)$$

So in summer, MPPT circuits provide an average increase of 27.4% transferred powers.
Making the same calculations for the winter, we get the following chart:

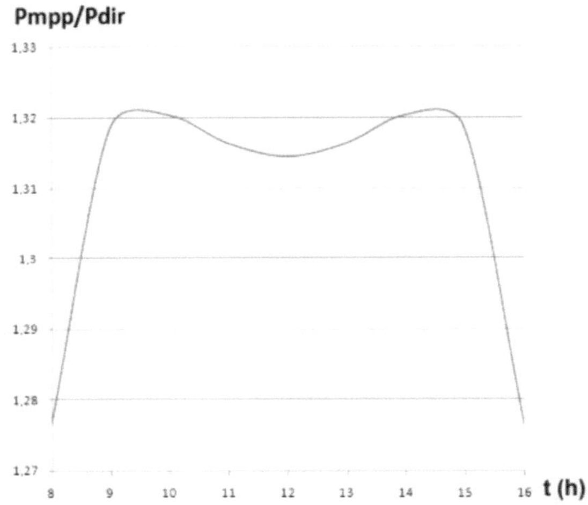

Fig. 8. Power ratio under winter conditions

If we evaluate the average MPPT power gain for winter, we get:

$$\overline{\frac{P_{MPPw}}{P_{DIRw}}} = 1{,}276 + \frac{2\cdot\left(\frac{0{,}04}{2} + 2\cdot 0{,}04 + 2\cdot\frac{0{,}003}{2} + 0{,}04\right)}{8} = 1{,}276 + \frac{0{,}143}{4} = 1{,}312\,, \quad (20)$$

It follows an average power gain of 31.2%, higher than in summer.
Therefore, the minimum power gain, obtained by MPP tracking, is 27.4%.

4. Proposed MPPT algorithms

The maximum power point (MPPT) tracking algorithms use the I(V) characteristic and the P(V) characteristic. All over the world there are many studies concerning the maximum power point (MPP) tracking. The performance of various types of MPPT algorithms (Chenni et al., 2007), (Faranda et al., 2007), (Santos et al., 2006), (Hui, 2008) is always measured by precision tracking of MPP and responsiveness to changes in the power curve.
In several studies we have addressed MPP tracking algorithms (Zafiu et al., 2009) and have designed new solutions, original circuits and applications for this purpose (Milea, 2010).
To determine the MPP we used previous mathematical relationships to create the algorithms presented in this section.
The proposed MPP tracking algorithms stands upon the relation *dP/dV=0* and involves the continuous adjustment of impedance adaption circuit (increasing or decreasing). Initially we choose two points so that *dP/dV<0,* respectively *dP/dV>0,* and the MPP is estimated to the medium value of these two points. Then, while *dP/dV>err,* the distance between points will be decreases and the MPP will take again the medium value of these points. The principle of MPP determination follows two phases. Firstly there is the measurement of three successive points (according to *V*) in the coordinates of I and V. Secondly, starting from these three points, the control circuit will decide the next adjustment to achive the MPP.

4.1 General presentation of the algorithms principle

This algorithm considers the values of three points (the last measured values) and Δ the distance between points. The measured value is in the middle and the other points are equidistant positioned on the left and on the right side of middle point (Fig. 9.).

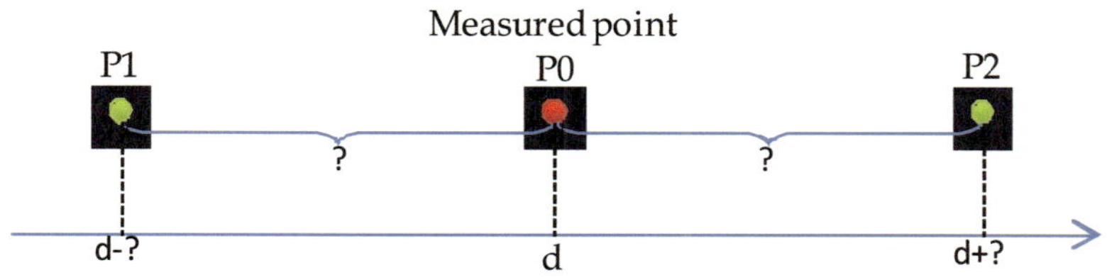

Fig. 9. Point's relative position

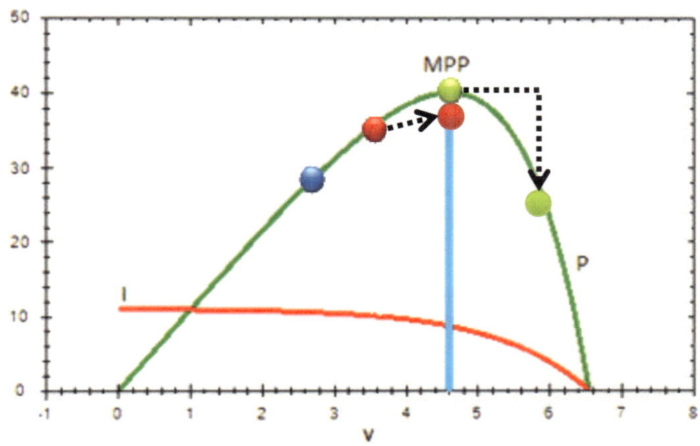

Fig. 10. Example of adjusting *d* and *Δ* so that MPP is between *d-Δ* and *d+Δ*

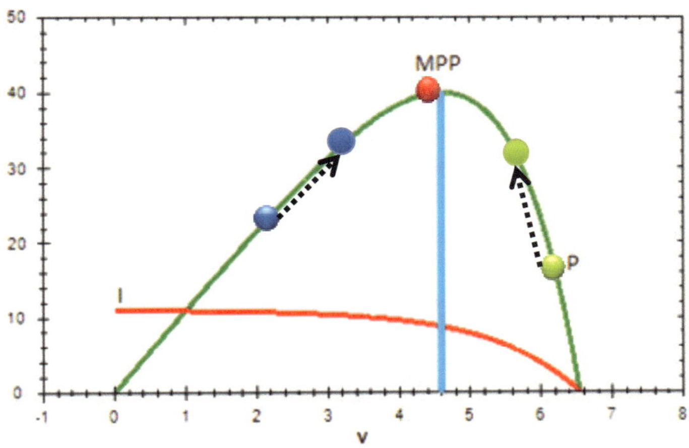

Fig. 11. Example where *Δ* is progressively reduced

It is considered that the three points are equidistant. Variable *d* corresponds to the value of the middle range (maximum power point value). Boundary values of the range are given by *d-Δ* (first point) and *d+Δ* (last point).
At each step of the algorithm, values of *d* and *Δ* are adjusted so that the middle point is located on top of the curve describing the power given by the photovoltaic module. If the points come to be on the same slope of the curve, distance *Δ* is increased so that move the middle point at the peak of the curve. If MPP was not peaked and the three points are in the pattern $P_1 < P_0$ and $P_2 < P_0$ then Δ will be reduced.
The domain of values used to represent voltage points is building using an n-bit representation. The number of bits determines the adjustment quantum and the algorithm error. The intermediate values couldn't be represented.

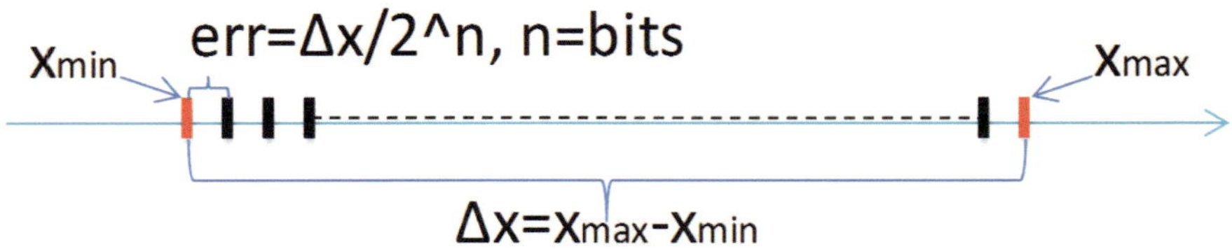

Fig. 12. The principle of building range of values using the n-bit representation

The values and errors are computed by starting from minimum and maximum values. The interval is spitted into 2^n intervals. The subinterval lentgh is $err = (x_{\max} - x_{\min}) / 2^n$. This length is also used as error value.
Following studies and tests, we designed two original algorithms to determine the maximum operating point.

4.2 Algorithm with three equidistant points

This algorithm is applied to a photovoltaic cell module in which the two cells are used as ends of range. All calculations are made with some calculation error fixed in advance.
Detailed algorithm is as follows:

- computing the point of short circuit and the range in which to find this point: as long as the voltage of the second edge is less than zero, the first edge (point) will receive the value of the second edge (point), and the second edge will receive a higher value; as long as the "distance" between the two points previously obtained is higher than the acceptable error, one of the two points will close to the other point with the average of points' values, finally achieving the short circuit point.
- compute the open circuit point and the range in which to find this point: as long as the voltage of the second point is higher than zero, the first point will receive the second point value and the second point will receive another point value; as long as the "distance" between the two points is higher than the acceptable error, the two points will approach each other with the average of points' values, thereby achieving the open circuit point;
- Determine the maximum power point as follows: *d* is initialized with the average of short circuit and open circuit values, Δ is initialized with the difference between *d* and the point of short circuit; the three points are taken to determine the MPP's and are initialized with: *d-Δ* to the point P_1, *d* to the point P_0, *d+Δ* to the point P_2; if the "distance" between P_1 and P_0 in addition to the "distance" between P_0 and P_2 is greater than the calculation error, are treated the following cases:

- If $P_1 < P_0$ then
  - If $P_0 < P_2$ then $d = d + \Delta$ and $\Delta = \Delta * 2$;
  - If $P_0 = P_2$ then $d = d + \Delta$ and $\Delta = \Delta * 1.5$;
  - Else $\Delta = \Delta / 2$;
- If $P_1 = P_0$ then
  - If $P_0 < P_2$ then $d = d - \Delta$ and $\Delta = \Delta * 1.5$;
  - If $P_0 = P_2$ then $\Delta = \Delta * 2$;
  - Else $d = d - \Delta$ and $\Delta = \Delta / 2$;
- Else
  - If $P_0 < P_2$ then $\Delta = \Delta * 2$;
  - If $P_0 = P_2$ then $d = d - \Delta$ and $\Delta = \Delta * 1.5$;
  - Else $d = d - \Delta$ and $\Delta = \Delta * 1.5$;

After these calculations are made the following adjustments: $d = \max(d, P_{sc} + err)$, $d = \min(d, P_{oc} - err)$, $\Delta = \min(\Delta, \min(d - P_{sc}, P_{oc} - d))$, and points P_1, P_2 and P_0 take the appropriate values for $d - \Delta$, $d + \Delta$, respectively d.

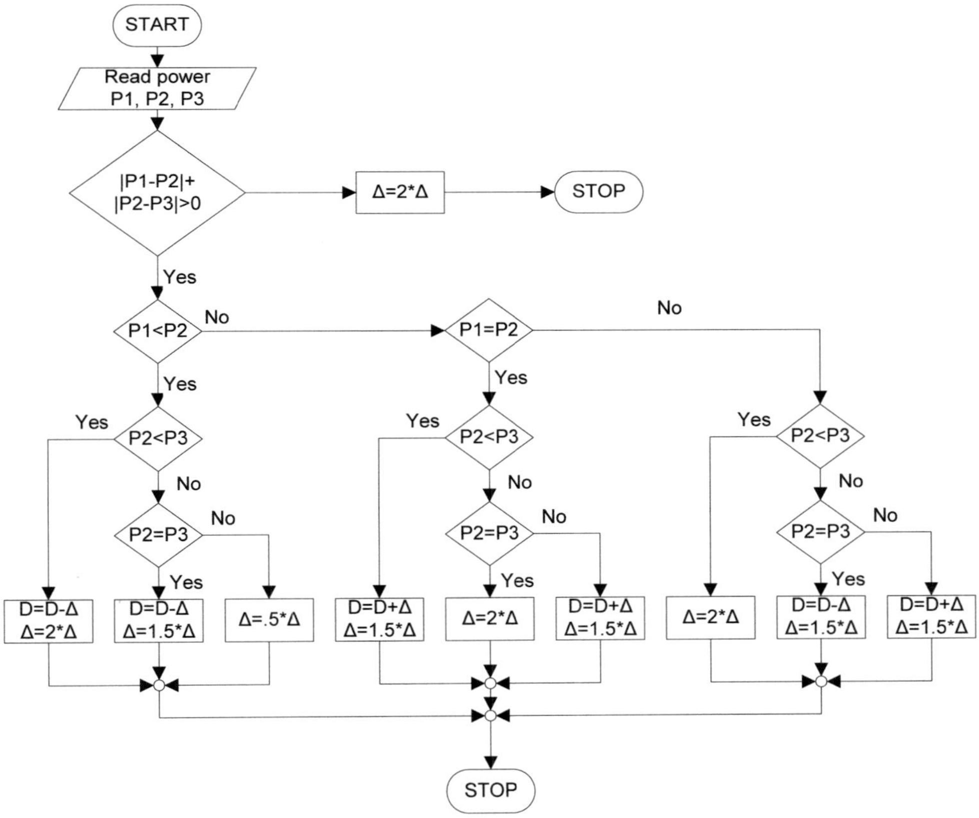

Fig. 13. The algorithm for MPP's calculation with three points

4.3 Computing MPP with a three dynamic step method

The proposed method consists in the adjustment of the pulse width which control the DC/DC converter, with dynamic step, based on the last three previous obtained successive (I, V) pairs. By comparing the values of that pairs, the control algorithm decides the step and direction of the next adjustment.

We consider that the first points are equidistant. Value D corresponds to the middle point. First point corresponds to value D-Δ and last point corresponds to the value D+Δ. At each step, the values D and Δ are adjusted.

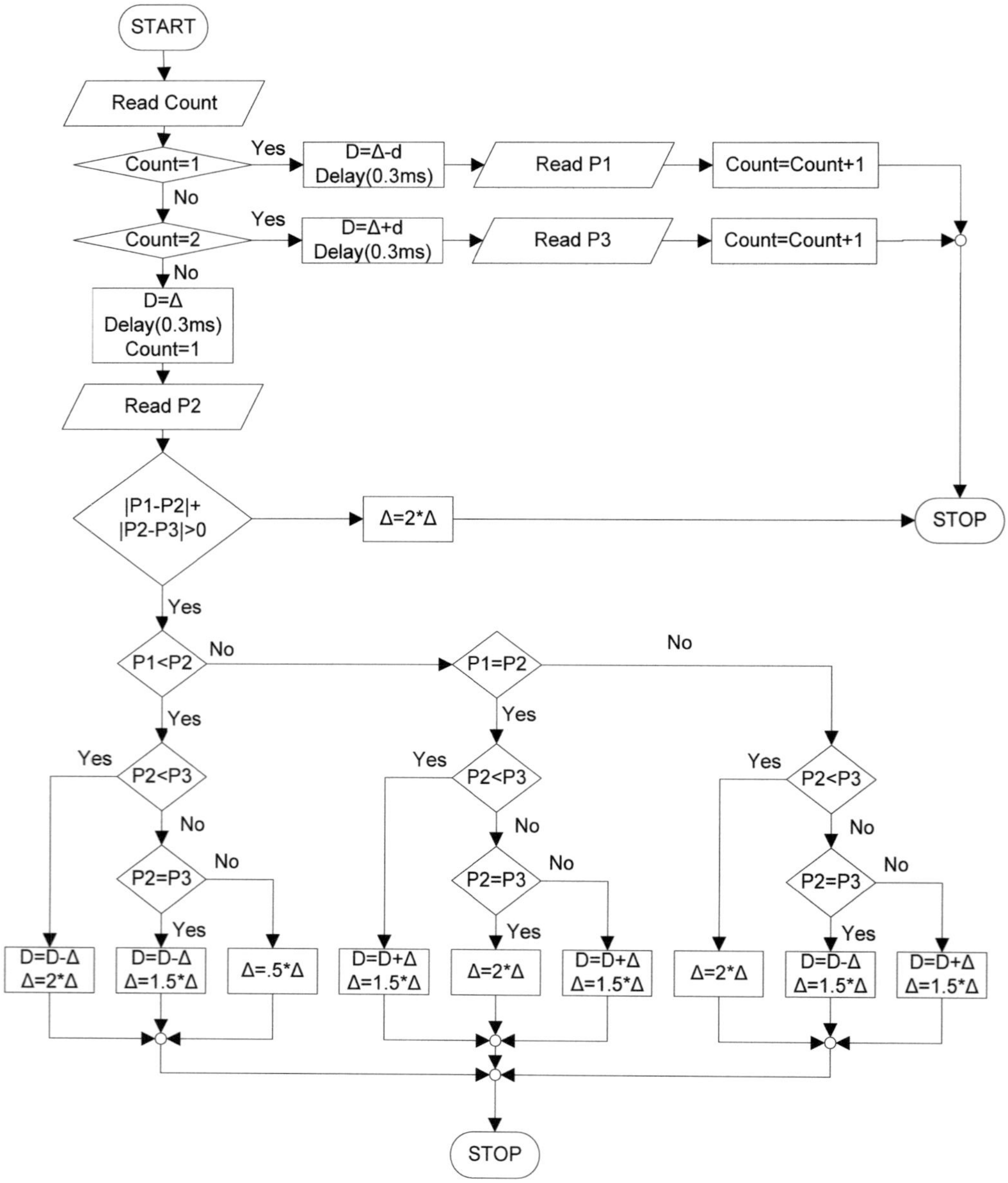

Fig. 14. The algorithm with three dynamic distanced points

5. Software application designed for mathematical model simulation and algorithm testing

For the algorithms testing we conceived and realized a software application for the modeling and simulation of cells and solar panel. Fig. 15. presents the interface for application settings, and Fig. 16. presents an example of characteristics tracing.

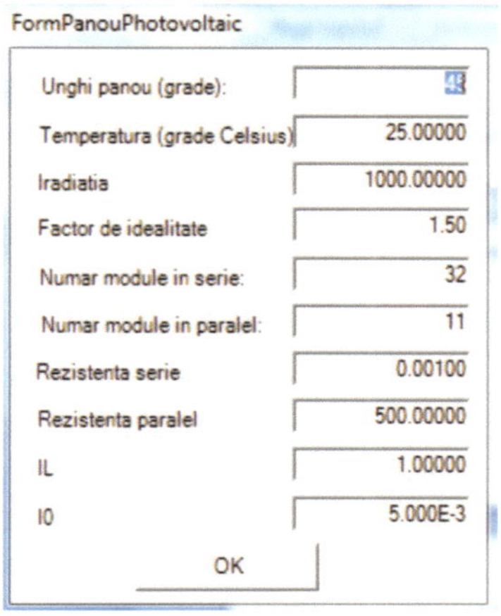

Fig. 15. Panel settings

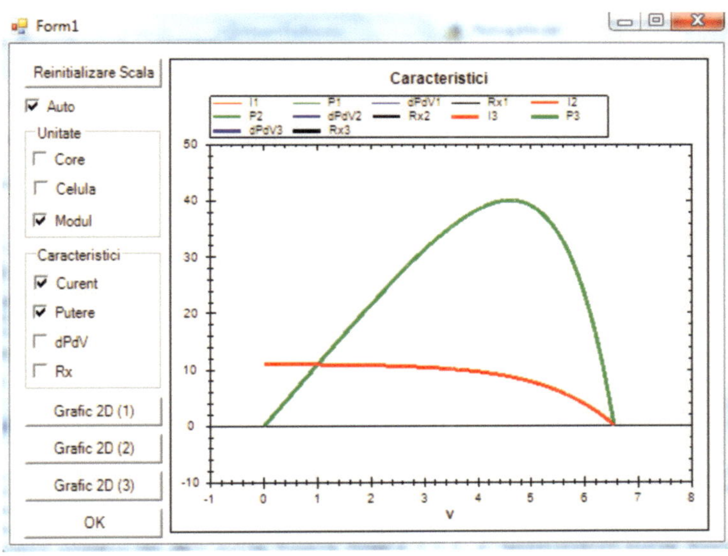

Fig. 16. The characteristic *I(V)* and *P(V)*

C# was used to design the application. The application was designed to be integrated, optionally, into an application for assessment of solar radiation intensity (Milea, 2010).

The first module represents the part in which is developed the MPP tracking algorithm for a photovoltaic cell (graphs were made using "ZedGraph library" - a set of classes written in C # for drawing 2D graphics based on arbitrary data sets; these classes provide a high degree of flexibility in that almost every aspect of the graph may be modified; ZedGraph includes a

"User Control" interface allowing editing of "drag and drop" type in the forms of Visual Studio, plus access from other languages such as C + + and Visual Basic).
Main classes implement photovoltaic cell behavior and algorithms for finding maximum power point (classes "PV" and "Algorithm").
Here you can set the angle of the panel, the temperature (if you want to see the effect of temperature on the photovoltaic cells characteristic), the solar irradiation (irradiation effect on the characteristic), the ideality factor, the number of cells connected in series, the number of cells connected in parallel, the series resistance and the parallel resistance of the equivalent circuit.
Current-voltage and power-voltage characteristics of the panel are shown in Fig. 16.
Buttons - 2D Graphics (1), 2D Graphics (2) and 2D Graphics (3) helped us to achieve the effect of temperature, irradiation and other graphs.

6. Simulations and comparative results

On the software simulation, we tested the algorithm presented in Fig. 13, on 8, 10 and 12 bits. The results are compared with the results of other algorithms and all of them are compared with the results of an ideal MPPT. The MPP tracking algorithm on 8 bits was used as reference to compare the algorithms variants. The bits number determines the discrete values that the duty cycle D can take. The newly developed algorithm uses a second variable Δ, for duty cycle adjustment, which values are dependent on the bits number, too.
The two variables can take the following values:

$$\{1/2^B, 2/2^B, 3/2^B, ..., 2^B - 1/2^B,\} \tag{21}$$

were B is the used number of bits.
As any tracking algorithm has no problem in estimating the maximum power if the variations are very small or non-existent (corresponding to a sunny day without clouds), the images show the behavior of our algorithm in the case of a cloudy day, where relatively high variations in G and T.
Figure 17 shows the energy output of our cell array by using our tracking algorithm on 8, 10 and 12 bits compared to the literature 8 bits method (Santos et al., 2006). The predicted duty cycle values will get the device very close to the computed theoretical maximal power point, obtaining tracking efficiencies near the maximum (100%).

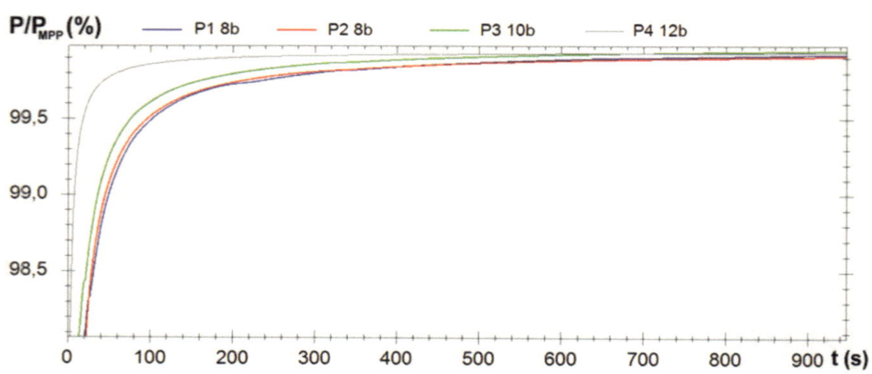

Fig. 17. The algorithm behavior: energy output
(Algorithm result for MPP calculation using $dP/dV=0$ relation)

There is a small loss in energy in our algorithm for the first ~20ms until it tracks the correct values. This difference is only felt once at the algorithm's startup and is recuperated very rapidly (in ~2 seconds our algorithm becomes more efficient: 99,94% at 8 bits, 99,98% at 10 bits, while literature algorithm (Santos et al., 2006) is at 99,935%).

Fig. 18.a) shows the power output of our cell array by using our tracking algorithm on 8, 10 and 12 bits compared to the (Santos et al., 2006) method on 8 bits and Fig. 18.b) shows a detail of the power tracking behavior.

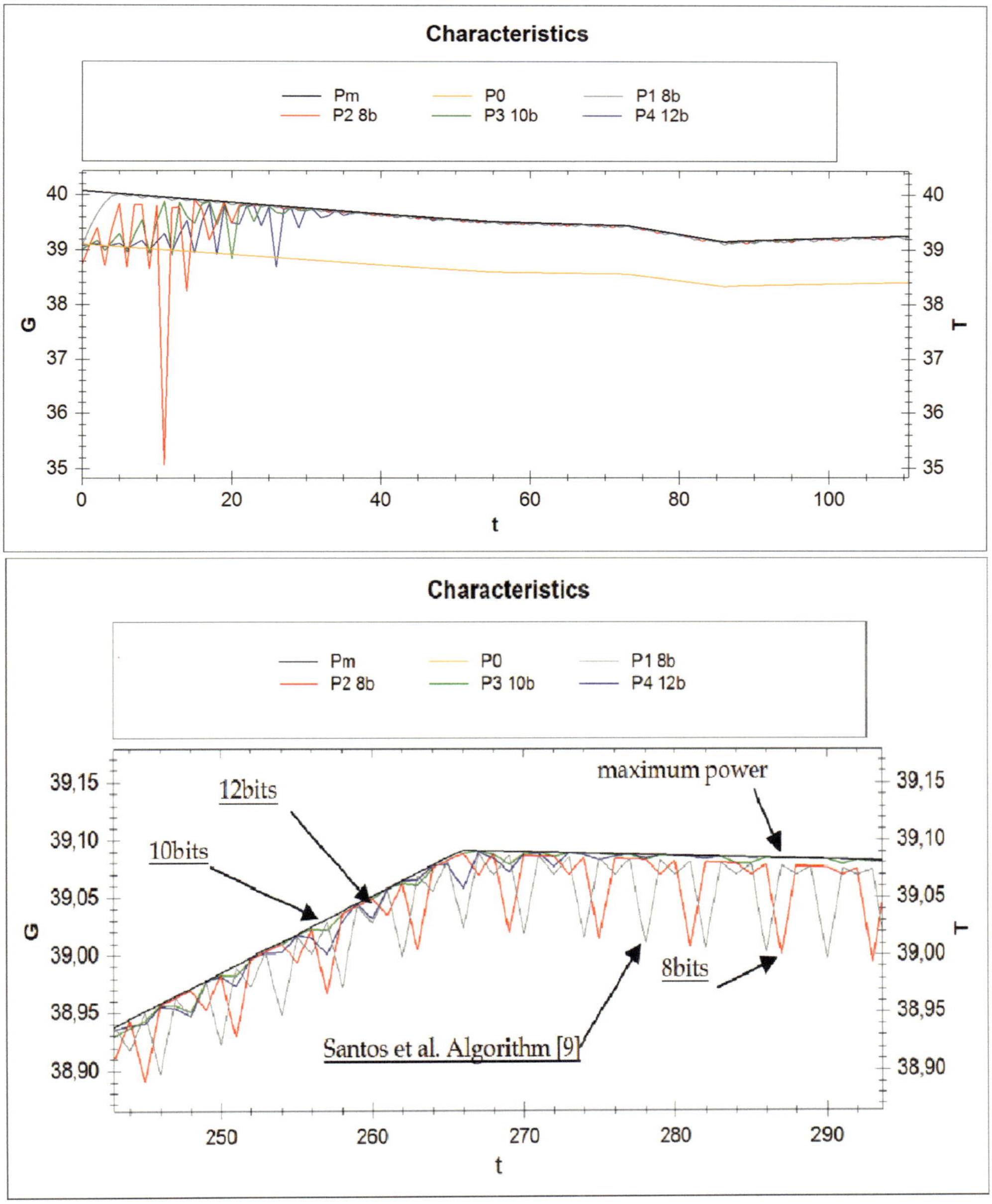

Fig. 18. The algorithm behavior: power output (a) and detail on power tracking (b)

The theoretical maximal power as seen in the figure is around 40 W. Our algorithm behaves best on 10 and 12 bits by staying as close as possible to the maximum power, while on 8 bits our algorithm's performance is still comparable to (Santos et al., 2006) algorithm.
Fig. 19. shows that the algorithm needs about 50 ms (less than 15 steps) to calibrate the variables D and Δ.

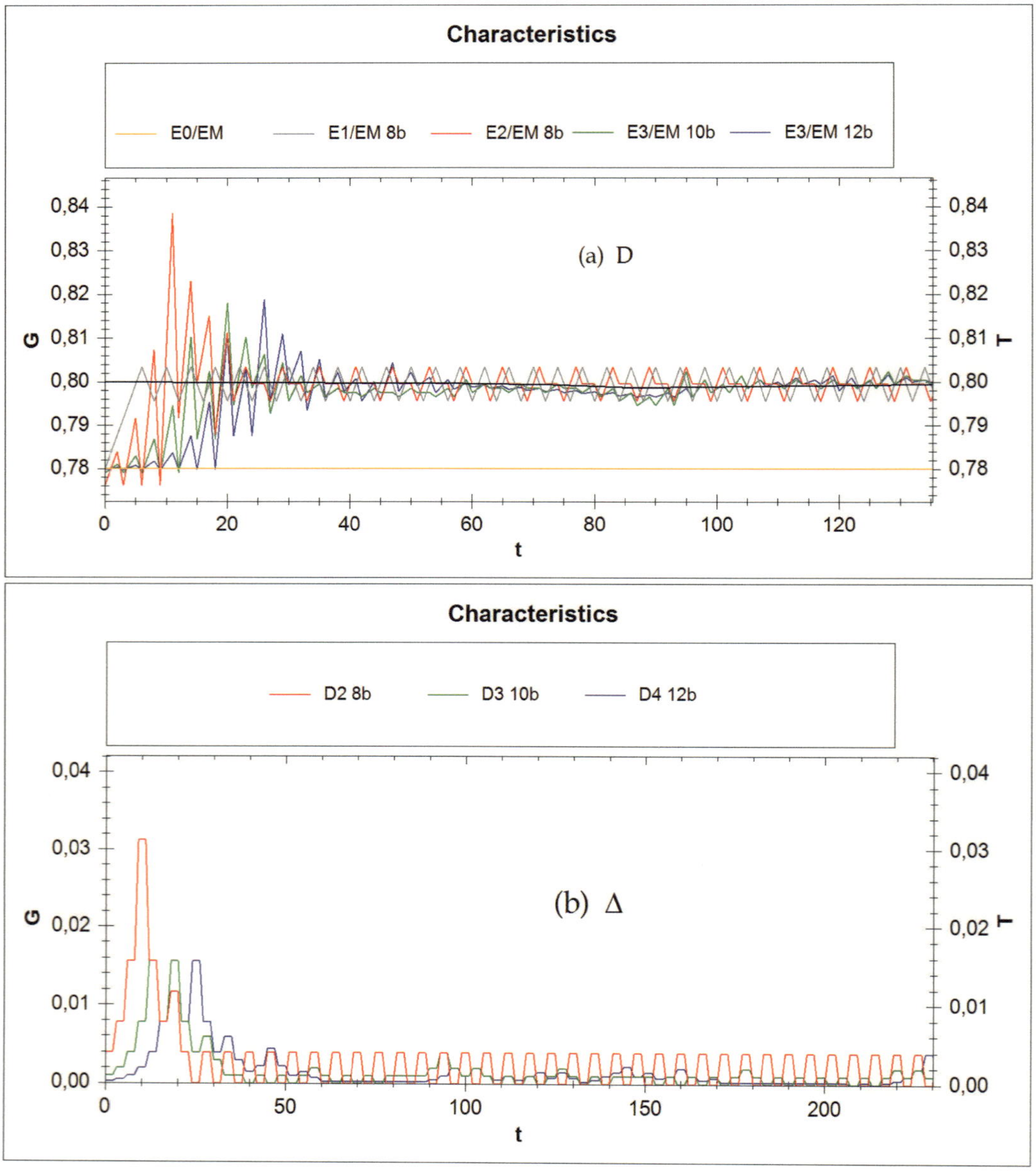

Fig. 19. The algorithm behavior: (a) D variation and (b) Δ variation

Using this algorithm we reached the following result, represented graphically:

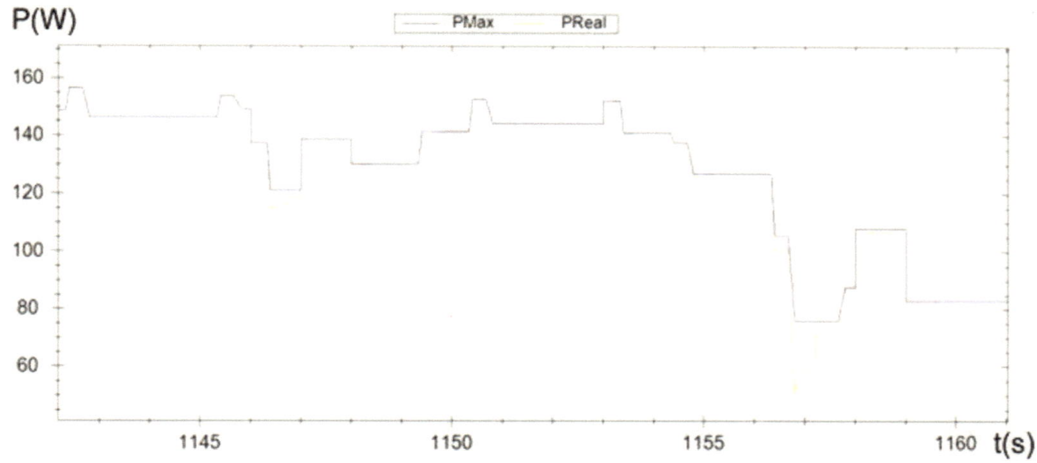

Fig. 20. Algorithm result for MPP's calculation with three points

Note that the power produced (P_{Real}) seeks the maximum power (P_{Max}) with a delay of 5 seconds. The algorithm needs 5 seconds to calibrate the variables Δ and d. The difference between the captured power and the available power is less than 5%.

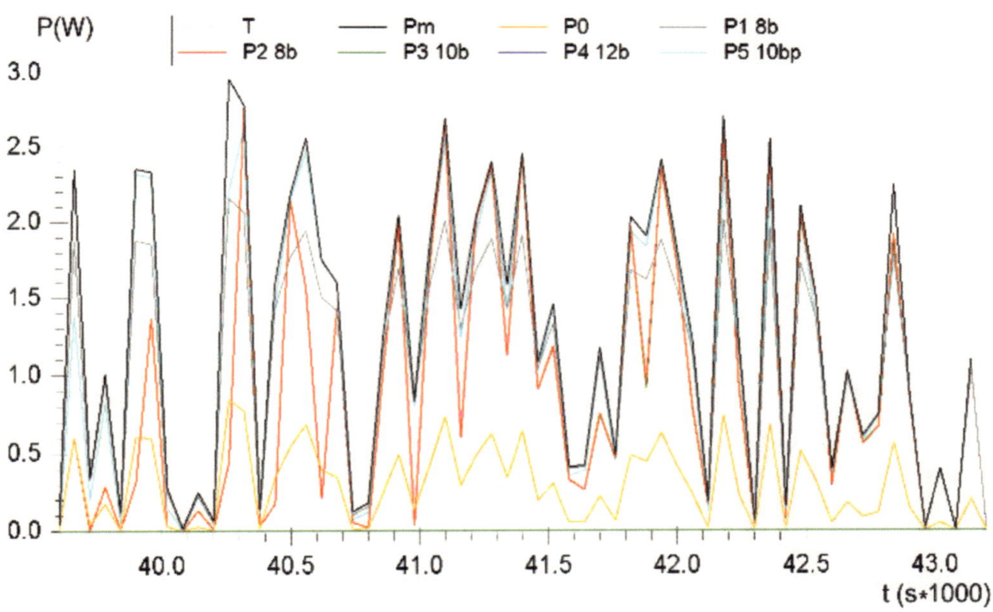

Fig. 21. Tracing of the maximum power through different algorithms, for high lightness and temperature variations

We made test implementations of this algorithm, on 8, 10 and 12 bits, and we compared their results with a set obtained from an 8 bits reference algorithm (P_{18b}) (Santos et al., 2006) and with one without MPPT (P_0), for very fluctuant meteorological conditions (simulated by our application - figure 5).

MPP determination using the reference algorithm (Santos 2007) is shown in Fig. 22.

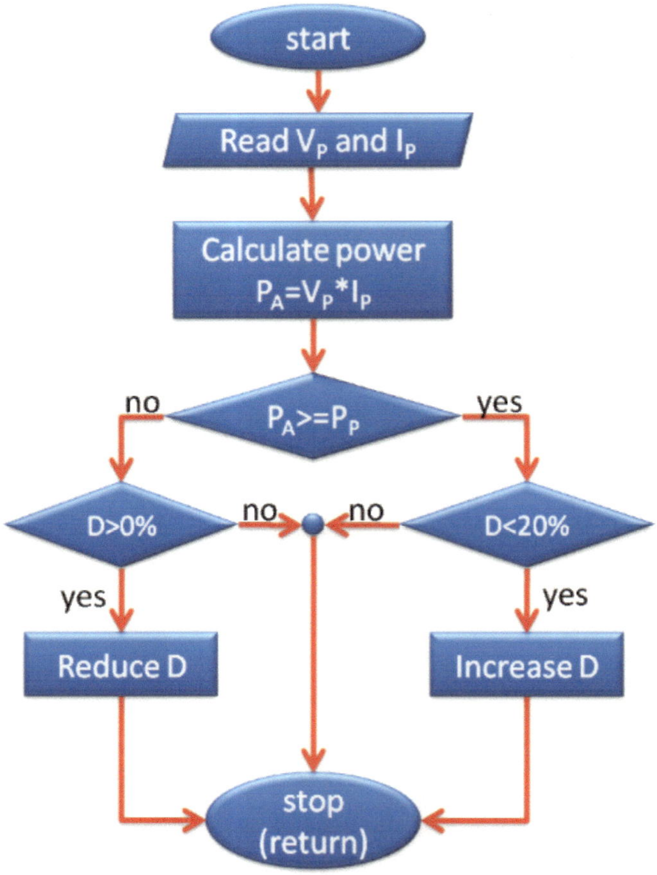

Fig. 22. J.L. Santos algorithm

7. Comparative determination between experimental and simulated results

To experimentally determine the advantages of MPPT chargers over the simple ones, we measured the output voltages and currents of these chargers, connected to two identical batteries, of *12V/80Ah*, and finally compared them with simulated results.

7.1 Experimental comparative determination between simple and MPPT charging

To measure load currents, we connected a resistor (R_I=*0.1Ω*) in series with each battery to the ground. Voltages on the resistors and on the battery-resistor assemblies were monitored for 16 hours with a four channel data aquisition device. To prevent the charge controllers to become out of service, following full charging of the batteries, we occasionally coupled load resistors at the battery terminals.

We realized the same measurements in two summer days: one predominant clear and a cloudy one.

We used the diagram below:

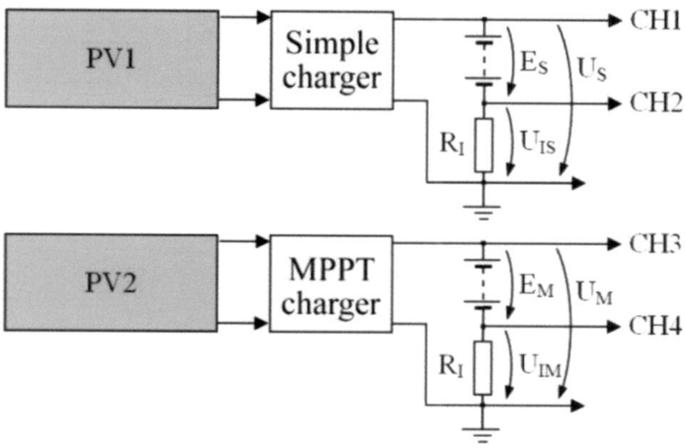

Fig. 23. The measurements diagram

The results are averaged over the hours shown in the tables below, together with calculated currents and powers. We also calculated the difference of power supplied by two regulators, *Pm-Ps*. The two powers and their difference were plotted in the figures below, and hourly average power data is presented in the table.

Ora	Us (V)	Uis (V)	Is (A)	Ps (W)	Um (V)	Uim (V)	Im (A)	Pm (W)	Pm-Ps (W)
4	11,92	0,05	0,50	5,96	11,97	0,07	0,70	8,38	2,42
5	11,93	0,13	1,30	15,51	12,09	0,17	1,70	20,55	5,04
6	12,12	0,20	2,00	24,24	12,33	0,25	2,50	30,83	6,59
7	12,33	0,28	2,80	34,52	12,35	0,35	3,50	43,23	8,70
8	12,47	0,33	3,30	41,15	12,53	0,40	4,00	50,12	8,97
9	12,70	0,39	3,90	49,53	12,69	0,47	4,70	59,64	10,11
10	13,04	0,42	4,20	54,77	12,90	0,49	4,90	63,21	8,44
11	13,13	0,42	4,20	55,15	13,04	0,48	4,80	62,59	7,45
12	12,98	0,44	4,40	57,11	12,91	0,51	5,10	65,84	8,73
13	12,90	0,38	3,80	49,02	12,86	0,44	4,40	56,58	7,56
14	12,80	0,33	3,30	42,24	13,01	0,39	3,90	50,74	8,50
15	12,63	0,29	2,90	36,63	13,10	0,33	3,30	43,23	6,60
16	12,60	0,20	2,00	25,20	13,14	0,24	2,40	31,54	6,34
17	12,57	0,13	1,30	16,34	12,87	0,15	1,50	19,31	2,96
18	12,36	0,05	0,50	6,18	12,61	0,07	0,70	8,83	2,65
19	12,30	0,01	0,10	1,23	12,49	0,01	0,10	1,25	0,02
20	12,26	0,00	0,00	0,00	12,32	0,00	0,00	0,00	0,00

Table 1. Electrical quantities measured and calculated for a clear day of summer

Energy provided by the two panels with associated chargers was determined as average hourly values and has daily values:

$$E_{S1} = 514{,}78Wh\,, \qquad E_{M1} = 615{,}86Wh \tag{22}$$

By using the MPPT charger, we get the following power gain:

$$A_1 = \frac{E_{M1}}{E_{S1}} = \frac{615,86}{514,78} = 1,1964 \tag{23}$$

So, in a bright summer day, tracking the MPP produced an energy surplus of 19.64%.

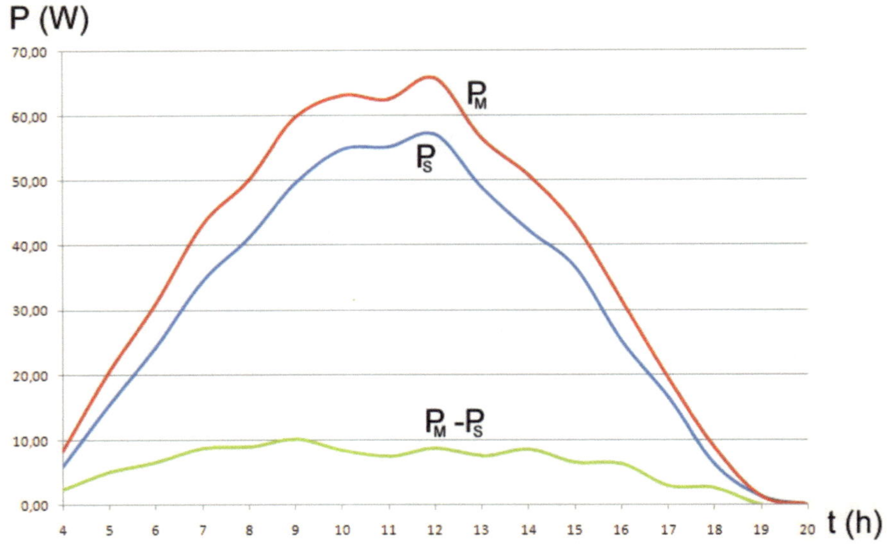

Fig. 24. Power transferred into batteries in a clear summer day

Ora	Us (V)	Uis (V)	Is (A)	Ps (W)	Um (V)	Uim (V)	Im (A)	Pm (W)	Pm-Ps (W)
4	11,90	0,03	0,30	3,57	11,93	0,03	0,30	3,58	0,01
5	11,86	0,06	0,60	7,12	11,99	0,07	0,70	8,39	1,28
6	12,00	0,08	0,80	9,60	12,18	0,10	1,00	12,18	2,58
7	12,14	0,09	0,90	10,93	12,12	0,12	1,20	14,54	3,62
8	12,27	0,13	1,30	15,95	12,29	0,16	1,60	19,66	3,71
9	12,47	0,16	1,60	19,95	12,42	0,20	2,00	24,84	4,89
10	12,76	0,14	1,40	17,86	12,58	0,17	1,70	21,39	3,52
11	12,89	0,18	1,80	23,20	12,78	0,22	2,20	28,12	4,91
12	12,71	0,17	1,70	21,61	12,61	0,21	2,10	26,48	4,87
13	12,64	0,12	1,20	15,17	12,57	0,15	1,50	18,86	3,69
14	12,61	0,14	1,40	17,65	12,80	0,18	1,80	23,04	5,39
15	12,46	0,12	1,20	14,95	12,91	0,14	1,40	18,07	3,12
16	12,49	0,09	0,90	11,24	13,01	0,11	1,10	14,31	3,07
17	12,49	0,05	0,50	6,25	12,78	0,06	0,60	7,67	1,42
18	12,34	0,03	0,30	3,70	12,57	0,03	0,30	3,77	0,07
19	12,30	0,01	0,10	1,23	12,49	0,01	0,10	1,25	0,02
20	12,27	0,01	0,10	1,23	12,33	0,01	0,10	1,23	0,01

Table 2. Electrical quantities measured and calculated for a cloudy summer day

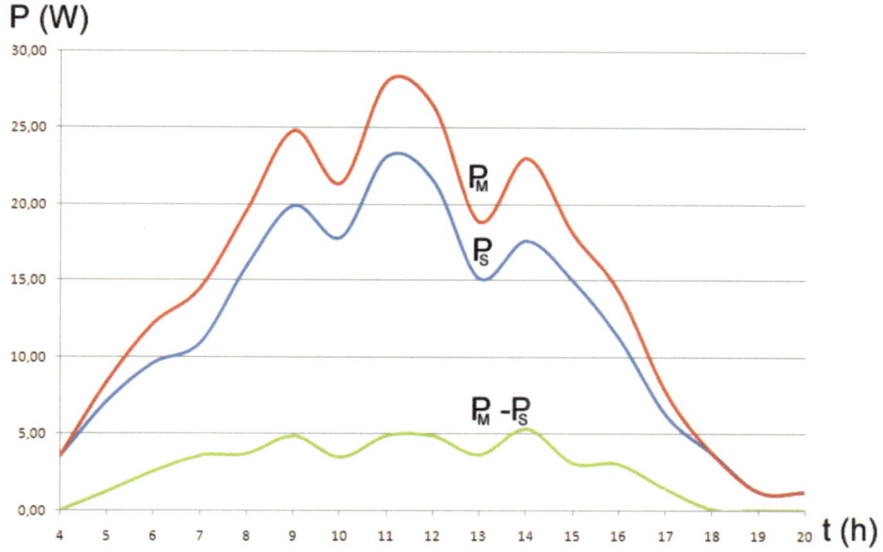

Fig. 25. The powers transferred to the batteries, in a cloudy summer day

The energy provided by the two panels during the cloudy day, via associated chargers, was computed based on average values of hourly and has daily values:

$$E_{S2} = 201{,}21Wh\,, \qquad E_{M2} = 247{,}38Wh \tag{24}$$

And the power gain caused by MPPT:

$$A_2 = \frac{E_{M2}}{E_{S2}} = \frac{201{,}21}{247{,}38} = 1{,}2295 \tag{25}$$

So, in a cloudy day of summer, using the MPPT charger, we obtain an energy surplus of 22.95%.

7.2 Experimental verification of the model and software

In parallel with the experimental determinations, we used the software application to estimate the energy increase caused by MPPT.

We modeled the *I*(*V*) characteristic of the practically used panels, using the following parameters, which was able to produce a characteristic very similar with the one given by panel catalog:

a	R_P (Ω)	R_S (mΩ)	I_L (A)	I_0 (μA)	N_S	N_P
1,5	500	1	1,1	1,04	42	5

Table 3. Modelling parameters

Using this program we modeled the few specific experimental determinations made previously. For this evaluation we chose areas of interest $G \in \{100 \div 900\}$ *W/m*^{2} and $T \in \{20 \div 60\}$ ^{0}C, sugested by the software.

The program window with the modelled *I(V)* characteristic is presented below:

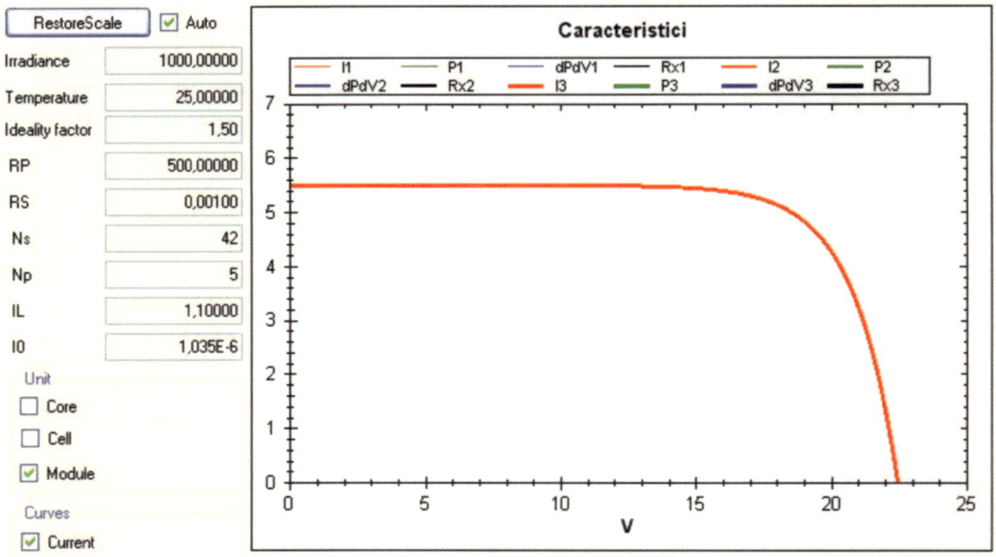

Fig. 26. The *I(V)* characteristic and model parameters of photovoltaic panel

P_{MPP} values are listed in the table below:

G \ T	**20°C**	**30°C**	**40°C**	**50°C**	**60°C**
100 W/m²	7,45	7,28	7,06	6,78	6,45
150 W/m²	11,67	11,45	11,15	10,76	10,31
200 W/m²	16,02	15,76	15,39	14,92	14,34
250 W/m²	20,48	20,18	19,75	19,19	18,51
300 W/m²	25,01	24,68	24,20	23,56	22,78
350 W/m²	29,60	29,26	28,73	28,02	27,13
400 W/m²	34,25	33,89	33,32	32,54	31,57
450 W/m²	38,95	38,58	37,97	37,13	36,06
500 W/m²	43,69	43,31	42,66	41,76	40,62
550 W/m²	48,47	48,08	47,41	46,45	45,22
600 W/m²	53,28	52,89	52,19	51,18	49,87
650 W/m²	58,13	57,74	57,01	55,95	54,57
700 W/m²	63,01	62,62	61,87	60,76	59,30
750 W/m²	67,91	67,53	66,75	65,60	64,07
800 W/m²	72,84	72,46	71,67	70,47	68,88
850 W/m²	77,79	77,42	76,62	75,38	73,72
900 W/m²	82,77	82,41	81,59	80,31	78,58

Table 4. The MPP power under different weather conditions

In this table we have marked with a gray background the P_{MPP} values which are representative for the times of day in which experimental measurements were made.

For G and T values in the table above, we determined the appropriate powers for direct connection of panel to the battery, P_S, from $P(V)$ characteristics, as shown graphically below.

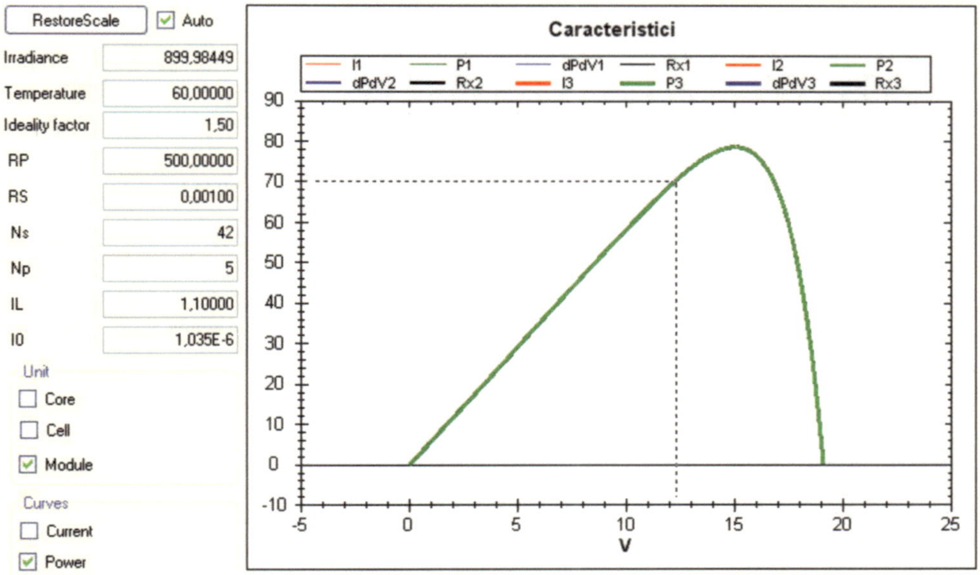

Fig. 27. The $P(V)$ characteristic of the panel model, at $T=60^0C$ and $G=900W/m^2$

The determined values (P_S), together with those of MPP (P_{MPP}), are listed in the table bellow:

G (W/m²)	T (⁰C)	P_{MPP} (W)	P_S (W)	P_{MPP}/P_S
100	20	7,45	6	1,2417
300	30	24,68	20	1,2340
500	40	42,66	36	1,1850
700	50	60,76	52	1,1685
900	60	78,58	71	1,1226

Table 5. MPP power, direct power and power gain, in different weather conditions

From this table we see that in most cases, power ratio is similar to situations experimentally determined, with errors under *1%*, correspondent to average illuminations of $500W/m^2$ (sunny day), respectively $300W/m^2$ (cloudy day).

8. Conclusions

A general approach on modeling photovoltaic modules is presented. The proposed MPP tracking computational method is based on a DC/DC converter control with an original algorithm. The theoretical evaluations of the MPPT advantages, based on the proposed model, sugest that the power gain, obtained by MPP tracking, is higher than 27%.

We developed, simulated and evaluated two MPPT algorithms, based on presented model. The proposed algorithms are independent of the used solar panel type, which could be considered a "black box".

The algorithms starts from the known fact that the power curve at given conditions have a global maximum. This algorithm is useful for any kind of electrical energy source, where this condition is covered. The performance of the algorithm is limited by the precision of measurement blocks and by the word size of used microcontroller.
Algorithm has relatively short response times, covering the difference between the extracted power and MPP in less than 15 steps (~ 50ms). These results are very good compared with some widely-adopted MPPT algorithms (Faranda et al., 2007). The power loss depends on the working frequency of the control module.
We made some experimental measurements on an implemented MPPT charging circuit and a simple one. The resulted MPPT power gain was lower with about 7% than the theoreticaly one, due to different efficiency of the charging circuits. The experimental results were compared with the simulated ones, for the same conditions and panel parameters. This comparison reveals that the differences between experimental data and simulated characteristics were less than *1%*.
We intend to tune and develop the presented method to improve the response time and efficiency and to apply it to other kind of unconventional energy sources.

9. Acknowledgement

The presented research work was done within the SINERG project (PNCDI2 22-140/2008), funded by the Romanian National Center for Programs Management.

10. References

Chenni, R.; Makhlouf, M.; Kerbache, T.; Bouzid, A. (2007). A detailed modeling method for photovoltaic cells, *Energy*, Vol. 32, no 9, 2007, pp. 1724-1730, ISSN 0360-5442

Faranda, R.; Leva, S.; Maugeri, V. (2007). Comparative study of ten Maximum Power Point Tracking algorithms for Photovoltaic System, *U.P.B. Scientific Bulletin*, Series C, Vol. 69, No. 4, 2007, pp. 271-278, ISSN 1454-234x

Hui, J. (2008). *An Adaptive Control Algorithm for Maximum Power Point Tracking for Wind Energy Conversion Systems - MSc thesis*, Queen's University Kingston, December 2008, Ontario, Canada

Milea, L.; Franti, E.; Dragulinescu, M.; Oltu, O.; Dascalu, M. (2008). Residential photovoltaic energetic system, optimized with an FPGA based control unit, *Proceedings of The 3rd International Conference on ENERGY & ENVIRONMENT (EE'08)*, pp. 518-522, ISBN: 978-960-6766-43-5, February 2008, WSEAS, Cambridge

Milea, L.; (2010). *Sources, Systems and Circuits for Unconventional Energies - PhD Thesis*, University POLITEHNICA of Bucharest, March 2010, Bucharest, Romania

Santos, J.L.; Antunes, F.; Chehab, A.; Cruz, C. (2006). A maximum power point tracker for PV systems using a high performance boost converter, *Solar Energy*, Vol. 80, Issue 7, July 2006, pp. 772–778, ISSN: 0038-092X

Yang, H.; Zhou, W.; Lu, L.; Fang, Z. (2008). Optimal sizing method for stand-alone hybrid solar–wind system with LPSP technology by using genetic algorithm, *Solar Energy*, Vol. 82, Issue 4, April 2008, Pages 354-367, ISSN: 0038-092X

Zafiu, A.; Ionescu, V.; Bizon, N.; Ghita, C.; Oproescu, M. (2009). A detailed model for PV Simulation and MPP Tracking with three points, *Proceedings of the International Conference on ELECTRONICS, COMPUTERS and ARTIFICIAL INTELLIGENCE – ECAI' 09*, ISSN – 1843 – 2115, July 2009, Pitesti, Romania

10

Maximum Power Point Tracker Applied in Batteries Charging with Photovoltaic Panels

José António Barros Vieira[1] and Alexandre Manuel Mota[2]
[1]*Instituto Politecnico de Castelo Branco, Escola Superior de Tecnologia de Castelo Branco, Departamento de Engenharia Electrotécnica*
[2]*Universidade de Aveiro, Departamento de Electrónica Telecomunicações e Informática, Instituto de Engenharia Electrónica e Telemática de Aveiro*
Portugal

1. Introduction

Recently, the concern for environmental issue has been rising in the world such as global warming by exhausting carbon dioxide (CO2) and breaking of ozone layer by freon gas. On December 1997, during the Kyoto Conference on Climate Change (COP3), it was agreed that by the year 2012 the developed countries would reduce at least 5% of the green house gases compared with year 1990 (Riza et al., 2003). Moreover, the global energy shortage and the need for sustainable energy systems enforce the development of power supply structures that are based mainly on renewable resources.

Photovoltaic (PV) system is gaining increased importance as a renewable source due to advantages such as the absence of fuel cost, little maintenance and no noise and wear due to the absence of moving parts. But there are still two principal barriers to the use of photovoltaic systems: the high installation cost and the low energy conversion efficiency.

A PV panel is a non-linear power source, i.e. its output current and voltage (power) depends on the terminal operating point. The maximum power generated by the PV panel changes with the intensity of the solar radiation and the operating temperature. To increase the ratio output power/cost of the installation it is important that PV panel operates in the maximum output power point (MPP).

This chapter describes a lead acid battery charger using a PV panel with high efficiency. It starts by introducing the PV panel characteristics and describing the DC/DC converter need to implement the MPPT algorithm. This is the most important part of this work and shows how to charge and discharge correctly a lead acid battery.

The developed prototype uses the perturbation and observation (P&O) MPPT algorithm with the objective to maximize energy storage in the battery. The MPPT algorithm is integrated in one of the main stages of charge of lead-acid batteries making an autonomous and intelligent system that can be used to feed any remote load or small application. It is very important to respect the correct battery charge curves because it will prolong its correct operation and live (Vieira and Mota, 2009).

The P&O MPPT algorithm is used to control the maximum transfer power from a PV panel to the battery. This algorithm is executed by a microcontroller using the PV voltage and current measurements to define the duty cycle of a pulse width modulation signal applied to the DC/DC converter. The schematic and design of the DC/DC converter are explained. The DC/DC converter used is of the SEPIC topology because it easily adapts any PV output voltage to any battery input voltage.

One of the most frequently used MPPT methods is the perturbation and observation algorithm, although this algorithm has some converging problems with rapidly insolation changes. This problem can be solved using the solution presented in (Sera et al., 2006). In this work, PV voltage and current are measured in the middle of the sampling interval, making possible to determine if the verified changes are due to perturbation algorithm or to shadows that cover the PV panel. Another popular MPPT algorithm is the incremental conductance method (IncCond) (Hussein et al., 1995). The authors developed the incremental conductance MPPT algorithm avoiding the drawbacks of the P&O MPPT algorithm. It is based on the fact that the derivative of the output power P with respect to the panel voltage V is equal to zero at the MPP. The solar panel's P-V characteristics presented in figure 2 show further that the derivative is greater than zero to the left of the MPP and less than zero to the right of the MPP. This algorithm shows that enough information is gathered to determine the relative location of the MPP by measuring only the incremental and instantaneous panel conductance's dI/dV and I/V, respectively.

In this work, perturbation and observation MPPT algorithm was chosen, due to its simplicity and to its low computational power needs (Knop, 1999).

To implement a correct charger for a lead-acid battery, the correct charge curves are presented and the four charging stages are described. This work also explains that if the correct charge curves are respected in the charge periods the correct operation period of time of the lead acid battery will be longer. In the developed system only two of the four charging stages are implemented. The first stage is avoided by the discharge supervisor implemented algorithm and the fourth stage is made extending the third stage. The discharge supervision algorithm simply monitors the battery voltage and if it goes bellow a minimum value the load is disconnected waiting for a new charge.

Finally, experimental results of the performance of the designed P&O MPPT algorithm, corresponding to the 2° stage of lead acid battery charge, are presented and compared with the results achieved with the direct connection of the PV panel to the battery.

The remainder of this chapter is organized as follows: section 2 presents the photovoltaic panel characteristics, section 3 presents the DC/DC SEPIC schematic and design, section 4 shows the perturbation and observation maximum power point tracking algorithm. Section 5 presents the algorithm proposed to the different stages of the lead-acid battery charging process, section 6 shows the implemented prototype board, section 7 discuss the experimental results of charging with the P&O MPPT algorithm and with out it, ending with the conclusions presented in section 8.

2. Solar array characteristics

The maximum power point of a solar panel changes in accordance with changes in the solar irradiance intensity, angle and panel temperature. The typical characteristic curves of current versus voltage, power versus voltage at different levels of solar irradiation and power versus voltage at different temperatures, are illustrated in figure 1, figure 2 and figure 3 respectively.

Figure 1 illustrates the operating characteristic of the panel under several given solar insolations. It consists of two regions: one is the current source region, and the other is the voltage source region. In the voltage source region, the internal impedance of the panel is low. That region is the right side of the current-voltage curve. The current source region, in which the internal impedance of the panel is high, is at the left side of the current-voltage curve. The MPP of the panel is located at the knee of the current-voltage curve. According to the maximum power transfer theory, the power delivered to the load is maximum when the source internal impedance matches the load impedance. Thus, the impedance seen from the converter input side (can be adjusted by PWM control signal) needs to match the internal impedance of the panel if the system is required to operate at or near the MPP of the solar array. If the system operates on the voltage source region (namely low impedance region) of panel characteristic curve, the panel terminal voltage will collapse (Hua & Lin, 2003).

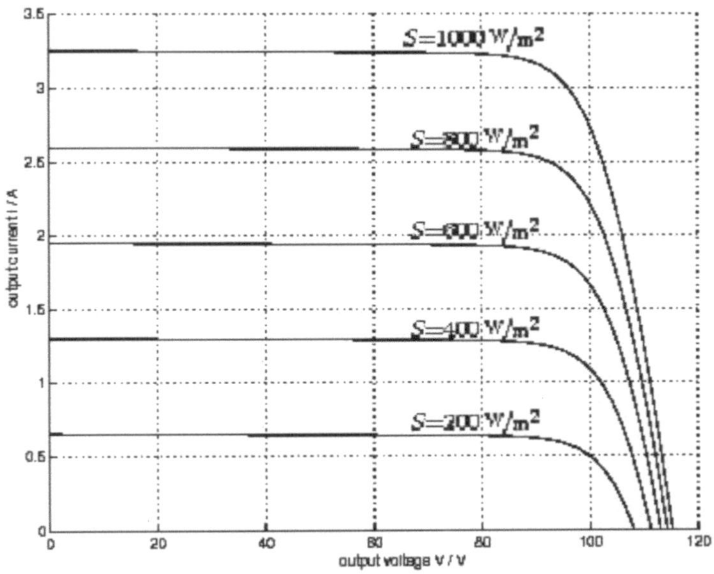

Fig. 1. I-V characteristics of a photovoltaic panel for different values of irradiance S at a temperature of 25 °C.

From figure 2 and Figure 3, it can be observed that each curve has a maximum power point, which is the optimal point for the efficient use of the panel. This point depends of the values of irradiance and working temperature. The main function of a MPPT is to adjust the panel output voltage to a value which the panel supplies the maximum energy to the load (Torres, 1998).

Thus, a DC/DC converter will be used to match the source internal impedance with the load impedance achieving the MPP. The applied MPPT algorithm will be explained in detailed in section 4.

3. DC/DC SEPIC converter

To implement the P&O MPPT algorithm a SEPIC (Single-Ended Primary Inductance Converter) is used. This DC/DC type of converter is an increasingly popular topology,

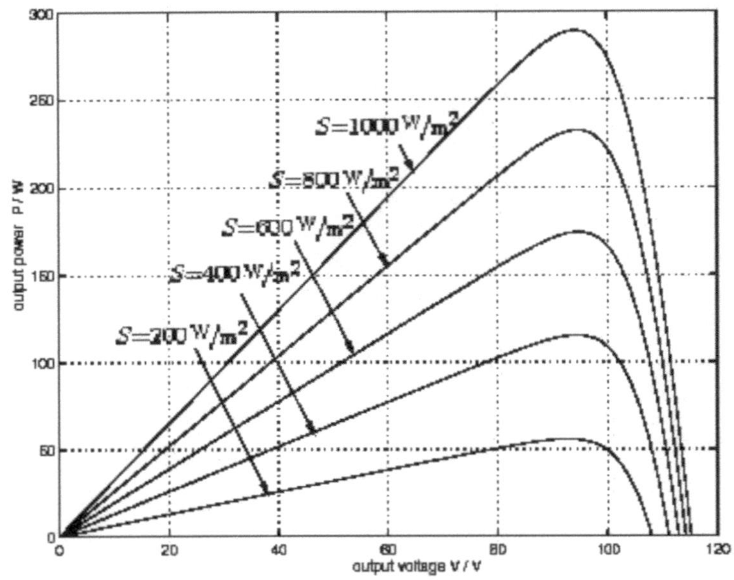

Fig. 2. P-V characteristics of a photovoltaic panel for different values of irradiance S at a temperature of 25 °C.

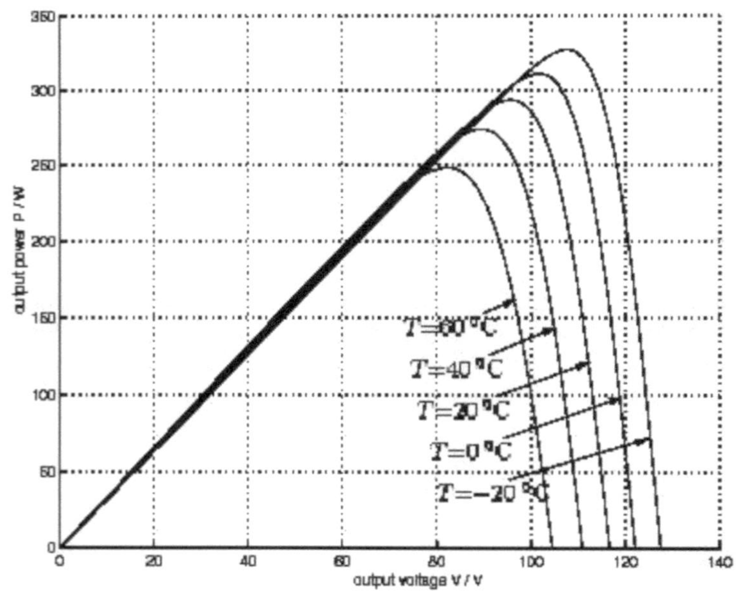

Fig. 3. P-V characteristics of a photovoltaic panel for different values of temperature T at irradiance of 1000W/m².

particularly in battery powered applications, because the input voltage can be higher or lower than the output voltage. This topology presents obvious design and working advantages. In this work, for the implementation of the maximum power point tracker the SEPIC, working in continuous conduction mode, is used as the power-processing unit. The PWM is controlled with a switching frequency of 125 kHz that actuates the Mosfet switch M1. The power flow is controlled by adjusting the on/off duty-cycle. Figure 4 shows the schematic of the DC/DC converter implemented. It has one Mosfet, one diode, two inductances and tree capacitors.

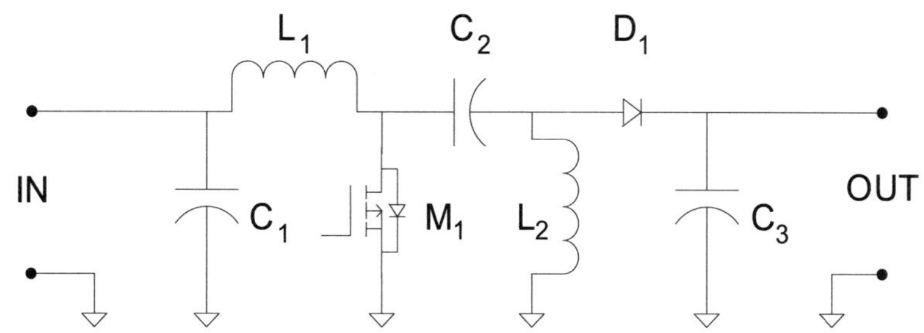

Fig. 4. SEPIC DC/DC converter circuit.

Using a PV panel with the following characteristics: maximum power P_{max}=9.31 W, maximum voltage V_{mp}=17.4 V, maximum current I_{mp}=0.54 A, open circuit voltage V_{oc} = 21.2 V, short-circuit current I_{cc}=0.66 A, the DC/DC converter design starts with the selection of the two separate inductors L_1 and L_2. For a general working point with:

Input voltage (V_{in})	12V - 16V
Output (V_{out} & I_{out})	12V, 0.6A
Switching frequency (F_s)	125kHz
Expected efficiency	90%

First it is need to calculate the duty cycle; $D = V_{out}/(V_{out} + V_{in})$; The worst case condition for inductor ripple current is at maximum input voltage so; D = 12/(12 + 16) = 0.429;
Calculating the value of L_2:

$$V = L\ di/dt \tag{1}$$

Where V is the voltage applied to the inductor, L in the inductance, di is the inductor peak to peak ripple current and dt is the duration the voltage applied. Hence:

$$L = V.dt/di \tag{2}$$

$$dt = 1/F_s \times D \tag{3}$$

$$dt = 1/(125 \times 10^3) \times 0.428 = 3.42\ \mu s \tag{4}$$

$V = V_{in}$ during the switch ON time so:

$$L_2 = 16 \times (3.42 \times 10^{-6}/0.4) \tag{5}$$

$$L_2 = 136.8\ \mu H \tag{6}$$

Using the nearest preferred value would lead to the selection of a 150 μH inductor. It is common practice to select the same value for both input and output inductors in SEPIC designs although when two separate parts are being used it is not essential.
Having selected the inductance value we now need to calculate the required RMS and peak current ratings for both inductors. For input inductor L_1:

$$I_{rms} = (V_{out} \times I_{out})/(V_{in}\,(min) * efficiency) \tag{7}$$

$$I_{rms} = (12 \times 0.6)/(12 \times 0.9) = 0.667A \tag{8}$$

$$I_{peak} = I_{rms} + (0.5 \times I_{ripple}) \tag{9}$$

Although worst case ripple current is at maximum input voltage the peak current is normally highest at the minimum input voltage.

$$I_{ripple} = (V.dt)/L \tag{10}$$

$$I_{ripple} = (12 \times 3.42 \times 10^{-6})/\ 150 \times 10^{-6} = 0.27A \tag{11}$$

$$I_{peak} = 0.667 + 0.135 = 0.804A \tag{12}$$

So a 150μH, 0.667Arms and 0.804Apk rated inductor is required. For the output inductor L_2:

$$I_{rms} = I_{out} = 0.6A \tag{13}$$

$$I_{ripple} = (16 \times 3.42 \times 10^{-6})/\ 150 \times 10^{-6} = 0.365A \tag{14}$$

$$I_{peak} = 0.6 + 0.182 = 0.782A \tag{15}$$

So a 150μH, 0.6Arms and 0.782Apk rated inductor is required.
Finally, the SEPIC components used are:

L_1 = 150μH and L_2 = 150μH	with I_{sat} = 4.0A because if V_{in} = 6V I_{rms} will be near to the double of the calculated current for V_{in} = 12V
C_1 = 47 μF, C_2 = 47μF ,C_3 = 47μF	with V_{max} = 40V
M_1 of I_{max} = 4.0A	switching Mosfet at 125kHz
D_1 of I_{max} 4.0A	Schotkky Diode

4. The P&O maximum power point tracker algorithm

The P&O MPPT is one of the so called 'hill-climbing' methods, which are based on the fact that in case of the V-P characteristic, on the left of the MPP the variation of the power against voltage dP/dV > 0, while at the right, dP/dV < 0 (Weidong & Dunford, 2004).
In Figure 2, if the operating voltage of the PV panel is perturbed in a given direction and dP/dV > 0, it is known that the perturbation moved the panel's operating point toward the MPP. The P&O algorithm would then continue to perturb the PV panel voltage in the same direction. If dP/dV < 0, then the change in operating point moved the PV panel away from the MPP, and the P&O algorithm reverses the direction of the perturbation (Hohm & Ropp, 2000).

The main advantage of the P&O method is its implementation simplicity and its low computational demand. However it shows some limitations, like oscillations around the MPP in steady state operation, slow response speed, and tracking in wrong way under rapidly changing atmospheric conditions (Hohm & Ropp, 2000), (Femia et al., 2004), (Brambilla et al., 1999). To reduce the presented limitations it will be useful to use a small sampling rate. In this work it was used a sampling rate of 100 ms.
Using a SEPIC with current and voltage resistance sensors illustrated in Figure 5, the P&O MPPT algorithm was implemented. The algorithm needs only the PV voltage and current information to work correctly, the battery voltage and current information will be used to control the battery charging stages and supervise its discharge.

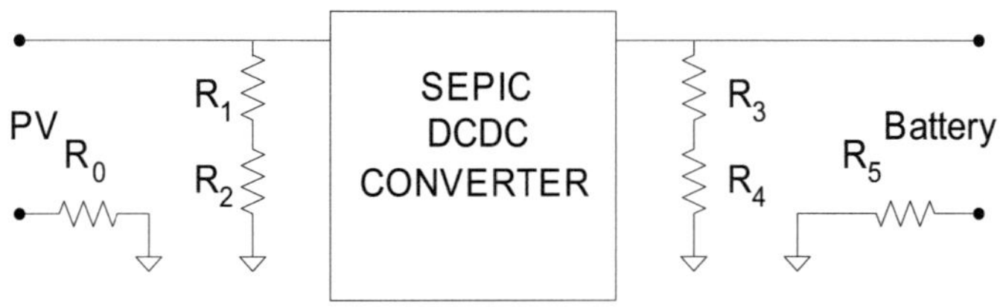

Fig. 5. Voltage and current resistive sensors for MPPT and battery charging algorithm.

$R_0=R_5=0.01\Omega$ for current measurements and $R_1=R_3=910k\Omega$ and $R_2=R_4=150k\Omega$ for voltage measurements.
The flow chart of the P&O MPPT implemented algorithm is illustrated in figure 6. The parameter K is the step given to the PWM signal. This parameter can vary depending of the working point of the DC/DC converter. To get a faster convergence we need big K values and to avoid big oscillations near to the MPPT working point we need small K values.
The P&O MPPT control algorithm is implemented in a microcontroller (ATTINY861V) that has ten 10-bits analogue-to-digital (A/D) converters and two fast PWM mode signals with 10-bits of resolution. The control circuit compares the PV output power before and after a change in the duty-cycle of the DC/DC converter control signal and acts in conformity. It is expected that the algorithm shows a small constant oscillation in the MPP working point inherent to the is working principle. The PWM_old is the sample of the PWM signal in the previous iteration of the algorithm and ΔP_{PV} is the variation of the delivered power to the battery.

5. Battery charging algorithm

The complete battery charging demands a complex control strategy, in which it would be possible to charge the battery, between its limits, in the faster possible way because the daily period of energy generation of the PV panel is limited (Galdino & Ribeiro, 1994).
To achieve a fast, safe and complete battery lead-acid charge, some of the manufacturers recommend dividing the charging process in four stages that are designated by: (i) trickle charge, (ii) bulk charge, (iii) over charge and (iv) float charge (Hesse, 1997) and (Rosemback, 2004). Figure 7 show the curves of current and voltage applied to the battery during a correct charging cycle.

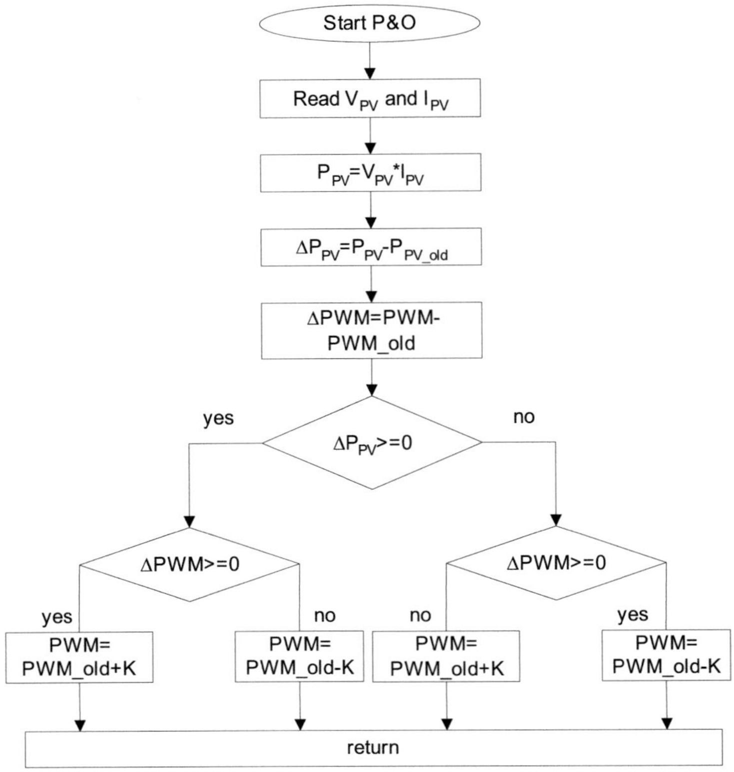

Fig. 6. P&O MPPT algorithm.

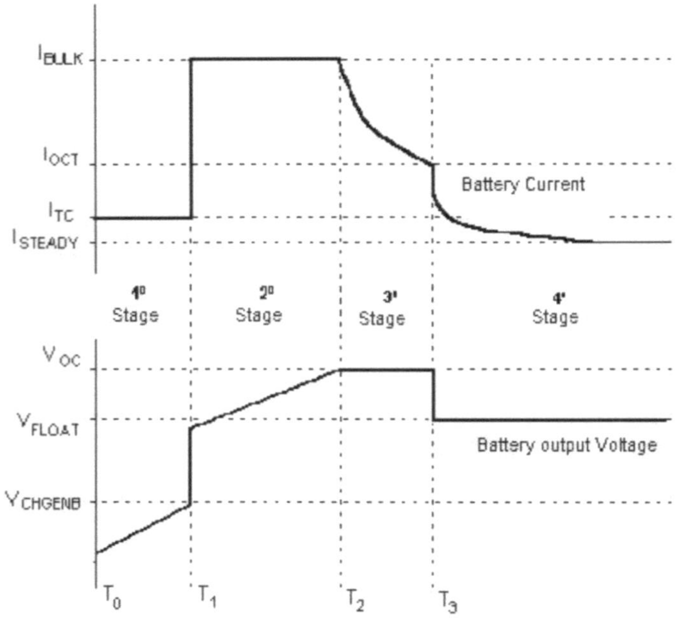

Fig. 7. Current and voltage curves in the four stages of battery charge.

5.1 Trickle charge - 1º stage (from T0 to T1)

This first stage is active when the battery voltage is below the value $V_{CHGENB.}$ This voltage value, specified for the manufacturers, shows that the battery arrives at its critical discharge capacity. In this condition the battery should receive a small charge current defined by I_{TC} that has a typical value of C/100 where C is the normal battery capacity with a 10 hours charging process.

This small current I_{TC} is applied until the battery voltage reaches the value of V_{CHGENB}. This stage also avoids that some accident could happens in the case of the one battery element is in curt circuit, therefore if this really happens the battery voltage will not grow and then the battery charging process does not pass to the next stage.

5.2 Bulk charge - 2º stage (from T1 to T2)

After the battery voltage reaches the value V_{CHGENB} it should be applied to the battery a constant current I_{BULK}. The I_{BULK} is the maximum charge current that battery supports without a big water losing, and its value is specified by the manufacturers. This current is applied until the battery voltage reaches the maximum value of over charge voltage, defined by V_{OC}, and also specified by the manufacturers. In this stage the prototype implemented board will run the P&O MPPT algorithm but the I_{BULK} is never exceeded. The maximum power of the PV panel should be correctly chosen.

5.3 Over charge - 3º stage (from T2 to T3)

During this stage the control algorithm should regulate the battery voltage V_{OC} until the complete charge has been reached. When the charging current fall down to a pre-established value I_{OTC} and the voltage stays in the value V_{OC}, the charge process should go to the next, and final, stage. The value of I_{OCT} is around 10% of the I_{BULK}.

5.4 Float charge - 4º stage (from T3 until the end)

In this stage the control algorithm will apply to the battery a constant voltage V_{FLOAT} which is specified by the battery manufacturers. This voltage is applied to the battery with the objective of avoiding its auto-discharge. During the discharging process the battery voltage will fall down and when it reaches 0.9 V_{FLOAT} the control algorithm will execute again the 2º stage providing the I_{BULK} current.

The control algorithm only returns to the 2º stage if the PV panel is capable of delivering energy. If it is not the case the battery will continue the discharge process. If the voltage goes below the value V_{CHGENB} the control algorithm should restart the charging process in 1º stage as soon as the PV panel is capable of delivering energy.

In this work some simplifications have been introduced in the implementation of the four different charging stages of a lead-acid battery. The 1º stage was not implemented because the discharge batteries voltage, with this prototype board, does not go below V_{FLOAT}. The possible applied load is disconnected from the battery by the control algorithm avoiding reaching the critical discharge.

The 4º stage was not implemented but the 3º stage is continued until the charge current reach I_{STEADY} and finally the charging process is ended. When the PV panel has energy to delivery and the battery voltage is below the V_{OC}, the control algorithm executes the 2º stage.

The battery charging algorithm can be seen in figure 8. Values V_b and I_b are the battery voltage an delivered current and T_b is the battery temperature. The maximum value of the V_{OC}

depends of the battery temperature. The temperature of the battery T_b is measured using a NTC temperature sensor and its linearisation is made in software using a conversion table. From figure 8 it is clear that only the 2° and the 3° stages are implemented from the four stages proposed in (Hesse, 1997) and (Rosemback, 2004).

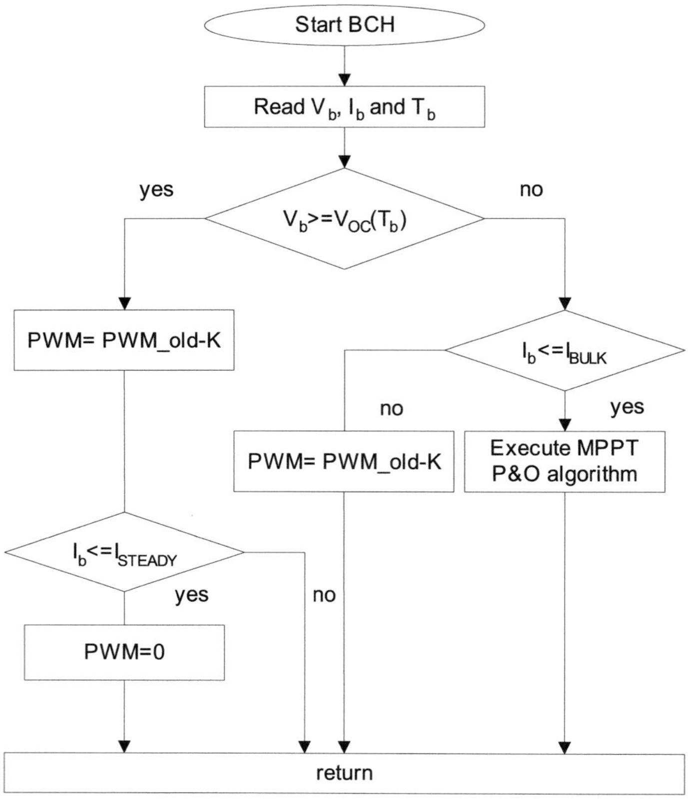

Fig. 8. Battery charging algorithm with two main stages.

6. Implemented prototype board

The implemented prototype board is illustrated in figure 9. It can be seen the PV panel connection in the right side of the photo (IN) and the connection to the battery (B) and to the possible load (L) both in the left side.
The described charging process of lead-acid batteries is executed with the P&O MPPT algorithm integrated to make an autonomous system that can be used to feed any autonomous load application. This board is also prepared to feed led light autonomous signalisation systems and could be used in any other remote small application.
The board also monitors the discharge of the battery. There is a minimum battery voltage, depending of the battery temperature, that shouldn't be over crossed. If that happens the system disconnects the load until a new charge.

Fig. 9. Photo of the MPPT and battery charger prototype board.

7. P&O MPPT experimental results

The experimental results of battery charging using the P&O MPPT algorithm are divided in two separated tests each one divided in two phases. In the first phase the Photo Voltaic panel is directly connected to the battery element (first 85 samples) and in the second phase the panel is connected to the battery element using the developed board running the P&O MPPT algorithm (from samples 85 to the end).

In the first test a PV panel with a Pmax = 9.31 W (Vmp = 17.4 V, Imp = 0.54 A) connects to one lead-acid battery of V = 12 V (Imax = 7.5 Ah). In the second test the same PV panel is connected to a bank of four lead-acid batteries of V = 6 V (Imax = 1.8 Ah) connected in parallel. The tests results are illustrated in figures 10 and 11.

From first test it can be seen that charging the 12V battery with the direct connection of the PV panel to the battery, the absorbed power from the PV panel is around 7W and with the P&O MPPT algorithm the absorbed power from the PV panel is around 8W. The MPPT algorithm presents small oscillations around the maximum power point as expected. The algorithm takes about 60 samples to go from zero to the maximum power point.

From the second test results it can be seen that with the direct connection of the PV panel to the bank of four batteries the absorbed power from the PV panel is around 4.5W and charging the batteries with the P&O MPPT algorithm, corresponding to the second charge stage, the absorbed power from the PV panel is around 7.5W. The algorithm takes about 40 samples from zero to the maximum power point.

The experimental setup using the P&O MPPT always gives more delivered energy to the battery than the direct connection. The P&O MPPT has increased the PV panel capacity of supply energy in 12.5% using a 12V battery and 40% using four 6V batteries connected in parallel.

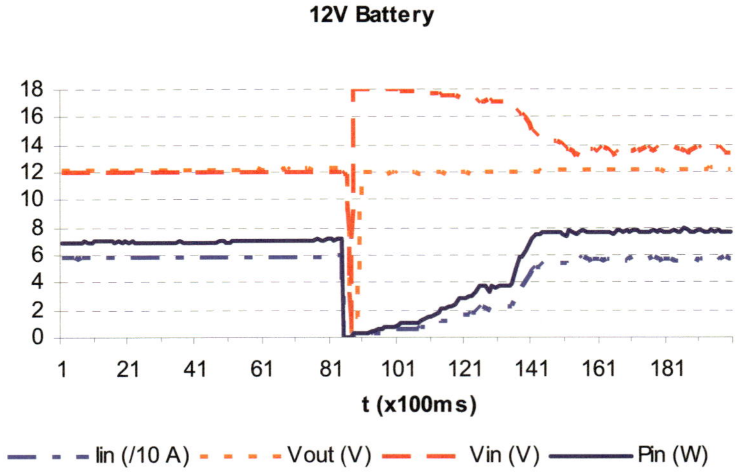

Fig. 10. Experimental results of the P&O MPPT algorithm power with a 12V battery (first phase direct connection and second phase using the developed board).

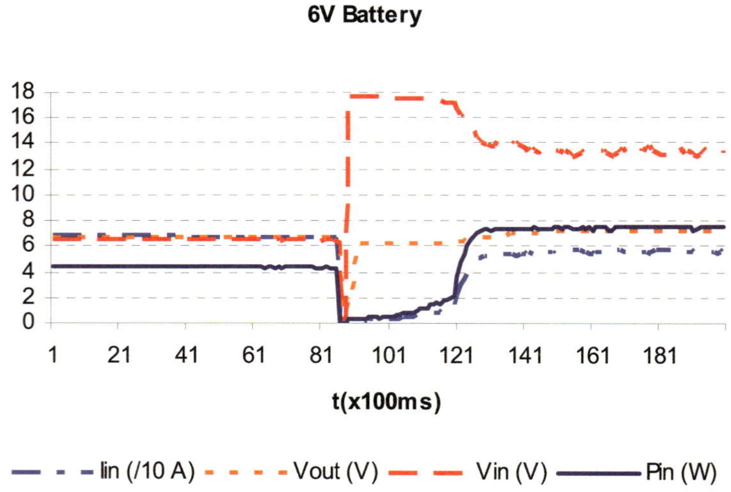

Fig. 11. Experimental results of the P&O MPPT algorithm power with four 6V batteries (first phase direct connection and second phase using the developed board).

8. Conclusions

This work presents a prototype board based in a small microcontroller that controls the lead acid battery charging process and also the correct use of the lead-acid battery supervising its discharge. The control algorithm executes the P&O maximum power point tracking function allowing, according to solar irradiance and temperature, transfer the maximum energy generated by photovoltaic panel to the battery. This P&O algorithm increase the efficiency power transference in comparison to systems that have not a MPPT (direct connection), reducing the size and the cost of the PV panel. The use of a SEPIC converter has some advantages because it easily adapts any PV output voltage to any input battery voltage as showed in the presented experiments.

This board enables the fast, safe and complete battery lead-acid charging process and also monitor its discharge. For future work the complete charging process should be analysed to compare with another system working with out P&O MPPT algorithm. From these results it is expect that the charging process using the MPPT algorithm will be faster and more efficient. These results will prolong for more time the correct operation of the lead acid battery.

For future work, it would be interesting to apply the P&O MPPT algorithm to a thermal solar panels to absorb the maximum thermal power from the irradiated solar energy. The thermal energy could be transferred to water tanks for future utilizations for domestic or industrial use. To implement this system we will need the measurement of the water flow and increase of the temperature of the water (output subtracted to input temperatures of the water of the thermal solar panel) to calculate the water thermal power that is the product of the two referred measurements. Finally, with the control of the water flow it should be possible to impose the MPP in a solar thermal system optimizing the efficiency of the thermal energy transference.

9. References

Brambilla A., Gambarara M., Garutti A. and Ronchi F.(1999). New Approach to Photovoltaic Arrays Maximum Power Point Tracking. *Proceedings of Power Electronics Specialists Conference, PESC 99*, vol. 2, pp 632–637, 27 June-1 July 1999.

Femia N., Petrone G., Spagnuolo G. and Vitelli M. (2004). Optimizing Sampling Rate Of P&O MPPT Technique, *Proceedings of Power Electronics Specialists Conference, PESC04*, vol. 3, pp 1945-1949, 20-25 June 2004.

Galdino M. A. E., Ribeiro C. M. (1994). An Intelligent Battery Charge Controller for Small Scale PV Panel, *Proceedings of 12th European Photovoltaic Solar Energy Conference and Exhibition*, 1994.

Hesse K. (1997). An Off-Line Lead -Acid Charger Based on the UC3909, *Technical report*, Unitrode Company, 1997.

Hohm D.P, Ropp M. E. (2000). Comparative Study of Maximum Power Point Tracking Algorithms Using an Experimental, Programmable, Maximum Power Point Tracking Test Bed, *Proceedings of Photovoltaic Specialists Conference*, Conference Record of the Twenty-Eighth IEEE, pp 1699 - 1702, 15-22 Septembre 2000.

Hua C. and Lin J. (2003). An On-Line MPPT Algorithm for Rapidly Changing Illuminations of Solar Arrays, *Proceedings of Renewable Energy*, vol. 28, pp 1129–1142, 2003.

Hussein K. H., Muta I., Hoshino T. and Osakada M. (1995). Maximum Photo- Voltaic Power Tracking: an Algorithm for Rapidly Changing Atmospheric Conditions, *In IEE Proceedings Generation, Transmission and Distribution*, vol. 142(1), pp 59-64, IEE Steven age, Herts, U.K., January, 1995.

Knop H. (1999). *Analysis, Simulation, and Evaluation Of Maximum Power Point Tracking (MPPT) Methods for a Solar Powered Vehicle*, Mater of Science Thesis in Electrical and Computer Engineering, Portland State University, 1999.

Riza M., et al. (2003). A Maximum Power Point Tracking For Photovoltaic-Spe System Using A Maximum Current Controller, *Proceedings of Solar Energy Materials & Solar Cells*, vol.75 , pp 697–706, 2003.

Rosemback R. H. (2004). *Conversor Cc-Cc Bidirecional Buck-Boost Atuando Como Controlador De Carga De Baterias Em Um Sistema Fotovoltaico*, Mater of Science Thesis in Electrical Engineering, University Federal de Juiz de Fora, 2004.

Sera D., Kerekes T., Teodorescu R., and Aalborg F. Blaabjerg (2006). Improved MPPT Method For Rapidly Changing Environmental Conditions, *IEEE International Symposium on Industrial Electronics*, vol. 2, pp 1420 - 1425, 9-13 July, 2006.

Torres A. M. (1998). *Aproveitamento Fotovoltaico Controlado por Redes Neurais Artificiais Interligado ao Sistema Elétrico*, MSc Thesis, GPEC - DEE - UFC, Septembre, 1998.

Weidong X., Dunford W.G. (2004). A Modified Adaptive Hill Climbing MPPT Method For Photovoltaic Power Systems, *Proceedings of Power Electronics Specialists Conference PESC 04*, vol. 3, 20-25, pp 1957–1963, June 2004.

Vieira J. A., Mota A. M. (2009). Maximum Power Point Tracker Applied in Batteries Charging with Thermoelectric Generator Using the Waste Energy from a Gas Water Heater, *Proceedings of the Conference on Control Applications*, St. Petersburg, Russia, vol. 1, pp. 1477-1482, July 2009.

11

Titanium Dioxide Nanomaterials: Basics and Design, Synthesis and Applications in Solar Energy Utilization Techniques

Fuqiang Huang, Yaoming Wang, Jianjun Wu and Xujie Lü
Shanghai Institute of Ceramics, Chinese Academy of Sciences
People's Republic of China

1. Introduction

Titanium dioxide (TiO_2) nanomaterials have been extensively studied in the last two decades. Due to their versatile properties, TiO_2 nanomaterials have possessed themselves vast applications, including paint, toothpaste, UV protection, photocatalysis, photovoltaics, sensing, electrochromics, as well as photochromics. An in-depth study of the basic material properties, electrical transport-favored nano/micro-structure design and processing of TiO_2 nanomaterials will be present in this chapter, focusing on solar energy utilization efficiency enhancement.

2. Basics and design

2.1 A criterion for ranking the charge separation abilities of semiconductors

Nanomaterials used for gathering solar energy inevitably involve charge transport process, and solar energy utilization efficiency often comes down due to the difficulty of charge separation in many material systems, TiO_2 nanomaterials are not exceptional. How to evaluate the charge separation/transport abilities of TiO_2 and other semiconductors is an urgent question to be answered. Solving this problem will give an insight into intrinsic nature of compounds and bring great convenience to material & device design.

Here we have developed the packing factor (PF) concept to evaluate inherently existing internal fields that can be used to rank the charge separation abilities among oxide materials (Lin et al., 2009). The concept is based on the idea that lower elastic stiffness can promote distortion, which promotes internal field, and it can be easily implemented using the packing factor. This packing factor model is a broadly applicable criterion for ranking charge seperation/transport and photocatalytic ability of the materials with similar chemistry or structure. Lower PF value results in lower elastic stiffness, higher internal field, more efficient light-induced electron-hole separation and transport, and higher photocatalytic activity.

PF of a compound was computed by dividing the sum of spherical volumes by the unit cell volume, as seen in the equation of $PF = Z\,(xV_A+yV_B+zV_C)/V_{cell}$, where Z is the number of the formula unit in one unit cell of a semiconductor ($A_xB_yC_z$); V_A, V_B and V_C are ion volumes calculated by assuming spherical ions with a Shannon radius that depends on the coordination number; and V_{cell} is the cell volume. The different compounds are attributed to the atoms to be packed in their preferred ways to gain the lowest total energy in light of physics. Therefore, the crystal packing factor is not only related to mass density, packing manners, bonding habits, etc. in the crystal structure, but also related to charge density, band width, band gap, carrier mobility, etc. in the electronic structure.

As the two most investigated phases of TiO_2, anatase is widely reported more photocatalytically active than rutile (Yu & Wang, 2007). Meanwhile in our experiments, the different representative organic pollutants (methyl orange, methyl blue, and phenol) for photocatalysis were used to test the activity, but the measured activity trend remains the same, anatase (PF= 0.6455) > rutile (PF= 0.7045). Besides organic pollutant photodegradation, photoinduced water splitting over TiO_2 is also adopted as primary evaluation means to scale the photocatalytic activity. The same activity sequence is obtained, the same as that for dye degradation and mineralization described above. As known, anatase TiO_2 (density = 3.90 g/cm^3, PF = 0.6455) is a more loosely packed structure compared to rutile (density = 4.27 g/cm^3, PF = 0.7045). The loosely packed structure of anatase TiO_2 is favorable for photocatalytic activity.

Based on the lifetime and mobility of electrons and holes, we can give a full explanation from the packing factor model. It is conceivable that photocatalytically active ion in a lower PF structure is more polarizable, therefore its exciton radius is larger as are the lifetimes of electrons and holes. In addition, a lower PF structure is more deformable, which lowers the activation (hopping) barrier for polarons (e.g., those associated with O^-) thus increasing their mobility. The band dispersion often associated with low PF structures may additionally increase the dispersion at the edges of CBM and VBM, thus decreasing the effective mass of electrons and holes. This would further contribute to a higher mobility. These generic mechanisms may operate in a broad range of structures and at selected sites where photoelectrons and holes are generated and transported. Consequently, they could lead to wide applicability of the PF model.

The packing factor model — lower PF value results in more efficient light-induced electron-hole separation and transport, can also explains that anatase TiO_2 with a better charge transport ability than rutile TiO_2 has been broadly used as the sensitized electrode of dye-sensitized solar cells (DSC). Meanwhile, the PF model also gained wide supports from the literatures covering compounds of d^0 cations (Ti^{4+}, V^{5+}, Nb^{5+}, Ta^{5+}, Cr^{6+}, Mo^{6+} and W^{6+}) and d^{10} cations (Ag^+, Zn^{2+}, Cd^{2+}, Ga^{3+}, In^{3+}, Sn^{4+} and Sb^{5+}). So far, the PF model has been proven by over 60 systems covering about 120 photocatalysts (Lin et al., 2009). The finding not only provides a new focus on ranking the charge separation and transport abilities for DSC electrode materials, but also discloses insights for developing new photocatalysts with high UV- and/or visible-light responsive activities.

2.2 Electrical transport and charge separation favored nano/micro-structure design

Charge transport is of great importance for the performance of electronic devices, especially for those solar energy gathering devices, such as solar cells, photocatalysts, and chlorophylls in photosynthesis, etc. On one hand, the transport behavior of sensitized anode electrode TiO_2 for DSCs or new concept solar cells is attributed to the carrier (electron) concentration and mobility. The high electron mobility in TiO_2 relies on the high crystallinity of the lattice, while the crystallinity is closely related to the preparation condictions. We have successfully controlled the crystallinity of TiO_2 via varying the reaction temperature and solvents. The effect of the crystallinity on charge transport and separation has been also fully discussed. On the other hand, the charge transport properties of single and/or conventional materials may not be sufficient. Nano/micro-structure materials design offers a powerful approach for tailoring the transport property and charge separation ability, and great enhancement in performance can be expected. We successfully designed two kinds of nano/micro-structure configurations as sensitized anode for DSCs, one is *electron-transport favored semiconductor*, and the other is *composite structure of* TiO_2 *| semimetal | semiconductor*.

The sensitized anode for DSCs is preferred to be an excellent electron conductor, and its conduction band should match the dye's LUMO (the lowest unoccupied molecular orbitals). Furthermore, a tightly chemical binding interface is necessary for electron-transfer from dye to TiO_2 and between the TiO_2 particles. Nb-doped TiO_2 has also appeared to have promising applications on transparent conducting oxide (TCO) (Furubayashi et al., 2005), antistatic material, and gas sensor (Sharma et al., 1998). However, few studies have been reported on the positive roles of Nb-doped TiO_2 nanoparticles applied as the photoanode material of DSCs, and the mechanism of the effects by ion doping is still controversial. In this chapter, the Nb-doped TiO_2 nanocrystalline powders were demonstrated to be an *electron-injection and transport favored semiconductor* to enhance the performance of dye-sensitized solar cells. The improvement was ascribed to the enhanced electron injection and transfer efficiency caused by positive shift of flat-band potential (V_{fb}) and increased powder conductivity (Lü etal., 2010).

A new composite structure of TiO_2 *| semimetal | semiconductor* have been investigated to promote charge separation and electron transport. In general, such heterojunction structure requires (1) an alignment of the conduction band of the *semiconductor* with that of TiO_2, (2) little solubility of the *semiconductor* in TiO_2, (3) a highly conductive *semimetal* interface such as transparent conducting oxide (TCO), and (4) a high electron mobility in the *semiconductor*. One example is TiO_2|ZnO:Ti|ZnO, in which ZnO has a similar band structure but much higher electron mobility (205–300 $cm^2\ V\ s^{-1}$) than TiO_2 (0.1–4 $cm^2\ V\ s^{-1}$) (Zhang et al., 2009), Zn^{2+} has very low solubility in TiO_2 (Bouchet et al., 2003), and the Ti-doped ZnO (ZnO:Ti) is a TCO with a high conductivity (up to 1.5×10^3 $S\ cm^{-1}$) that depends on the doping level and microstructure (Chung etal., 2008). In this chapter, the new composite construct with a hollow spherical geometry with a hybrid TiO_2/ZnO composition is proposed for solar energy utilization. The hybrid TiO_2/ZnO spheres exhibit enhanced energy-conversion efficiency for the DSC. These improvements are ascribed to the enhanced charge-separation and electron-transport efficiencies made possible by the nano-heterojunction structure of TiO_2|ZnO:Ti|ZnO.

3. Synthesis and applications

3.1 Crystallinity control and solvent effect

As a bottom-up method, solvothermal method is a facile route for direct synthesis of nano-TiO_2. However, the main attention is often directed toward control over the structure and morphology only by varying the reaction temperature, duration, additive, and pH value during solvothermal treatment, while the solvent has rarely been deliberately selected to achieve different well-crystallized nanostructures. Initial failures in the solvothermal growth of a specific compound are usually the result of lack of proper data on the type of solvents, the solubility, and solvent-solute interaction. Solubility is a vital physicochemical and technological parameter which strongly influences the rate of dissolution, the degree of the supersaturation, thus the rate of nuclei formation. Solubility depends upon the nature of the substance, its aggregate state, temperature, pressure and a series of other factors, among which, the dielectric constant has a crucial effect on the solubility of precursor due to the diverse solvation energy. We have studied the formation of well-crystallized nano-TiO_2 on the basis of a one-pot solvothermal route. The effect of the dielectric constant on the solubility of the precursor, the nucleation and the crystal growth was discussed in detail. Moreover, the photocatalytic activity of the samples was also fully investigated in close conjunction with crystallinity (Wu et al., 2009).

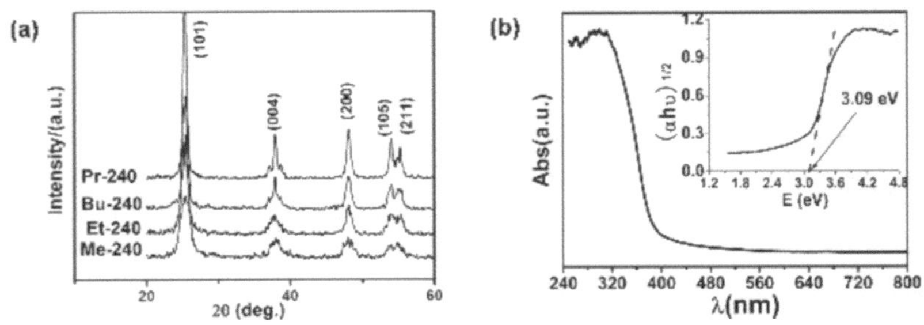

Fig. 1. (a) XRD patterns for samples at 240 °C. Et here shows the first two letters of the solvent (ethanol). Me, Pr and Bu are for methanol, 2-propanol and n-butanol, respectively. (b) UV-Vis spectrum for a typical nano-TiO_2

Fig. 1a presents the XRD patterns for the powders synthesized in the four different alcohols. Hereafter, Et-240 was denoted for nano-TiO_2 treated at 240 °C for 6 h with ethanol as solvent. All of the powders belong to the anatase type of TiO_2 (JCPDS No. 21-1272). Moreover, Pr-240 obtained the sharpest peaks when the temperature was set at 240 °C, indicating the relatively high crystallinity was obtained by these two samples. A typical UV-Vis spectrum for the obtained nano-TiO_2 was shown in the Fig. 1b. To obtain more precise optical band gap, plots of $(\alpha h\nu)^{1/2}$ vs the energy of absorbed is used to obtain the band gap because of its indirect transition nature (Tian et al., 2008). Eg was determined to be 3.09 eV.

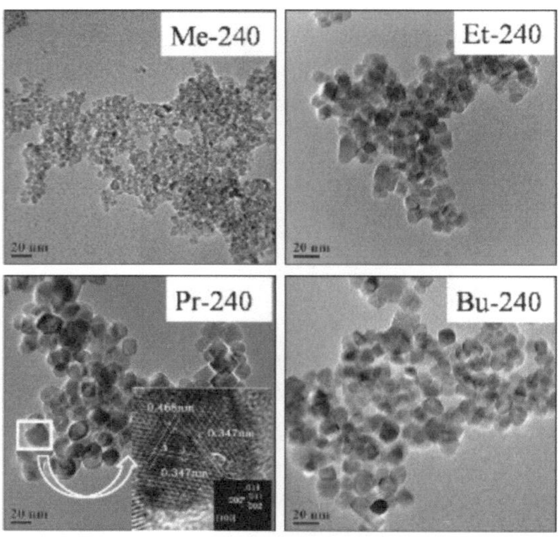

Fig. 2. TEM images for the TiO_2 nanoparticles at 240 °C

The TEM images for samples obtained at 240 °C were presented in Fig. 2. The crystallite size and shape strongly depend on the type of the solvent employed. Particles with amorphous shape are severely agglomerated and poor-crystallized in the case of methanol. While for Pr-240, the crystallinity is greatly enhanced and the shape tends to exhibit equiaxed geometry bounded by crystallographic facets. Additionally, HRTEM observation confirms the anatase structure for Pr-240. The inset shows the lattice image of a TiO_2 grain and its FFT diffractogram which is consistent to a [100]-projected diffraction pattern of the anatase TiO_2. Among the all four powders obtained at 240 °C, Pr-240 has obtained the largest crystallite size of about 15 nm determined from the corresponding TEM image. Considering that the samples prepared in the present work are synthesized under the same conditions, *i.e.*, temperature and time, the varied morphology and XRD patterns of the powders should originate from the different solvents for their distinct physicochemical properties.

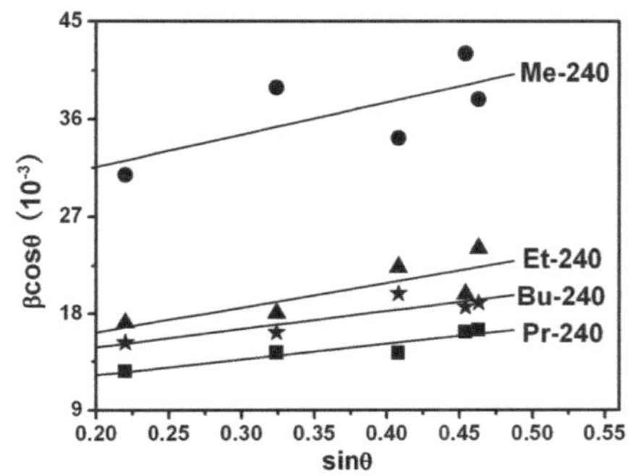

Fig. 3. The relation between βcosθ and sinθ for the samples

Crystallite size (D) and lattice strain (ε) are calculated via the Williams and Hall equation, $\beta\cos\theta = K\lambda / D + 2\varepsilon \sin\theta$, plots of βcosθ against sinθ based on the XRD patterns (Fig. 1a) are shown in Fig. 3. For Et-240, Bu-240 and Pr-240, it shows relatively good linearity, which gives reliable values of D and ε. Table 1 depicts the quantitative values of D and ε for each sample. Crystallinity enhances, *i.e.*, the growth of crystallite and the decrease in lattice strain, in the order: Me-240, Et-240, Bu-240 and Pr-240, indicating that the crystallinity for the nano-TiO_2 has a strong dependence on the solvent used

Catalyst	D (nm)	ε (10^{-3})
Me-240	5.7	14.94
Et-240	11.6	11.87
Bu-240	12.2	8.56
Pr-240	14.8	7.27

Table 1. The obtained D and ε based on the data shown in Fig. 3

Solvents with different physicochemical properties have a pronounced effect on the crystallinity and morphology of the final nanocrystals by influencing the solubility, reactivity, diffusion behavior and the crystallization kinetics (crystal nucleation and growth rate). Here, we give a closer look on the effect of dielectric constant on the crystallinity of the obtained nano-TiO_2. The crystallization for nanoparticles generally consists of two processes (Sirachaya et al., 2006): nucleation and crystal growth. The nucleation rate, J_N, can be expressed as follows with a pre-factor, J_0:

$$J_N = J_0 \exp\left(\frac{-16\pi V_m^2 \gamma^3}{3(RT)^3 (\ln S)^2}\right)$$

Where V_m is the molar volume of the solid material, S is the supersaturation degree, and $S = C_l / C_s$. C_l the precursor concentration, C_s the solubility of the solid phase, J_0 the frequency of collisions between precursor molecules, γ the interfacial tension, R the gas constant, and T the temperature. Hence, it can be concluded that the nucleation rate is expected to increase strongly with increasing supersaturation. The solubility of an inorganic salt decreases with a decrease in the dielectric constant of the solvent, due to the decreased solvation energy. Meanwhile, during the process of the crystal growth, larger particles grow at the expense of the smaller ones owing to the energy difference between the larger particles and the smaller ones of a higher solubility based on the Gibbs-Thompson law. This refers to the "Ostwald ripening" process applied and confirmed in numbers of papers (Li et al., 2007). In methanol, as Table 2 shows, a higher dielectric constant (η = 32.35) invites a higher solubility of the solid metal oxide and a lower supersaturation degree in this system, which predicts less nuclei numbers, inadequate nutriments-supply and slower crystal-growth rates (Hua et al., 2006), thus lower crystallinity. As mentioned above, the crystallinity (concerning two part: crystallite size and lattice strain) of the obtained nano-TiO_2 should be foretold in the enhanced order: Me-240 < Et-240 < Pr-240 < Bu-240. However, the present data show some unexpected results, *i.e.*, Pr-240 obtains a better crystallization than Bu-240, demonstrating that other properties of the solvent, such as viscosity, saturated vapor pressure, coordinating ability and steric hindrance should be taken into account (Zhang et al., 2002). In other words, crystallinity depends on dielectric constant of the solvent to a great extent, not in all the range.

Solvent	Methanol	Ethanol	2-Propanol	n-butanol
η	32.35	25.00	18.62	17.50

Table 2. Dielectric constant for the alcohols used, η refers to dielectric constant, and the values are provided by (Moon et al., 1995).

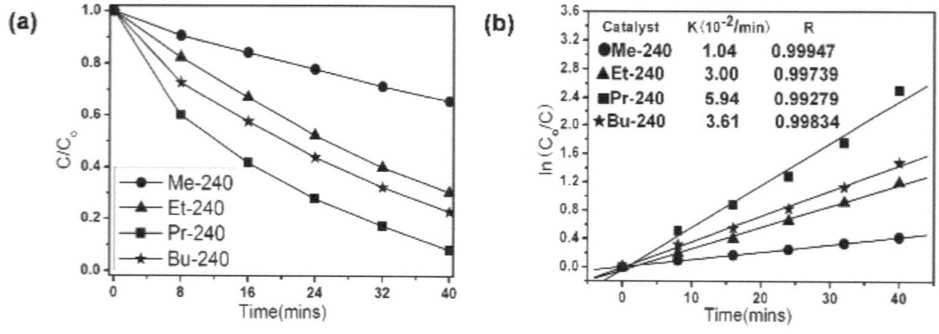

Fig. 4. MO photodegradation over samples under UV-light irradiation

Fig. 4 depicts the result of the photocatalytic degradation of methyl orange (MO) for nano-TiO_2. The photocatalysis efficiency decreases gradually in the order: Pr-240 > Bu-240 > Et-240 > Me-240, in an agreement with the tendency of the crystallite size, as shown in Table 1. In other words, the photocatalytic efficiency increased in the order: Me-240 < Et-240 < Bu-240 < Pr-240, simultaneously with an increase of the crystallinity, *i.e.*, the increase in crystallite size and the decrease in lattice strain, as Fig. 5 shows, confirming the dependence of the photocatalysis on the crystallinity.

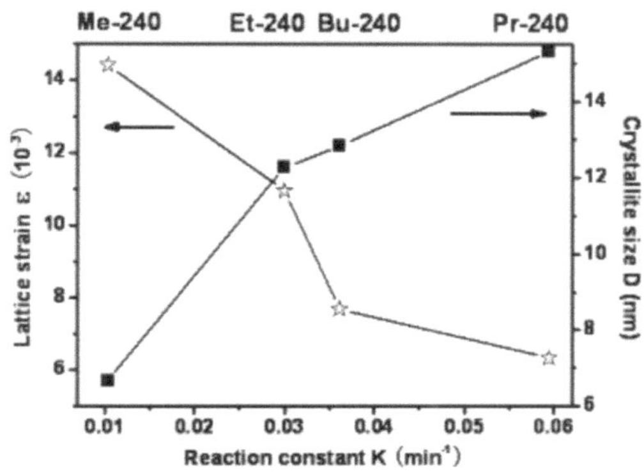

Fig. 5. The effect of the crystallinity on the reaction constant K

Crystallinity was proved to have an indispensible effect on the two most important processes of the photocatalysis: charges separation and charges transport, as follows (Chen & Mao, 2007): (1) the highly crystallized anatase can promote the charges transfer from particle center to surface. The residual strain of the poor-crystallized TiO_2 lattice leads to disorder and distortion of the TiO_2 matrix, which have a severe scattering effect on the charges transport. Furthermore, an electron and a hole can migrate a longer distance in a crystal of larger crystallite size than in a smaller one, separating more the reducing and oxidizing sites on the surface of the crystal. So the volume recombination may occur less frequently; (2) it eliminates the crystal defects, *i.e.*, impurities, dangling bonds, and microvoids, which behave as recombination centers for the e^-/h^+ pairs, thus the surface recombination is greatly suppressed. It is, thus, no wonder that Pr-240 of which the crystallite size is about 14.8 nm and lattice strain about 7.27×10^{-3} holds the maximum in the reaction constant K of MO decomposition, *i.e.*, about 6 times of that for Me-240.

3.2 Synthesis and solar-spectrum tunable TiO_2: Eu

Extensive research interests are focused in photocatalysis, but investigations and applications for the photoluminescence (PL) properties of TiO_2 have not been simultaneously satisfied. As we konw, high-energy photons (UV, etc.) in the solar spectrum are harmful to the components of DSCs (dye dissociation) and silicon solar cells (overheated silicon). Based on our recent study of TiO_2: Eu (Wu etal., 2010), through the excition at 394 nm (UV) and 464 nm (blue light), it shows intense emissions at 592 nm (yellow) and 612 nm (red). In other words, TiO_2: Eu can be used as a solar-spectrum tunable photoluminescent material to convert high-energy photons to low-energy photons, *i.e.*, from UV and/or blue to yellow or red light. The PL process of TiO_2: Eu comprises the intrinsic excitation resulted from the *f-f* inner-shell transitions and the host excitation ascribed to the charge transfer band (CTB) from O–Ti to Eu^{3+} ions. It requires a perfect lattice of TiO_2 for charges transfer, in order to avoid space charge regions and e-h recombination. So the crystallinity of the TiO_2 lattice is to have a pronounced effect on the PL process, which should be further investigated.

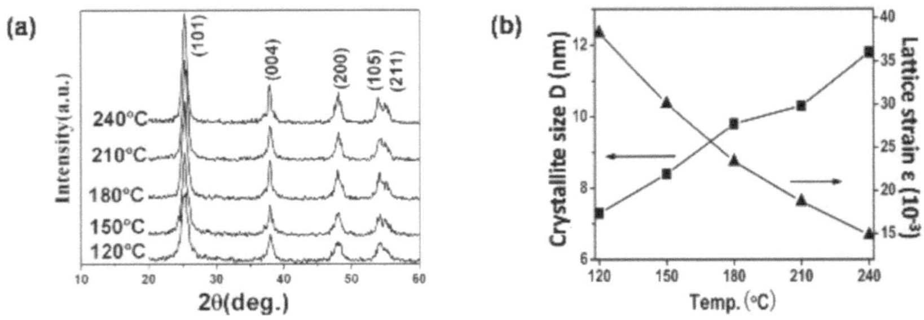

Fig. 6. (a) XRD patterns and (b) the corresponding crystallite size D and lattice strain ε for the TiO_2: Eu nanoparticles on the hydrothermal temperature

Based on the Williams and Hall Equation, D increases from 7.3 nm to 11.8 nm and ε decreases from 38.25×10^{-3} to 14.82×10^{-3} for the TiO_2: Eu samples when increasing the hydrothermal temperature (Fig. 6). The growth of crystallite and the decrease in lattice strain, indicating that the crystallinity of the nanoparticles has been enhanced, and that various structural defects, such as small displacement of atoms neighboring, non-uniform strain and residual stress of the lattice, have been gradually eliminated. These defects were reasonably supposed to influence the PL performance.

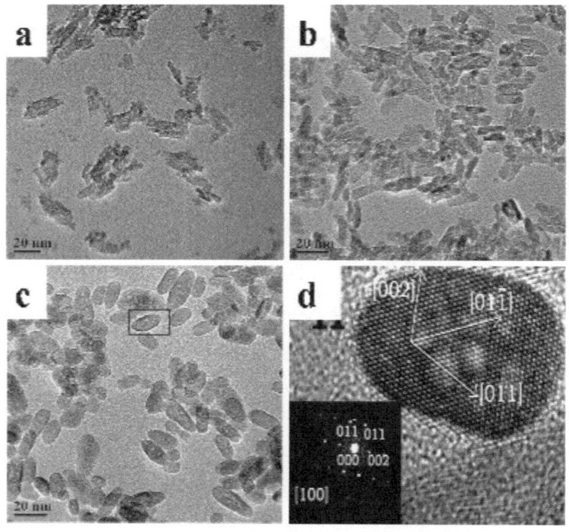

Fig. 7. The TEM images of (a) Eu^{3+}/TiO_2-120, (b) Eu^{3+}/TiO_2-180, (c) Eu^{3+}/TiO_2-240, (d) HRTEM of Eu^{3+}/TiO_2-240, Fast-Fourier Transformed diffractogram of Eu^{3+}/TiO_2-240 (inset)

The morphology of the nanoparticles changes from polyhedron to rod-like with Eu^{3+} doping (Fig. 7), which implies that the Eu^{3+} doping plays an important effect on the crystallographic orientation of TiO_2 nanocrystal. Eu^{3+} hinders the growth of specific facets of anatase TiO_2 based on the "oriented attachment" mechanism (Ghosh & Patra, 2007). The similar case was also observed in Er^{3+}-doped TiO_2. And HRTEM of a representative rod also shows its anatase structure, and the corresponding FFT diffractogram demonstrate its single crystal nature (Fig. 7d).

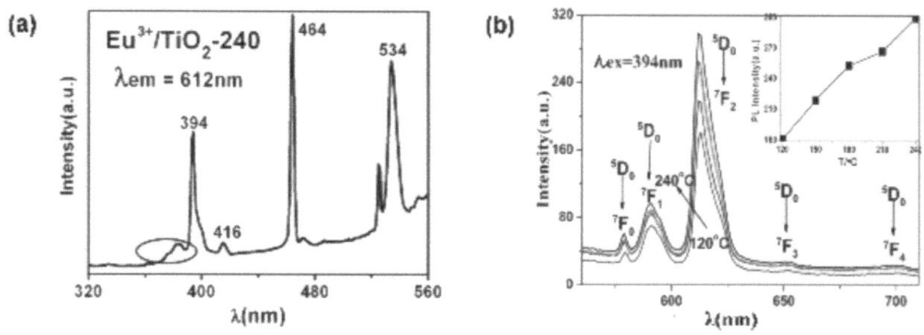

Fig. 8. (a) The excitation spectrum of Eu^{3+}/TiO_2-240 (λ_{em} = 612 nm), (b) the emission spectra (λ_{ex} = 394 nm) of the TiO_2:Eu^{3+} samples, where their maximum emission (λ_{em} = 612 nm) intensities at 612 nm in the inset

Fig. 8a depicts the typical excitation spectrum of the Eu^{3+}/TiO_2-240. By monitoring the emission line of 612 nm, the excitation lines appear at 394, 416, 464, and 534 nm are ascribed to the *f-f* inner-shell transitions within the Eu^{3+} $4f^{6}$ configuration. Besides, a new band appears in the range from 320 to 380 nm, although it's not obvious. Based on the previous papers, the new wide band can be attributed to the host excitation and assigned to the charge transfer band (CTB) from O–Ti to the Eu^{3+} ions. Similar broad band has also been observed and attributed to the CTB from O–Ti to Eu^{3+} ions in the previous works (You & Nogami, 2004).

Sample	I [$^5D_0 \rightarrow {}^7F_2$] (a.u.)	I [$^5D_0 \rightarrow {}^7F_1$] (a.u.)	R
Eu^{3+}/TiO_2-120	2.324	0.901	2.58
Eu^{3+}/TiO_2-150	2.793	1.054	2.65
Eu^{3+}/TiO_2-180	3.228	1.117	2.89
Eu^{3+}/TiO_2-210	3.415	1.149	2.97
Eu^{3+}/TiO_2-240	3.822	1.258	3.05

Table 3. The integrated intensity ratio of $^5D_0 \rightarrow {}^7F_2$/ $^5D_0 \rightarrow {}^7F_1$ of the samples. **R:** Integrated intensity ratio of $^5D_0 \rightarrow {}^7F_2$ and $^5D_0 \rightarrow {}^7F_1$

The five characteristic peaks at 579, 592, 612, 651, 699 nm corresponding to $^5D_0 \rightarrow {}^7F_0$, $^5D_0 \rightarrow {}^7F_1$, $^5D_0 \rightarrow {}^7F_2$, $^5D_0 \rightarrow {}^7F_3$, $^5D_0 \rightarrow {}^7F_4$ transitions of Eu^{3+} ion, respectively, are observed for all the Eu^{3+} doped samples at the excitation wavelength of 394 nm in Fig. 8b. It can be seen that the $^5D_0$ emission is intensified with the increment in temperature accompanied with gradually enhanced crystallnity. For $^5D_0 \rightarrow {}^7F_2$ transition, the PL intensity was quantitatively analysed and tabulated in the inset of Fig. 8b. The intensity ratio (R) of $^5D_0 \rightarrow {}^7F_2$ (612 nm) to

${}^5D_0 \rightarrow {}^7F_1$ (592 nm) increases as the degree of Eu–O covalence increases, so R is widely used to investigate the bonding environment of the Eu^{3+} ions. The integrated intensity ratio (R) of the samples obtained at different temperature are shown in Table 3.Note that R increases with hydrothermal temperature, accompanied with the promoted crystallinity, indicating that the covalence degree of the Eu^{3+} ions increases.

On the other hand, the great mismatch of ionic radius between Eu^{3+} (0.95 Å) and Ti^{4+} (0.68 Å) makes the doping Eu^{3+} hardly enter into the TiO_2 lattice (Lin & Yu, 1998), but inclined to distribute in the crystallite surface or interstitials of TiO_2 nanocrystals. For the poor-crystallized TiO_2 matrix, the Eu^{3+} has a tendency to form clusters due to the reduction of Eu^{3+}–Eu^{3+}distances (Stone et al., 1997). The clusters are undesirable which lead to an enhanced interparticle contact of the Eu–Eu pairs, thus quench its luminescence through cross relaxation. As the crystallinity enhances, the gradual formation of Eu^{3+}–O^{2-}–Ti^{4+} bonding leads to reducing the extent of the Eu^{3+} clusters, suppressing the cross relaxation and intensifying the luminescence effectively. Furthermore, the great elimination of the crystal defects, as quenching centers for luminescence, can diminish the undesired nonradiative recombination routes for electrons and holes (Ikeda et al., 2008), contributing to the enhanced luminescence.

3.3 Synthesis and application of TiO_2: Nb in DSCs

The highly crystallized Nb-doped TiO_2 nanoparticles were prepared by a one-step hydrothermal process and applied as the photoanode materials in DSCs, which facilitate electron injection and transfer, contributing to the significant improvement of energy conversion efficiency of the DSCs. The mechanism of the improvement caused by Nb doping was discussed in detail.

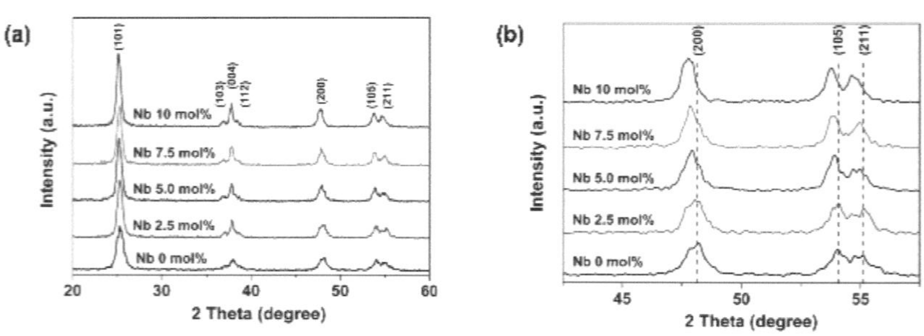

Fig. 9. (a) XRD patterns of as-prepared samples with different Nb contents; (b) Details of the XRD patterns around 48° and 54° 2θ values

Fig. 9 shows the XRD patterns of the Nb-doped TiO_2 with different Nb contents. All peaks of the as-prepared samples can be assigned to the anatase phase, indicating that the anatase nanocrystalline structure is retained after doping. The diffraction peaks shift to lower theta values with increasing Nb content, due to the larger radius of Nb^{5+} (0.64 Å) than Ti^{4+} (0.61 Å) according to the Bragg equation of $2d\sin\theta = \lambda$ (Fig. 9b). Furthermore, the intensity of the diffraction peaks strengthens gradually with the increasing Nb content. Consequently, as the superiority of the new method, the higher ordered nature of the TiO_2 nanoparticles introduced by the Nb doping would be in favor of electron transfer, resulting in the increased photocurrent. The HRTEM images in Fig. 10 indicate the high crystallinity of the TiO_2 nanoparticles.

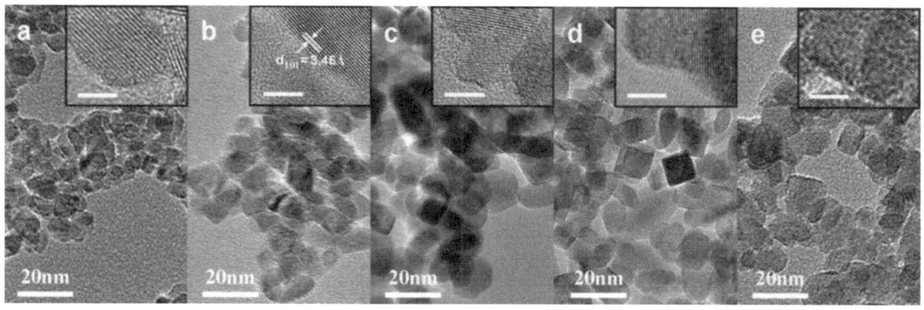

Fig. 10. TEM images of the as-prepared TiO_2 nanoparticles with different Nb contents (a) 0 mol%, (b) 2.5 mol%, (c) 5.0 mol%, (d) 7.5 mol%, and (e) 10.0 mol%. Inset shows the corresponding HRTEM image of each sample (Scale bar 5 nm)

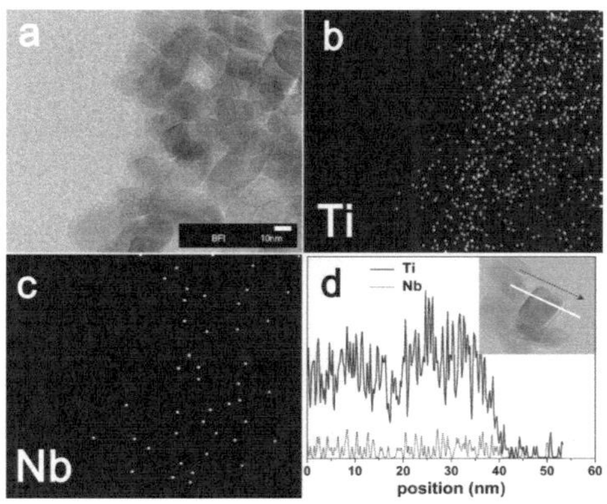

Fig. 11. (a) Bright-field STEM image of 5.0 mol% Nb-doped TiO_2; (b, c) the corresponding elemental mapping of Ti (b) and Nb (c); (d) line-scanning analysis across the nanoparticles indicated by the line as shown in the inset

Fig. 11 shows the STEM image of 5.0 mol% Nb-doped TiO_2 nanoparticles, and the corresponding elemental mapping, revealing the homogeneous spatial distribution of Nb. The uniform distribution of Nb in the TiO_2 lattice was also confirmed by the line-scanning analysis (Fig. 11d).

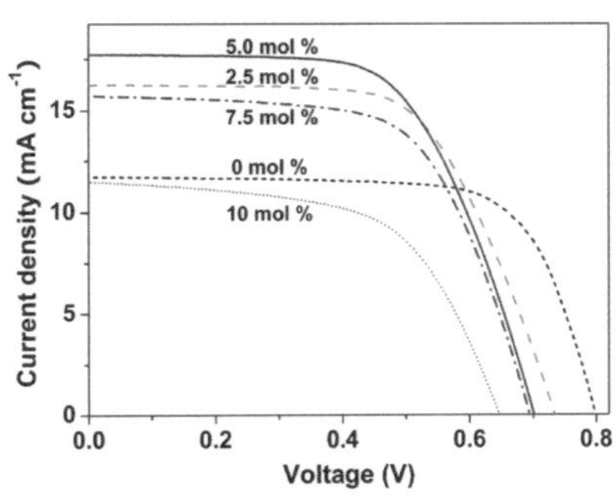

Fig. 12. Current - voltage curves of dye-sensitized solar cells based on the undoped and Nb-doped TiO_2 electrodes

Fig. 12 shows the current-voltage curves of the open cells based on the Nb-doped and undoped TiO_2 photoelectrodes. The performance characteristics are summarized in Table 4.

A pronounced increase in the photocurrent for the DSCs based on the Nb-doped TiO_2 was observed by the Nb doping between 2.5 - 7.5 mol%. As a result, an improved energy conversion efficiency of 7.8% was achieved for DSC based on the 5.0 mol% Nb-doped TiO_2, which was 18.2% higher than that of the undoped one. Whereas, the influence on the open circuit potential (V_{oc}) by the doping of Nb is negative. It is evident that the conduction band edge has been changed by the Nb doping.

DSCs	J_{sc} [mA cm^{-2}]	V_{oc} [V]	FF [%]	η [%]	amount of dye [a] [mol cm^{-2}] × 10^{-8}	film thickness [b] [μm]
0 mol%	11.87 ± 0.26	0.79 ± 0.01	70 ± 1	6.6 ± 0.1	5.2 ± 0.7	5.5 ± 0.2
2.5 mol%	15.75 ± 0.51	0.74 ± 0.01	64 ± 1	7.5 ± 0.3	5.5 ± 0.6	5.4 ± 0.3
5.0 mol%	17.67 ± 0.19	0.70 ± 0.01	63 ± 1	7.8 ± 0.2	5.4 ± 0.3	5.4 ± 0.2
7.5 mol%	15.91 ± 0.22	0.69 ± 0.01	63 ± 2	6.9 ± 0.2	5.7 ± 0.2	5.5 ± 0.2
10.0 mol%	11.79 ± 0.57	0.65 ± 0.01	57 ± 3	4.4 ± 0.2	5.1 ± 0.9	5.4 ± 0.5

Table 4. Performance characteristics of dye-sensitized solar cells based on the undoped and Nb-doped TiO_2 electrodes

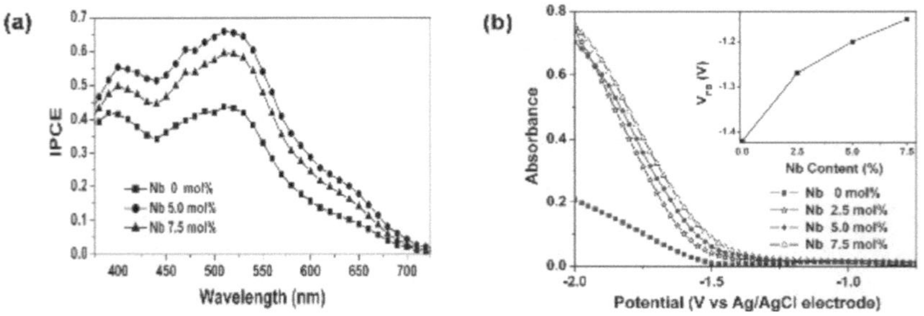

Fig. 13. (a) Action spectra of the dye-sensitized solar cells based on the undoped and Nb-doped TiO_2 electrodes. (b) Optical absorbance at 870 nm of undoped and Nb-doped TiO_2 films measured as a function of applied potential. Inset shows the flat-band potential of the samples as a function of the Nb contents

The reasons leading to a higher photocurrent for the solar cells based on Nb-doped TiO_2 are revealed according to the measurements on photocurrent action spectra and flat-band potential (V_{fb}). The action spectra are shown in Fig. 13, which present a significant enhancement in the IPCE of the DSCs based on the Nb-doped TiO_2 electrodes compared with that of the undoped one. The improvement can be attributed to the enhanced electron injection and charge transfer efficiency as well as the slightly higher amount of dye absorption as listed in Table 4. It has been reported that when the dye uptake increased 1.2 times, the IPCE only increased approximately 3% (Redmond & Fitzmaurice, 1993). Thus, the intrinsic increase in the photocurrent and IPCE are primarily due to the enhanced electron injection and transfer ability of the Nb-doped TiO_2. The effects caused by the Nb doping on electron injection, transfer and recombination of the DSCs would be discussed via the studies of flat-band potential and electrochemical impedance spectra as follows.

Photocurrent generation depends on electron injection, charge transfer, and charge recombination processes. Here the effect of the Nb doping on the above factors is qualitatively discussed. The different positions of the excited energy level of the dye and the conduction band minimum (CBM) of the semiconductor are essential to the electron injection. Central to an understanding of the band energetics of a semiconductor electrode is the determination of flat-band potential (V_{fb}). As shown in Fig. 13b, the results indicate a positive shift of the flat-band potential with the increasing of Nb content. Consequently, the driving force for electron injection, E_{fb} – LUMO (the lowest unoccupied molecular orbital energy level) (Kron et al., 2003), is increased by the Nb doping, which correspondingly makes contribution to the enhancement of electron injection efficiency. Meanwhile, the open current potential (V_{oc}) of DSCs is dependent on the difference of the flat-band potential of TiO_2 and the redox potential of I^-/I_3^- couple. Therefore, the V_{oc} of the DSCs would decrease due to the positive shift of V_{fb}, as shown in Fig. 12 and Table 4. By optimally selecting the photoanode material, dye and electrolyte, the photocurrent density can be improved without significantly lowering the V_{oc}. One approach to increase V_{oc} is to adjust the redox potential to a more positive value (Han et al., 2004), while the dye's ground state potential should be positive enough comparing with the redox potential to make sure the efficient dye regeneration rate. Another approach is to choose a more efficient sensitizer, and then more electrons are injected to the photoanode, raising the Fermi level of the oxide and thus shift its potential.

The J_{sc} improvement is also related to the charge transfer ability. After the Nb doping, the charge compensation of Nb^{5+} in substitution to Ti^{4+} is achieved either by the creation of one Ti cation vacancy per four Nb introduced or by the stoichiometric reduction of Ti^{4+} to Ti^{3+} per Nb introduced.

$$\frac{1}{2}Nb_2O_5 + Ti_{Ti}^{x} \rightarrow Nb_{Ti}^{\bullet} + \frac{1}{4}V_{Ti}^{''''} + TiO_2 + \frac{1}{4}O_2 \quad (1)$$

$$\frac{1}{2}Nb_2O_5 + Ti_{Ti}^{x} \rightarrow Nb_{Ti}^{\bullet} + Ti_{Ti}^{'} + \frac{5}{4}O_2 \quad (2)$$

The occurrence of one or the other of two scenarios depends on the synthetic conditions and Nb concentration. High oxidative synthetic condition and low Nb content might play in favor of the scenario corresponding to Equation 1 because cations would be maintained in their higher oxidation state, whereas scenario corresponding to Equation 2 should be considered in low oxidative synthesis condition and high Nb concentrations. Here, the reactions occurred in a sealed autoclave with a rather low oxidative circumstance and the Nb contents are quite high (>2.5 mol%), thus the occurrence here is in favor of the scenario corresponding to Equation 2 and this has been demonstrated by Hirano and Matsushima (Nakamura et al., 2003). Consequently, one excess electron in the Ti 3d orbital due to each Nb^{5+} substituting for Ti^{4+} raises the electron concentration.

The enhancement of electron transfer ability was discussed on the basis of theoretical model of the electrical conductivity, which is based on the equation of $\sigma = ne\mu$, where e is elementary charge, n denotes the concentration of electrons, and μ is the electron mobility. The increasing of the electron concentration enhances the electron conductivity, and the improved electron transport efficiency results in the increase of the photocurrent density. However, the electron mobility decreased rapidly at high defect concentration due to the electron scattering by the defects. The severe defects increase charge recombination and that would become the dominant factor when the Nb content reaches a high level. The mechanism for electron transport through mesoporous TiO_2 is still a hotly debated topic. Deducing the exact mechanism through experimental and theoretical investigations is complicated, partly because of the apparent inability to systematically vary individual parameters without influencing others. Fortunately, there have been much experimental and theoretical evidence that supports the notion that the electron transport is governed by a trapping-detrapping process of electrons from the sub-bandgap states (Longo et al., 2002). In the DSC system, one dye molecule transfers one electron to Ti^{4+} $3d^0$ of TiO_2, and then one Ti^{3+} $3d^1$ is generated. The energy gap between the Ti^{4+} $3d^0$ band and the Ti^{3+} $3d^1$ energy level is rather shallow, and the electron at Ti^{3+} $3d^1$ is easy to be transferred to the neighboring Ti^{4+} instead of being trapped to form space charge. This wonderful feature makes the loose-packed anatase TiO_2 be an excellent dye-sensitized electrode material. By doping Nb into the TiO_2 in this work, the Ti^{3+} $3d^1$ states existing in the nanocrystals increase the electron concentration, and these Ti^{3+} $3d^1$ states plus Nb^{5+} $4d^0$ make the band structure near conduction band minimum (CBM) more dispersed to enhance the mobility of the excited electrons. However, Ti^{3+} can also be the electron traps, when the TiO_2 has a very poor crystallinity or excessive imperfects. Furthermore, the results shown in Fig. 14 indicate that the resistance of powder drops sharply at the beginning of doping and changes slightly when the Nb content exceeds 5.0 mol%. This result certifies the reason of J_{sc} improvement discussed above.

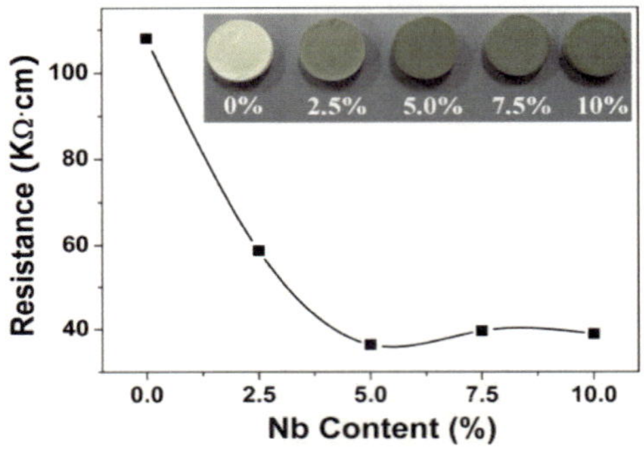

Fig. 14. Powder resistance of the as-prepared undoped and Nb-doped TiO_2. Inset shows the color change after Nb-doping

The internal resistances of DSCs were studied via electrochemical impedance spectroscopy (EIS) in the frequency range of 0.1 Hz – 100 kHz, and with alternating current amplitude of 10 mV. Fig. 15 shows the EIS results at forward bias of the open-circuit voltage under light irradiation and the results were represented as Nyquist plots. The responses in the frequency regions around 10^4, 10^3, 10 and 0.1 – 1 Hz are assigned to charge transfer processes occurring at the Pt/electrolyte interface, TiO_2 /TiO_2 particles interface, TiO_2/dye/electrolyte interface and the Nernst diffusion within the electrolyte, respectively. The relative low resistance between Pt/electrolyte interface results in an unobvious semicircle at the frequency ω_1 = 14.7 kHz. The border between the arcs of ω_2 and ω_3 was vague for the undoped TiO_2 electrode with the severe overlap between ω_2 and ω_3 resulting from the relative high resistance between TiO_2 particles. In contrast, the borders of the Nb-doped samples are clear. Obviously, the second semicircle at the frequency ω_2 = 1.2 kHz become smaller with increasing of Nb content (see Fig. 9b), owing to the enhanced electron conductivity. The third semicircle at the frequency ω_3 = 4.5 Hz expanded with the Nb content increasing from 2.5 mol% to 7.5 mol%. The raise of resistance at the TiO_2/dye/electrolyte interface is beneficial for suppressing the charge recombination at the interface, which can compensate the drop of V_{oc} caused by the positive shift of flat-band potential. In Fig. 12, the V_{oc} of the cell based on the 7.5 mol% Nb-doped TiO_2 is close to than of the 5.0 mol% one, due to the greater compensation. However, the result at the Nb content of 10.0 mol% is abnormal which may because the severe defects became the recombination centers and hindered the charge transfer. The EIS results mentioned above confirm the mechanism of improvement.

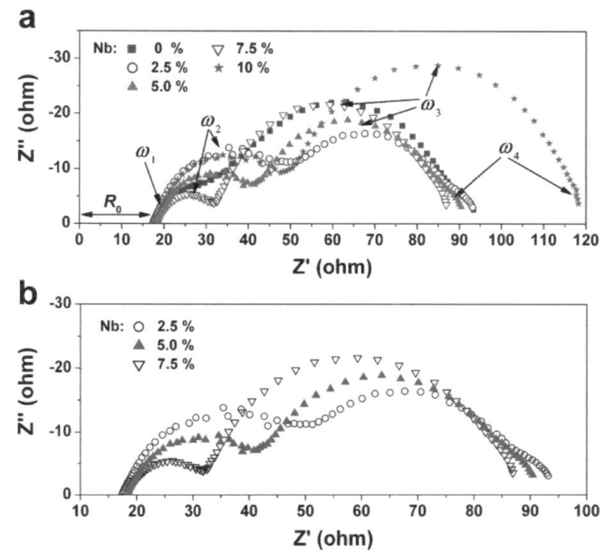

Fig. 15. Electrochemical impedance spectra of dye-sensitized solar cells based on the undoped and Nb-doped TiO_2 electrodes

In this section, the Nb-doped TiO_2 nanocrystalline powders were demonstrated to be an *electron-injection and transport favored semiconductor* to enhance the performance of dye-sensitized solar cells. The improvement was ascribed to the enhanced electron injection and transfer efficiency caused by positive shift of flat-band potential (V_{fb}) and increased powder conductivity, and the mechanism was verified by powder resistance and EIS analyses. Such systematic investigation on the effect of the Nb doping will provide valuable insight on designing the high-performing DSCs.

3.4 Synthesis and application of TiO_2 | ZnO: Ti | ZnO in photocatalysis and DSCs

TiO_2 hollow spheres with a hybrid composition were prepared by a hydrated-salt assisted solvothermal (HAS) strategy. In this method, a metallorganic Ti source reacts with the water that is slowly released from a hydrated salt of another metal, and hybrid metal oxides are obtained forming the desired nano-heterojunction structure of *semiconductor | semimetal | semiconductor* (*e.g.* TiO_2|ZnO:Ti|ZnO). We also report the photocatalytic activity and photovoltaic efficiency of a DSC fabricated with TiO_2/ZnO spheres demonstrating improved performance.

The hollow spherical morphology of the sample has been revealed by transmission electron microscopy (TEM). Also evident from Fig. 16b is the nanocrystallites in the shell of spheres, and the selected area electron diffraction (SAED) image (Fig. 16c) indicates the nanocrystallites are random in orientation. Energy-dispersive X-ray spectroscopy (EDS) shown in Fig. 16d determines the Zn content in the product to be 1.1 atomic%. According to

EXAFS spectroscopy, Zn^{2+} segregation occurs when the Zn concentration is above 0.1 atomic% in nanocrystalline anatase TiO_2 (Bouchet et al., 2003), which is reasonable in view of the large mismatch in the charge and the ionic radius between Ti^{4+} (0.61 Å) and Zn^{2+} (0.74 Å). Therefore, Zn^{2+} apparently has difficulty in entering the TiO_2 lattice and is likely to form very small crystallites that are incorporated into the TiO_2/ZnO composite in the hollow spheres. Such ZnO nanocrystals located between TiO_2 nanocrystals are expected to have a beneficial effect on electron mobility and charge separation.

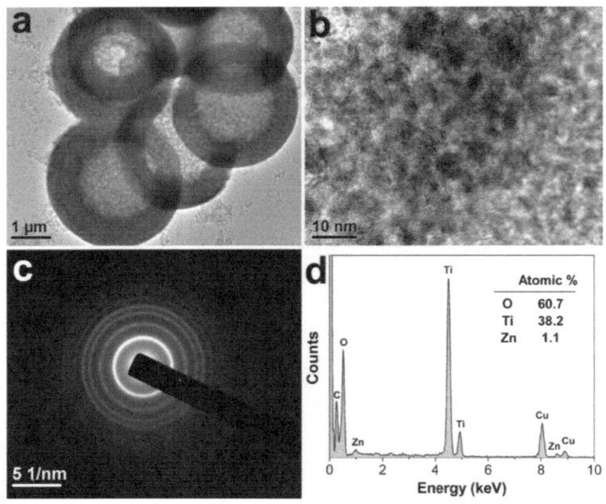

Fig. 16. (a, b) TEM images, (c) selected area electron diffraction (SAED) image, and (d) energy-dispersive X-ray spectroscopy (EDS) of the TiO_2/ZnO spheres

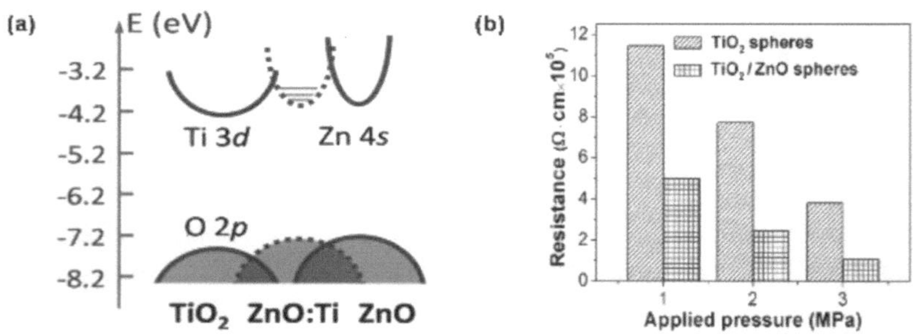

Fig. 17. (a) Schematic band structure of TiO_2 | ZnO:Ti | ZnO heterojunction, (b) Powder resistances of the TiO_2/ZnO spheres and TiO_2 hollow spheres

As mentioned above, ZnO and TiO_2 have similar band structures, and charge can be easily transferred at their interface. As is well known, the smaller effective mass (m^*) of electrons implies the higher electron mobility (μ). Since the conduction band of TiO_2 originates from the *d*-orbital, which has a narrow bandwidth and a large m^* (~10 m_e), whereas the conduction band of ZnO has an *s*-orbital character giving rise to a much smaller m^* (~0.2 m_e) (Roh et al., 2006). Therefore, ZnO has a much higher electron mobility than TiO_2, which should have a beneficial effect on electron transport in the hybrid TiO_2/ZnO spheres. Moreover, although Zn^{2+} has a very small solubility in TiO_2, Ti^{4+} can dissolve up to 4 mol% in ZnO (Lin et al., 2005). Therefore, in the hybrid spheres, there is likely to exist a TiO_2/ZnO interface, Ti-doped ZnO (ZnO:Ti), which is a well-known TCO. Overall, the hybrid composite could achieve the schematic band structure configuration shown in Fig. 17a. Such a construct of TiO_2 | ZnO:Ti | ZnO is suitable for charge separation and electron transport, so enhanced performance in both photocatalysis and DSCs can be expected. To demonstrate the beneficial effect of ZnO addition on electron transport, we compared the resistance of powder compacts of TiO_2/ZnO spheres and TiO_2 hollow spheres with comparable radius and shell thickness (These spheres have a comparable morphology and surface area as the hybrid TiO_2/ZnO spheres.). These compacts were cold-pressed under various pressures. As shown in Fig. 17b, regardless of compaction pressures, the TiO_2/ZnO hybrid compacts are always less resistive than nonhybrid TiO_2 compacts.

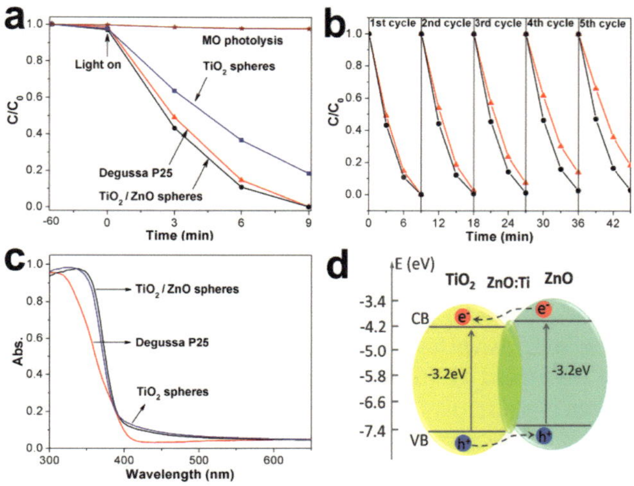

Fig. 18. (a) photocatalytic degradation of MO (10 mg L^{-1}) over TiO_2/ZnO spheres (●), TiO_2 hollow spheres (■), Degussa P25 (▲) and without catalyst (★), (b) cycling experiments of MO degradation over TiO_2/ZnO spheres (●) and Degussa P25 (▲), (c) UV-vis diffuse reflectance spectra of the TiO_2/ZnO spheres, TiO_2 hollow spheres and Degussa P25, and (d) schematic illustration of the band structure and charge separation in TiO_2/ZnO hybrid

To demonstrate the beneficial effect of ZnO addition on photocatalysis, the photocatalytic activity of the hybrid TiO_2/ZnO spheres is compared with similar TiO_2 hollow spheres using the methyl orange (MO) assay. Degussa P25, a highly effective photocatalyst often considered as the gold standard in this field, is also used as the reference. As shown in Fig. 18a, after UV irradiation for 9 min, MO was totally bleached over the TiO_2/ZnO spheres, whereas only 80% of MO was degraded over TiO_2 hollow spheres. The hybrid spheres also compared favorably with P25, and are more robust than P25 for repeated reuse (Fig. 18b). The superior performance of the hybrid spheres compared to P25 is probably attributed to a higher specific surface area (150 m^2 g^{-1} vs. 50 m^2 g^{-1}) and more efficient light harvesting by the hollow spheres. On the other hand, since TiO_2 spheres and hybrid TiO_2/ZnO spheres have very similar UV-vis absorption and surface area, their different photocatalytic activities must be attributed to the differences in charge separation and electron transport caused by ZnO. According to the schematic band diagram of the TiO_2 | ZnO:Ti | ZnO heterojunction (Fig. 18d), electrons created in the conduction bands (CB) of TiO_2 and ZnO and holes in the valence bands (VB) can be separated at the heterojunctions due to the favorable energy bias between the two sides (Zhang et al., 2009). This reduces electron-hole recombination and maintains the requisite electron/hole populations required for photocatalytic reactions with organic dyes . In addition, the lower resistance caused by ZnO addition (Fig. 17b) indicates that electron/hole transport is facilitated which should also favor photocatalytic activity. Incidentally, the similar absorption spectra of TiO_2 hollow spheres and TiO_2/ZnO spheres provide further evidence that few Zn^{2+} ions enter the TiO_2 lattice. Otherwise, aliovalent substitution would have created substitutional and charge-compensating point defects that affect optical absorption.

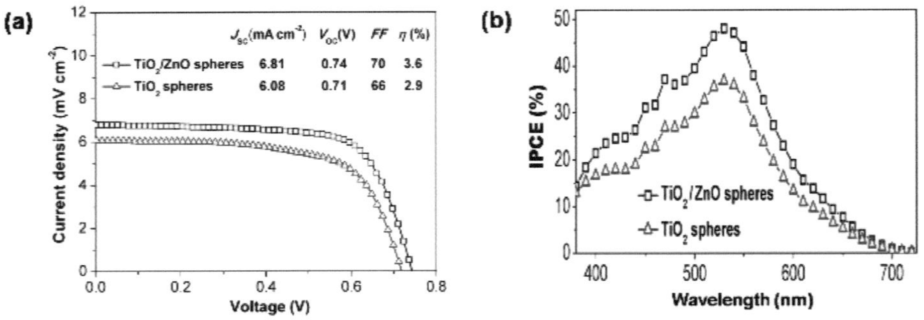

Fig. 19. (a) Photocurrent density-voltage curves, (b) action spectra of the DSCs with anodes made of TiO_2/ZnO spheres and TiO_2 hollow spheres

When used as the anode material to fabricate DSCs, enhanced performance can also been achieved. The photocurrent density-voltage (*J-V*) curves are shown in Fig. 19a. The energy-conversion efficiency increased from 2.9 % for TiO_2 hollow spheres to 3.6 % for hybrid TiO_2/ZnO spheres. This is primarily due to the increased photocurrent density, as well as the higher photovoltage and fill factor, which is not always easy to achieve by impurity doping only. In this case, the inhibition of electron back transfer from TiO_2 to the redox electrolyte (I_3^-) by the heterojunctions may contribute to the improvement in the photovoltage and fill factor (Kay & Gratzel, 2002). As shown in Fig. 19b, the incident-photon-to-current efficiency (IPCE) of the cell with a hybrid electrode is higher than that with a TiO_2 (hollow spheres) electrode at all wavelengths. Since there is only a slight difference in the dye adsorption between these two electrodes, and the influence of dye adsorption is known to be relatively minor (Ma et al., 2005), the main reason for the increase in the photocurrent density and IPCE in the cells with hybrid electrodes may be attributed to their enhanced electron transport efficiency. Under the solar illumination, the injected electrons in the Ti^{4+} 3*d* states transfer easily to the Zn^{2+} 4*s* states in the composite structure of TiO_2|ZnO:Ti|ZnO. Such a band-structure-matched heterojunction can be imaged as the "bridge" for electrons to transport from here to there. The enhanced electron transport efficiency raises the photocurrent density, results in the improvement of energy-conversion efficiency.

In conclusion, a new composite construct of *TiO_2 | semimetal | semiconductor* with a hollow spherical geometry with a hybrid TiO_2/ZnO composition is proposed for solar energy utilization. The hybrid TiO_2/ZnO spheres exhibit a higher photocatalytic activity and enhanced energy-conversion efficiency for the DSC. These improvements are ascribed to the enhanced charge-separation and electron-transport efficiencies made possible by the nano-heterojunction structure of TiO_2|ZnO:Ti|ZnO.

4. Summary

Over the past decades, the tremendous effort put into TiO_2 nanomaterials has resulted in a rich database for their synthesis, properties, modifications, and solar applications. The synthesis and modifications of TiO_2 nanomaterials have brought new properties and new applications with improved performance via solar energy utilization techniques in our lab. Meanwhile, TiO_2 nanomaterials also exhibit size-dependent as well as shape- and structure-dependent optical, electronic, thermal, and structural properties, as reported by other groups. TiO_2 nanomaterials have continued to be highly active in photocatalytic and photovoltaic applications, and they also demonstrate new applications including electrochromics, sensing, and hydrogen storage. This steady progress has demonstrated that TiO_2 nanomaterials are playing and will continue to play an important role in the protections of the environment and in the search for renewable and clean energy technologies.

5. References

Bouchet, R. ; Weibel, A. & Knauth, P. (2003). EXAFS study of dopant segregation (Zn, Nb) in nanocrystalline anatase (TiO_2). *Chem. Mater.*, 15, 26, 4996-5002, 0897-4756

Chen, X.; Mao, S. S. (2007). Titanium dioxide nanomaterials: Synthesis, properties, modifications, and applications. *Chem. Rev.* 107, 7, 2891–2959, 0009-2665

Chung, L.; Chen, J. C. & Tseng, C. J. (2008). Preparation of TiO_2-doped ZnO films by radio frequency magnetron sputtering in ambient hydrogen-argon gas, *Appl. Surf. Sci.*, 255, 5, 2494-2499, 0169-4332

Furubayashi, Y.; Hitosugi, T. & Yamamoto, Y. (2005). A transparent metal: Nb-doped anatase TiO_2. *Appl. Phys. Lett.*, 86, 25, 252101, 0003-6951

Ghosh, G.; Patra, A. (2007). Influence of surface coating on physical properties of TiO_2/Eu^{3+} nanocrystals. *J. Phys. Chem. C,* 111, 19, 7004-7010, 1932-7447

Han, L.; Koide, N. & Chiba, Y. (2004). Modeling of an equivalent circuit for dye-sensitized solar cells. *Appl. Phys. Lett.*, 84, 13, 2433-2435, 0003-6951

Hua, Z. L.; Wang, X. M. & Shi, J. L. (2006). Solvent effect on microstructure of yttria-stabilized zirconia (YSZ) particles in solvothermal synthesis. *J. Eur Ceram. Soc.*, 26, 12, 2257–2264, 0955-2219

Ikeda, M.; Li, J. G. & Kobayashi, N. (2008). Phase formation and luminescence properties in Eu^{3+}-doped TiO_2 nanoparticles prepared by thermal plasma pyrolysis of aqueous solutions. *Thin Solid Films,* 516, 19, 6640–6644, 0040-6090

Kay, A.; Gratzel, M. (2002). Dye-sensitized core-shell nanocrystals: Improved efficiency of mesoporous tin oxide electrodes coated with a thin layer of an insulating oxide. *Chem. Mater.*, 14, 7, 2930-2935, 0897-4756

Kron, G.; Rau, U. & Werner, J. H. (2003). Influence of the Built-in Voltage on the Fill Factor of Dye-Sensitized Solar Cells. *J. Phys. Chem. B*, 107, 48, 13258-13261, 1520-6106

Li, J.; Zeng, H. C. (2007). Hollowing Sn-doped TiO_2 nanospheres via Ostwald ripening, *J. Am. Chem. Soc.*, 129, 51, 15839–15847, 0002-7863

Lin, J.; Yu, J. C. (1998). An investigation on photocatalytic activities of mixed TiO_2-rare earth oxides for the oxidation of acetone in air. *J. Photochem. & Photobio. A: Chem.*, 116, 1, 63-67, 1010-6030

Lin, S. S.; Huang, J. L. & Sajgalik, P. (2005). The properties of Ti-doped ZnO films deposited by simultaneous RF and DC magnetron sputtering. *Surface and Coatings Technology*, 191, 3, 286-292, 0257-8972

Lin, X. P.; Wu, J. J. & Huang, F. Q. (2009). Novel antimonate photocatalysts MSb_2O_6 (M= Ca, Sr and Ba): a correlation between packing factor and photocatalytic activity. *Phys.Chem.Chem.Phys.*, 11, 43, 10047–10052, 1463-9076

Longo, C.; Nogueira, A. F. & Cachet, H. (2002). Solid-state and flexible dye-sensitized TiO_2 solar cells: a study by electrochemical impedance spectroscopy. *J. Phys. Chem. B*, 106, 23, 5925-5930, 1520-6106

Lü, X. J.; Mou, X. L. & Huang, F. Q. (2010). Improved-performance dye-sensitized solar cells using Nb-doped TiO_2 electrodes: efficient electron injection and transfer. *Adv. Funct. Mater.*, 20, 3, 209-515, 1616-301X

Ma, T. L.; Akiyama, M. & Abe, E. (2005). High-efficiency dye-sensitized solar cell based on a nitrogen-doped nanostructured titania electrode. *Nano. Lett.*, 5, 12, 2543-2547, 1530-6984

Moon, Y. T.; Park, H. K. & Seog, I. S. (1995). Preparation of monodisperse and spherical zirconia powders by heating of alcohol-aqueous salt-solutions. *J. Am. Ceram. Soc.* 1995, 78, 10, 2690–2694, 0002-7820

Nakamura, R.; Imanishi, A. & Murakoshi, K. (2003). In situ FTIR studies of primary intermediates of photocatalytic reactions on nanocrystalline TiO_2 films in contact with aqueous solutions. *J. Am. Chem. Soc.*, 125, 24, 7443-7450, 0002-7863

Redmond, G.; Fitzmaurice, D. (1993). Spectroscopic determination of flat band potentials for polycrystalline TiO_2 electrodes in nonaqueous solvents. *J. Phys. Chem.*, 97, 7, 1426-1430, 0022-3654

Roh, S.; Mane, R. & Han, S. (2006). Achievement of 4.51% conversion efficiency using ZnO recombination barrier layer in TiO_2 based dye-sensitized solar cells. *Appl. Phys. Lett.*, 89, 25, 253512, 0003-6951

Sharma, R. K.; Bhatnagar, M. C. & Sharma, G. L. (1998). Mechanism in Nb doped titania oxygen gas sensor. *Sensors and Actuators B: Chem.*, 46, 3, 194-201, 0925-4005

Sirachaya, K. N. A.; Okorn M. & Piyasan P. (2006). Solvothermal synthesis of ZnO with various aspect ratios using organic solvents. *Cryst. Growth & Des.*, 6, 11, 2446–2450, 1528-7483

Stone, B. T.; Costa, V. C. & Bray, K. L. (1997). In situ dehydroxylation in Eu^{3+}-doped sol-gel silica. *Chem. Mater.*, 9, 11, 2592-2598, 0897-4756

Tian, G. H.; Fu, H. G. & Xin, B. F. (2008). Preparation and characterization of stable biphase TiO_2 photocatalyst with high crystallinity, large surface area, and enhanced photoactivity. *J. Phys. Chem. C*, 112, 8, 3083–3089, 1932-7447

Wu, J. J.; Lü, X. J. & Huang, F. Q. (2009). Dielectric constant-controlled solvothermal synthesis of photocatalyst TiO_2 with tunable crystallinity: A strategy for solvent-selection. *Eur. J. Inorg. Chem.*, 2009, 19, 2789–2795, 1434-1948

Wu, J. J.; Lü, X. J. & Huang, F. Q. (2010). Crystallinity control on photocatalysis and photoluminescence of TiO_2-based nanoparticles. *J. Alloy. Compd.*, 496, 1, 234-240, 0925-8388

You, H. P.; Nogami, M. (2004). Optical properties and local structure of Eu^{3+} ions in sol-gel TiO_2-SiO_2 glasses. *J. Phys. Chem. B*, 108, 32, 12003-12008, 1520-6106

Yu, J. G. & Wang, G. H. (2007). Effects of hydrothermal temperature and time on the photocatalytic activity and microstructures of bimodal mesoporous TiO_2 powders. *Appl. Catal. B: Environ*, 69, 3, 171–180, 0926-3373

Zhang, J.; Sun, L. D. & Yan, C. H. (2002). Control of ZnO morphology via a simple solution route, *Chem. Mater.*, 14, 10, 4172–4177, 0897-4756

Zhang, L. S.; Wong, K. H. & Wong, P. K. (2009). AgBr-Ag-Bi_2WO_6 nanojunction system: A novel and efficient photocatalyst with double visible-light active components. *Appl. Catal. A: Gen.*, 363, 2, 221-229, 0926-860X

Zhang, Q. F.; Dandeneau, C. S. & Zhou, X. Y. (2009). ZnO nanostructures for dye-sensitized solar cells. *Adv. Mater.*, 21, 41, 4087-4108, 0935-9648

12

Sensorless Control of a Polar-Axis Photovoltaic Tracking System

John T. Agee and Adisa A. Jimoh
Tshwane University of Technology
P. Bag X680 Pretoria 0001,
South Africa

1. Introduction

Photovoltaic solar power installations can be broadly classified as static (non-tracking), single-axis tracking, polar axis-tracking and two-axis tracking installations (Agee et al., 2006). In general, tracking photovoltaic systems have higher percentage energy recovery per Kilowatt of installed capacity than static solar power systems (Ed. Kusoke et al., 2003). A key component of existing photovoltaic tracking systems is the solar position sensor and associated conditioning circuitry, which provides the information with which the tracking angle is updated. These sensors add to the overall cost of installed photovoltaics. For example, in South Africa where the average installed cost of photovoltaics is ZAR 29.00/Watt (Greenology, 2010), the percentage sensor(s) cost for installed photovoltaic wattage is shown in Figure 1, based on an average sensor cost of USD110.0. It is evident from Figure 1 that for low power solar photovoltaic applications, the percentage sensor cost motivates the exploitation of alternative tracking strategies that are devoid of sensors. Sensor-less tracking offer a cost effective solution in such low power applications. Sensor-less tracking has been reported in literature (Ibrahim et al., 2004; Cheng & Wong, 2009; Power from the Sun, 2010; Chen et al., 2006; Stine & Harrington, 1988) concerning solar-thermal systems. These rely on the use of well established astronomical formulae to extract the direction of sunrays as a function of the local clock time, after due compensation for any differences between the local clock time and the solar time. The equation of time (EOT) and the local longitude compensation are factored into the derivation of the final local time equation. EOT is an equation that evaluates the difference between the local clock time and the solar hour. In the discourse presented in the current chapter, the sensor-less tracking of a polar-axis solar tracker is reported. The concepts of differential flatness (Fliess et al.; Fliesss et al.; Levine & Nguyen, 2003; Bitaud, 1990, 1997, 2003) is used for embedding the equations of the direction of sunrays into the feedback loop of the controller.

In the rest of the chapter, the physical structure of the polar-axis solar tracker and the derivation of its dynamic equations are described in section two. The concepts of differential flatness and the derivation of the flat output for the polar-axis solar tracker is presented in section three. Controller design is contained in section four. A derivation of the relationship between the local clock time and the direction of sunrays with respect to an observer (or the photovoltaics platform) at a given location, together with the integration of time-based values of the sunrays angle for sensor-less tracking is presented in section five of the chapter. Illustrative simulations and results presentation and discussion form section six of the chapter. Conclusions are presented in section seven. A list of references is included at the end of the chapter.

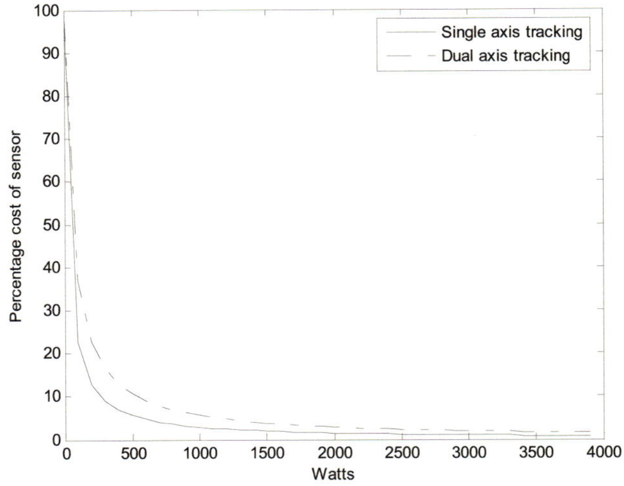

Fig. 1. Percentage sensor cost as a function of installed wattage of photovoltaics

2. The polar-axis photovoltaic solar tracker

The platform carries ten Shott 300W photovoltaic panels. In addition, two smaller, Shell SQ 80W solar panels are provided, to compensate for the energy looses in the electrical installation. The detailed design of the 3KW platform is presented elsewhere (de Lazzer, 2005). The standing 3KW platform is shown in Figure 2. The drive system consists of a d.c motor linked to the platform through a gear train having a gear ratio of 800. Additional provision was made for the occasional manual adjustment of the elevation of the platform for the purposes of field experimentation.

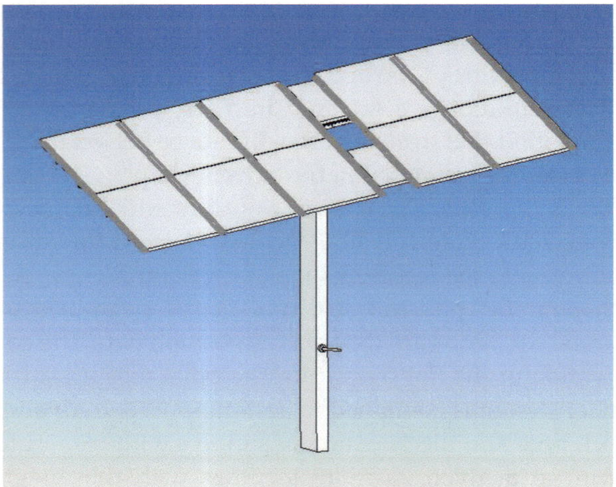

Fig. 2. The 3KW polar-axis solar power platform

2.1 Mathematical modelling of the 3 KW solar power platform

The block diagram representation of the platform in the east-west direction is shown in Figure 3. Where: $\theta_s(t)$ is the instantaneous direction of sunlight and $\theta_p(t)$ the instantaneous position of the platform. Following (de Lazzer, 2005; Agee et al., 2006; Agee & Jimoh, 2007), for the D.C motor we can write:

$$e_a = R_a i_a + L_a \frac{di_a}{dt} + K_b \frac{d\theta_m}{dt} \tag{1}$$

where $e_a(t)$: armature voltage (V); $i_a(t)$: armature current (A); R_a: armature resistance (Ω); L_a: armature inductance (H); K_b: back-emf constant (V/rad/s) and $\theta_m(t)$: rotor displacement (rad.). Similarly, the mechanical torque developed by the motor is given by

$$T_m = K_m i_a \tag{2}$$

where $T_m(t)$ is torque(N.m.) and K_m the torque constant (N.m/A).
Furthermore, the mechanical torque is written as in equation (3).

$$T_m = J_t \frac{d^2\theta_m}{dt^2} + B\frac{d\theta_m}{dt} + K\theta_m \tag{3}$$

where $J_t = J_m + N^2 J_l$ and J_m : moment of inertia of the motor ($kg.m^2$); J_l : moment of inertia of the load($kg.m^2$); N : gear-train ratio between motor and load; B: viscous-friction coefficient of the motor ($kg.m.s^{-1}$); K: spring constant ($kg.m^2.s^{-2}$).
The physical-variables state-space description of the platform, $\dot{x} = Ax + Bu$, could thus be written as:

$$\begin{aligned}
\dot{x}_1 &= x_2 \\
\dot{x}_2 &= -\frac{K}{J_t}x_1 - \frac{B}{J_t}x_2 + \frac{K_m}{J_t}x_3 \\
\dot{x}_3 &= -\frac{K_b}{L_a}x_2 - \frac{R_a}{L_a}x_3 + \frac{1}{L_a}u \\
y &= x_1 = \theta_m ; u = e_a \\
x^T &= [x_1, x_2, x_3] = [\theta_m, \dot{\theta}_m, i_a] \\
B^T &= [0, 0, 1/L_a]
\end{aligned} \tag{4}$$

3. Differential flatness of platform

By definition, a linear system given by:

$$\begin{aligned}
\dot{x} &= Ax + Bu \\
x &\in R^n, u \in R^m; n \geq m+1
\end{aligned} \tag{5}$$

is said to be differentially flat (or simply *flat*) if it is equipped with a set of variables h_1, called the flat output (Levine & Nguyen, 2003), such that for some integer r,

$$h_1 = g(x, u, \dot{u}, \ddot{u}, \ldots\ldots, u^{(r)}), 0 < r \leq \infty; h_1 \in R^m \tag{6}$$

such that every state $x_i, i = 1,2,\ldots n$ of the linear system, together with its input u can be described completely in terms of the flat output and its derivatives as in equation (7).

$$\begin{aligned} x_i &= p_i(h_1, \dot{h}_1, \ddot{h}_1, \ldots\ldots, h_1^{(q)}) \\ u &= Q(h_1, \dot{h}_1, \ddot{h}_1, \ldots\ldots, h_1^{(q+1)}) \end{aligned} \tag{7}$$

Where q is a finite integer, such that the initial equations $\dot{x} = Ap(h_1, \dot{h}_1, \ddot{h}_1, \ldots\ldots, h_1^{(q)}) + BQ(h_1, \dot{h}_1, \ddot{h}_1, \ldots\ldots, h_1^{(q+1)})$, where $\alpha^T = [\alpha_1, \alpha_2, \ldots\ldots \alpha_n]$, are identically satisfied. We shall thus show that every state variable of the physical model of the platform could be written in terms of a set of variables, the flat variable, and a finite number of its derivatives.

3.1 Derivation of the flat output for a linear system with a scalar input

For the given linear system, re-write the dynamics in the formal variable **s** as:

$$\begin{aligned} A_1(s)X(s) &= Bu(s) \\ A_1(s) &= sI - A \end{aligned} \tag{8}$$

The formal derivation of the flat output for (8) follows the method of Levine and Nguyen; and requires that there be a matrix C, of rank ***n-m***, orthogonal to $\boldsymbol{B}$ (Levine & Nguyen, 2003) such that,

$$C^T B = 0 \tag{9}$$

$$C^T A_1(s)P(s) = 0 \tag{10}$$

$$Q(s) = (B^T B)^{-1} B^T A_1(s)P(s) \tag{11}$$

hence, for a given linear system for which $A_1(s)$ and B(s) are known, **C** can be evaluated from (9). P(s) is then evaluated from equation (10), and finally, Q(s) is evaluated from equation (11)

3.2 Derivation of the flat output for the polar-axis-type photovoltaic solar power platform

The detailed derivation of the flat output for the polar-axis solar tracker is presented in (Agee & Jimoh, 2010). Key result is summarised as follows:

$$\begin{aligned} p_1(s) &= \frac{K_m}{J_t} h_1(s) \\ p_2(s) &= s p_1(s) \\ p_3(s) &= \left\{ s^2 + \frac{b}{J_t} s + \frac{K}{J_t} \right\} h_1(s) \end{aligned} \tag{12}$$

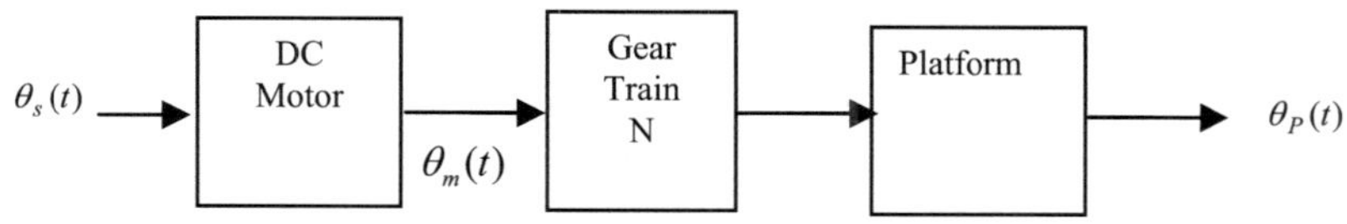

Fig. 3. Block diagram of the open-loop 3KW solar power platform

and $Q(s)=(B^TB)^{-1}B^TA_1(s)P(s)$ yields

$$Q(s)=\left\{\frac{R_aK}{J_t}+\frac{\{R_aB+K_bK_m+L_aK\}}{J_t}s+\frac{\{L_aB+R_aJ\}}{J_t}s^2+L_as^3\right\}h_1(s) \tag{13}$$

It is evident from equation (12) that each of the states of the platform could now be written in terms of the flat output $h_1(t)$ and its derivatives. The input u(t) could be written from equation (13). Hence,

$$\begin{aligned}\theta_m(t)&=\frac{K_m}{J_t}h_1(t)\\ \dot{\theta}_m(t)&=\frac{K_m}{J_t}\dot{h}_1(t)\\ \ddot{\theta}_m(t)&=\frac{K_m}{J_t}\ddot{h}_1(t)\\ i_a(t)&=\ddot{h}_1+\frac{B}{J_t}\dot{h}_1+\frac{K}{J_t}h_1(t)\end{aligned} \tag{14}$$

$$\begin{aligned}\theta_p(t)&=\frac{K_m}{800J_t}h_1(t)\\ \dot{\theta}_P(t)&=\frac{K_m}{800J_t}\dot{h}_1(t)\end{aligned} \tag{15}$$

$$u(t)=e_a=\frac{R_aK}{J_t}h_1+\frac{\{R_aB+K_bK_m+L_aK\}}{J_t}\dot{h}_1+\frac{\{L_aB+R_aJ\}}{J_t}\ddot{h}_1+L_ah_1^{(3)} \tag{16}$$

Alternatively,

$$\begin{aligned}h_1(t)&=\frac{J_t}{K_m}\theta_m(t)\\ \dot{h}_1(t)&=\frac{J_t}{K_m}\dot{\theta}_m(t)\\ \ddot{h}_1&=i_a(t)-\frac{K}{K_m}\theta_m(t)-\frac{B}{K_m}\dot{\theta}_m(t)\end{aligned} \tag{17}$$

Notice also that, if the desired trajectories of motion are either know apriori, or given, the reference values of the flat output and it derivatives $h_1^*,\dot{h}_1^*,\ddot{h}_1^*,h_1^{(3)*}$, could be described.

3.3 Alternative Description of the Dynamics of the Tracker

The flat description of the systems dynamics, as in equations (14)-(16), enables an alternative

$$\begin{aligned} \dot{h}_1 &= h_2 \\ \dot{h}_2 &= h_3 \\ \dot{h}_3 &= -\frac{R_a K}{J_t L_a} h_1 - \frac{\{R_a B + K_b K_m + L_a K\}}{J_t L_a} h_2 - \frac{\{L_a B + R_a J\}}{J_t L_a} h_3 + \frac{1}{L_a} u(t) \end{aligned} \tag{18}$$

presentation of the dynamics of the platform, in terms of the flat output. Hence, using equation (16), a representation of the plant in terms of the flat output could be presented as in equation (18). The presentation in equation (8) is particularly suitable for controller design.

4. Controller design

The differential flatness property allows to exploit sensor-less control of the platform, in which loop closure is with respect to the time derivation of the angular position of the sun with respect to an earth-based observer. Details are presented in section five of the chapter.

4.1 Controller tuning

For the design of a three-term controller in the flat variables, substitute equation (19) in equation (18),

$$\dot{h}_3(t) = \dot{h}_3^* - K_1(h(t)_1 - h_1^*(t)) - K_2(h_2(t) - h_2^*(t)) - K_3(h_3 - h_3^*(t)) \tag{19}$$

and the steering input is thus given by:

$$\begin{aligned} u_{1c}(t) &= K_1 L_a h_1^*(t) + K_2 L_a h_2^*(t) + K_3 L_a h_3^*(t) + (\beta_1 - K_1) h_1(t) \\ &+ (\beta_2 - K_2 L_a) h_2(t) + (\beta_3 - K_3 L_a) h_3(t) + L_a \dot{h}_3(t) \end{aligned} \tag{20}$$

Where,

$$\beta_1 = \frac{R_a K}{J_t}; \beta_2 = \frac{R_a B + K_b K_m + L_a K}{J_t}; \beta_3 = \frac{L_a B + R_a J}{J_t} \tag{21}$$

the substitution resulting also in the following controlled system in equation (22):

$$\begin{aligned} \dot{h}_1 &= h_2 \\ \dot{h}_2 &= h_3 \\ \dot{h}_3 &= \dot{h}_3^* - K_1 e - K_2 \dot{e} - K_3 \ddot{e} \\ e &= h_1 - h_1^*, \dot{e} - h_2 - h_2^*, \ddot{e} = h_3 - h_3^*, \dddot{e} = \dot{h}_3 - \dot{h}_3^* \end{aligned} \tag{22}$$

with the equivalent characteristics equation given by equation (23).

$$[s^3 + K_3 s^2 + K_2 s + K_1] E(s) = 0 \tag{23}$$

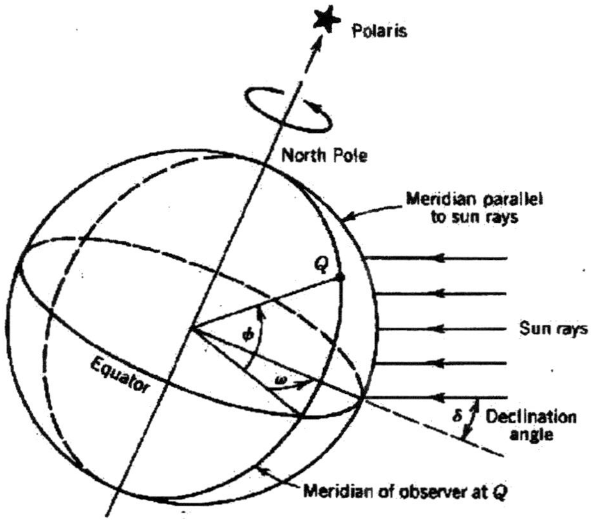

Fig. 4. Measurement of direction of sunrays, in the earth-centred coordinate systems

To obtain the controller parameters K_1, K_2, K_3, the platform is tuned to yield the same dominant pole-pair as in (Agee et al., 2006); with the dominant poles given as $s_1, s_2 = -2.324 + j2.34$ and the third pole given by $s_3 = -33.506$. This system of closed-loop poles leads to the closed-loop dynamics of equations (24) and (25).

$$\begin{aligned} \dot{h}_1 &= h_2 \\ \dot{h}_2 &= h_3 \\ \dot{h}_3 &= -355.7276(h_1 - h_1^*(t)) - 165.0125(h_2 - h_2^*(t)) - 38.114(h_3 - h_3^*(t)) \\ \therefore K_1 &= 335.7276, K_2 = 165.0125, K_3 = 38.114 \end{aligned} \tag{24}$$

$$\begin{aligned} u(t) = u^*(t) &+ \left(\frac{R_a K}{J_t} - 355.7276\right)(h - h^*{}_1) + \left(\frac{R_a B + K_b K_m + L_a K}{J_t}\right. \\ &\left. -165.0125\right)(h_2 - h_2^*) + \left(\frac{L_a B + R_a J}{J_t} - 38.114\right)(h_3 - h_3^*) + L_a \dddot{h}_1 \end{aligned} \tag{25}$$

In equation (25), it has also been assumed that the reference acceleration $\dot{h}_3^*(t) = 0$, for bravity.

4.2 Classical control of polar-axis tracker with a solar position sensor

In the classical control of systems, feedback loop closure is with respect to the expected steady-state values of state. Suppose that the tracker is at rest at an initial position $\theta_p(0)$; for generality, use the notation

$$\theta_p(0) = 0 \tag{26}$$

Also at this rest position, all velocities and accelerations are zero. Hence, $\dot{\theta}_P(0) = \ddot{\theta}_P(0) = 0....$ It also follows from equations (14)-(17) that:

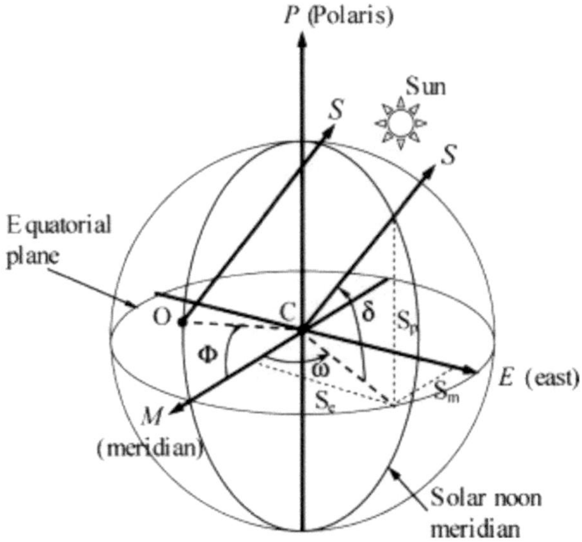

Fig. 5. Relationship between earth-centred solar coordinates and the perpendicular coordinates used on the surface of the earth

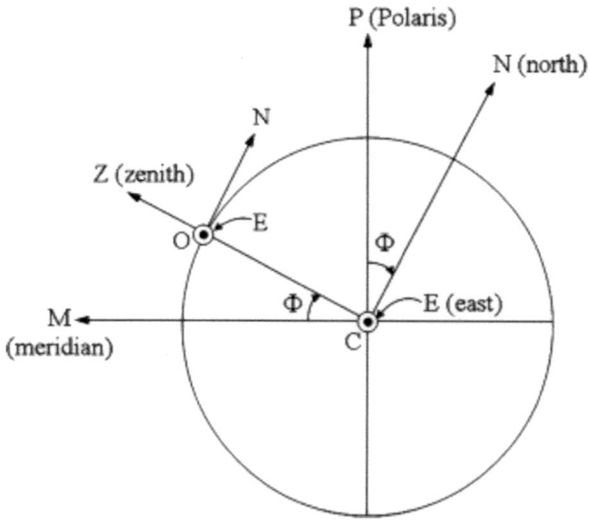

Fig. 6. Reference from equatorial plane to the plane of observer latitude

$$\begin{aligned} &\theta_m(0) = h_1(0) = 0 \\ &\dot{\theta}_m(0) = \dot{h}_1(0) = \ddot{\theta}_m(0) = \ddot{h}_1(0) = 0 \end{aligned} \tag{27}$$

And the classical control strategy is implemented as in equation (29):
Suppose that the sensor updates its output to a non-zero step change in the direction of sunrays $\dot{\theta}_P(t)$. The tracker will respond and track this new direction, which becomes the desired steady state direction of the sunrays; or $\dot{\theta}_P(t) = \theta_P(\infty)$, then for the new angular position of the sun:

$$h_1^*(t) = h_1(\infty) = \frac{800J_t}{K_m}\dot{\theta}_P(\infty), h_2^*(t) = h_2(\infty) = 0, h_3^*(t) = h_3(\infty) = 0 \quad (28)$$

$$\begin{aligned}
\dot{h}_1 &= h_2 \\
\dot{h}_2 &= h_3 \\
\dot{h}_3 &= -355.7276(h_1 - h_1^*(\infty)) - 165.0125h_2 - 38.114h_3 \\
u(t) &= (\frac{R_a K}{J_t} - 355.7276)\{\ h_1(t) - \frac{800J_t}{K_m}\theta_P(\infty)\ \} \\
&+ (\frac{R_a B + K_b K_m + L_a K}{J_t} - 165.0125)h_2 + (\frac{L_a B + R_a J}{J_t} - 38.114)h_3
\end{aligned} \quad (29)$$

4.3 Design of controller with trajectories of sunrays

The flatness property enables to integrate the mathematical formula for the trajectory of sunrays into the controller structure. These trajectories are derived in the sequel.

5. Derivation of the direction of sun rays as a function of local clock time

Sensorless solar tracking has been applied in solar-thermal systems (Ibrahim et al., 2004; Cheng & Wong, 2009; Power from the Sun, 2010; Chen et al., 2006; Stine & Harrington, 1988), and uses the concept of solar time and solar angle to relate the time of the day and time of the year to the position of the sun. For the derivation of the mathematical relationships employed in sensor-less solar tracking, consider first Figure 4, which shows the traditional measurement of the direction of sunrays, in a coordinate system with its origin as the centre of the earth. This coordinate system has one axis pointing toward the poles of the earth, and the other being the equatorial plane of the earth. Consider also the sunrays having an instantaneous declination δ with respect to the equatorial plane, through a meridian that differ from the meridian **O**, of the observer (the observer here being the polar-axis solar platform) by an angle (the solar angle) ω. The solar angle being the difference between the current meridian (or longitude) of sunrays and the observer meridian.

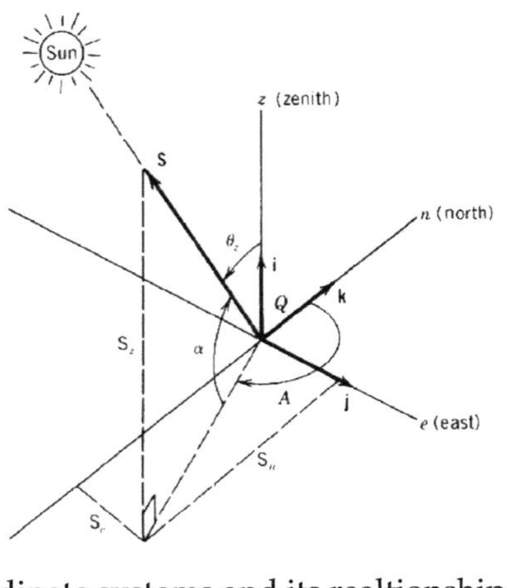

Fig. 7. Collector-centred coordinate systems and its realtionship to the earth-centred co-ordinates at the observer location

On the earth surface however, an observer uses a set of co-ordinates wherein, one of the cardinal axes points vertically upward, and the remaining two point north-south and east-west respectively. Figure 5 shows the relative orientations of the original solar co-ordinates and those used by an observer at the surface of the earth. The directional cosines relating the two co-ordinate systems [S], at the equator, are given by (Ibrahim et al., 2004; Cheng & Wong, 2009; Power from the Sun, 2010; Chen et al., 2006; Stine & Harrington, 1988) as:

$$[S]=\begin{bmatrix} S_m \\ S_e \\ S_p \end{bmatrix}=\begin{bmatrix} \cos\delta\cos\omega \\ -\cos\delta\sin\omega \\ \sin\delta \end{bmatrix} \tag{30}$$

Referred to the observer latitude Φ at position ***O***, as shown in Figure 6, a further set of angular transformations is given by equation (31):

$$[\Phi]=\begin{bmatrix} \cos\Phi & 0 & \sin\Phi \\ 0 & 1 & 0 \\ -\sin\Phi & 0 & \cos\Phi \end{bmatrix} \tag{31}$$

where is Φ the latitude angle.
For a tracking collector mounted on a stand, motion is only possible in two axes. Hence, for the collector surface located at ***O***, the solar position is measured in terms of an observer-centred cordinate system, consisting of a vertical (OV) or zenith (OZ) axis, and a horizontal (OH) axis. The elavation angle α (or its complement, the zenth angle θ_Z) and the azimuth angle β are therefore sufficient descriptors of the colloctor orientation in the OV-OH plane. This collector-centred co-ordinate system compares with the earth-surface co-ordinates as shown in Figure 7. The collector orientation is shown in Figure 8, and the orientation admits the representation in equation (32):

$$S^{/}=\begin{bmatrix} S_V \\ S_H \\ S_R \end{bmatrix}=\begin{bmatrix} \sin\alpha \\ \cos\alpha\sin\beta \\ \cos\alpha\cos\beta \end{bmatrix} \tag{32}$$

The collector elavation α, and its azimuth β, are the required tracking angles. In an ideal azimuth-elavation system, OV, OH and OR axes of the collector-centred frame are parallel to the OZ, OE and ON axes of the earth-surface frame, as shown in Figure 9. Generally, this coincidence may not apply, and the two co-ordinates are rotated from each other. The three possible scenarios are as illustrated in Figure 10. The associated transformation angles of the three orientations are given by equations (33)-(35):

$$[\varphi]=\begin{bmatrix} 1 & 0 & 0 \\ 0 & \cos\phi & -\sin\phi \\ 0 & \sin\phi & \cos\phi \end{bmatrix} \tag{33}$$

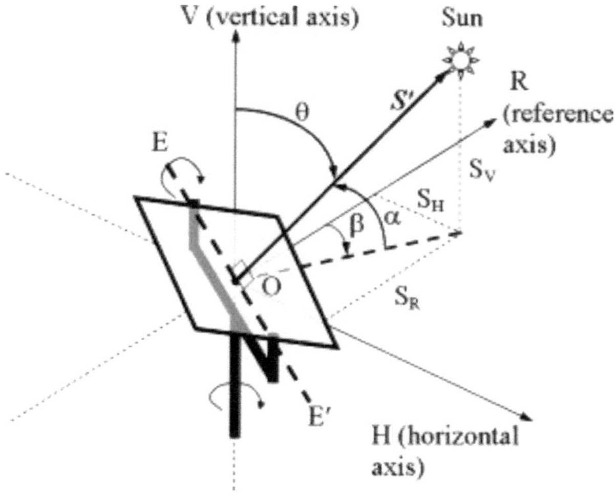

Fig. 8. Orientation of the collector in the OV-OH-OR axes

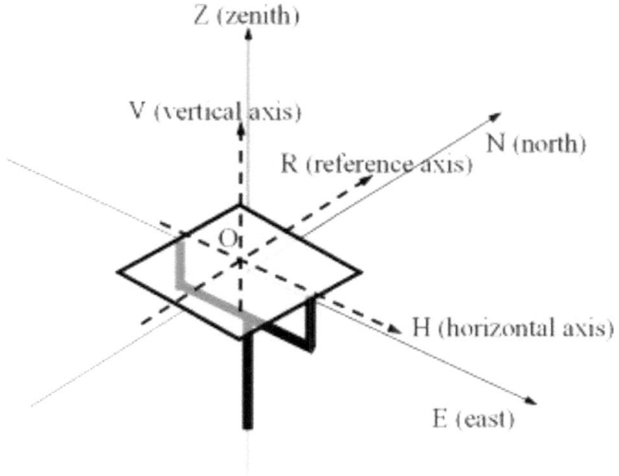

Fig. 9. Ideal orientation of collector in the OV-OH-OR axes

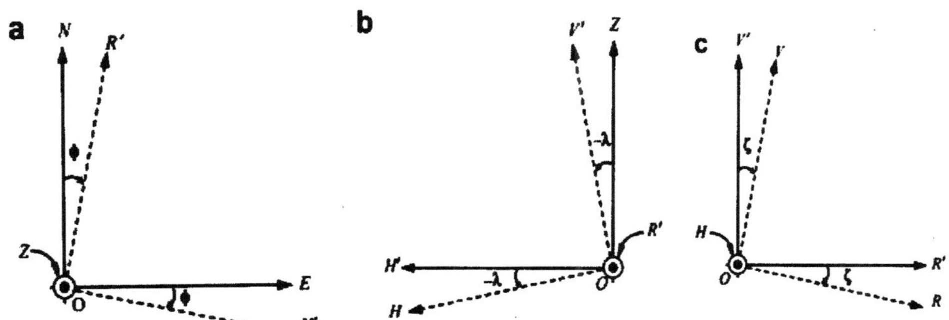

Fig. 10. Orientations of the OV-OH-OR axes relative to the OZ-OE-ON axes

$$[\lambda]=\begin{bmatrix}\cos\lambda & -\sin\lambda & 0\\ \sin\lambda & \cos\lambda & 0\\ 0 & 0 & 1\end{bmatrix} \tag{34}$$

$$[\zeta]=\begin{bmatrix}\cos\zeta & 0 & \sin\zeta\\ 0 & 1 & 0\\ -\sin\zeta & 0 & \cos\zeta\end{bmatrix} \tag{35}$$

The overall cosine angles are threfore given by the combined transformation as in equation (36):

$$\begin{bmatrix}S_V\\ S_H\\ S_R\end{bmatrix}=\begin{bmatrix}\sin\alpha\\ \cos\alpha\sin\beta\\ \cos\alpha\cos\beta\end{bmatrix}=[\zeta][\lambda][\phi][\Phi]\begin{bmatrix}\cos\delta\cos\omega\\ -\cos\delta\sin\omega\\ \sin\delta\end{bmatrix} \tag{36}$$

Solving the above matrix equation for the solar altitude angle in the collector-centred frame, we have for α, the elevation angle of the sun with respect to the orientation of the platform, given by:

$$\begin{aligned}&\alpha=\sin^{-1}\{ab+c+d\}\\ &a=\cos\delta\cos\omega\\ &b=\cos\zeta\cos\lambda\cos\Phi-\cos\zeta\sin\lambda\sin\phi\sin\Phi-\sin\zeta\cos\phi\sin\Phi\\ &c=-\cos\delta\sin\omega[\sin\zeta\sin\phi-\cos\zeta\sin\lambda\cos\phi]\\ &d=\sin\delta[\cos\zeta\cos\lambda\sin\Phi+\cos\zeta\sin\lambda\sin\phi\cos\Phi+\sin\zeta\cos\phi\cos\Phi]\end{aligned} \tag{37}$$

Or the zenith angle

$$\theta_Z=\frac{\pi}{2}-\alpha \tag{38}$$

Similarly, the azimuth angle is given by:

$$\begin{aligned}&\beta=\begin{cases}\sin^{-1}\left\{\dfrac{ae+m+f)}{\cos\alpha}\right\};\beta\geq 0\\ \pi-\sin^{-1}\left\{\dfrac{ae+m+f)}{\cos\alpha}\right\};\beta<0\end{cases}\\ &a=\cos\delta\cos\omega\\ &e=\sin\lambda\cos\Phi+\cos\lambda\sin\phi\sin\Phi\\ &m=-\cos\delta\sin\omega\cos\lambda\cos\phi\\ &f=\sin\delta[\sin\lambda\sin\Phi-\cos\lambda\sin\phi\cos\Phi]\end{aligned} \tag{39}$$

and because β can exist in any of the four quadrants, depending on the observer location on the earth, time of the day and season of the year, the following two evaluations must be jointly made to determine the quadrant of β, hence its actual value.

$$\begin{aligned}
&\sin\beta = \frac{g+h+i}{\cos\alpha} \\
&g = \cos\delta\cos\omega[\sin\lambda\cos\Phi + \cos\lambda\sin\phi\sin\Phi] \\
&h = -\cos\delta\sin\omega\cos\lambda\cos\phi \\
&i = \sin\delta[\sin\lambda\sin\Phi - \cos\lambda\sin\phi\cos\Phi]
\end{aligned} \tag{40}$$

$$\begin{aligned}
&\cos\beta = \frac{aj+k+l}{\cos\alpha} \\
&a = \cos\delta\cos\omega \\
&j = -\sin\zeta\cos\lambda\cos\Phi + \sin\zeta\sin\lambda\sin\phi\sin\Phi - \cos\zeta\cos\phi\sin\Phi \\
&k = -\cos\delta\sin\omega[\sin\zeta\sin\lambda\cos\phi + \cos\zeta\sin\phi] \\
&l = \sin\delta[-\sin\zeta\cos\lambda\sin\Phi - \sin\zeta\sin\lambda\sin\phi\cos\Phi + \cos\zeta\cos\phi\cos\Phi]
\end{aligned} \tag{41}$$

5.1 Special cases of observer solar angles

Special cases of the sun angles from the collector are presented by (Ibrahim et al., 2004). For example, for elavation-azimuth tracking, set the angles $\phi = \pi$, $\lambda = 0$ and $\zeta = \Phi - \pi/2$ in the general formulas. For this case, the general tracking formula can be then simplified to

$$\begin{aligned}
&\theta_Z = \pi / 2 - \delta \\
&\beta = \omega
\end{aligned} \tag{42}$$

Where θ_Z is the zenith angle, and β is the azimuth angle. In polar-axis tracking, the zenith angle is fixed (or seassonally fixed), as a function of the local lattitude angle. The azimuth anlge information is then available for one-axis tracking, from sunrise to sunset. From equation (42), the azimuth angle at any instant of time has the value of the sun hour angle, ω. Now, from (Power from the Sun, 2010),

$$\beta = \omega = 15(t_s - 12)^o; -180^o \le \omega \le 180^o \tag{43}$$

and t_s the solar hour, ω the solar angle is zero degrees when the sun is directly overhead, or when the solar hour is 12.00hrs.

5.2 Relating the azimuth angle of sunrays to the local clock time

The solar hour may differ from the local clock time (LCT). This difference is quantified by the equation-of-time (EOT). A version of the EOT which is accurate to within 0.63 seconds is given by (Power from the Sun, 2010):

$$EOT = \sum_{K=0}^{5} \left\{ A_K \cos\left(\frac{360Kn}{365.25}\right) + B_K \sin\left(\frac{360Kn}{365.25}\right) \right\} [hours] \tag{44}$$

Where the A_K's and B_K's are given in Table 1 and ***n*** is the number of days into a leap year cycle, with ***n=1*** being January of each leap year and ***n=1461*** corresponding to 31st December of the 4th year in a leap year cycle. The complete relationship between solar time and the local clock time is given in equation (45):

$$t_s = LCL + EOT - LC - D[hours] \tag{45}$$

K	A_K(hr)	B_K(hr)
0	2.0870×10^{-4}	0
1	9.2869×10^{-3}	$- 1.2229 \times 10^{-1}$
1	-5.2258×10^{-2}	$- 1.5698 \times 10^{-1}$
3	$- 1.3077 \times 10^{-3}$	$- 5.1602 \times 10^{-3}$
4	$- 2.1867 \times 10^{-3}$	$- 2.9823 \times 10^{-3}$
5	$- 1.5100 \times 10^{-4}$	$- 2.3463 \times 10^{-4}$

Table 1. Coefficients for conversion between solar time and clock time

D is 1 hour where daylight saving time is used, otherwise, D=0. In South Africa, D=0. The local longitude correction (LC) is given by:

$$LC = \frac{LL - LoLTz}{15}[hours] \tag{46}$$

where $\boldsymbol{LC}$: the longitude collection, $\boldsymbol{LL}$: local logitude at the location of the local time clock, and $\boldsymbol{LoLTz}$: Longitude of the standard time zone meridian. Summarising equations (43-46), the azimuth direction of sunrays, at a given location, as a function of time in seconds could be re-written as in equation (47).

$$\begin{gathered}\beta_s(t) = \frac{15}{3600}t - 180 \quad [\deg rees] \\ 0 \le t \le 86400 \sec\end{gathered} \tag{47}$$

And local clock time in seconds have been used in equation (47), instead of LCT. Given that no tracking is needed before sunrise, and after sunset, equation (47), may be written as

$$\begin{gathered}\beta_s(t) = 0.00416667(t - t_R) + \beta_R(0) - 180 \quad [\deg rees] \\ t_R \le t \le t_S\end{gathered} \tag{48}$$

Where $\boldsymbol{t_R}$ is the sunrise time (seconds) at the location, $\boldsymbol{\beta_R}(0)$=0.004166 $\boldsymbol{t_R}$ is the sunrise angle, and $\boldsymbol{t_s}$ is the sunset time.

5.3 Trajectories generation for sensorless tracking

To generate trajectories of motion for sensorless control, substitute equation (48) into equation (17) to obtain equation (49), for which the reference trajectories for feedback are given by:

$$\begin{aligned} h_1^*(t) &= \frac{K_m}{J_t}\beta_s(t) \\ h_2^*(t) &= \frac{K_m}{J_t}\dot{\beta}_s(t) \\ h_3^*(t) &= \frac{K_m}{J_t}\ddot{\beta}_s(t) \end{aligned} \tag{49}$$

Which, by substitution yields:

$$h_1^*(t) = 0.00416667\frac{K_m}{J_t}t; h_1^*(0) = 0.00416667\frac{K_m}{J_t}t_R$$
$$h_2^*(t) = 0.00416667\frac{K_m}{J_t}; h_2^*(0) = 0 \quad (50)$$
$$h_3^*(t) = 0, h_3^*(0) = 0$$

6. Simulations, results presentation and discussion

Simulation results present the dynamic response of the open-loop platform, the impact of feedback on platform performance, and the performance of the platform in tracking the direction of sunrays through sensorless control.

6.1 System data

The data used for the simulation of the platform systems is shown in Table 2.

R_a=5Ω	L_a=0.003H	B=3.95.10^{-6} Kg.ms^{-1}
K_b=0.0636V/rad/s	K_m=0.00711 Kgm/A	K=0.01Kgm^{2}s^{-2}
J_M=7.72.10^{-6} Kg m^2	J_L=970Kgm^2	N=1/n=1/800

Table 2. Platform system parameters

6.2 Dynamics of uncontrolled platform

The dynamics of the uncontrolled platform is shown in Figure 11. Overshoots of 96% are confirmed in the rotor position and speed. The settling time is 102seconds. The results confirm those presented in (de Lazzer, 2005; Agee et al., 2006).

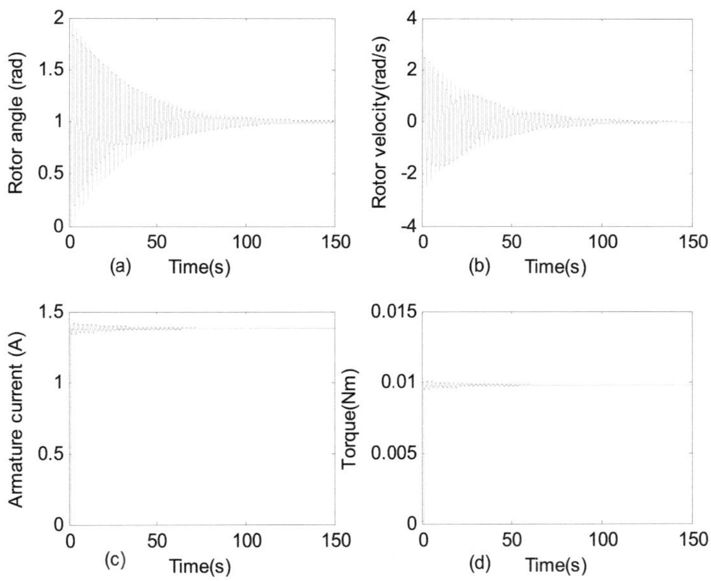

Fig. 11. Dynamics of open-loop platform

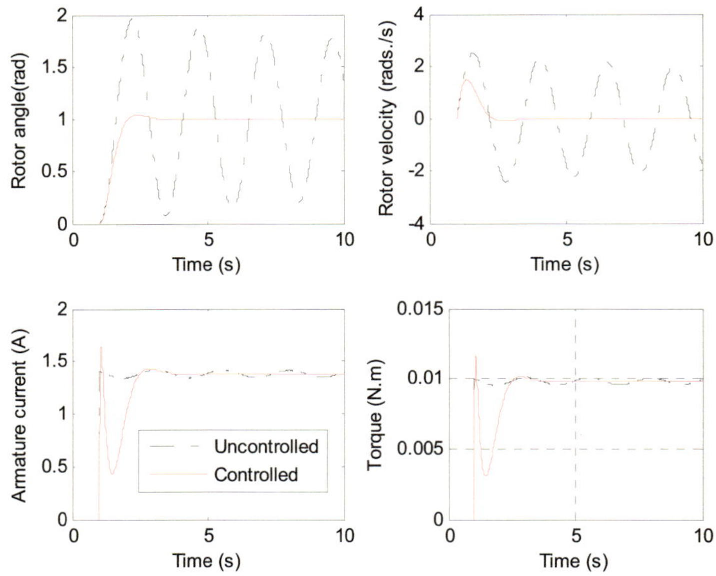

Fig. 12. Comparison of dynamics for controlled and uncontrolled platform

6.3 Dynamics of platform controlled by classical three-term controller

The effect of including a classical linear three-term controller is shown in Figure 12. Compared to the dynamics of the uncontrolled platform, the controller brings the systems to settle in about 2 seconds. Overshoots of the rotor angle and velocity are within the acceptable limit of 17% (Kuo and Galnoraghi, 2003). However, the overshoots in the motor armature current and the delivered torque have been worsened by the inclusion of the classical controller. Current overshoots and torque overshoots have implications in the choice of the rating and costs of the drive systems, as well as the extra installed PV capacity required for the drive hardware.

6.4 Performance of controlled platform with trajectories of motion

Figures 13-15 show the dynamics of the platform in sensorless tracking. In Figure 13, we see that the tracking of the direction of the sunrays is achieved within a second, and without oscillations. From Figures 14 and 15, we could conclude that the transient velocity (maximum value of about 0.003.8deg./sec) and maximum acceleration (0.02deg./sec^2) were not excessive. As such, the energy required for sensorless tracking is low. This is a major factor of merit when considering also the cost of supplying the tracker drive system.

7. Conclusions

In conclusion, a flat model of the solar tracker was presented and used for controller design. Mathematical derivation of the direction of sunrays as a function of the local clock time was given. It was also shown how the flatness property could be combined with the mathematical formulation of the direction of sunrays to generate trajectories of motion for sensorless tracking. Results also showed that sensorless tracking was achieved without oscillations, at modest velocities and accelerations. The low energy requirement in

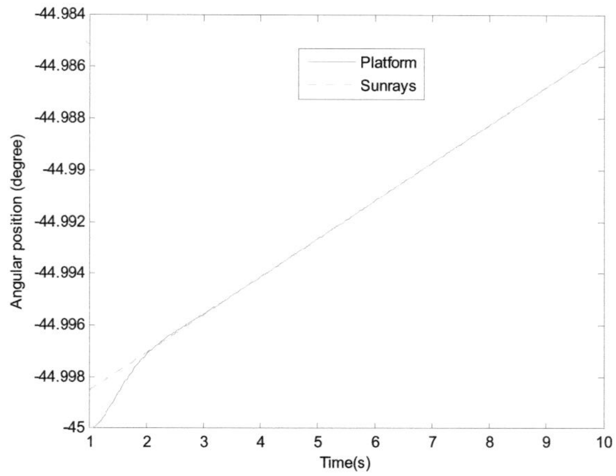

Fig. 13. Polar axis tracking of angle of sunrays

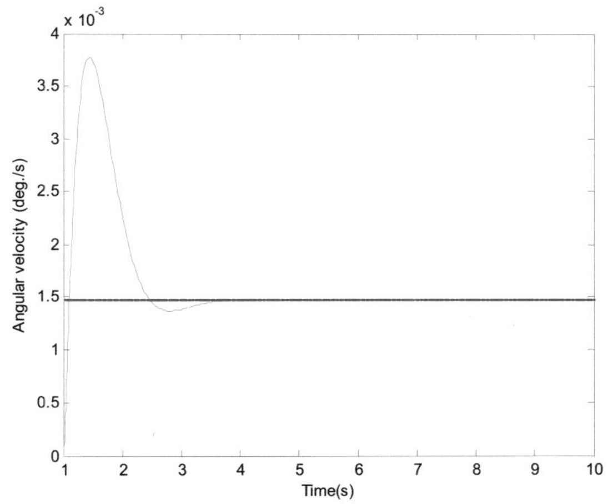

Fig. 14. Velocity response of tracker

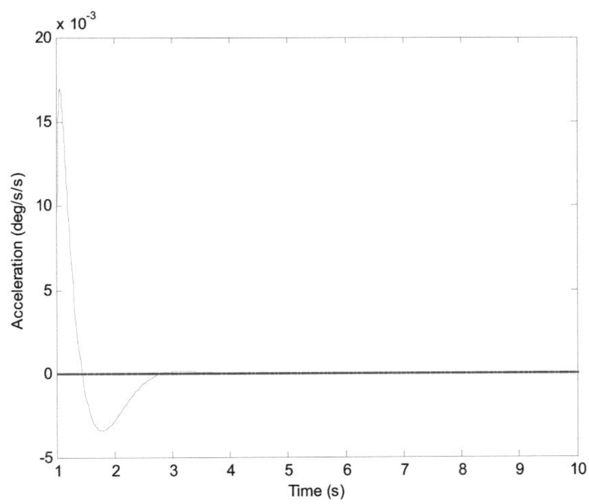

Fig. 15. Comparison of acceleration

sensorless tracking could be beneficial in reducing the rating requirements of auxiliary photovoltaic power, required for the tracker drive system. Combined with the elimination of sensor cost, the reduced drive energy requirement could lead to significant reductions in the overall cost of photovoltaic hardware.

8. References

Agee, J. T. Obok-Opok, A. and de Lazzer, M. (2006). "Solar tracker technologies: market trends and field applications. *Int. Conf. on Eng. Research and Development: Impact on Industries.* 5-7th September, 2006.

Agee, J. T. , de Lazzer, M. an Yanev, M. K. "A Pole cancellation strategy for stabilising a 3KW solar power platform. *Int. Conf. Power and Energy Systems(EuroPES 2006)*, Rhodes, Greece. June 26-28.

Agee, J. T. and Jimoh, A. A. (2010) Flat Control of a Polar-Axis Photovoltaic Solar power Platform. Submitted.

Greenology (2010). Available on http://www.greeology.co.za, 25th June, 2010

Bitaud, L., Fliess, M. and Levine, J. (2003).A Flatness-Based Control Synthesis of Linear Systems and Application to Wind Sheild Wiper. Proceedings of the European Control Conference (ECC'97), Brussels. Pp. 1-6.

Cheng, K. K. and Wong, C. W. (2009). General Formula for Ones-axis Tracking Systems and its Application in Improving Tracking Accuracy of Solar Collectors. Solar Energy vol. 83, Issue 3, pp. 298-305.

Chen, Y. T., Lim, B. H. and Lim, C. S (2006). General Sun Tracking Formula for Heliostats with Arbitrary Oriented Axes, Journal of Solar Energy, vol 128, pp. 245-250.

De Lazzer, M. *Positioning System for an Array of Solar Panels* (M.Eng Thesis (Unpublished). University of Botswana and Ecoles de Saint Cyr, France, 2005).

Energy from the Desert (2003): Practical Proposals for Large Scale Photovoltaic Systems. Edited bt: Kusoke Korokawa, Keiichi Komoto, Peter van der Vlueten and David Faiman. Pp. 150.

Fliess, M., Levine, J, Martan, P., Ollivier, F. and Rouchon, P. (1997).Controlling Nonlinear Systems by Flatness. In Systems and Control in the Twenty-first Century (Progress in Systems and Control Theory); ed. Byrness, C. I., Datta, B. N., Gilliam, S. And Martin, C. F. Birhauser, Boston. Pp. 137-154.

Fliess, M, Levine, J, Martin, P and Rouchon, P. (1990). A Lie-Backland Approach to Equivalence and Flatness of Nonlinear Systems. IEEE Transactions in Automatic Control; vol. 44, no.5, pp. 922-937.

Kuo, B. C. and Golnaraghi, F. *Automatic Control Systems* (eight edition, John Wiley and Sons, Inc., 2003).

Stine, W. B. And Harringan, R. W. (1985) Solar Energy Fundamentals and design (First ed.). Willey Interscience, New York. Pp. 38-69.

The Suns position. Available on http://www.powerfromthesun.net/chapter3/chapter3word.htm, 25th June, 2010

13

General Formula for On-Axis Sun-Tracking System

Kok-Keong Chong, Chee-Woon Wong
Universiti Tunku Abdul Rahman
Malaysia

1. Introduction

Sun-tracking system plays an important role in the development of solar energy applications, especially for the high solar concentration systems that directly convert the solar energy into thermal or electrical energy. High degree of sun-tracking accuracy is required to ensure that the solar collector is capable of harnessing the maximum solar energy throughout the day. High concentration solar power systems, such as central receiver system, parabolic trough, parabolic dish etc, are the common in the applications of collecting solar energy. In order to maintain high output power and stability of the solar power system, a high-precision sun-tracking system is necessary to follow the sun's trajectory from dawn until dusk.

For achieving high degree of tracking accuracy, sun-tracking systems normally employ sensors to feedback error signals to the control system for continuously receiving maximum solar irradiation on the receiver. Over the past two decades, various strategies have been proposed and they can be classified into the following three categories, i.e. open-loop, closed-loop and hybrid sun-tracking (Lee et al., 2009). In the open-loop tracking approach, the control program will perform calculation to identify the sun's path using a specific sun-tracking formula in order to drive the solar collector towards the sun. Open-loop sensors are employed to determine the rotational angles of the tracking axes and guarantee that the solar collector is positioned at the right angles. On the other hand, for the closed-loop tracking scheme, the solar collector normally will sense the direct solar radiation falling on a closed-loop sensor as a feedback signal to ensure that the solar collector is capable of tracking the sun all the time. Instead of the above options, some researchers have also designed a hybrid system that contains both the open-loop and closed-loop sensors to attain a good tracking accuracy. The above-mentioned tracking methods are operated by either a microcontroller based control system or a PC based control system in order to trace the position of the sun.

Azimuth-elevation and tilt-roll tracking mechanisms are among the most commonly used sun-tracking methods for aiming the solar collector towards the sun at all times. Each of these two sun-tracking methods has its own specific sun-tracking formula and they are not interrelated in many decades ago. In this chapter, the most general form of sun-tracking formula that embraces all the possible on-axis tracking approaches is derived and presented in details. The general sun-tracking formula not only can provide a general mathematical solution, but more significantly, it can improve the sun-tracking accuracy by tackling the installation error of the solar collector. The precision of foundation alignment during the installation of solar collector becomes tolerable because any imprecise configuration in the tracking axes can be easily compensated by changing the parameters' values in the general sun-tracking formula. By integrating the novel general formula into the open-loop sun-tracking system, this strategy is definitely a cost effective way to be capable of remedying the installation error of the solar collector with a significant improvement in the tracking accuracy.

2. Overview of sun-tracking systems

2.1 Sun-tracking approaches

A good sun-tracking system must be reliable and able to track the sun at the right angle even in the periods of cloud cover. Over the past two decades, various types of sun-tracking mechanisms have been proposed to enhance the solar energy harnessing performance of solar collectors. Although the degree of accuracy required depends on the specific characteristics of the solar concentrating system being analyzed, generally the higher the system concentration the higher the tracking accuracy will be needed (Blanco-Muriel et al., 2001).

In this section, we would like to briefly review the three categories of sun-tracking algorithms (i.e. open-loop, closed-loop and hybrid) with some relevant examples. For the closed-loop sun-tracking approach, various active sensor devices, such as CCD sensor or photodiode sensor are utilized to sense the position of the solar image on the receiver and a feedback signal is then generated to the controller if the solar image moves away from the receiver. Sun-tracking systems that employ active sensor devices are known as closed-loop sun trackers. Although the performance of the closed-loop tracking system is easily affected by weather conditions and environmental factors, it has allowed savings in terms of cost, time and effort by omitting more precise sun tracker alignment work. In addition, this strategy is capable of achieving a tracking accuracy in the range of a few milli-radians (mrad) during fine weather. For that reason, the closed-loop tracking approach has been traditionally used in the active sun-tracking scheme over the past 20 years (Arbab et al., 2009; Berenguel et al., 2004; Kalogirou, 1996; Lee et al., 2006). For example, Kribus et al. (2004) designed a closed-loop controller for heliostats, which improved the pointing error of the solar image up to 0.1 mrad, with the aid of four CCD cameras set on the target. However, this method is rather expensive and complicated because it requires four CCD cameras and four radiometers to be placed on the target. Then the solar images captured by CCD cameras must be analysed by a computer to generate the control correction feedback for correcting tracking errors. In 2006, Luque-Heredia et al. (2006) presented a sun-tracking error monitoring system that uses a monolithic optoelectronic sensor for a concentrator photovoltaic system. According to the results from the case study, this monitoring system achieved a tracking accuracy of better than 0.1°. However, the criterion is that this tracking system requires full clear sky days to operate, as the incident sunlight has to be above a certain threshold to ensure that the minimum required resolution is met. That same year, Aiuchi et al. (2006) developed a heliostat with an equatorial mount and a closed-loop photo-sensor control system. The experimental results showed that the tracking error of the heliostat was estimated to be 2 mrad during fine weather. Nevertheless, this tracking method is not popular and only can be used for sun trackers with an equatorial mount configuration, which is not a common tracker mechanical structure and is complicated because the central of gravity for the solar collector is far off the pedestal. Furthermore, Chen et al. (2006, 2007) presented studies of digital and analogue sun sensors based on the optical vernier and optical nonlinear compensation measuring principle respectively. The proposed digital and analogue sun sensors have accuracies of 0.02° and 0.2° correspondingly for the entire field of view of ±64° and ±62° respectively. The major disadvantage of these sensors is that the field of view, which is in the range of about ±64° for both elevation and azimuth directions, is rather small compared to the dynamic range of motion for a practical sun tracker that is about ±70° and ±140° for elevation and azimuth directions, respectively. Besides that, it is just implemented at the testing stage in precise sun sensors to measure the position of the sun and has not yet been applied in any closed-loop sun-tracking system so far.

Although closed-loop sun-tracking system can produce a much better tracking accuracy, this type of system will lose its feedback signal and subsequently its track to the sun position when the sensor is shaded or when the sun is blocked by clouds. As an alternative method to overcome the limitation of closed-loop sun trackers, open-loop sun trackers were introduced by using open-loop sensors that do not require any solar image as feedback. The open-loop sensor such as encoder will ensure that the solar collector is positioned at pre-calculated angles, which are obtained from a special formula or algorithm. Referring to the literatures (Blanco-Muriel et al., 2001; Grena, 2008; Meeus, 1991; Reda & Andreas, 2004; Sproul, 2007), the sun's azimuth and elevation angles can be determined by the sun position formula or algorithm at the given date, time and geographical information. This tracking approach has the ability to achieve tracking error within ±0.2° when the mechanical structure is precisely made as well as the alignment work is perfectly done. Generally, these algorithms are integrated into the microprocessor based or computer based controller. In 2004, Abdallah and Nijmeh (2004) designed a two axes sun tracking system, which is operated by an open-loop control system. A programmable logic controller (PLC) was used to calculate the solar vector and to control the sun tracker so that it follows the sun's trajectory. In addition, Shanmugam & Christraj (2005) presented a computer program written in *Visual Basic* that is capable of determining the sun's position and thus drive a paraboloidal dish concentrator (PDS) along the East-West axis or North-South axis for receiving maximum solar radiation.

In general, both sun-tracking approaches mentioned above have both strengths and drawbacks, so some hybrid sun-tracking systems have been developed to include both the open-loop and closed-loop sensors for the sake of high tracking accuracy. Early in the 21st century, Nuwayhid et al. (2001) adopted both the open-loop and closed-loop tracking methods into a parabolic concentrator attached to a polar tracking system. In the open-loop scheme, a computer acts as controller to calculate two rotational angles, i.e. solar declination and hour angles, as well as to drive the concentrator along the declination and polar axes. In the closed-loop scheme, nine light-dependent resistors (LDR) are arranged in an array of a circular-shaped "iris" to facilitate sun-tracking with a high degree of accuracy. In 2004, Luque-Heredia et al. (2004) proposed a novel *PI* based hybrid sun-tracking algorithm for a concentrator photovoltaic system. In their design, the system can act in both open-loop and closed-loop mode. A mathematical model that involves a time and geographical coordinates function as well as a set of disturbances provides a feed-forward open-loop estimation of the sun's position. To determine the sun's position with high precision, a feedback loop was introduced according to the error correction routine, which is derived from the estimation of the error of the sun equations that are caused by external disturbances at the present stage based on its historical path. One year later, Rubio et al. (2007) fabricated and evaluated a new control strategy for a photovoltaic (PV) solar tracker that operated in two tracking modes, i.e. normal tracking mode and search mode. The normal tracking mode combines an open-loop tracking mode that is based on solar movement models and a closed-loop tracking mode that corresponds to the electro-optical controller to obtain a sun-tracking error, which is smaller than a specified boundary value and enough for solar radiation to produce electrical energy. Search mode will be started when the sun-tracking error is large or no electrical energy is produced. The solar tracker will move according to a square spiral pattern in the azimuth-elevation plane to sense the sun's position until the tracking error is small enough.

2.2 Types of sun trackers

Taking into consideration of all the reviewed sun-tracking methods, sun trackers can be grouped into one-axis and two-axis tracking devices. Fig. 1 illustrates all the available types of sun trackers in the world. For one-axis sun tracker, the tracking system drives the collector about an axis of rotation until the sun central ray and the aperture normal are coplanar. Broadly speaking, there are three types of one-axis sun tracker:

1. **Horizontal-Axis Tracker** – the tracking axis is to remain parallel to the surface of the earth and it is always oriented along East-West or North-South direction.
2. **Tilted-Axis Tracker** – the tracking axis is tilted from the horizon by an angle oriented along North-South direction, e.g. Latitude-tilted-axis sun tracker.
3. **Vertical-Axis Tracker** – the tracking axis is collinear with the zenith axis and it is known as azimuth sun tracker.

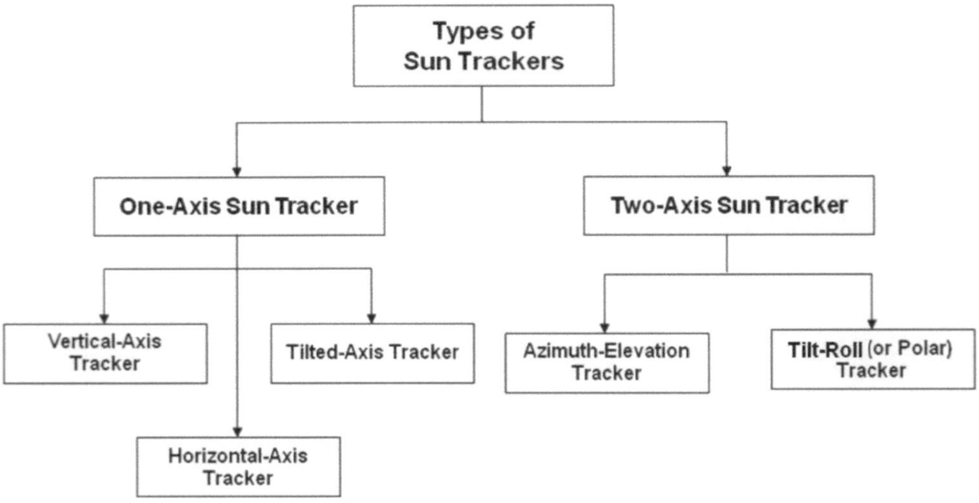

Fig. 1. Types of sun trackers

In contrast, the two-axis sun tracker, such as azimuth-elevation and tilt-roll sun trackers, tracks the sun in two axes such that the sun vector is normal to the aperture as to attain 100% energy collection efficiency. Azimuth-elevation and tilt-roll (or polar) sun tracker are the most popular two-axis sun tracker employed in various solar energy applications. In the azimuth-elevation sun-tracking system, the solar collector must be free to rotate about the azimuth and the elevation axes. The primary tracking axis or azimuth axis must parallel to the zenith axis, and elevation axis or secondary tracking axis always orthogonal to the azimuth axis as well as parallel to the earth surface. The tracking angle about the azimuth axis is the solar azimuth angle and the tracking angle about the elevation axis is the solar elevation angle. Alternatively, tilt-roll (or polar) tracking system adopts an idea of driving the collector to follow the sun-rising in the east and sun-setting in the west from morning to evening as well as changing the tilting angle of the collector due to the yearly change of sun path. Hence, for the tilt-roll tracking system, one axis of rotation is aligned parallel with the earth's polar axis that is aimed towards the star Polaris. This gives it a tilt from the horizon equal to the local latitude angle. The other axis of rotation is perpendicular to this polar axis. The tracking angle about the polar axis is equal to the sun's hour angle and the tracking angle about the perpendicular axis is dependent on the declination angle. The advantage of tilt-roll tracking is that the tracking velocity is almost constant at 15 degrees per hour and therefore the control system is easy to be designed.

2.3 The challenges of sun-tracking systems

In fact, the tracking accuracy requirement is very much reliant on the design and application of the solar collector. In this case, the longer the distance between the solar concentrator and the receiver the higher the tracking accuracy required will be because the solar image becomes more sensitive to the movement of the solar concentrator. As a result, a heliostat or off-axis sun tracker normally requires much higher tracking accuracy compared to that of on-axis sun tracker for the reason that the distance between the heliostat and the target is normally much longer, especially for a central receiver system configuration. In this context, a tracking accuracy in the range of a few miliradians (mrad) is in fact sufficient for an on-axis sun tracker to maintain its good performance when highly concentrated sunlight is involved (Chong et al, 2010). Despite having many existing on-axis sun-tracking methods, the designs available to achieve a good tracking accuracy of a few mrad are complicated and expensive. It is worthwhile to note that conventional on-axis sun-tracking systems normally adopt two common configurations, which are azimuth-elevation and tilt-roll (polar tracking), limited by the available basic mathematical formulas of sun-tracking system. For azimuth-elevation tracking system, the sun-tracking axes must be strictly aligned with both zenith and real north. For a tilt-roll tracking system, the sun-tracking axes must be exactly aligned with both latitude angle and real north. The major cause of sun-tracking errors is how well the aforementioned alignment can be done and any installation or fabrication defect will result in low tracking accuracy. According to our previous study for the azimuth-elevation tracking system, a misalignment of azimuth shaft relative to zenith axis of 0.4° can cause tracking error ranging from 6.45 to 6.52 mrad (Chong & Wong, 2009). In practice, most solar power plants all over the world use a large solar collector area to save on manufacturing cost and this has indirectly made the alignment work of the sun-tracking axes much more difficult. In this case, the alignment of the tracking axes involves an extensive amount of heavy-duty mechanical and civil works due to the requirement for thick shafts to support the movement of a large solar collector, which normally has a total collection area in the range of several tens of square meters to nearly a hundred square meters. Under such tough conditions, a very precise alignment is really a great challenge to the manufacturer because a slight misalignment will result in significant sun-tracking errors. To overcome this problem, an unprecedented on-axis general sun-tracking formula has been proposed to allow the sun tracker to track the sun in any two arbitrarily orientated tracking axes (Chong & Wong, 2009). In this chapter, we would like to introduce a novel sun-tracking system by integrating the general formula into the sun-tracking algorithm so that we can track the sun accurately and cost effectively, even if there is some misalignment from the ideal azimuth-elevation or tilt-roll configuration. In the new tracking system, any misalignment or defect can be rectified without the need for any drastic or labor-intensive modifications to either the hardware or the software components of the tracking system. In other words, even though the alignments of the azimuth-elevation axes with respect to the zenith-axis and real north are not properly done during the installation, the new sun-tracking algorithm can still accommodate the misalignment by changing the values of parameters in the tracking program. The advantage of the new tracking algorithm is that it can simplify the fabrication and installation work of solar collectors with higher tolerance in terms of the tracking axes alignment. This strategy has allowed great savings in terms of cost, time and effort by omitting complicated solutions proposed by other researchers such as adding a closed-loop feedback controller or a flexible and complex mechanical structure to level out the sun-tracking error (Chen et al., 2001; Luque-Heredia et al., 2007).

3. General formula for on-axis sun-tracking system

A novel general formula for on-axis sun-tracking system has been introduced and derived to allow the sun tracker to track the sun in two orthogonal driving axes with any arbitrary orientation (Chong & Wong, 2009). Chen et al. (2006) was the pioneer group to derive a general sun-tracking formula for heliostats with arbitrarily oriented axes. The newly derived general formula by Chen et al. (2006) is limited to the case of off-axis sun tracker (heliostat) where the target is fixed on the earth surface and hence a heliostat normal vector must always bisect the angle between a sun vector and a target vector. As a complimentary to Chen's work, Chong and Wong (2009) derive the general formula for the case of on-axis sun tracker where the target is fixed along the optical axis of the reflector and therefore the reflector normal vector must be always parallel with the sun vector. With this complete mathematical solution, the use of azimuth-elevation and tilt-roll tracking formulas are the special case of it.

3.1 Derivation of general formula

Prior to mathematical derivation, it is worthwhile to state that the task of the on-axis sun-tracking system is to aim a solar collector towards the sun by turning it about two perpendicular axes so that the sunray is always normal relative to the collector surface. Under this circumstance, the angles that are required to move the solar collector to this orientation from its initial orientation are known as sun-tracking angles. In the derivation of sun-tracking formula, it is necessary to describe the sun's position vector and the collector's normal vector in the same coordinate reference frame, which is the collector-centre frame. Nevertheless, the unit vector of the sun's position is usually described in the earth-centre frame due to the sun's daily and yearly rotational movements relative to the earth. Thus, to derive the sun-tracking formula, it would be convenient to use the coordinate transformation method to transform the sun's position vector from earth-centre frame to earth-surface frame and then to collector-centre frame. By describing the sun's position vector in the collector-centre frame, we can resolve it into solar azimuth and solar altitude angles relative to the solar collector and subsequently the amount of angles needed to move the solar collector can be determined easily.

According to Stine & Harrigan (1985), the sun's position vector relative to the earth-centre frame can be defined as shown in Fig. 2, where CM, CE and CP represent three orthogonal axes from the centre of earth pointing towards the meridian, east and Polaris respectively. The unified vector for the sun position **S** in the earth-centre frame can be written in the form of direction cosines as follow:

$$\boldsymbol{S} = \begin{bmatrix} S_M \\ S_E \\ S_P \end{bmatrix} = \begin{bmatrix} \cos\delta\cos\omega \\ -\cos\delta\sin\omega \\ \sin\delta \end{bmatrix} \tag{1}$$

where δ is the declination angle and ω is hour angle are defined as follow (Stine & Harrigan, 1985): The accuracy of the declination angles is important in navigation and astronomy. However, an approximation accurate to within 1 degree is adequate in many solar purposes. One such approximation for the declination angle is

$$\delta = \sin^{-1}\{0.39795 \cos [0.98563 (N-173)]\} \quad \text{(degrees)} \tag{2}$$

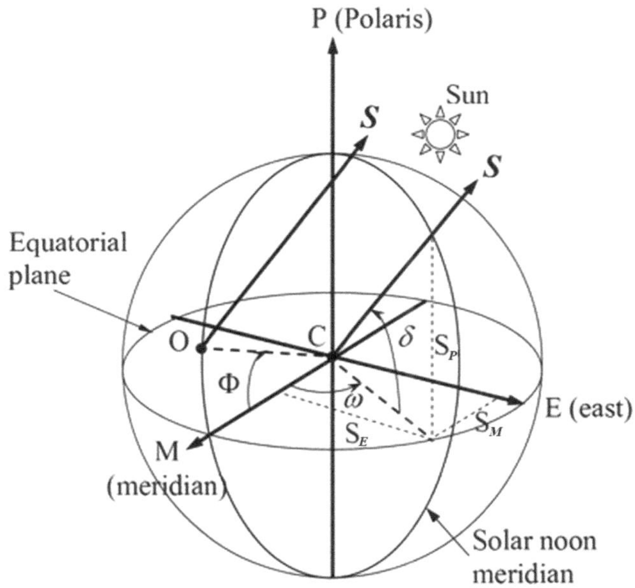

Fig. 2. The sun's position vector relative to the earth-centre frame. In the earth-centre frame, CM, CE and CP represent three orthogonal axes from the centre of the earth pointing towards meridian, east and Polaris, respectively

where N is day number and calendar dates are expressed as the N = 1, starting with January 1. Thus March 22 would be N = 31 + 28 + 22 = 81 and December 31 means N = 365.
The hour angle expresses the time of day with respect to the solar noon. It is the angle between the planes of the meridian-containing observer and meridian that touches the earth-sun line. It is zero at solar noon and increases by 15° every hour:

$$\omega = 15(t_s - 12) \quad \text{(degrees)} \tag{3}$$

where t_s is the solar time in hours. A solar time is a 24-hour clock with 12:00 as the exact time when the sun is at the highest point in the sky. The concept of solar time is to predict the direction of the sun's ray relative to a point on the earth. Solar time is location or longitudinal dependent. It is generally different from local clock time *(LCT)* (defined by politically time zones)

Fig. 3 depicts the coordinate system in the earth-surface frame that comprises of OZ, OE and ON axes, in which they point towards zenith, east and north respectively. The detail of coordinate transformation for the vector *S* from earth-centre frame to earth-surface frame was presented by Stine & Harrigan (1985) and the needed transformation matrix for the above coordinate transformation can be expressed as

$$[\Phi]=\begin{bmatrix}\cos\Phi & 0 & \sin\Phi\\ 0 & 1 & 0\\ -\sin\Phi & 0 & \cos\Phi\end{bmatrix} \tag{4}$$

where Φ is the latitude angle.

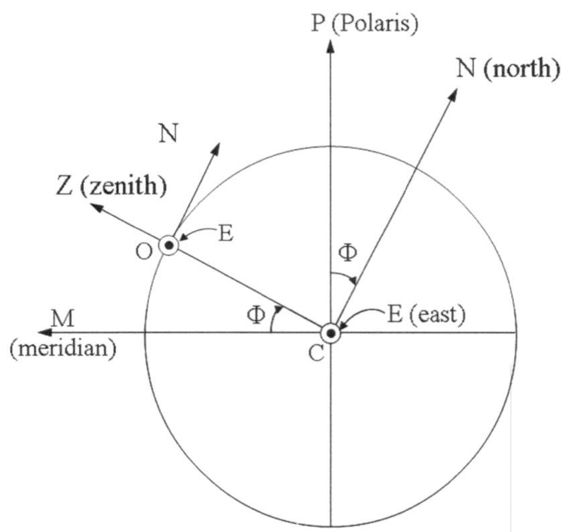

Fig. 3. The coordinate system in the earth-surface frame that consists of OZ, OE and ON axes, in which they point towards zenith, east and north respectively. The transformation of the vector S from earth-centre frame to earth-surface frame can be obtained through a rotation angle that is equivalent to the latitude angle (Φ)

Now, let us consider a new coordinate system that is defined by three orthogonal coordinate axes in the collector-centre frame as shown in Fig. 4. For the collector-centre frame, the origin O is defined at the centre of the collector surface and it coincides with the origin of earth-surface frame. OV is defined as vertical axis in this coordinate system and it is parallel with first rotational axis of the solar collector. Meanwhile, OR is named as reference axis in which one of the tracking angle β is defined relative to this axis. The third orthogonal axis, OH, is named as horizontal axis and it is parallel with the initial position of the second rotational axis. The OR and OH axes form the level plane where the collector surface is driven relative to this plane. Fig. 4 also reveals the simplest structure of solar collector that can be driven in two rotational axes: the first rotational axis that is parallel with OV and the second rotational axis that is known as EE′ dotted line (it can rotate around the first axis during the sun-tracking but must always be perpendicular with the first axis). From Fig. 4, θ is the amount of rotational angle about EE′ axis measured from OV axis, whereas β is the rotational angle about OV axis measured from OR axis. Furthermore, α is solar altitude angle in the collector-centre frame, which is equal to $\pi/2-\theta$. In the collector-centre frame, the sun position S' can be written in the form of direction cosines as follow:

$$S' = \begin{bmatrix} S_V \\ S_H \\ S_R \end{bmatrix} = \begin{bmatrix} \sin\alpha \\ \cos\alpha\sin\beta \\ \cos\alpha\cos\beta \end{bmatrix} \tag{5}$$

In an ideal azimuth-elevation system, OV, OH and OR axes of the collector-centre frame are parallel with OZ, OE and ON axes of the earth-surface frame accordingly as shown in Fig. 5. To generalize the mathematical formula from the specific azimuth-elevation system to any arbitrarily oriented sun-tracking system, the orientations of OV, OH and OR axes will be described by three tilted angles relative to the earth-surface frame. Three tilting angles have been introduced here because the two-axis mechanical drive can be arbitrarily oriented about any of the three principal axes of earth-surface frame: ϕ is the rotational angle about zenith-axis if the other two angles are null, λ is the rotational angle about north-axis if the other two angles are null and ζ is the rotational angle about east-axis if the other two angles are null. On top of that, the combination of the above-mentioned angles can further generate more unrepeated orientations of the two tracking axes in earth-surface frame, which is very important in later consideration for improving sun-tracking accuracy of solar collector.

Fig. 6(a) – (c) show the process of how the collector-centre frame is tilted step-by-step relative to the earth-surface frame, where OV′, OH′ and OR′ axes represent the intermediate position for OV, OH and OR axes, respectively. In Fig. 6(a), the first tilted angle, $+\phi$, is a rotational angle about the OZ axis in clockwise direction. In Fig. 6(b), the second tilted angle, $-\lambda$, is a rotational angle about OR′ axis in counter-clockwise direction. Lastly, in Fig. 6(c), the third tilted angle, $+\zeta$, is a rotational angle about OH axis in clockwise direction. Fig. 7 shows the combination of the above three rotations in 3D view for the collector-centre frame relative to the earth-surface frame, where the change of coordinate system for each axis follows the order: $Z \rightarrow V' \rightarrow V$, $E \rightarrow H' \rightarrow H$ and $N \rightarrow R' \rightarrow R$. Similar to the latitude angle, in the direction representation of the three tilting angles, we define positive sign to the angles, i.e. ϕ, λ, ζ, for the rotation in the clockwise direction. In other words, clockwise and counter-clockwise rotations can be named as positive and negative rotations respectively.

As shown in Fig. 6(a) – (c), the transformation matrices correspond to the three tilting angles (ϕ, λ and ζ) can be obtained accordingly as follow:

$$[\phi] = \begin{bmatrix} 1 & 0 & 0 \\ 0 & \cos\phi & -\sin\phi \\ 0 & \sin\phi & \cos\phi \end{bmatrix} \tag{6a}$$

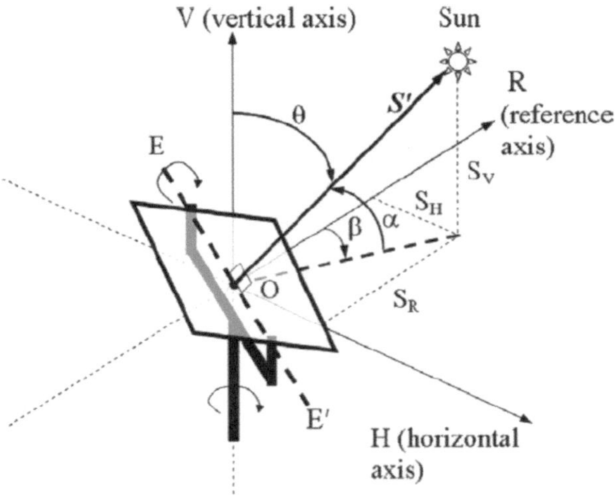

Fig. 4. In the collector-centre frame, the origin O is defined at the centre of the collector surface and it coincides with the origin of earth-surface frame. OV is defined as vertical axis in this coordinate system and it is parallel with first rotational axis of the solar collector. Meanwhile, OR is named as reference axis and the third orthogonal axis, OH, is named as horizontal axis. The OR and OH axes form the level plane where the collector surface is driven relative to this plane. The simplest structure of solar collector that can be driven in two rotational axes: the first rotational axis that is parallel with OV and the second rotational axis that is known as EE' dotted line (it can rotate around the first axis during the sun-tracking but must always remain perpendicular with the first axis). From the diagram, θ is the amount of rotational angle about EE' axis measured from OV axis, whereas β is the amount of rotational angle about OV axis measured from OR axis. Furthermore, α is solar altitude angle in the collector-centre frame, which is expressed as $\pi/2$ - θ

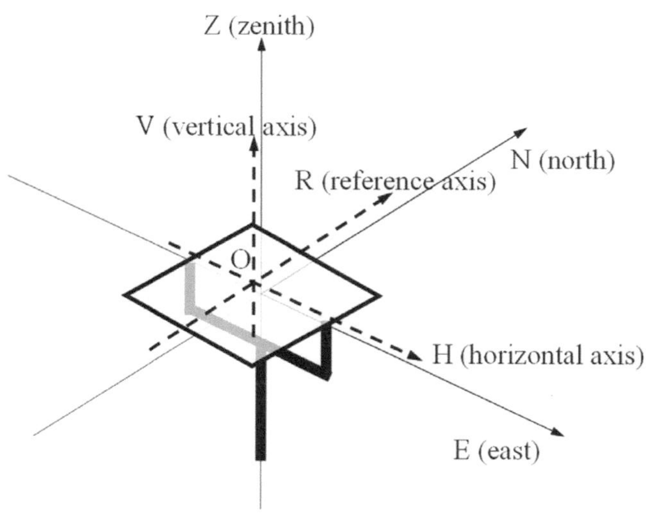

Fig. 5. In an ideal azimuth-elevation system, OV, OH and OR axes of the collector-centre frame are parallel with OZ, OE and ON axes of the earth-surface frame accordingly

(a)

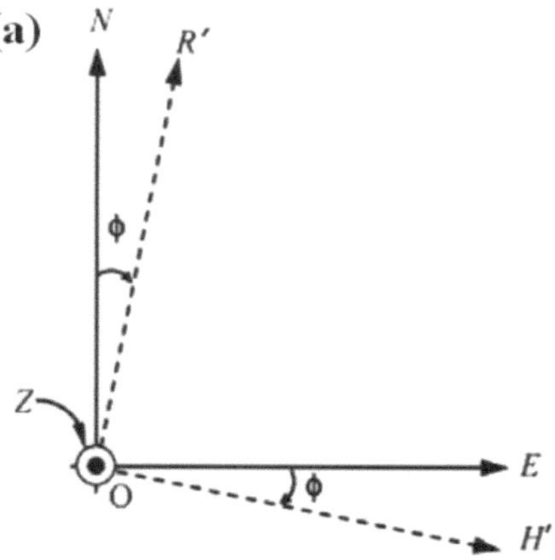

$$\begin{bmatrix} Z \\ H' \\ R' \end{bmatrix} = \begin{bmatrix} 1 & 0 & 0 \\ 0 & \cos\phi & -\sin\phi \\ 0 & \sin\phi & \cos\phi \end{bmatrix} \begin{bmatrix} Z \\ E \\ N \end{bmatrix}$$

(b)

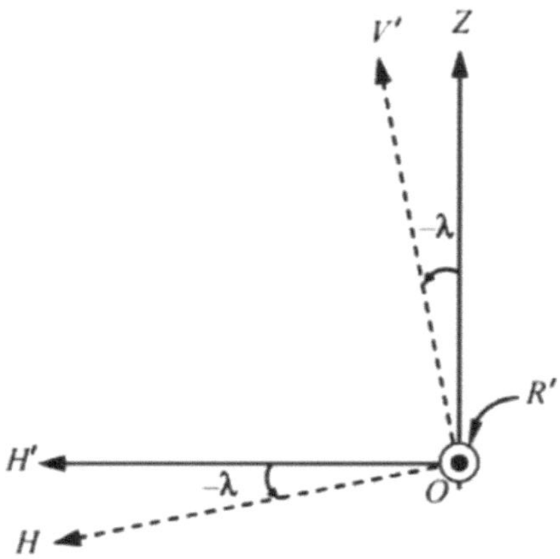

$$\begin{bmatrix} V' \\ H \\ R' \end{bmatrix} = \begin{bmatrix} \cos\lambda & -\sin\lambda & 0 \\ \sin\lambda & \cos\lambda & 0 \\ 0 & 0 & 1 \end{bmatrix} \begin{bmatrix} Z \\ H' \\ R' \end{bmatrix}$$

(c)

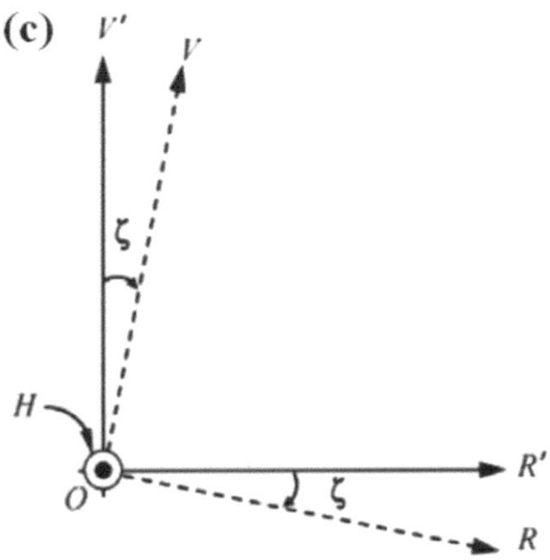

$$\begin{bmatrix} V \\ H \\ R \end{bmatrix} = \begin{bmatrix} \cos\zeta & 0 & \sin\zeta \\ 0 & 1 & 0 \\ -\sin\zeta & 0 & \cos\zeta \end{bmatrix} \begin{bmatrix} V' \\ H \\ R' \end{bmatrix}$$

Fig. 6. The diagram shows the process of how the collector-centre frame is tilted step-by-step relative to the earth-surface frame, where OV′, OH′ and OR′ axes represent the intermediate position for OV, OH and OR axes, respectively. (a) The first tilted angle, $+\phi$, is a rotational angle about OZ-axis in clockwise direction in the first step of coordinate transformation

$$[\lambda]=\begin{bmatrix}\cos\lambda & -\sin\lambda & 0\\ \sin\lambda & \cos\lambda & 0\\ 0 & 0 & 1\end{bmatrix} \tag{6b}$$

$$[\zeta]=\begin{bmatrix}\cos\zeta & 0 & \sin\zeta\\ 0 & 1 & 0\\ -\sin\zeta & 0 & \cos\zeta\end{bmatrix} \tag{6c}$$

The new set of coordinates *S′* can be interrelated with the earth-centre frame based coordinate *S* through the process of four successive coordinate transformations. It will be first transformed from earth-centre frame to earth-surface frame through transformation matrix $[\Phi]$, then from earth-surface frame to collector-centre frame through three subsequent coordinate transformation matrices that are $[\phi]$, $[\lambda]$ and $[\zeta]$. In mathematical expression, *S′* can be obtained through multiplication of four successive rotational transformation matrices with *S* and it is written as

$$\begin{bmatrix}S_V\\ S_H\\ S_R\end{bmatrix} = [\zeta][\lambda][\phi][\Phi]\begin{bmatrix}\cos\delta\cos\omega\\ -\cos\delta\sin\omega\\ \sin\delta\end{bmatrix},$$

$$\begin{bmatrix}\sin a\\ \cos a\sin\beta\\ \cos a\cos\beta\end{bmatrix} = \begin{bmatrix}\cos\zeta & 0 & \sin\zeta\\ 0 & 1 & 0\\ -\sin\zeta & 0 & \cos\zeta\end{bmatrix}\times\begin{bmatrix}\cos\lambda & -\sin\lambda & 0\\ \sin\lambda & \cos\lambda & 0\\ 0 & 0 & 1\end{bmatrix}\times\begin{bmatrix}1 & 0 & 0\\ 0 & \cos\phi & -\sin\phi\\ 0 & \sin\phi & \cos\phi\end{bmatrix} \tag{7}$$
$$\times\begin{bmatrix}\cos\Phi & 0 & \sin\Phi\\ 0 & 1 & 0\\ -\sin\Phi & 0 & \cos\Phi\end{bmatrix}\times\begin{bmatrix}\cos\delta\cos\omega\\ -\cos\delta\sin\omega\\ \sin\delta\end{bmatrix}$$

Solving the above matrix equation for the solar altitude angle (α) in collector-frame, we have

$$\alpha=\arcsin\begin{bmatrix}\cos\delta\cos\omega(\cos\zeta\cos\lambda\cos\Phi-\cos\zeta\sin\lambda\sin\phi\sin\Phi-\sin\zeta\cos\phi\sin\Phi)\\ -\cos\delta\sin\omega(\sin\zeta\sin\phi-\cos\zeta\sin\lambda\cos\phi)\\ +\sin\delta(\cos\zeta\cos\lambda\sin\Phi+\cos\zeta\sin\lambda\sin\phi\cos\Phi+\sin\zeta\cos\phi\cos\Phi)\end{bmatrix} \tag{8}$$

Thus, the first tracking angle along EE′ axis is

$$\theta=\frac{\pi}{2}-\arcsin\begin{bmatrix}\cos\delta\cos\omega(\cos\zeta\cos\lambda\cos\Phi-\cos\zeta\sin\lambda\sin\phi\sin\Phi-\sin\zeta\cos\phi\sin\Phi)\\ -\cos\delta\sin\omega(\sin\zeta\sin\phi-\cos\zeta\sin\lambda\cos\phi)\\ +\sin\delta(\cos\zeta\cos\lambda\sin\Phi+\cos\zeta\sin\lambda\sin\phi\cos\Phi+\sin\zeta\cos\phi\cos\Phi)\end{bmatrix} \tag{9}$$

from earth-surface frame to collector-centre frame. (b) The second tilted angle, $-\lambda$, is a rotational angle about OR′ axis in counter-clockwise direction in the second step of coordinate transformation from earth-surface frame to collector-centre frame. (c) The third tilted angle, $+\zeta$, is a rotational angle about OH axis in clockwise direction in the third step of coordinate transformation from earth-surface frame to collector-centre frame

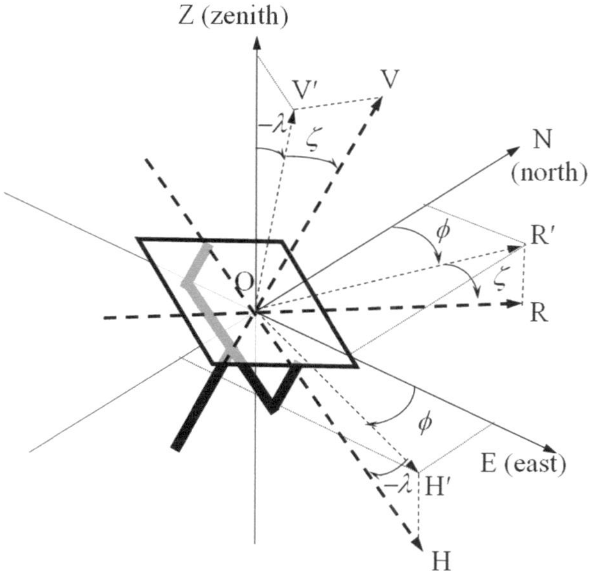

Fig. 7. The combination of the three rotations in 3D view from collector-centre frame to the earth-surface frame, where the change of coordinate system for each axis follows the order: $Z \rightarrow V' \rightarrow V$, $E \rightarrow H' \rightarrow H$ and $N \rightarrow R' \rightarrow R$

Similarly, the other two remaining equations that can be extracted from the above matrix equation expressed in cosine terms are as follow:

$$\sin\beta = \frac{\begin{bmatrix}\cos\delta\cos\omega(\sin\lambda\cos\Phi + \cos\lambda\sin\phi\sin\Phi) - \cos\delta\sin\omega\cos\lambda\cos\phi \\ + \sin\delta(\sin\lambda\sin\Phi - \cos\lambda\sin\phi\cos\Phi)\end{bmatrix}}{\cos\alpha} \tag{10}$$

$$\cos\beta = \frac{\begin{bmatrix}\cos\delta\cos\omega(-\sin\zeta\cos\lambda\cos\Phi + \sin\zeta\sin\lambda\sin\phi\sin\Phi - \cos\zeta\cos\phi\sin\Phi) \\ - \cos\delta\sin\omega(\sin\zeta\sin\lambda\cos\phi + \cos\zeta\sin\phi) \\ + \sin\delta(-\sin\zeta\cos\lambda\sin\Phi - \sin\zeta\sin\lambda\sin\phi\cos\Phi + \cos\zeta\cos\phi\cos\Phi)\end{bmatrix}}{\cos\alpha} \tag{11}$$

In fact, the second tracking angle along OV axis, β, can be in any of the four trigonometric quadrants depending on location, time of day and the season. Since the arc-sine and arc-cosine functions have two possible quadrants for their result, both equations of $\sin\beta$ and $\cos\beta$ require a test to ascertain the correct quadrant. Consequently, we have either

$$\beta = \arcsin\begin{bmatrix}\dfrac{\cos\delta\cos\omega(\sin\lambda\cos\Phi + \cos\lambda\sin\phi\sin\Phi) - \cos\delta\sin\omega\cos\lambda\cos\phi + \sin\delta(\sin\lambda\sin\Phi - \cos\lambda\sin\phi\cos\Phi)}{\cos\alpha}\end{bmatrix} \tag{12}$$

when $\cos\beta \geq 0$

or

$$\beta = \pi - \arcsin\left[\frac{\cos\delta\cos\omega(\sin\lambda\cos\Phi + \cos\lambda\sin\phi\sin\Phi) - \cos\delta\sin\omega\cos\lambda\cos\phi + \sin\delta(\sin\lambda\sin\Phi - \cos\lambda\sin\phi\cos\Phi)}{\cos\alpha}\right] \tag{13}$$

when $\cos\beta < 0$.

3.2 General formula for on-axis solar collector

The derived general sun-tracking formula is the most general form of solution for various kinds of arbitrarily oriented on-axis solar collector on the earth surface. In overall, all the on-axis sun-tracking systems fall into two major groups as shown in Fig. 1: (i) two-axis tracking system and (ii) one-axis tracking system. For two-axis tracking system, such as azimuth-elevation and tilt-roll tracking system, their tracking formulas can be derived from the general formula by setting different conditions to the parameters, such as ϕ, λ and ζ. In the case of azimuth-elevation tracking system, the tracking formula can be obtained by setting the angles $\phi = \lambda = \zeta = 0$ in the general formula. Thus, we can simplify the general formula to

$$\theta = \tfrac{\pi}{2} - \arcsin\left[\sin\delta\sin\Phi + \cos\delta\cos\omega\cos\Phi\right] \tag{14}$$

$$\beta = \arcsin\left[-\frac{\cos\delta\sin\omega}{\cos\alpha}\right] \tag{15}$$

when $\cos\beta \geq 0$
or

$$\beta = \pi - \arcsin\left[-\frac{\cos\delta\sin\omega}{\cos\alpha}\right] \tag{16}$$

when $\cos\beta < 0$
On the other hand, polar tracking method can also be obtained by setting the angles $\phi = \pi$, $\lambda = 0$ and $\zeta = \Phi - \pi/2$. For this case, the general tracking formula can be then simplified to

$$\theta = \pi/2 - \delta \tag{17}$$

$$\beta = \omega, \text{ when } -\pi/2 < \omega < \pi/2 \tag{18}$$

For one-axis tracking system, the tracking formula can be easily obtained from the full tracking formula by setting one of the tracking angles, which is either θ or β, as a constant value. For example, one of the most widely used one-axis tracking systems is to track the sun in latitude-tilted tracking axis. Latitude-tilted tracking axis is derived from tilt-roll tracking system with θ to be set as $\pi/2$ and the solar collector only tracks the sun with the angle $\beta = \omega$.

3.3 Application of general formula in improving sun-tracking accuracy

General sun-tracking formula not only provides the general mathematical solution for the case of on-axis solar collector, but also gives the ability to improve the sun-tracking accuracy by compensating the misalignment of the azimuth axis during the solar collector installation work. According to the general formula, the sun-tracking accuracy of the system is highly reliant on the precision of the input parameters of the sun-tracking algorithm: latitude angle (Φ), hour angle (ω), declination angle (δ), as well as the three orientation angles of the tracking axes of solar concentrator, i.e., ϕ, λ and ζ. Among these values, local latitude, Φ, and longitude of the sun tracking system can be determined accurately with the latest technology such as a global positioning system (GPS). On the other hand, ω and δ are both local time dependent parameters as shown in the Eq. (2) and Eq. (3). These variables can be computed accurately with the input from precise clock that is synchronized with the internet timeserver. As for the three orientation angles (ϕ, λ and ζ), their precision are very much reliant on the care paid during the on-site installation of solar collector, the alignment of tracking axes and the mechanical fabrication. Not all these orientation angles can be precisely obtained due to the limitation of measurement tools and the accuracy of determination of the real north of the earth. The following mathematical derivation is attempted to obtain analytical solutions for the three orientation angles based on the daily sun-tracking error results induced by the misalignment of sun-tracking axes (Chong et al., 2009b).

From the Eq. (7), the unit vector of the sun, $\mathbf{S}'$, relative to the solar collector can be obtained from a multiplication of four successive coordinate transformation matrices, i.e., [Φ], [ϕ], [λ] and [ζ] with the unit vector of the sun, $\mathbf{S}$, relative to the earth. Multiply the first three transformation matrices [ϕ], [λ] and [ζ], and then the last two matrices [Φ] with $\mathbf{S}$ as to obtain the following result:

$$\begin{bmatrix} \sin\alpha \\ \cos\alpha\sin\beta \\ \cos\alpha\cos\beta \end{bmatrix} = \begin{bmatrix} \cos\zeta\cos\lambda & -\cos\zeta\sin\lambda\cos\phi+\sin\zeta\sin\phi & \cos\zeta\sin\lambda\sin\phi+\sin\zeta\cos\phi \\ \sin\lambda & \cos\lambda\cos\phi & -\cos\lambda\sin\phi \\ -\sin\zeta\cos\lambda & \sin\zeta\sin\lambda\cos\phi+\cos\zeta\sin\phi & -\sin\zeta\sin\lambda\sin\phi+\cos\zeta\cos\phi \end{bmatrix} \times \begin{bmatrix} \cos\Phi\cos\delta\cos\omega+\sin\Phi\sin\delta \\ -\cos\delta\sin\omega \\ -\sin\Phi\cos\delta\cos\omega+\cos\Phi\sin\delta \end{bmatrix} . \tag{19}$$

From Eq. (19), we can further break it down into Eq. (20):

$$\sin\alpha = (\cos\Phi\cos\delta\cos\omega+\sin\Phi\sin\delta)(\cos\zeta\cos\lambda)+(-\cos\delta\sin\omega)(-\cos\zeta\sin\lambda\cos\phi+\sin\zeta\sin\phi) + (-\sin\Phi\cos\delta\cos\omega+\cos\Phi\sin\delta)(\cos\zeta\sin\lambda\sin\phi+\sin\zeta\cos\phi) \tag{20a}$$

$$\cos\alpha\sin\beta = (\cos\Phi\cos\delta\cos\omega+\sin\Phi\sin\delta)(\sin\lambda)+(-\cos\delta\sin\omega)(\cos\lambda\cos\phi) + (-\sin\Phi\cos\delta\cos\omega+\cos\Phi\sin\delta)(-\cos\lambda\sin\phi) \tag{20b}$$

$$\cos\alpha\cos\beta = (\cos\Phi\cos\delta\cos\omega+\sin\Phi\sin\delta)(-\sin\zeta\cos\lambda)+(-\cos\delta\sin\omega)(\sin\zeta\sin\lambda\cos\phi+\cos\zeta\sin\phi) + (-\sin\Phi\cos\delta\cos\omega+\cos\Phi\sin\delta)(-\sin\zeta\sin\lambda\sin\phi+\cos\zeta\cos\phi) \tag{20c}$$

The time dependency of ω and δ can be found from Eq. (20). Therefore, the instantaneous sun-tracking angles of the collector only vary with the angles ω and δ. Given three different local times LCT_1, LCT_2 and LCT_3 on the same day, the corresponding three hours angles ω_1, ω_2 and ω_3 as well as three declination angles δ_1, δ_2 and δ_3 can result in three elevation angles α_1, α_2 and α_3 and three azimuth angles β_1, β_2 and β_3 accordingly as expressed in Eqs. (20a)–(20c). Considering three different local times, we can actually rewrite each of the Eqs. (20a)–(20c) into three linear equations. By arranging the three linear equations in a matrix form, the Eqs. (20a)–(20c) can subsequently form the following matrices

$$\begin{bmatrix}\sin\alpha_1\\ \sin\alpha_2\\ \sin\alpha_3\end{bmatrix}=\begin{bmatrix}\cos\Phi\cos\delta_1\cos\omega_1+\sin\Phi\sin\delta_1 & -\cos\delta_1\sin\omega_1 & -\sin\Phi\cos\delta_1\cos\omega_1+\cos\Phi\sin\delta_1\\ \cos\Phi\cos\delta_2\cos\omega_2+\sin\Phi\sin\delta_2 & -\cos\delta_2\sin\omega_2 & -\sin\Phi\cos\delta_2\cos\omega_2+\cos\Phi\sin\delta_2\\ \cos\Phi\cos\delta_3\cos\omega_3+\sin\Phi\sin\delta_3 & -\cos\delta_3\sin\omega_3 & -\sin\Phi\cos\delta_3\cos\omega_3+\cos\Phi\sin\delta_3\end{bmatrix}\times\begin{bmatrix}\cos\zeta\cos\lambda\\ -\cos\zeta\sin\lambda\cos\phi+\sin\zeta\sin\phi\\ \cos\zeta\sin\lambda\sin\phi+\sin\zeta\cos\phi\end{bmatrix}. \tag{21a}$$

$$\begin{bmatrix}\cos\alpha_1\sin\beta_1\\ \cos\alpha_2\sin\beta_2\\ \cos\alpha_3\sin\beta_3\end{bmatrix}=\begin{bmatrix}\cos\Phi\cos\delta_1\cos\omega_1+\sin\Phi\sin\delta_1 & -\cos\delta_1\sin\omega_1 & -\sin\Phi\cos\delta_1\cos\omega_1+\cos\Phi\sin\delta_1\\ \cos\Phi\cos\delta_2\cos\omega_2+\sin\Phi\sin\delta_2 & -\cos\delta_2\sin\omega_2 & -\sin\Phi\cos\delta_2\cos\omega_2+\cos\Phi\sin\delta_2\\ \cos\Phi\cos\delta_3\cos\omega_3+\sin\Phi\sin\delta_3 & -\cos\delta_3\sin\omega_3 & -\sin\Phi\cos\delta_3\cos\omega_3+\cos\Phi\sin\delta_3\end{bmatrix}\times\begin{bmatrix}\sin\lambda\\ \cos\lambda\cos\phi\\ -\cos\lambda\sin\phi\end{bmatrix}. \tag{21b}$$

$$\begin{bmatrix}\cos\alpha_1\cos\beta_1\\ \cos\alpha_2\cos\beta_2\\ \cos\alpha_3\cos\beta_3\end{bmatrix}=\begin{bmatrix}\cos\Phi\cos\delta_1\cos\omega_1+\sin\Phi\sin\delta_1 & -\cos\delta_1\sin\omega_1 & -\sin\Phi\cos\delta_1\cos\omega_1+\cos\Phi\sin\delta_1\\ \cos\Phi\cos\delta_2\cos\omega_2+\sin\Phi\sin\delta_2 & -\cos\delta_2\sin\omega_2 & -\sin\Phi\cos\delta_2\cos\omega_2+\cos\Phi\sin\delta_2\\ \cos\Phi\cos\delta_3\cos\omega_3+\sin\Phi\sin\delta_3 & -\cos\delta_3\sin\omega_3 & -\sin\Phi\cos\delta_3\cos\omega_3+\cos\Phi\sin\delta_3\end{bmatrix}\times\begin{bmatrix}-\sin\zeta\cos\lambda\\ \sin\zeta\sin\lambda\cos\phi+\cos\zeta\sin\phi\\ -\sin\zeta\sin\lambda\sin\phi+\cos\zeta\cos\phi\end{bmatrix}. \tag{21c}$$

where the angles Φ, ϕ, λ and ζ are constants with respect to the local time.

In practice, we can measure the sun tracking angles i.e. (α_1, α_2, α_3) and (β_1, β_2, β_3) during sun-tracking at three different local times via a recorded solar image of the target using a CCD camera. With the recorded data, we can compute the three arbitrary orientation angles (ϕ, λ and ζ) of the solar collector using the third-order determinants method to solve the three simultaneous equations as shown in Eqs. (21a)–(21c). From Eq. (21b), the orientation angle λ can be determined as follows:

$$\lambda=\sin^{-1}\left(\frac{\begin{vmatrix}\cos\alpha_1\sin\beta_1 & -\cos\delta_1\sin\omega_1 & -\sin\Phi\cos\delta_1\cos\omega_1+\cos\Phi\sin\delta_1\\ \cos\alpha_2\sin\beta_2 & -\cos\delta_2\sin\omega_2 & -\sin\Phi\cos\delta_2\cos\omega_2+\cos\Phi\sin\delta_2\\ \cos\alpha_3\sin\beta_3 & -\cos\delta_3\sin\omega_3 & -\sin\Phi\cos\delta_3\cos\omega_3+\cos\Phi\sin\delta_3\end{vmatrix}}{\begin{vmatrix}\cos\Phi\cos\delta_1\cos\omega_1+\sin\Phi\sin\delta_1 & -\cos\delta_1\sin\omega_1 & -\sin\Phi\cos\delta_1\cos\omega_1+\cos\Phi\sin\delta_1\\ \cos\Phi\cos\delta_2\cos\omega_2+\sin\Phi\sin\delta_2 & -\cos\delta_2\sin\omega_2 & -\sin\Phi\cos\delta_2\cos\omega_2+\cos\Phi\sin\delta_2\\ \cos\Phi\cos\delta_3\cos\omega_3+\sin\Phi\sin\delta_3 & -\cos\delta_3\sin\omega_3 & -\sin\Phi\cos\delta_3\cos\omega_3+\cos\Phi\sin\delta_3\end{vmatrix}}\right) \tag{22a}$$

Similarly, the other two remaining orientation angles, ϕ and ζ can be resolved from Equation (21b) and Equation (21c) respectively as follows:

$$\phi = -\sin^{-1}\left(\frac{\begin{vmatrix} \cos\Phi\cos\delta_1\cos\omega_1 + \sin\Phi\sin\delta_1 & -\cos\delta_1\sin\omega_1 & \cos\alpha_1\sin\beta_1 \\ \cos\Phi\cos\delta_2\cos\omega_2 + \sin\Phi\sin\delta_2 & -\cos\delta_2\sin\omega_2 & \cos\alpha_2\sin\beta_2 \\ \cos\Phi\cos\delta_3\cos\omega_3 + \sin\Phi\sin\delta_3 & -\cos\delta_3\sin\omega_3 & \cos\alpha_3\sin\beta_3 \end{vmatrix}}{\begin{vmatrix} \cos\Phi\cos\delta_1\cos\omega_1 + \sin\Phi\sin\delta_1 & -\cos\delta_1\sin\omega_1 & -\sin\Phi\cos\delta_1\cos\omega_1 + \cos\Phi\sin\delta_1 \\ \cos\Phi\cos\delta_2\cos\omega_2 + \sin\Phi\sin\delta_2 & -\cos\delta_2\sin\omega_2 & -\sin\Phi\cos\delta_2\cos\omega_2 + \cos\Phi\sin\delta_2 \\ \cos\Phi\cos\delta_3\cos\omega_3 + \sin\Phi\sin\delta_3 & -\cos\delta_3\sin\omega_3 & -\sin\Phi\cos\delta_3\cos\omega_3 + \cos\Phi\sin\delta_3 \end{vmatrix}} \times \frac{1}{\cos\lambda} \right) \quad (22b)$$

$$\zeta = -\sin^{-1}\left(\frac{\begin{vmatrix} \cos\alpha_1\cos\beta_1 & -\cos\delta_1\sin\omega_1 & -\sin\Phi\cos\delta_1\cos\omega_1 + \cos\Phi\sin\delta_1 \\ \cos\alpha_2\cos\beta_2 & -\cos\delta_2\sin\omega_2 & -\sin\Phi\cos\delta_2\cos\omega_2 + \cos\Phi\sin\delta_2 \\ \cos\alpha_3\cos\beta_3 & -\cos\delta_3\sin\omega_3 & -\sin\Phi\cos\delta_3\cos\omega_3 + \cos\Phi\sin\delta_3 \end{vmatrix}}{\begin{vmatrix} \cos\Phi\cos\delta_1\cos\omega_1 + \sin\Phi\sin\delta_1 & -\cos\delta_1\sin\omega_1 & -\sin\Phi\cos\delta_1\cos\omega_1 + \cos\Phi\sin\delta_1 \\ \cos\Phi\cos\delta_2\cos\omega_2 + \sin\Phi\sin\delta_2 & -\cos\delta_2\sin\omega_2 & -\sin\Phi\cos\delta_2\cos\omega_2 + \cos\Phi\sin\delta_2 \\ \cos\Phi\cos\delta_3\cos\omega_3 + \sin\Phi\sin\delta_3 & -\cos\delta_3\sin\omega_3 & -\sin\Phi\cos\delta_3\cos\omega_3 + \cos\Phi\sin\delta_3 \end{vmatrix}} \times \frac{1}{\cos\lambda} \right) \quad (22c)$$

Fig. 8 shows the flow chart of the computational program designed to solve the three unknown orientation angles of the solar collector: ϕ, λ and ζ using Eqs. (22a)–(22c). By providing the three sets of actual sun tracking angle α and β at different local times for a

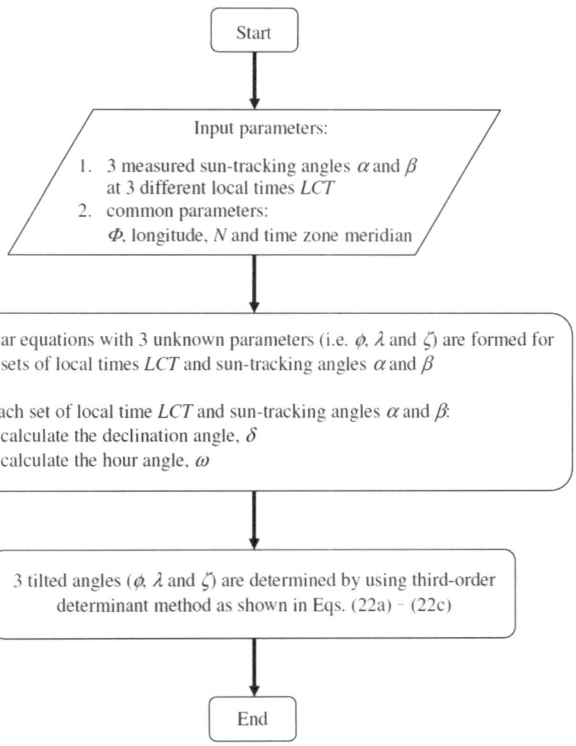

Fig. 8. The flow chart of the computational program to determine the three unknown orientation angles that cannot be precisely measured by tools in practice, i.e. ϕ, λ and ζ

particular number of day as well as geographical information i.e. longitude and latitude (Φ), the computational program can be executed to calculate the three unknown orientation angles (ϕ, λ and ζ).

4. Integration of general formula into open-loop sun-tracking system

4.1 Design and construction of open-loop sun-tracking system

For demonstrating the integration of general formula into open-loop sun-tracking control system to obtain high degree of sun-tracking accuracy, a prototype of on-axis Non-Imaging Planar Concentrator (NIPC) has been constructed in the campus of Univesiti Tunku Abdul Rahman (UTAR), Kuala Lumpur, Malaysia (located at latitude 3.22° and longitude 101.73°). A suitable geographical location was selected for the installation of solar concentrator so that it is capable of receiving the maximum solar energy without the blocking of any buildings or plants. The planar concentrator, applies the concept of non-imaging optics to concentrate the sunlight, has been proposed in order to achieve a good uniformity of the solar irradiation with a reasonably high concentration ratio on the target (Chong et al., 2009a; Chong et al., 2010). Instead of using a single piece of parabolic dish, the newly proposed on-axis solar concentrator employs 480 pieces of flat mirrors to form a total reflective area of about 25 m^2 with adjustable focal distance to concentrate the sunlight onto the target (see Fig. 9). The target is fixed at a focal point with a distance of 4.5 m away from the centre of solar concentrator frame.

Fig. 9. A prototype of 25m^2 on-axis Non-Imaging Planar Concentrator (NIPC) that has been constructed at Universiti Tunku Abdul Rahman (UTAR)

This planar concentrator is designed to operate on the most common two-axis tracking system, which is azimuth-elevation tracking system. The drive mechanism for the solar concentrator consists of stepper motors and its associated gears. Two stepper motors, with 0.72 degree in full step, are coupled to the shafts, elevation and azimuth shafts, with gear ratio of 4400 yielding an overall resolution of 1.64×10^{-4} °/ step. A Windows-based control program has been developed by integrating the general formula into the open-loop sun-tracking algorithm. In the control algorithm, the sun-tracking angles, i.e. azimuth (β) and elevation (α) angles, are first computed according to the latitude (Φ), longitude, day numbers (N), local time (LCT), time zone and the three newly introduced orientation angles (ϕ, λ and ζ). The control program then generate digital pulses that are sent to the stepper motor to drive the concentrator to the pre-calculated angles along azimuth and elevation movements in sequence. Each time, the control program only activates one of the two stepper motors through a relay switch. The executed control program of sun-tracking system is shown in Fig. 10.

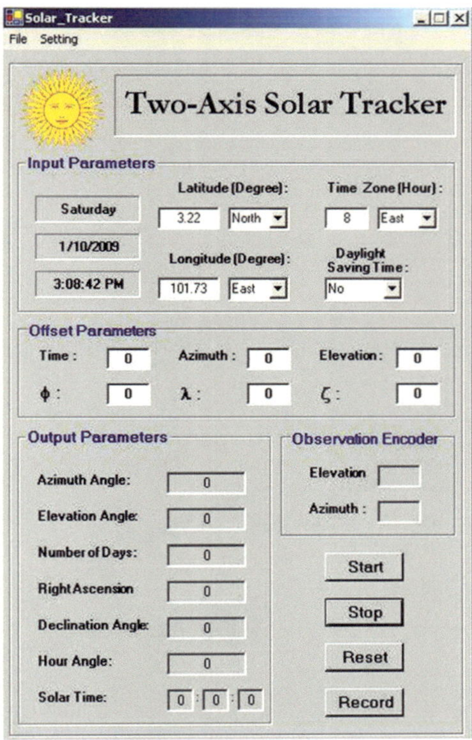

Fig. 10. A Windows-based control program that has been integrated with the on-axis general formula

An open-loop control system is preferable for the prototype solar concentrator to keep the design of the sun tracker simple and cost effective. In our design, open-loop sensors, 12-bit absolute optical encoders with a precision of 2,048 counts per revolution, are attached to the shafts along the azimuth and elevation axes of the concentrator to monitor the turning angles and to send feedback signals to the computer if there is any abrupt change in the encoder reading [see the inset of Fig. 11(b)]. Therefore, the sensors not only ensure that the instantaneous azimuth and elevation angles are matched with the calculated values from the general formula, but also eliminate any tracking errors due to mechanical backlash, accumulated error, wind effects and other disturbances to the solar concentrator. With the optical encoders, any discrepancy between the calculated angles and real time angles of solar concentrator can be detected, whereby the drive mechanism will be activated to move the solar concentrator to the correct position. The block diagram and schematic diagram for the complete design of the open-loop control system of the prototype are shown in Fig. 11 (a), (b) respectively.

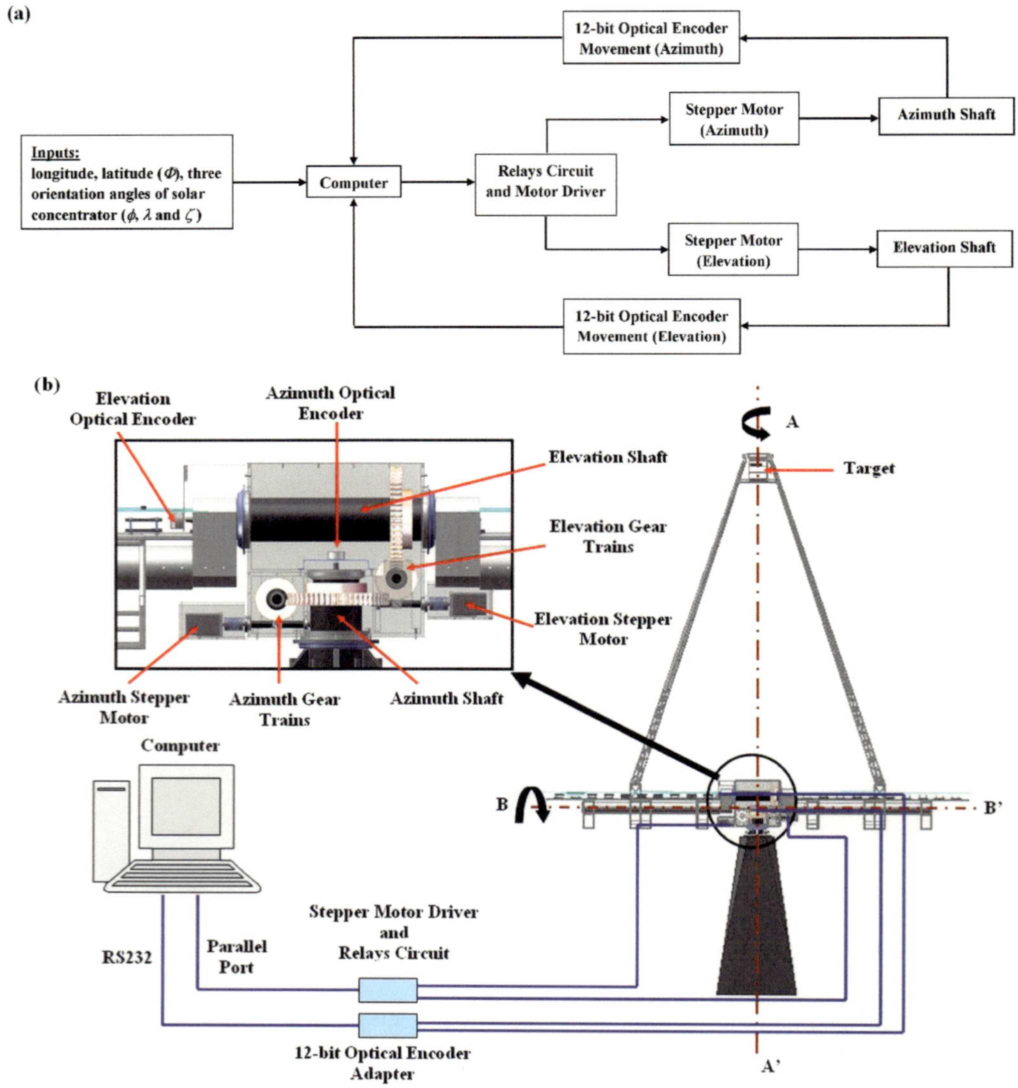

Fig. 11. (a) Block diagram to show the complete open-loop feedback system of the solar concentrator. (b) Schematic diagram to show the detail of the open-loop sun-tracking system of the prototype planar concentrator where AA′ is azimuth-axis and BB′ is elevation-axis.

4.2 Energy consumption

The estimated total electrical energy produced by the prototype solar concentrator and the total energy consumption by the sun-tracking system are also calculated. Taking into account of the total mirror area of 25 m^2, optical efficiency of 85%, and the conversion efficiency from solar energy to electrical energy of 30% for direct solar irradiation of 800 W/m^2, we have obtained the generated output energy of 35.7 kW-h/day for seven hours daily sunshine. Table 1 shows the energy consumption of 1.26 kW-h/day for the prototype includes the tracking motors, motor driver, encoders and computer. It corresponds to less than 3.5 % of the rated generated output energy. Among all these components, computer consumes the most power (more than 100W) and in future microcontroller can be used to replace computer as to reduce the energy consumption.

Total rotational angles of Elevation axis (degree/ day)	240
Total rotational angles of Azimuth axis (degree/ day)	540
Motor's rotational speed (rpm)	120
Gear ratio	1: 4400
Solar concentrator's angular speed (degree per second)	0.16
Total time for Elevation axis rotation (hour/ day)	0.41
Total time for Azimuth axis rotation (hour/ day)	0.92
Total operating time:10am-5pm (hour/ day)	7
Elevation motor's power consumption (watt)	99
Azimuth motor's power consumption (watt)	66
Power consumption of computer, encoders & motor driver (watt)	165
Energy Consumption of the Elevation motor (kW-h/day)	0.04
Energy Consumption of the Azimuth motor (kW-h/day)	0.06
Energy Consumption of computer, encoder & driver (kW-h/day)	1.16
Total Energy Consumption of the motors (kW-h/day)	1.26

Table 1. Specification and energy consumption of prototype sun-tracking system

5. Performance study and results

Before the performance of sun-tracking system was tested, all the mirrors are covered with black plastic (see Fig. 9), except the one mirror located nearest to the centre of the concentrator frame. To study the performance of the sun-tracking system, a CCD camera with 640 × 480 pixels resolution is utilized to capture the solar image cast on the target, which has a dimension of 60 cm × 60 cm and with a thickness of 1 cm steel plate, drawn with

28 cm × 26 cm target area. The camera is connected to a computer via a Peripheral Component Interconnect (PCI) video card as to have a real time transmission and recording of solar image. For the sake of accuracy, the CCD camera is placed directly facing the target to avoid the Cosine Effect. By observing the movement of the solar image via CCD camera, the sun-tracking accuracy can be analysed and recorded in the computer database every 30 minutes from 10 a.m. to 5 p.m. local time. Three different performance studies were executed in the year of 2009.

Study no. 1: First performance study has been carried out on 13 January 2009. Initially, we assume that the alignment of solar concentrator is perfectly done relative to real north and zenith by setting the three orientation angles as $\phi = \lambda = \zeta = 0°$ in the control program. According to the recorded results as shown in Fig. 12, the recorded tracking errors, ranging from 12.12 to 17.54 mrad throughout the day, have confirmed that the solar concentrator is misaligned relative to zenith and real north. Fig. 13 illustrates the recorded solar images at different local times.

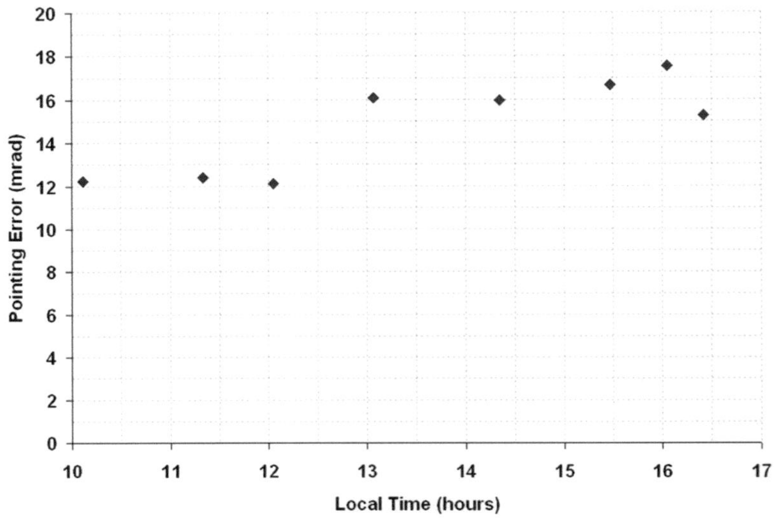

Fig. 12. The plot of pointing error (mrad) versus local time (hours) for the parameters, i.e. $\phi = \lambda = \zeta = 0°$, on 13 January 2009

Study no. 2: To rectify the problem of the sun-tracking errors due to imperfect alignment of the solar concentrator during the installation, we have to determine the three misaligned angles, i.e. ϕ, λ, ζ and then insert these values into the edit boxes provided by the control program as shown in Fig. 10. Thus, the computational program using the methodology as described in Fig. 8 was executed to compute the three new orientation angles of the prototype based on the data captured on 13 January 2009. The actual sun-tracking angles, i.e. (α_1, α_2 , α_3) and (β_1, β_2, β_3) at three different local times, can be determined from the central point of solar image position relative to the target central point by using the ray-tracing method. Three sets of sun-tracking angles at three different local times from the previous data were used as the input values to the computational program for simulating the three unknown parameters of ϕ, λ and ζ. The simulated results are $\phi = -0.1°$, $\lambda = 0°$, and $\zeta = -0.5°$.

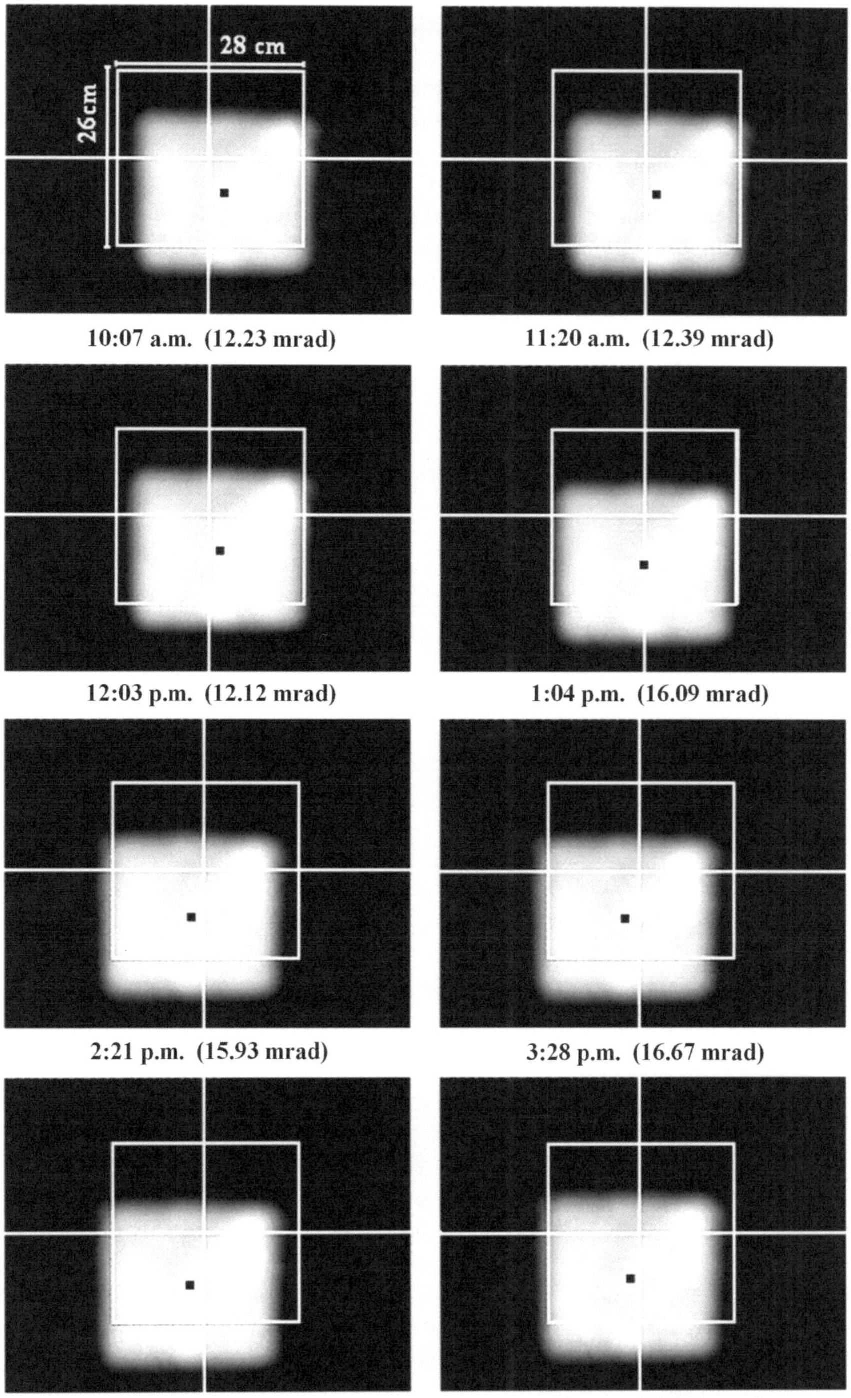

Fig. 13. The recorded solar images cast on the target of prototype solar concentrator using a CCD camera from 10:07 a.m. to 4:25 p.m. on 13 January 2009 with $\phi = \lambda = \zeta = 0°$

To substantiate the simulated results, these values were then used in the next session of sun-tracking that was performed on 16 January 2009. With the new orientation angles, the performance of the prototype in sun-tracking has been successfully improved to the accuracy of below 2.99 mrad, as shown in Fig. 14. It has reached the accuracy limit of the prototype as the optical encoder resolution that corresponds to 4.13 mrad, unless higher resolution of encoder is used for giving feedback signal. Fig. 15 shows the recorded solar images at the target for different local times ranging from 10 a.m. to 5 p.m. on 16 January 2009. In order to confirm the validation of the sun-tracking results, the sun-tracking system has been tested by running it for a period of more than six months.

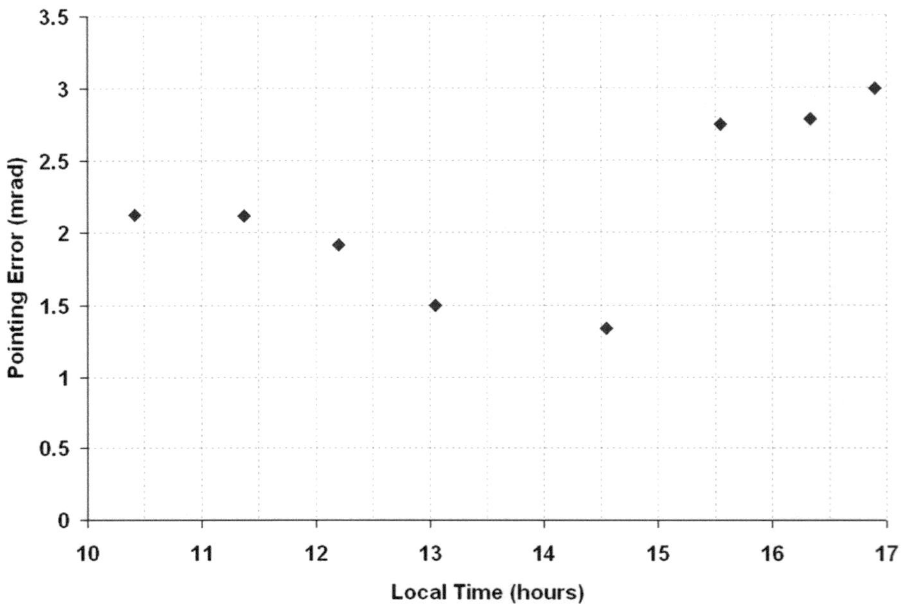

Fig. 14. The plot of pointing error (mrad) versus local time (hours) for the parameters, i.e. $\phi = -0.1°$, $\lambda = 0°$, and $\zeta = -0.5°$, on 16 January 2009

Study no. 3: On top of that, additional effort has been made to improve the sun-tracking accuracy beyond the resolution of the optical encoder by including the step count of stepper motor, 1.64×10^{-4} °/step, in fine-tuning the position of the prototype solar concentrator. Referring to the algorithm flow as shown in Fig. 16, the initial concentrator's azimuth and elevation angles are first defined with the reading of optical encoders, which are mounted on the azimuth and elevation axes respectively. Subsequently, the sun position angles, i.e. azimuth and elevation angles, are computed according to the general formula. The control program in succession compares the calculated sun angles with the current encoders' reading. If the absolute difference between the calculated azimuth or elevation angle and the encoder reading for azimuth or elevation axis (Δ_1) is larger than or equal to the encoder resolution, i.e. 0.176°, the control program then generates digital pulses that are sent to the stepper motor driver to drive the solar concentrator to the pre-calculated angles along azimuth and elevation axes in sequence with the use of relay switches, and stores the current reading of encoders as the latest concentrator's azimuth and elevation angles. In this

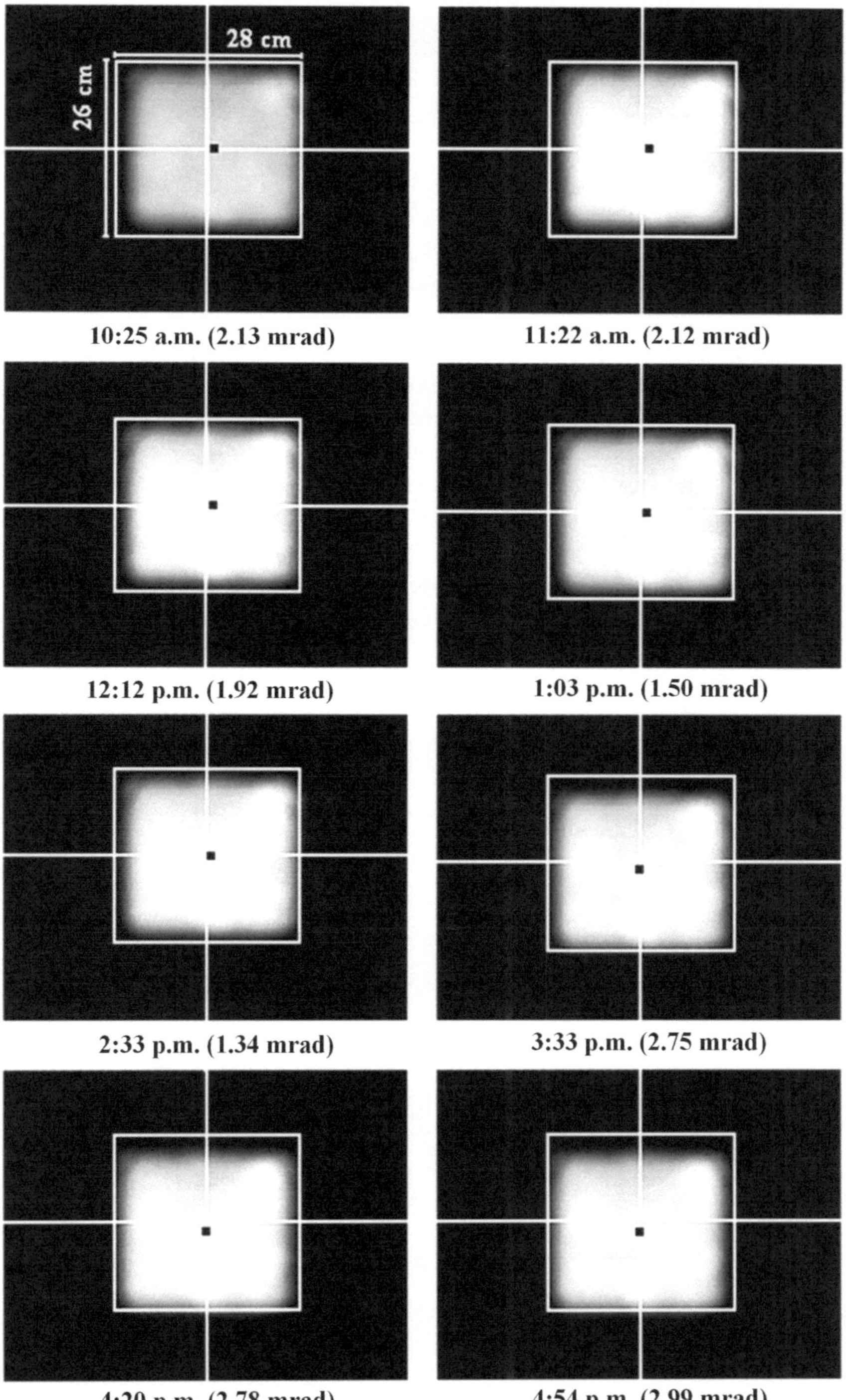

Fig. 15. The recorded solar images on the target of prototype solar concentrator using a CCD camera from 10:25 a.m. to 4:54 p.m. on 16 January 2009 with $\phi = -0.1°$, $\lambda = 0°$, and $\zeta = -0.5°$.

case, the program operates in feedback loop that is capable of making correction or compensating the disturbances like wind load and backlash so that the difference between the concentrator position angle and the calculated sun position angle is within the encoder resolution. Since the motor driving resolution (1.64 × 10^{-4} °/step) is far higher than the encoder resolution, a non-feedback loop has been introduced when the solar concentrator operates within the resolution of the optical encoder. In the non-feedback loop, we have included two assumptions in which backslash and step loss are negligible within the encoder resolution during the driving operation. When the absolute difference between the calculated azimuth or elevation angle and the concentrator's azimuth or elevation angles (Δ_2) is larger than or equal to 0.05° (this angle is sufficient for an on-axis solar concentrator to achieve a tracking accuracy of below 1 mrad), the control program will send the required pulses to motors for rotating the solar concentrator towards the sun along azimuth and elevation axes in order. After that, the timer is activated and the position of solar concentrator is updated with the sum of previous concentrator position angle and Δ_2. The solar concentrator is programmed to follow the sun at all times since the program is to run in repeated loops in every 10 seconds. Fig. 17 illustrates the pointing error (mrad) versus local time (hours) for different local times ranging from 10:00 a.m. to 4:10 p.m. on 6 August 2009 which has included the step count of stepper motor, 1.64 × 10^{-4} °/step. This strategy has further improved the tracking accuracy to 0.96 mrad on 6 August 2009 as shown in Fig. 18. Similarly, the performance of the sun-tracking has been observed for several months until end of year 2009.

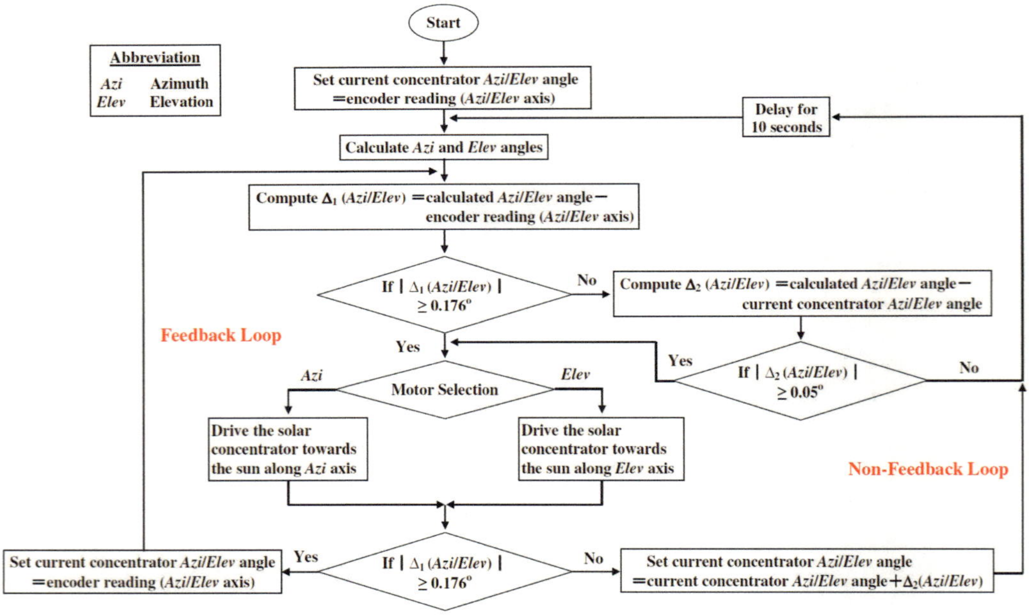

Fig. 16. The algorithm flow of the sun-tracking control program that including motor step count, 1.64 × 10^{-4} °/step, in fine-tuning the position of solar concentrator prototype and improving the sun tracking accuracy beyond the resolution of the optical encoder.

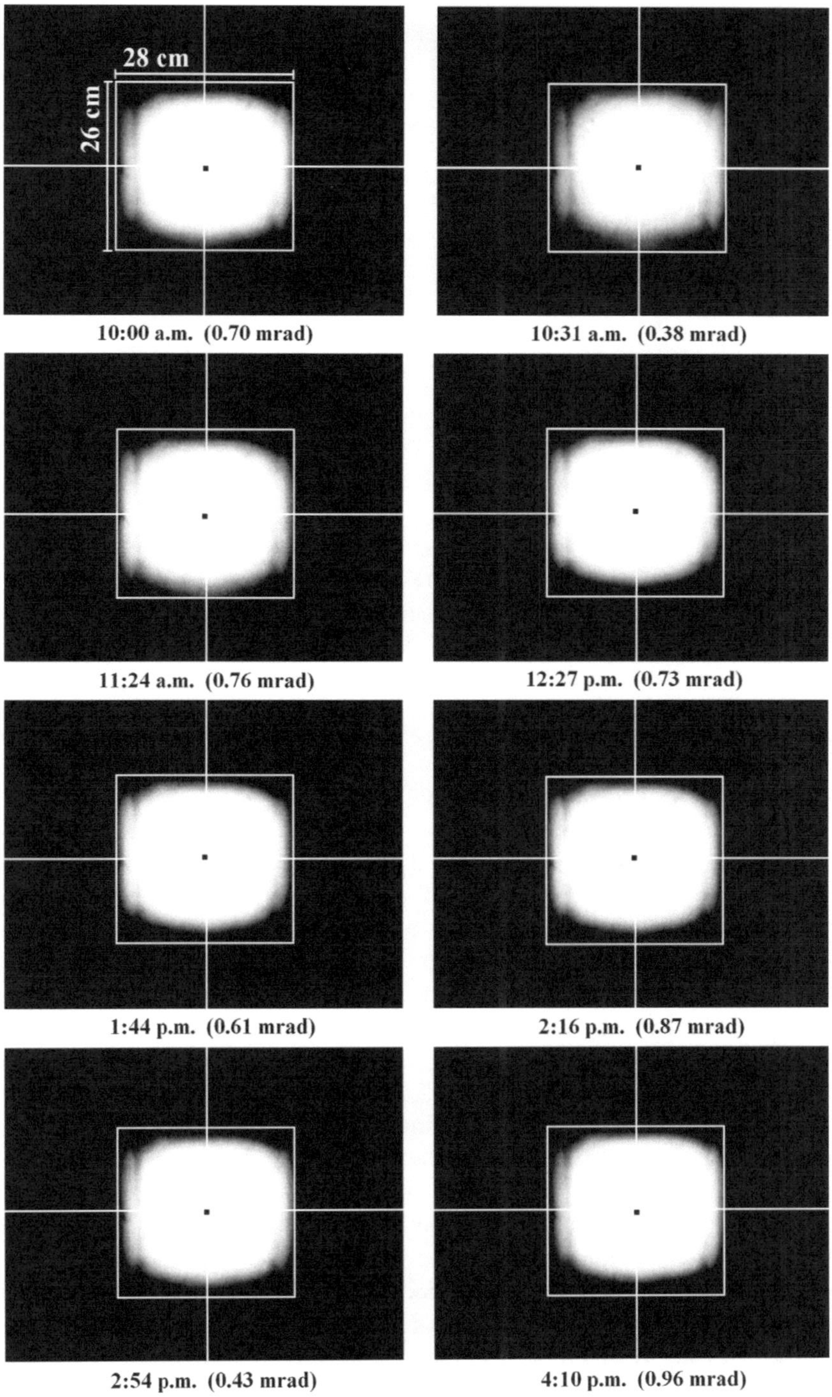

Fig. 17. The recorded solar images on the target of prototype solar concentrator using a CCD camera from 10:00 a.m. to 4:10 p.m. on 6 August 2009 with $\phi = -0.1°$, $\lambda = 0°$, and $\zeta = -0.5°$ and including the step count of stepper motor, 1.64×10^{-4} °/step

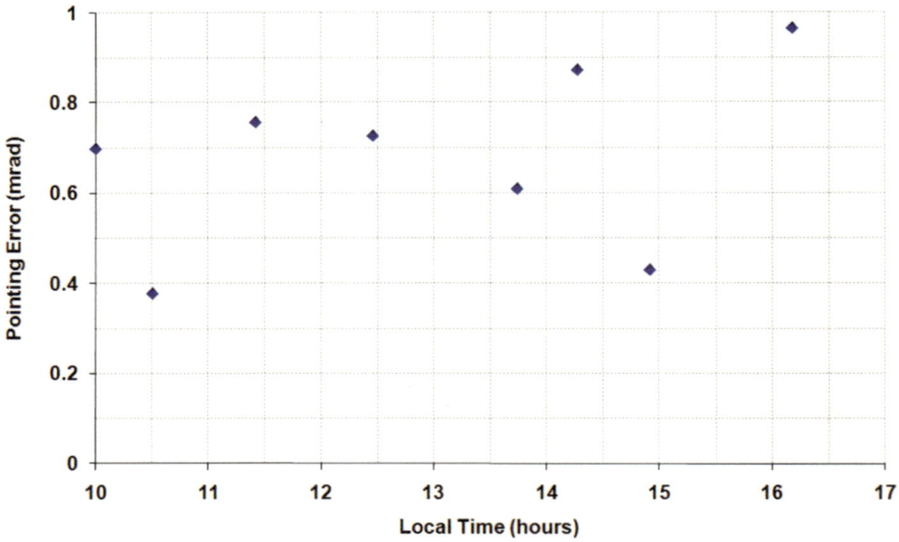

Fig. 18. The plot of pointing error (mrad) versus local time (hours) for the parameters, i.e. $\phi = -0.1°$, $\lambda = 0°$, and $\zeta = -0.5°$ on 6 August 2009 which has included the step count of stepper motor, 1.64×10^{-4} °/step

6. Conclusion

A novel on-axis general sun-tracking formula for various kinds of arbitrarily oriented on-axis solar collector has been derived using coordinate transformation method and integrated into the open-loop azimuth-elevation sun-tracking system intended for improving the tracking accuracy. In accordance with the experimental results, even though the misalignment on the azimuth axis relative to the zenith axis is within the range of 0.5 degree, the resulted sun-tracking error is significant, especially for the solar collector which requires high solar concentration and in particular for dense array concentrator photovoltaic (CPV) systems. With these results, the general sun-tracking formula is confirmed to be capable of rectifying the installation error of the solar concentrator with a significant improvement in the sun-tracking accuracy. In fact, there are many solutions of improving the tracking accuracy such as adding a closed-loop feedback system to the controller, designing a flexible mechanical platform that capable of two-degree-of-freedom for fine adjustment of azimuth shaft, etc. Nevertheless, all these solutions require a more complicated engineering design to the solar collector, which is also more complex and expensive. General sun-tracking formula allows the on-axis solar concentrator to track the sun accurately and simplifies the fabrication as well as the installation work of solar concentrator with higher tolerance in terms of the tracking axes alignment. Instead of using a complicated sun-tracking method, integrated on-axis general sun-tracking formula into the open-loop sun-tracking system is a clever method to get a reasonably high precision in sun-tracking with a much simple design and cost effective. This approach can significantly improve the performance and reduce the cost of solar energy collectors especially for high concentration systems.

7. References

Abdallah, S. & Nijmeh, S. (2004). Two axes sun tracking system with PLC control. *Energy Conversion and Management, Vol. 45, No. 11-12, (July* 2004) page numbers (1931-1939), ISSN 0196-8904

Aiuchi, K.; Yoshida, K.; Onozaki, M.; Katayama, Y.; Nakamura, M. & Nakamura, K. (2006). Sensor-controlled heliostat with an equatorial mount. *Solar Energy*, Vol.. *80, No. 9, (September 2006) page numbers (*1089-1097), ISSN 0038-092X

Arbab, H.; Jazi, B & Rezagholizadeh, M. (2009). A computer tracking system of solar dish with two-axis degree freedoms based on picture processing of bar shadow. *Renewable Energy, Vol. 34, No. 4, (April 2009) page numbers (*1114-1118), ISSN 0960-1481

Berenguel, M.; Rubio, F.R.; Valverde, A.; Lara, P.J.; Arahal, M.R.; Camacho, E.F. & Lopez, M. (2004). An artificial vision-based control system for automatic heliostat positioning offset correction in a central receiver solar power plant. *Solar Energy, Vol. 76, No. 5, (*2004) page numbers (563-575), ISSN 0038-092X

Blanco-Muriel, M.; Alarcon-Padilla, D.C.; Lopez-Moratalla, T. & Lara-Coira, M. (2001). Computing the solar vector. *Solar Energy, Vol. 70, No. 5, (*2001) page numbers (431-441), ISSN 0038-092X

Chen, F.; Feng, J. & Hong, Z. (2006). Digital sun sensor based on the optical vernier measuring principle. *Measurement Science and Technology, Vol. 17, No. 9, (September* 2006) page numbers (2494-2498), ISSN 0957-0233

Chen, F. & Feng, J. (2007). Analogue sun sensor based on the optical nonlinear compensation measuring principle. *Measurement Science and Technology, Vol. 18, No. 7, (July* 2007) number pages (2111-2115), ISSN 0957-0233

Chen, Y.T.; Chong, K.K.; Bligh, T.P.; Chen, L.C.; Yunus, J.; Kannan, K.S.; Lim, B.H.; Lim, C.S.; Alias, M.A.; Bidin, N.; Aliman, O.; Salehan, S.; Rezan S.A.H., S.A.; Tam, C.M. & Tan, K.K. (2001). Non-imaging, focusing heliostat. *Solar Energy, Vol. 71, No. 3, (*2001) page numbers (155-164), ISSN 0038-092X

Chen, Y.T.; Lim, B.H. & Lim, C.S. (2006). General sun tracking formula for heliostats with arbitrarily oriented axes. *Journal of Solar Energy Engineering,* Vol. 128, No. 2, (May 2006) page numbers (245-250), ISSN 0199-6231

Chong K.K. & Wong, C.W. (2009). General formula for on-axis sun-tracking system and its application in improving tracking accuracy of solar collector. *Solar Energy*, Vol. 83, No. 3, (March 2009) page numbers (298-305), ISSN 0038-092X

Chong, K.K.; Siaw, F.L.; Wong, C.W. & Wong G.S. (2009a). Design and construction of non-imaging planar concentrator for concentrator photovoltaic system. *Renewable Energy*, Vol. 34, No. 5, (May 2009) page numbers (1364-1370), ISSN 0960-1481

Chong, K.K.; Wong, C.W.; Siaw, F.L.; Yew, T.K.; Ng, S.S.; Liang, M.S.; Lim, Y.S. & Lau S.L. (2009b). Integration of an on-axis general sun-tracking formula in the algorithm of an open-loop sun-tracking system. *Sensors*, Vol. 9, No. 10, (September 2009) page numbers (7849-7865), ISSN 1424-8220

Chong, K.K.; Wong, C.W.; Siaw, F.L. & Yew, T.K. (2010). Optical characterization of nonimaging planar concentrator for the application in concentrator photovoltaic system. *Journal of Solar Energy Engineering,* Vol. 132, No. 1, (February 2010) page numbers (11011-11019), ISSN 0199-6231

Grena, R. (2008). An algorithm for the computation of the solar position. *Solar Energy,* Vol. 82, No. 5, (May 2008) page number (462-470), ISSN 0038-092X

Kalogirou, S.A. (1996). Design and construction of a one-axis sun-tracking system. *Solar Energy,* Vol. 57, No. 6, (December 1996) page number (465-469), ISSN 0038-092X

Kribus, A.; Vishnevetsky, I.; Yogev, A. & Rubinov, T. (2004). Closed loop control of heliostats. *Energy, Vol* 29, No. 5-6, (April-May 2004) page numbers (905-913), ISSN 0360-5442

Lee, C.D.; Yeh, H.Y.; Chen, M.H.; Sue, X.L. & Tzeng, Y.C. (2006). HCPV sun tracking study at INER. *Proceedings of the IEEE 4th World Conference on Photovoltaic Energy Conversion,pp. 718-720, ISBN* 1-4244-0017-1, Waikoloa Village, May 7-12 2006

Lee, C.Y.; Chou, P.C.; Chiang, C.M. & Lin, C.F. (2009). Sun Tracking Systems: A Review. *Sensors,* Vol. 9, No. 5, (May 2009) page numbers (3875-3890), ISSN 1424-8220

Luque-Heredia, I.; Gordillo, F.& Rodriguez, F. (2004). A *PI* based hybrid sun tracking algorithm for photovoltaic concentration. *Proceedings of the 19th European Photovoltaic Solar Energy Conversion,* Paris, France, June 7-14, 2004

Luque-Heredia, I.; Cervantes, R. & Quemere, G. (2006). A sun tracking error monitor for photovoltaic concentrators. *Proceedings of the IEEE 4th World Conference on Photovoltaic Energy Conversion,pp. 706-709, ISBN* 1-4244-0017-1, Waikoloa Village, USA, May 7-12 2006

Luque-Heredia, I.; Moreno, J.M.; Magalhaes, P.H.; Cervantes, R.; Quemere, G. & Laurent, O. (2007). Inspira's CPV sun tracking, In: *Concentrator Photovoltaics;* Luque, A.L. & Andreev, V.M., (Eds.), page numbers (221-251), Springer, ISBN 978-3-540-68796-2, Berlin, Heidelberg, Germany

Meeus, Jean. (1991). *Astronomical Algorithms,* Willmann-Bell, Inc., ISBN 0-943396-35-2, Virginia.

Nuwayhid, R.Y.; Mrad, F. & Abu-Said, R. (2001). The realization of a simple solar tracking concentrator for the university research applications. *Renewable Energy,* Vol. 24, No. 2, (October 2001) page numbers (207-222), ISSN 0960-1481

Reda, I. & Andreas, A. (2004). Solar position algorithm for solar radiation applications. *Solar Energy,* Vol. 76, No. 5, (2004) page number (577-589), ISSN 0038-092X

Rubio, F.R.; Ortega, M.G.; Gordillo, F. & Lopez-Martinez, M. (2007). Application of new control strategy for sun tracking. *Energy Conversion and Management, Vol. 48, No. 7, (July* 2007) page numbers (2174-2184), ISSN 0196-8904

Shanmugam, S. & Christraj, W. (2005). The tracking of the sun for solar paraboloidal dish concentrators. *Journal of Solar Energy Engineering,* Vol. 127, no. 1, (February 2005) page numbers (156-160), ISSN 0199-6231

Sproul, A.B. (2007). Derivation of the solar geometric relationships using vector analysis. *Renewable Energy,* Vol. 32, No. 7, (June 2007) page numbers (1187-1205), ISSN 0960-1481

Stine, W.B. & Harrigan, R.W. (1985). The sun's position. In: *Solar Energy Fundamentals and Design: With Computer Applications,* page numbers (38 – 69).John Wiley & Sons, Inc., ISBN 0-471-88718-8, New York

14

Self Powered Instrumentation Equipment and Machinery using Solar Panels

Federico Hahn
Universidad Autónoma Chapingo
México

1. Introduction

Energy and water are required by any human being in order to live decently. Most of the rural population of the developing world lives without access to formal electrification. Electricity is one of the prerequisites for significant sustainable economic growth, being a reliable and reasonably priced energy essential for value-added agricultural and post-harvest processes. Modern energy supply also enables more intensive agriculture by providing irrigation (pumps) and immediate post-harvest treatment (cooling) and storage. Solar radiation can be converted into electricity using photovoltaic panels. Industrialized countries present a trend towards grid-connected photovoltaic systems, as battery energy storage is not required and the electricity is supplied to the network. Therefore, it is more economically interesting to supply the electricity produced by a photovoltaic system to the electricity network than to use it to drive a chiller.

Providing a reliable water supply for both human water pumping systems and agricultural needs in rural areas is one of the main applications of PV energy. The least expensive method of pumping water using PV energy is by connecting a DC motor without batteries (Abidin & Yesilata, 2004). Battery-less systems that directly couple PV modules to variable speed DC pump motors seems to have high potential for energy efficient and cost effective reverse osmosis desalinization (Ghermandi & Messalem, 2009). A simple irrigation fuzzy logic model analyzed the pumping system together with the crop to obtain the time of year to irrigate for compensating the lack of water (Damak et al., 2009). The use of pumped water for energy storage is an innovative alternative to battery storage due to its unlimited storage duration (Manolakos et al., 2004).

Many machines have been constructed thinking in photovoltaic powering. As photovoltaic panels can provide excellent energy on places with daily radiation of 4-6 kWh/m^2 prototypes are being constructed so that life becomes easier. For example, in African countries milling the average daily consumption (2.5 kg of grain) takes three hours (Chinsman, 1985). A PV-driven stone mill was constructed using two 50 W PV-panels and a battery of 85 Ah. Feed costs on dairy farms accounts for approximately half the cost of producing milk (Gardner et al., 1995). The feed dispensed to animals was measured to be within 6% of the programmed ratio and the cows adapted to eat from the feeder with training. The solar panels worked efficiently charging the batteries to provide 2.5 days of

reserve capacity to 50% depth of discharge (DOD). VaxiCool mobile freezers have been constructed in order to store and transport vaccines and drugs in remote areas.
Wireless instrumentation sensors have reduced its size and power demand. All these low-cost and lightweight devices need a compact energy source or if possible eliminate battery use. Actually worldwide research is carried out on alternatives solutions to batteries in what is called power harvesting (Vullers et al. 2009). Harvesting energy from the ambient can be done using vibration energy, thermal energy and light.
Solar panels can be used for supplying energy to measuring instruments and for small machines used during food harvesting. Instruments for taking measurements on sites were no electricity is available are generally supplied from a solar panel. Batteries are used in order to provide the energy required by the instruments during the night and under cloudy days. In this chapter a simple monitoring system that samples chloride in rivers is presented. Food production equipments require electric motors to operate; so soft start motor controllers are required to avoid current peaks in squirrel cage motors reducing solar panel sizing. Its application for a greenhouse automatic moving shelter is illustrated. A final application which uses solar panels is for prickle pear fruit cauterization to increase its shelf-life.

2. Important

If PV systems are constructed with rudimentary electronics, it is possible that most of the power generated by the solar panels can be wasted and dissipated as heat by the system components. It is worthwhile to investigate modern and sophisticated means of managing electrical power in PV systems using power electronics. Furthermore, since solar energy systems are relatively expensive in comparison to other energy sources, it would be advisable to maximize energy efficiency.
Only 30% of PV applications apply energy to motors being 15% of them alternating current motors. Its main application is for pumping water in rural areas with direct current motors, which although are bigger than AC motors, are easily started. In applications where the motor is hanging like in the greenhouses or in other equipments, weight becomes an important feature to be considered. Actually at homes, DC motors are rarely used and AC motors move most of the appliances. Squirrel cage induction motors have increased its efficiency decreasing its weight while DC motors are used as variable speed drives; actually inverters control AC motor speed without reducing its torque. AC motors driven by PV systems which are properly managed by PWM (pulse width modulated) soft starters reduce starting current peaks, increase batteries life, reduce its recharge rate and solar panel size. A PWM (pulse width modulated) soft starter was developed and later in this chapter it is applied to control a greenhouse shade curtain.
Rapid population growth, land development along river basin, urbanization and industrialization increase rivers pollution and environmental deterioration. All developed countries have the decision to maintain the water quality of their rivers. River water quality has to be characterized to identify changes or trends in quality over time or emerging problems. Pollutants discharge into the rivers can be spotted, as well as their possible sources in order to make recommendations for improving water quality in rivers. Ecosystems depend upon high water quality for provision of drinking water, cycling of nutrients, and biodiversity maintenance. Nitrogen and phosphorus are essential plant

nutrients, but when streams become enriched with these nutrients changes to species composition can result.

Real-time monitoring platforms can collect valuable data with extremely high temporal resolution without missing data in risky events like storms. Water monitoring sensor technology has been evolving, and most monitoring systems can work for weeks and even months without maintenance or calibration. In-situ automatic calibration can be performed without personal and a chloride sensor photovoltaic driven will be analyzed latter on this chapter considering autonomy and in-situ calibration. As monitoring platforms operate unattended and require relatively little maintenance, continuous monitoring programs quickly become more cost-effective than programs that rely on data-gathering personal. Data transmission via radio, cellular and phone telemetry provides a real-time picture of the water quality and automatic reports via internet provide water information to everybody.

For all fresh produce, climatic conditions and growing practices affect the quality at harvest. Successful marketing of still alive fresh fruits and vegetables depends on maintaining the quality harvested. Worldwide postharvest fruit and vegetables losses are as high as 30 to 40% and even much higher in some developing countries. Reducing postharvest losses is very important ensuring that sufficient high quality food is available to every inhabitant in our planet. World production of vegetables amounted to 487 million ton, while that of fruits reached 393 million ton. Freshness is a very important quality attribute and can be achieved by storing fruits and vegetables for short periods (days) under the proper conditions. The last example in this chapter, analyzes a cauterization technique to avoid water loss in prickle pears. Simple equipment creates a diaphragm coating in the cauterized zone. Farms require of many equipments operating in parallel so a PV grid system should be used in the future.

3. Equipments working with PV panels

There are a many equipments working with PV panels. In this section three different applications were selected from hundreds of application as they will represent clear examples. The first one will talk about monitoring systems for water quality, the second one a controller for increasing PV system efficiency and the third a machine which cuts and cauterizes prickle pears increasing its shelf life.

3.1 Automatic chloride detection in rivers

Water quality monitoring can control and detect possible pollution sites on streams, rivers and effluents (Gray, 1999). Bakker et al. (1997) developed ion selective sensors by introducing a polymeric membrane sensible to ions or cations within the circuit. Dissolved oxygen, electrical conductivity and nitrates were monitored on the Wissahickon River (Kozul & Haas, 1999). Ion-sensitive sensors are cheap to construct and although their accuracy is in the order of parts per million, they can be used in environmental sensing of wastewater.

Chloride can destroy river habitats (Detwiler et al., 1991) and free chlorine concentrations from 0.03 to 0.05 mg/l killed fish depending on the exposure time (Augspurger et al., 2003). Chlorinated wastewaters have the potential to affect fish and mussel habitats (USEPA, 1985). Mussel exposure to concentrated chloride is harmful and should be less than 20 parts per billion. In the total residual chloride criterion, the 4-day average concentrations should not exceed 11 mg/l more than once every 3 yr on average (Tikkanen et al., 2004).

3.1.1 Block diagram of the monitoring system

An autonomous water quality monitoring system has to consider the variables to be evaluated, their calibration and sensor cleaning. If the sensors are positioned in a river where water flow is intense a special sensor should be used. Timing between researchers visits are important as calibration liquids can expire or rain effects on framing can become dangerous. Positioning of the sensor and acknowledgement of available radiation hours per day and per season is necessary for proper energy harvesting management. Energy was stored in batteries or capacitors where it is ready to use by the sensors, datalogging and transmitting equipment.

A chloride measuring valve was designed for automatic sampling, calibration and cleaning of an ion-sensitive chloride sensor for environmental detection on rivers and effluents. Valve optimization decreased contamination of the calibration (buffer) liquids to a minimum value. The system presents photovoltaic panels for energy harvesting and stores the remaining energy in batteries. The system avoids the use of electrical cables becoming a wireless sensor although its power consumption is relatively high when compared with wireless sensors. Sampling measurements were taken continuously for 7 days with the valve at the wastewater effluent at Nativitas on the river Texcoco near Mexico City (Hahn et al., 2006).

3.1.2 Valve development

A 10 cm diameter by 30 cm long stainless-steel valve was coupled with a pair of stepper motors (model Step-Syn 103G770, Sanyo Denki Co., Ltd, Japan) for its automatic operation. The stepper motors presented a resolution of 2°/step and consumed a power of 10w and required a battery of 12 v at 70 A/h, auto-charged by a solar cell. Valve operation presented three options: sensor cleaning, sensor calibration and effluent or river sampling and an ATM 89C51 microcontroller provided the control of both stepper motors.

The right stepper motor (Fig. 2) drives two bottles (calibration liquids) weighing 60 g each. The right lateral wall presented four holes connected to the plastic tubes; when the bottle necks and the wall holes coincided liquid began to flow into the cavity. Each hole served a function during calibration: filling the sensor cavity or refilling the bottle where the solution is stored, Fig. 3. A 70 mm long shaft with a hollow section centered by a bearing was fixed to the valve bottom by a structure and had the sensor inserted at its centre. Three holes made on the rotating tube control the liquid flow towards bottle 1, bottle 2 and for distilled and sampled water disposal. Four of the plastic tubes introduced liquids to the sensor cavity (reference liquid 1, reference liquid 2,distilled water and sampling water) and three plastic tubes removed liquids from the sensor cavity (reference liquid 1, reference liquid 2 and disposal water).

The left stepper motor (Fig. 2) drives the worm used by the brush cleaner, and works together with right stepper motor for introducing distilled water, reference 1, reference 2 or sample liquid to the sensing cavity. The left stepper motor is the responsible of keeping whichever of the four liquids in the sensing cavity as well as its disposal. Basically the three operations are briefly described, but more detailed information is presented by Hahn, 2004.

Sample measurement: The valve samples the liquid after introducing it to the sensor cavity through a spoon mechanism. The spoon delivers the liquid to the drain channel and then to the sensor cavity. At the end of the measurement the liquid remaining on the sensor cavity returns to the stream.

Washing operation: A control signal stops the right stepper motor movement, and distilled water flows towards the sensor cavity. Once the cavity is full the left stepper motor drives the brush washer (worm-operated) through the sensor cavity, and advances for 40 mm. The brush is moved in both directions three times to obtain a better cleaning operation. Once the brush cleaning action is finished, it returns to its initial position and the water gets out from the cavity.

Calibration operation: Two bottles carried by the main gear contain the calibration solutions. The bottles have to be synchronized with the right lateral wall holes. The bottle in front of the plastic tube starts filling the sensor cavity with the high concentrated sodium chloride liquid. One minute later, after the measurement is taken the first bottle is refilled. Later the second bottle is aligned to the plastic flexible tube repeating the same procedure done with bottle 1.

3.1.3 Energy management of the sensor

Table 1 shows the time required for each operation so that the energy required can be obtained in a watt-hour basis. It is assumed that after every measurement washing takes place and that a measurement is taken every ten minutes. Sampling the river chloride will take 3 W per hour. Washing uses a brush to clean the cavity and takes 3.25 W-hr for cleaning six times per hour. Calibration is only done once and could be done every week if liquid is always maintained on the cavity. Therefore it was noted that the washing liquid (distilled water) should stay on the cavity until the next sample is taken.

Although the most important part of the sampling is the sensor it requires of a data logger for storing data and a WI-FI transmission equipment in order to transmit the information to a place nearby the sensor. Currently, there are two dominant short range wireless standards frequently incorporated into mobile devices: WiFi and Bluetooth. WiFi: IEEE 802.11 offers high-bandwidth local-area coverage up to 100 meters (Pering et al., 2006; Crk & Gniady, 2009). However WiFi transmission consumes higher power in the order of 890 mW, compared to only 120 mW for Bluetooth due to simpler radio architecture (Agarwal et al, 2005).

The energy used for monitoring, data-logging and transmitting the information is shown in Table 2. A limit of the number of operations per hour is ten after considering that the sensor will be operating continuously as the measuring operation will take half an hour. The washing routine can be adjusted to work in the other half an hour. It can be noted that the energy consumed by the sensor increases with the number of samples and seems higher than the data logger and transmission energy consumption.

A low power consumption data logger was developed for storing the measurement with a resolution of 0.2% at any voltage ranging between 0-5V. The logger can record up to 8 channels of 10-bit data and stores data in a 512Kbyte memory. The logger uses 'D' sized batteries and its consumption was of 523μA, indicating that they have to be replaced every year. Its power consumption is in the order of 2.5 mW and when the data logger is not working it stays in sleep mode reducing the power supply voltage to a minimum. Campbell Scientific data logger CR800 acquires 6 analog inputs with a 13 bit AD converter with an accuracy of 0.06% of the voltage signal. During sleeping mode it consumes 0.6 mA and up to 28 mA using the RS232 communication port. The applied voltage can be from 9.6 up to 16 volts avoiding the use of a solar cell regulator. The worst case power will be 28mA x 16V= 0.448 W. These results are much higher than the ones obtained by the embedded system which never exceed 0.1W, but for a case in which another datalogger is used it was considered as 0.5 W-hr.

3.1.4 Selection of solar cells and storing elements

The selection of the batteries and solar panels was done for the highest energy consumption (ten operations per hour). A ship vessel was constructed so that the system floated in the river. The total weight of the monitoring system, batteries and solar cells accounted for 25 kg, considering aluminum surrounding the solar panels and thin coating glass to protect them. The vessel structure was elaborated with fiber glass and could hold up the weight, Fig. 4. Three solar cells were installed; one superior in flat position and the other two in the sides with the proper slope for optimum energy harvesting.

Tilting of the solar panel has been always a concern as it has been recognized that tilting helps in:

1. More power production results from better sun angle;
2. Rain water promotes module cleaning and;
3. Improved airflow cooling allows more energy output.

Horizontal modules increase energy production in the summer and decrease its production in the winter, spring and autumn; therefore a reduction in energy production over a one-year period is noted. Figure 5 shows a simulation throughout the year for three different angles and the horizontal position. The red line is the production energy; the violet represents the incident energy and the green the power on the horizontal plane. As tilting increases the red and green lines tend to separate further apart being the highest at the summer. The red and green lines were exactly equal without tilting, Fig. 5.a., but at a tilting angle of 19° the red line was higher than the horizontal installed panel from the beginning of the year to day 90 and from day 243 to the end of the year (Fig. 5.c.). In the rest of the year the flat panel produced more energy. At a tilting angle of 80° energy production decreased notably, Fig 5.d.

In this particular application the monitoring system floating in the river is continuously moving avoiding that dust is collected in the flat panel. A capacitor sensor providing a variable signal between 0-5 V, detected the floating vessel movement. As the movement frequency is in the order of one second and the air movement was of 1 ms^{-1} the dust did not accumulate in the panel surface.

The energy/day required for sampling ten times during the 24 hours is 279.84 W. Adding the energy required for calibration to this value gives the final 280.86 W/day. If two days of autonomy are needed due to cloudy days, the 12 V battery should manage 93.62 Ah for a battery discharge of 50%. Two batteries of 50 Ah can be used to provide the system requirement. The determination of the photovoltaic cells (PV) depends on the energy used per day which is 23.41 Ah; the effective amperes required from the batteries will be 27.54 Ah considering a lead battery efficiency of 0.85. Table 3 shows the panels required to provide the energy.

Lead acid batteries were selected as they present a low self-discharge rate, and low maintenance requirements. Its recharging is slow after deep discharges and at higher operating temperatures a shorter battery life is expected. The battery charger used a semiconductor switching element between the array and battery which switched on/off at a variable duty cycle to maintain the battery at or very close to the voltage regulation set point reducing power dissipation to a minimum. At noon the battery voltage reached the regulation voltage set point and the controller began to regulate the PV array current; the current followed the same profile as the solar irradiance. Towards the end of the sunlight hours the PV array current output decreased to a low value wherein regulation was not

required to limit the battery voltage below the regulation set point of the controller. Once the sun sets the battery voltage begins a gradual decrease to its open-circuit voltage.
Two different installation panel setups were tested and the current measured with an Elnet GR current datalogger. The first setup was the inverted V (Fig. 4) and the second one presented two parallel panels of 30 W each with a slope of 19° and a 25 W flat panel of 25 W. The latter presented a higher current after noon up to 6 AM as the slope angle with respect to the water level was directed towards east (tilted west). The total current obtained per day in the parallel red curve was of 51.66 A which can be used for all the day required only of 23.41 A. The black line with the inverted V shape produced 49.6 A during the day, while the parallel panels tilted east produced a minimum value of 47.56 A. The generated energy difference was not considerable different, but as sunny days appear in the mornings and rain in the afternoon, the inverse V was selected due to its stable and compact design.

3.2 Movable shade curtains controller in greenhouses

Movab1e curtain insulation systems (heat blankets, thermal screens) reduce heat radiation losses at night inside greenhouses, and decrease the energy load on your greenhouse crop during warm and sunny conditions. These structures covered with polypropylene, polyethylene, or composite fabrics (Bartok 2005) are known as retractable roof shade houses; screens should reflect as much near infrared radiation and transmit high photosynthetic radiation. Energy savings of up to 30% have been reported, ensuring a quick payback period based on today's fuel prices (Plaisier & Svensson, 2005).
Plants grow and pass from its initial transplanting state to its harvest state in 25 days requiring of a constant daily light integral, which can be achieved only through shading or supplemental lighting (Albright et al., 2000). Seginer et al., (2006) revealed that light control signals may use 3-day light integrals rather than a single-day integral. An embedded greenhouse shade curtain controller reduced solar radiation and temperature within the greenhouse (Droga et al., 2006).
Temperature inside the greenhouse is affected by the shade curtains. Soil temperatures were measured at three different times 11:00 a.m., 1:00 pm and 3:30 pm, Table 4. At 13:00 the soil temperature was 10°C warmer at the soil than the air temperature, and beneath the retractable roof the temperature difference was of 1°C.

3.2.1 Soft starter controller

Retractable shade structures with motorized roll-up systems (Fig. 8) were controlled by light sensors to regulate the amount of sunlight that reaches the plants (Pass & Mahrer, 1997). The retractable shade used a black Raschel woven shade with 60% transmission. The mechanism implemented used wires split 60 cm as guides for curtain sustain. A long tube drived by a single gear motor rolled the wire over it and moved the curtain in any direction. The retractable roof was opened seven times between 9:45 AM and 14:14 PM, and a 1/2 HP motor started fourteen times, Fig. 9. The energy used to move the shade curtain during the day was 3.95 Ah at full voltage application.
The soft start motor controller uses an inverter drived by an ATM89C51 microcontroller, Fig. 10. The inverter is composed by a 12-120 V transformer switched on and off by a pair of MOSFETS. The microcontroller reads the battery voltage in order to predict maximum voltage that can be applied to the motor; for example a battery voltage of 10 VDC can

provide 100V RMS (root mean square) to the motor. The PWM voltage presents many time delays (Fig. 10) stored as t_1-t_0, t_2-t_1, etc. in a look up table (LUT). At moment t_0 the first MOSFET will be turned on for the period given by the first data in the Table. After counting this time it will arrive to t_1 where it is turned off; the second data of the LUT will indicate the period that the MOSFET has to stay off, arriving to t_2. At this moment the next value of the table is acquired and the same MOSFET is turned on until t_3. The opposite MOSFET that operates as a switch will turn on and off during the negative part of the cycle.
The microprocessor program applies a starting voltage of 85 V RMS to the motor for a time period of 5 seconds, expecting it to turn. If it didn't rotate the voltage is increased by 5 volts for another 5 seconds and so on until it turns on, Fig. 11. Once rotating the increments take place every 30 seconds until full voltage is supplied. With an uncharged battery the maximum voltage that can be supplied decreases changing the PWM timing constants. The rest of the waveform can be reconstructed from the first quarter and its RMS value can be obtained from Equation 1.
The method of charging lead-acid batteries is with a constant voltage, current-limited source. That method allows a high initial charge current until the battery reaches full charge. The battery charger is also managed by the microcontroller. The PV system presented one battery of 50Ah with a maximum discharge capacity of 50% and 3 storage days in case cloudy days were present. When the system was drived by the soft start controller the energy usage per day decreased from 28.4 Ah to 23.02 Ah, Table 5. The motor load is disconnected when the battery voltage drops below 11V and reconnected when it gets back to 12.5V. The number of panels of 30W each decreased from seven to six, depending on the voltage applied. If the curtains are opened 12 times instead of 7 the 50W solar panels installed decreased to 6 when started with 84 V, Table 6.

3.3 Prickle pear cauterizer

Mexico is the first world producer with 79.4% of all the prickle pears, which are cultivated in 49,165 ha (Añorve et al., 2006). The prickle pear is rich in vitamin C and proteins, low in fat with high potentials of calcium, phosphorus and iron. The soluble solid content is high presenting fructose and glucose, meanwhile the pH is high and the acidity low (Pimienta, 1990). Fruits stored over nine days at ambient temperature presented a high incidence of spots and rots, and after 20 days almost 70% of the fruit gets damaged (Cervantes, 1998). Domínguez (1992) reported that freezing the pear at 10°C increased its shelf life to 16 days. Moisture control inside storage rooms can increase the shelf life up to 75 days (Barrios & Hernández, 2004). It is common to harvest prickle pears with a glove, rotating the fruit until it separates from the nopal leaf. Pear rots is a problem when deficient cuts are done under high moisture conditions. Pathogens as *Erwina caratovora, Agrobacterium tumefaciens, Fusarium oxisporum, Botrytis cinerea Pers.*, and *Colectotricum. Sp.*, attack the orifice left open in the fruit during harvest (Dominguez, 2006).

3.3.1 Energy consumption of the cutter

Electric pulses are applied to the heating resistance and its number determines the energy required for the cutting operation (Eqn. 2). The pulse number having a period of 100 ms is given by k; V represents the supplied voltage in volts and I the current in amperes. The temperature increased during the T_{on} period (Fig. 12) and decreased during the T_{off} period, Hahn (2009). By increasing the duty cycle (T_{on} /T) the voltage is applied for a longer time to the resistance and the set point temperature is obtained quicker. The heating resistance is of 125 Ω and the energy required is given by Eqn. 3. The voltage supplied by a 12 volts battery is given by Eqn. 4. Table 7 shows the effect of cauterization on rot production on prickle pears. Forty five days after harvest all non-cauterized prickle pears presented rots. After one month of storage, 79% of the cauterized fruits at 50°C were healthy meanwhile no pear was rot at 150°C. After two months of storage 78% of the pears were healthy.

If the prickle pear is cut in one second (10 periods) with a 50% duty cycle it will consume 2.88 W. A field worker takes 5 seconds to cut the next pear, when it is very close to him; otherwise it will take from 15 to 20 seconds. An average of 150 pears are cut per hour requiring a power of 432 W. The 5 W and 0.8 kg solar panel PV fixed on the worker cap was connected directly to the motor (without battery), Fig.13. Whenever, the sun hides or is cloudy it is important to have a 100 Ah lead acid battery which can be charged every 2 hours. Although a 100 Ah battery is the ideal it is too heavy to carry (38 Kg) so a 70 Ah deep cycle battery was selected as it is lighter (16.5 kg).

The cutting blade reached 150°C after 83 seconds, and the resistance was disconnected automatically when it arrived to this temperature. Heat loss after 90 seconds decreased the cutter temperature to 140°C. Cauterizing of fruits was the main work (Hahn, 2009), but it can be applied during nopal farm management just before the season starts, Fig. 14. The cauterized prickle pears are placed vertically in boxes with the cutting edge looking upwards. The boxes are closed to avoid any contamination and contact with the fruits before being transported from the field to the storage room. To reduce energy requirements, the resistance is now being substituted by a light focusing and converging system which can heat up to 130°C without using energy.

3.3.2 Farm mode analysis

Theoretically an interesting system has been developed but not very useful to the producers who own a farm, Fig. 15. The producer hires ten workers so the system has to be changed in order to be economically feasible. A grid system is required as stored energy can be used by any worker. If 1500 pears are cut per hour and the working day lasts eight hours 34.56 kW are required per day. The current demand per day is 1440 A using 24 VDC motors. As there are 6 hours with radiation of 1000W/m^2, and considering a battery efficiency of 0.84, the panels should manage 282 A. This current without considering any class of autonomy is provided by 23 panels of 300 Watts. Cutting pears on cloudy days with high humidity is undesirable and cauterization is not as effective, so only harvesting is done during sunny days. Also the working day is divided so that during the middle of the day when radiation peaks the solar panels charge the batteries. The 29 battery bank handles 100 Ah each and the DOD were limited to 0.5. Operationally, the farm had two options for transferring the electricity which are shown on Fig 15 and 16. The latter one used cooper tubes over the soil making a DC grid, with low resistance and automatic disconnection when no worker was there. The system only works well when the freeway presents no nopal leaves which avoids the connecting box (motor to power line) movement. The second system presents the grid over the nopal tree and it's easier to work with, Fig. 16.

4. Figures and tables

Fig. 1. Water quality autonomous sensing

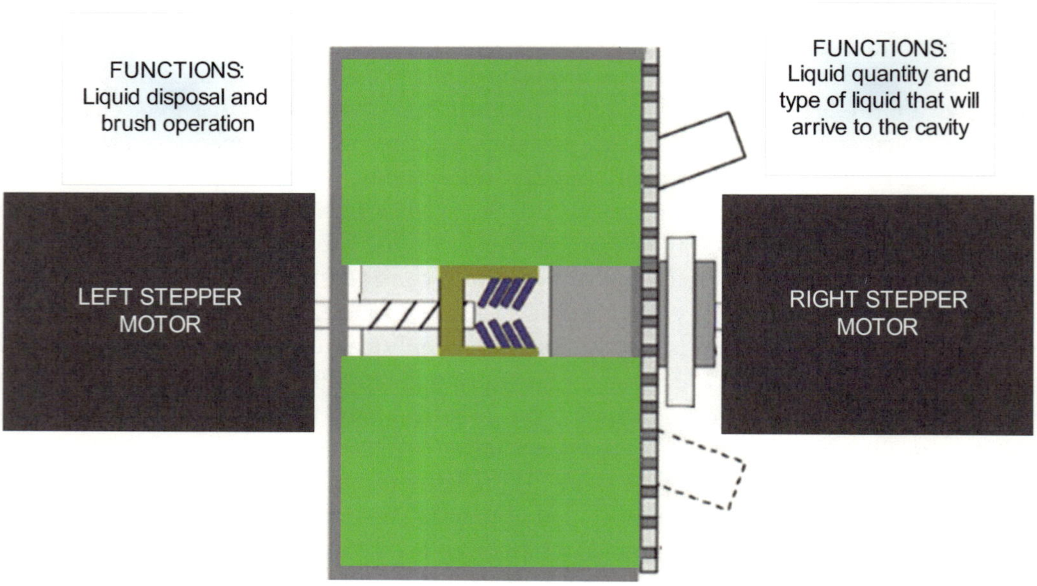

Fig. 2. Sensor simplest diagram design and motor functions

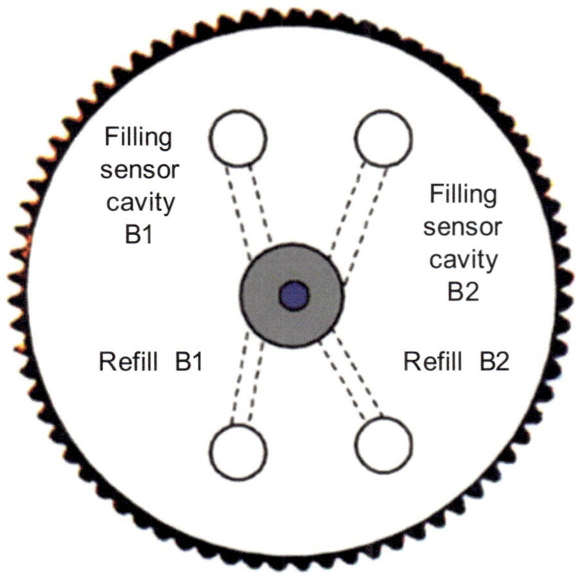

Fig. 3. Lateral view of the right wall and function realized per hole.

	Basic operation	Operation number	Time per operation, sec	Energy W-hr
Measurement	Hollow shaft filling by gravity, sensing time and sucking liquid	6/hr	180	3
Washing	Brush moving, positioning and distilled water management	6/hr	195	3.25
Calibration	Bottle positioning, refilling and water disposal	1/day	370	1.02

Table 1. Energy in watts-hour, time per operation for each of the sensor measuring, washing and calibration

	Four operation/hr	Six operation/hr	Ten operation/hr
Sensor	5.18	7.27	10.16
Data logger	0.5	0.5	0.5
Transmission	1	1	1

Table 2. Energy in watts-hour, time per operation for each of the sensor measuring, washing and calibration

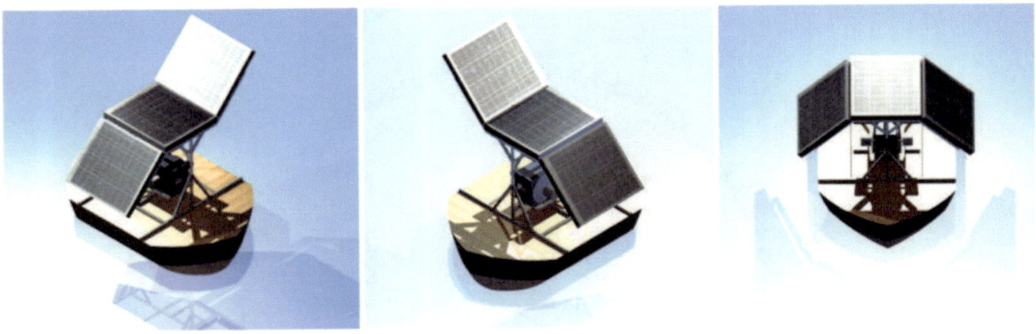

(a) (b) (c)

Fig. 4. Solar panels generate energy for chloride monitoring (a) facing east (b) facing west and (c) with flat panel in the top.

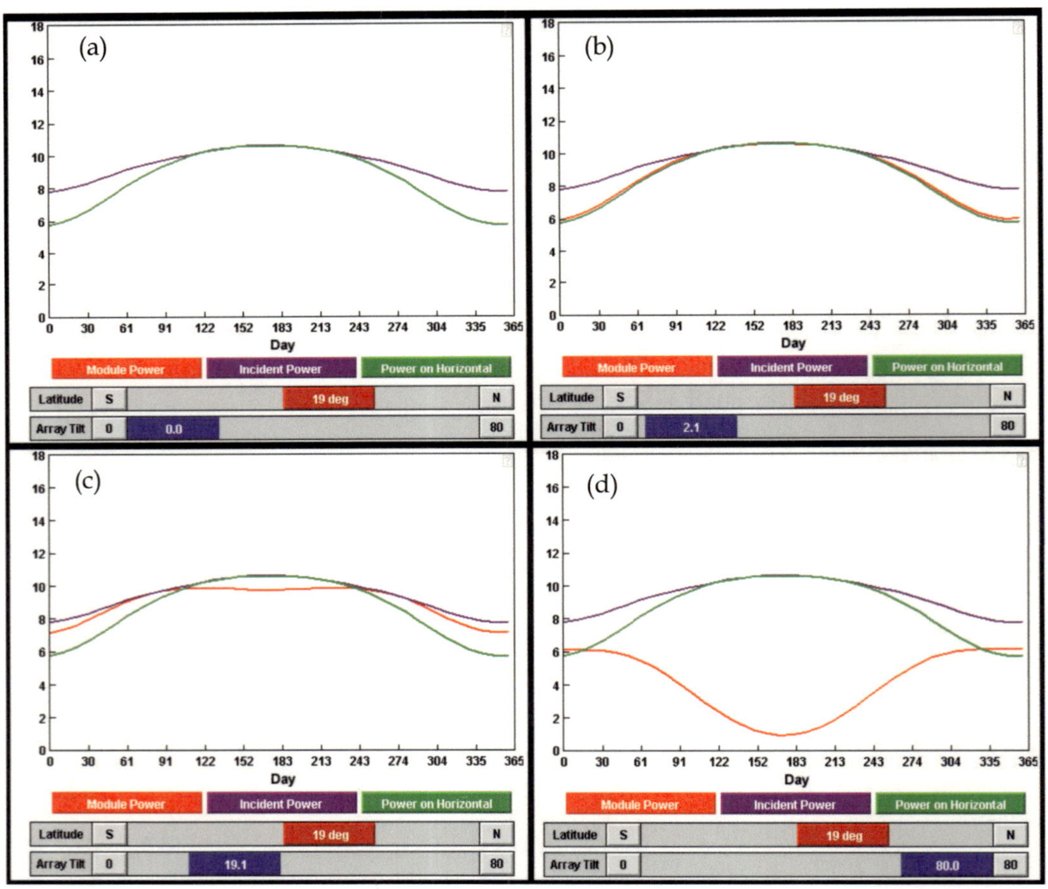

Fig. 5. Tilting in power production at (a) 0°, (b) 2.1°, (c) 19.1° and (d) 80° throughout all the year (http://energyworksus.com/solar_power_incident_angle.html).

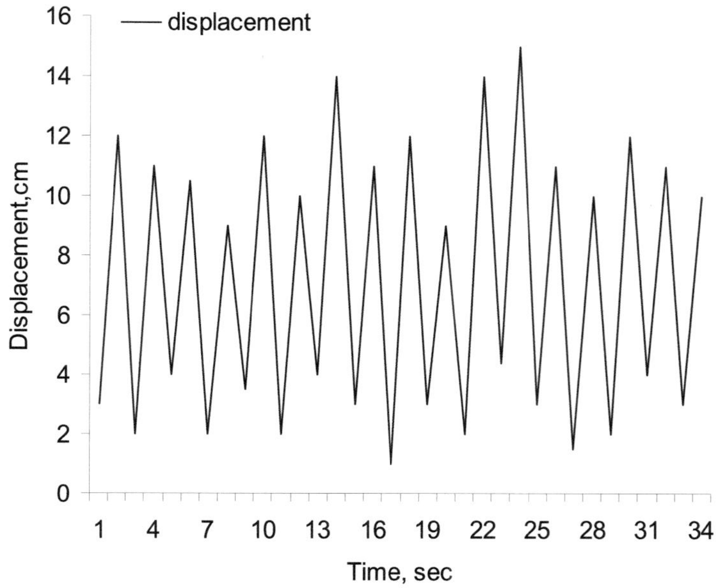

Fig. 6. Vessel vertical movement measured above the water surface.

Hours with high radiation	Current, A	Panel size	Number of panels	Weight, kg
5	5.51	90 W	3 of 30 W	11.7
6	4.59	85 W	2 of 30 W and 1 of 25W	10.6
7	3.93	70 W	2 of 20 W and 1 of 30W	8.5
8	3.44	60 W	3 of 20 W	6.0

Table 3. PV size required, amperes required to charge the battery and hours/day with a radiation over 1000 W/m^2.

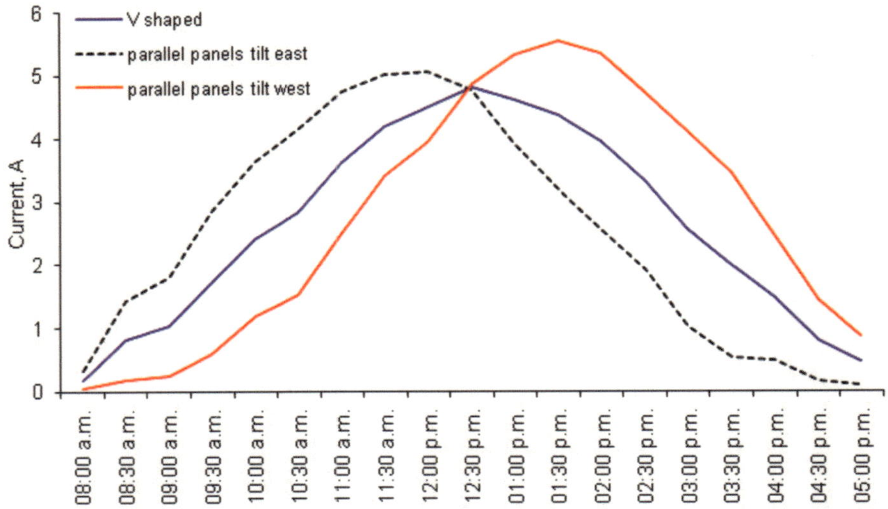

Fig. 7. Current produced for the same panels (two of 30 W and one of 25 W) fixed in different directions.

Fig. 8. Elements used for closing and opening the shade curtain including gear motor, pulleys and Raschel net.

	11:00 AM			1:00 PM			3:30 PM		
	OUT	GH	RET	OUT	GH	RET	OUT	GH	RET
Light level,	53	39	34	170	100	85	39	32	26
Air temp °C	26	30.4	27	29	34	30	24	28	25
Soil temp °C	35	26	25.5	39	29	29	33	24	23

RET: retractable; OUT: outside; GH inside the greenhouse

Table 4. Temperature and light intensity measured at different times outside and inside the greenhouse.

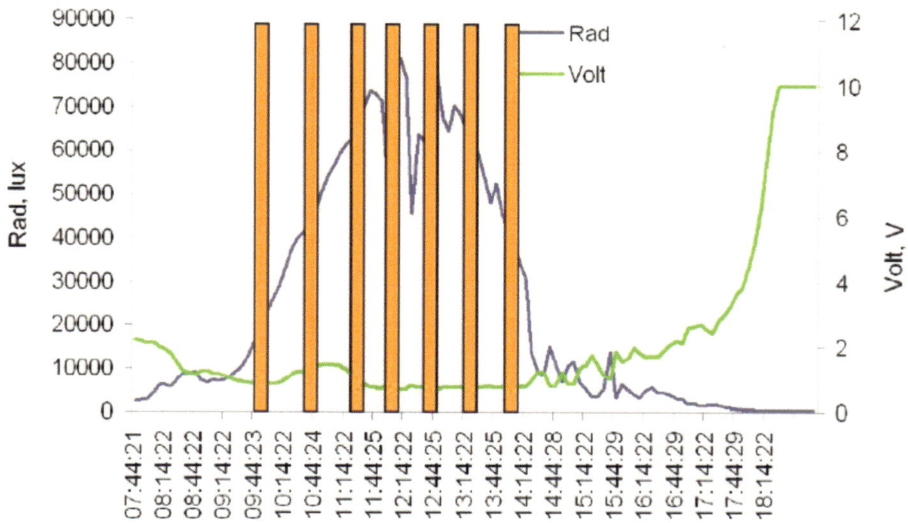

Fig. 9. Closing of the retractable roof and voltage obtained from the circuit.

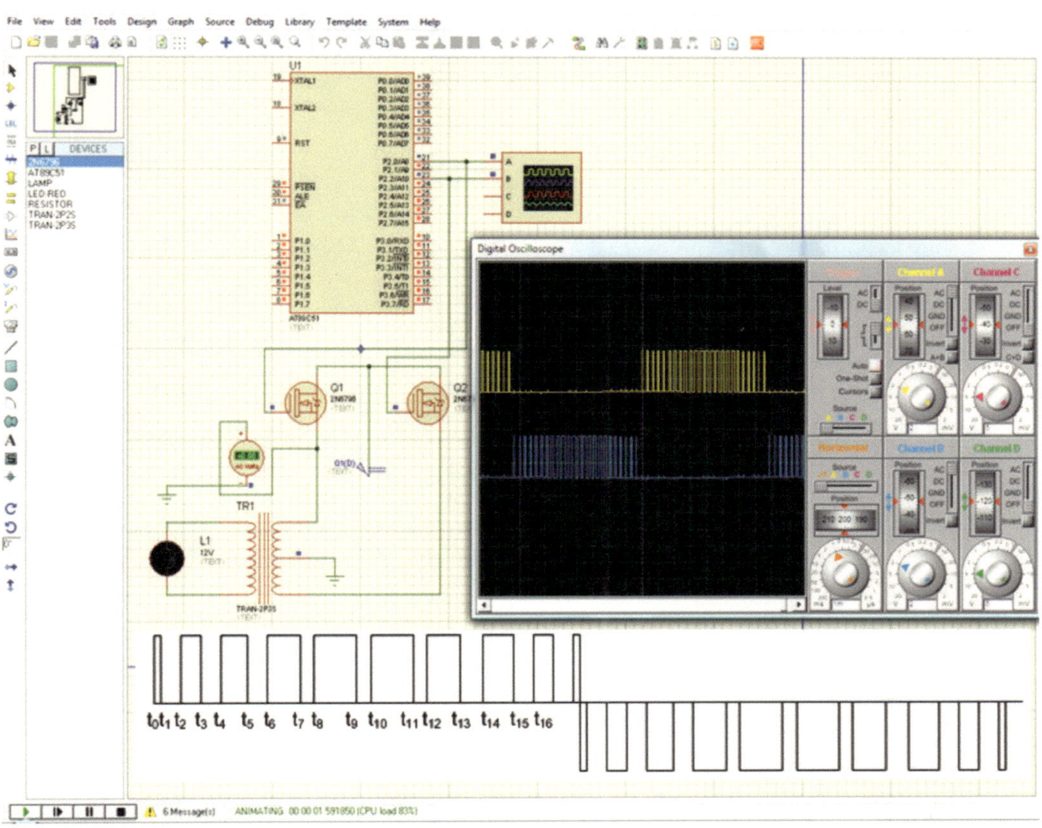

Fig. 10. Embedded closed loop PWM controller and PWM timing sequence.

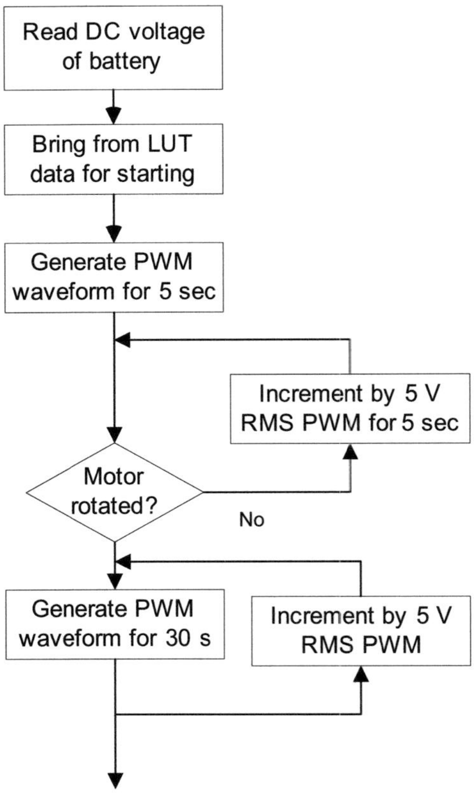

Fig. 11. Block diagram of the motor starter

	100%	90%	80%	70%
Daily current, Ah	3.95	3.73	3.5	3.2
Calculated current, Ah	28.4	26.88	25.2	23.02
Batteries	1	1	1	1
Panel @ 30W	7	7	6	6

Table 5. Current required daily for 7 openings under different starting voltages.

	100%	90%	80%	70%
Daily current, Ah	6.7	6.4	6	5.48
Calculated current, Ah	48.7	46.08	43.2	39.4
Batteries	1	1	1	1
Panel @ 50W	7	7	7	6

Table 6. Current required daily for 12 openings under different starting voltages.

Storage time (days)	Treatment efficiency, %			
	Without cauterization	50ºC	100ºC	150ºC
15	71	98	100	100
30	25	78	88	100
45	0	54	61	95
60	0	22	49	78

Table 7. Decrease in pricke pear rot during cauterization

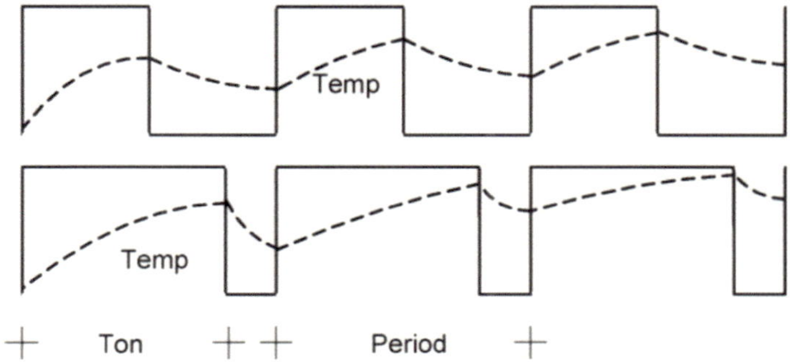

Fig. 12. Duty cycle and its effect on temperature increase.

Fig. 13. Lady cutting a pear wearing a PV cell in the cap

Fig. 14. Prickle pear (a) well cauterized, (b) beginning with rot; nopal (c) just cut and after (d) cauterization

Fig. 15. Prickle pear farm with aerial grid cables

Fig. 16. Conductor in the soil and box connection

5. Equations

$$RMS = \sqrt{\frac{V^2(t_1 - t_0) + V^2(t_2 - t_1) + V^2(t_3 - t_2) + \ldots + V^2(t_n - t_{n-1})}{(t_1 - t_0) + (t_2 - t_1) + (t_3 - t_2) + \ldots + (t_n - t_{n-1})}} \quad (1)$$

$$E = 0.5kVI \quad (2)$$

$$E = 0.004kV^2 \quad (3)$$

$$E = 0.576k \quad (4)$$

6. Conclusions

Photovoltaic systems provide electricity to everyone where sun shines and even satellites use these panels to provide the energy they require for their operation and its crew. Agriculture PV grids can provide energy to remote locations for different applications like pumping water for irrigation and human consumption, small equipments for agro industry and home appliance management. It is our responsibility to save energy and to use clean energy to share a better life quality.

Monitoring sensors are becoming wireless and batteries more efficient and smaller. Actual batteries show an increased efficiency and higher storage capacity than ten years ago. Since solar energy systems are relatively expensive in comparison to other energy sources, it is advisable to maximize energy efficiency. PV electronics systems dissipate minimum energy as heat by the system components optimizing harvested energy. A monitoring self calibrated PV driven chloride sensor was presented in this chapter; another sensor could be used being the information available even without being in the river bank.

The third project presented in the chapter shows how PV cells were used for a simple fruit cauterizer and how PV grids can be managed to operate on a farm contour. Cauterizing can even be simplified and its energy consumption reduced. Another machine PV driven projects can be implemented by our team to solve special needs and contribute to make of the Earth a wonderful planet.

7. Acknowledgments

I would like to acknowledge CONACYT and ICAMEX for funding the projects and to Armando Reyes Amador for helping me in drawing the pictures and Miss Nancy Rodriguez Perales for her help with the computer.

8. References

Abidin, Z. & Yesilata, B. (2004). New approaches on the optimization of directly coupled PV pumping systems. *Solar Energy*, 77(1), 81- 93

Agarwal, Y.; Schurgers C. & Gupta R. (2005). Dynamic Power Management using On Demand Paging for Networked Embedded Systems, In *Proc. of Asia-South Pacific Design Automation Conference (ASPDAC)*, January 18-21, 2005, Shanghai, China, ISBN: 0-7803-8737-6

Albright, L.; Both, A. & Chiu, A. (2000). Controlling greenhouse light to a consistent daily integral. *Trans. of the ASAE*, 43(2), 421-431

Añorve, J.; Aquino, E. & Mercado, E. (2006). Effect of controlled atmosphere on the preservation of minimally processed cactus pears. *Acta Horticulturae*, 728, 211-216

Augspurger, T.; Keller, A.; Black, M.; Cope, W. & Dwyer, F. (2003). Water quality guidance for protection of freshwater mussels (Unionidae) from ammonia exposure. *Environmental Toxicology and Chemistry*, 22, 2569–2575

Bakker, E.; Buhlmann, P. & Pretsch, E. (1997). Carrier-based ionselective electrodes and optodes. *Chemical Review*, 97, 3083–3132

Barrios, R. & Hernández, G. (2004). Cambios físicos y fisiológicos de tuna (*Opuntia spp.*) variedad burrona durante el almacenamiento en bodegas de reciente construcción. [Physical and physiological changes of prickle pear (*Opuntia spp.*) variety burrona during storage on new buildings] Tesis de Licenciatura. Departamento de Ingeniería Agroindustrial, Universidad Autónoma Chapingo, México.

Bartok, J. (2005). Retractable Roof Greenhouses and Shadehouses. *USDA Forest Service Proceedings* RMRS-P-35, 73-75, July 26–29 2004, Medford, Oregon, USA

Beshada, E.; Bux, M. & Waldenmaier, T. (2006). Construction and Optimization of a PV-Powered Grain Mill. Agricultural CIGR Ejournal. Manuscript FP 06 002, Vol. VIII, 1-11

Cervantes, A. (1998). Evaluación y caracterización poscosecha de tres variedades de tuna (*Opuntia spp.*). [Evaluation and post harvest characterization of three prickle pear varieties (*Opuntia spp.*)]. Tesis de Licenciatura. Instituto de Ciencias Agrícolas. Universidad de Guanajuato, México.

Crk, I. & Gniady, C. (2009). Understanding energy consumption of sensor enabled applications on mobile phones. Engineering in Medicine and Biology Society, 2009. EMBC 2009. Annual International Conference of the IEEE. 6885-6888

Damak, A.; Guesmi, A. & Mami, A. (2009). Modeling and fuzzy control of a photovoltaic-assisted watering system. *Journal of Engineering and Technology Research*, Vol.1(1), 7-13

Detwiler, R.; Kjellsen, K. & Gjorv, O. (1991). Resistance to chloride intrusion to concrete cured at different temperatures. *Journal American Concrete Institute Materials*, 88(1), 19–24

Dogra, A.; Parsad, K.; Kumar, N. & Suri, K. (2006). An Embedded Controller for Greenhouse Shade Curtain. *Acta Horticulturae*, 710: 121- 126

Domínguez, C. (2006). Evaluación financiera de una desespinadora y un empaque de tuna en el Cardonal, Hidalgo. [Financial evaluation of a spine-removing machine and a packinghouse at the Cardonal, Hidalgo] Tesis de Licenciatura. División de Ciencias Económicas Administrativas. Universidad Autónoma Chapingo, México.

Domínguez, J. (1992) Efectos de la incidencia de daños por frío sobre la fisiología y la calidad de frutos de tuna (*Opuntia amyclaea T*). [Effects on the incidence of freezing damage on the physiology and quality of prickle pear fruits (*Opuntia amyclaea T*).] Master Degree Theses. Colegio de Posgraduados, México.

Gardner, M.; Buckmaster, D. & Muller, L. (1995). Development of a mobile solar-powered dairy concentrate feeder. *Applied Engineering in Agriculture*, Vol 11(6), 785-790

Ghermandi, A. & Messalem, R. (2009). Solar-driven desalination with reverse osmosis: the state of art. *Desalination and Water Treatment*, 7, 285-296

Gray, F. (1999). Water Technology: An Introduction for Environmental Students. Elsevier Science & Technology Books, New York.

Hahn, F.; Manqueros, E. & García B. (2010). Controller for optimizing PV usage during motor starting in greenhouse applications. IX Congreso Latinoamericano y del Caribe de Ingeniería Agrícola CLIA 2010, 25-29 July 2010, Victoria, Brasil, To be published.

Hahn, F. (2009). Cactus pear cauterizer increases shelf life without cooling processes. *Computers and electronics in agriculture*, 65(1), 1-6

Hahn, F.; Miranda, G.; Perez, F.; Mayo, O.; Rojas, F. & Coras, P. (2006). Monitoreo de la calidad del agua del Rio Texcoco mediante sensores selectivos de iones. *Agrociencia*, 40, 277-287

Hahn, F. (2005) Novel valve for automatic calibration of a chloride sensor for river monitoring. *Biosystems Engineering*, 92(3), 275-284.

Kozul, J. & Haas, L. (1999). Uses and development of water quality monitoring technology. *Appalachian Rivers II conference*, West Virginia, USA. www.nett.doe.gov/publications. 18 Dic. 2005

Manolakos, D.; Papadakis, G.; Papantonis, D. & Kyritsis, S. (2004). A stand-alone photovoltaic power system for remote villages using pumped water energy storage. *Energy*, 29, 57–69

Pass, N. & Mahrer, Y. (1997). The effect of an external moveable screen on the greenhouse microclimate: an application of a one dimensional numerical model. *Acta Horticulturae*, 443, 111-118.

Pering, T.; Agarwal, Y.; Gupta, R. & Want, R. (2006). CoolSpots: Reducing the Power consumption of Wireless Mobile Devices with Multiple Radio Interfaces. *MobiSys'06*, June 19–22, Uppsala, Sweden.

Pimienta, B. (1990). El nopal tunero. [Prickle pear nopal]. First edition. Universidad de Guadalajara. Jalisco, México.

Plaisier, H. & Svensson, L. (2005). Use of Adapted Energy Screens in Tomato Production with Higher Water Vapour Transmission. *Acta Horticulturae*, 691, 583-588.

Seginer, I.; Albright, L. & Ioslovich, I. (2006). Improved strategy for a constant daily light integral in greenhouses. *Biosystems Engineering*, 93 (1), 69–80.

Tikkanen, M.; Schroeter, J.; Leong, L. & Ganesh, R. (2004). Guidance manual for the disposal of chlorinated water. Integra Chemical Company, Renton, WA, USA

USEPA (1985). Ambient water quality criteria for chlorine. EPA 440/5-84-030. United States Environmental Protection Agency, Office of Water, Criteria and Standards Division, Washington, DC

Vullers, R.; van Schaijk, R.; Doms, I.; Van Hoof, C. & Mertens, R. (2009). Micropower energy harvesting. *Solid State Electronics*, 53, 684–693

15

Artificial Intelligence Techniques in Solar Energy Applications

Soteris A. Kalogirou[1] and Arzu Şencan[2]
[1]Cyprus University of Technology, Department of Mechanical Engineering and Materials Sciences and Engineering, P. O. Box 50329, 3603 Limassol,
[2]Department of Mechanical Education, Technical Education Faculty, Süleyman Demirel University, 32260 Isparta,
[1]Cyprus
[2]Turkey

1. Introduction

Many human mental activities such as writing computer programs, doing mathematics, engaging in commonsense reasoning, understanding language, and even driving an automobile are said to demand "intelligence". Over the past few decades, several computer systems have been built that can perform tasks such as these. Specifically, there are computer systems that can diagnose diseases, plan the synthesis of complex organic chemical compounds, solve differential equations in symbolic form, analyze electronic circuits, understand limited amounts of human speech and natural language text, or write small computer programs to meet formal specifications. We might say that such systems possess some degree of artificial intelligence. Most of the work on building these kinds of systems has taken place in the field called Artificial Intelligence (AI) (Nilsson, 1980). Most AI programs are quite complex objects and mastering their complexity is a major research goal. A comprehensive study of the problems that exist in AI programs requires a precise formalization so that detailed analyses can be carried out so as satisfactory solutions can be obtained (Bourbakis, 1992).

The main objectives of AI research are (Akerkar, 2005):

- Understand human cognition
- Cost-effective automation replaces humans in intelligent tasks.
- Cost-effective intelligent amplification builds systems to help humans think better, and faster.
- Superhuman intelligence builds programs to exceed human intelligence.
- General problem-solving solves a broad range of problems.
- Coherent discourse communicates with people using natural language.
- Autonomy has intelligent systems acting on own initiative.
- Training of the system should be able to gather own data.
- Store information and know how to retrieve it.

The aim of this chapter is to introduce briefly the various AI techniques and to present various applications in solar energy applications. Solar energy applications include the estimation of solar radiation, solar heating, photovoltaic (PV) systems, sun tracking systems, solar air-conditioning systems and many others. Therefore, the possibilities of applying AI in solar energy applications will be shown.

2. AI techniques

AI techniques have the potential for making better, quicker and more practical predictions than any of the traditional methods. AI consists of several branches such as artificial neural network (ANN), fuzzy logic (FL), Adaptive Network based Fuzzy Inference System (ANFIS) and Data Mining (DM).

2.1 Artificial Neural Networks (ANN)

Neural networks are composed of simple elements operating in parallel. These elements are inspired by biological nervous systems. As in nature, the network function is determined largely by the connections between elements. A neural network can be trained to perform a particular function by adjusting the values of the connections (weights) between the elements. Commonly neural networks are adjusted, or trained, so that a particular input leads to a specific target output. Such a situation is shown in Fig 1. Here, the network is adjusted, based on the comparison between the output and the target, until the network output matches the target. Typically many such input/target output pairs are needed to train a network (MATLAB Neural Network Toolbox 4.0.4).

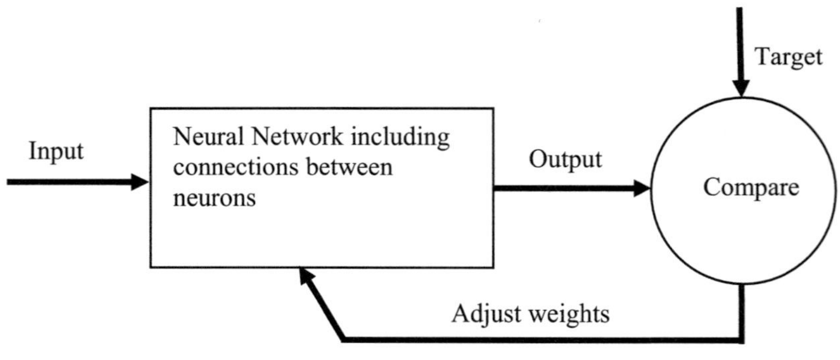

Fig. 1. Basic Principles of Artificial Neural Networks

ANNs have been applied successfully in a number of application areas. Some of the most important ones are (Kalogirou, 2000; 2001):

1. *Function approximation*. Mapping of a multiple input to a single output is established. Unlike most statistical techniques, this can be done with adaptive model-free estimation of parameters.
2. *Pattern association and pattern recognition*. This is a problem of pattern classification. ANNs can be effectively used to solve difficult problems in this field, for instance in sound, image, or video recognition. This task can even be made without an *a priori* definition of the pattern. In such cases the network learns to identify totally new patterns.
3. *Associative memories*. This is the problem of recalling a pattern when given only a subset clue. In such applications the network structures used are usually complicated, composed of many interacting dynamical neurons.

4. *Generation of new meaningful patterns*. This general field of application is relatively new. Some claims are made that suitable neuronal structures can exhibit rudimentary elements of creativity.

ANNs have been applied successfully in various fields of mathematics, engineering, medicine, economics, meteorology, psychology, neurology, and many others. Some of the most important ones are in pattern, sound and speech recognition, in the analysis of electromyographs and other medical signatures, in the identification of military targets and in the identification of explosives in passenger suitcases. They have also being used in weather and market trends forecasting, in the prediction of mineral exploration sites, in electrical and thermal load prediction, and in adaptive and robotic control. Neural networks are used for process control because they can build predictive models of the process from multidimensional data routinely collected from sensors (Kalogirou, 2000; 2001).

The network usually consists of an input layer, some hidden layers and an output layer. In its simple form, each single neuron is connected to other neurons of a previous layer through adaptable synaptic weights. Knowledge is usually stored as a set of connection weights (presumably corresponding to synapse efficacy in biological neural systems). Training is the process of modifying the connection weights in some orderly fashion using a suitable learning method. The network uses a learning mode, in which an input is presented to the network along with the desired output and the weights are adjusted so that the network attempts to produce the desired output. The weights after training contain meaningful information whereas before training they are random and have no meaning (Kalogirou, 2000; 2001). Figure 2 illustrates how information is processed through a single node. The node receives weighted activation from other nodes through its incoming connections. First, these are added up (summation). The result is then passed through an activation function; the outcome is the activation of the node. For each of the outgoing connections, this activation value is multiplied by the specific weight and transferred to the next node.

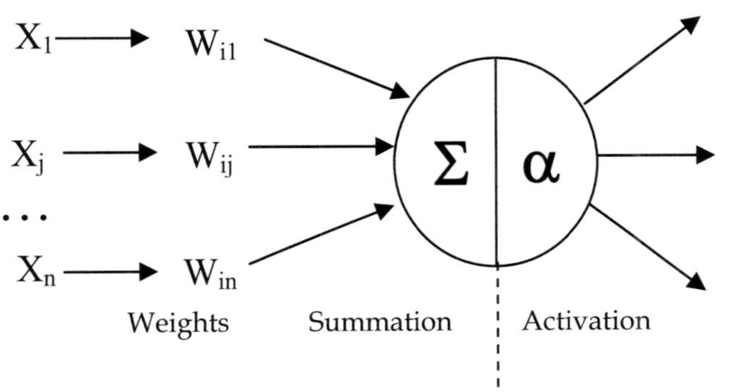

$$\alpha_i = f\left(\sum_{j=1}^{n} X_j W_{ij}\right)$$

Fig. 2. Information processing in a neural network unit

More details on neural networks can be found in Kalogirou (2000; 2001).

2.2 Fuzzy Logic (FL)

Fuzzy logic has two different meanings. In a narrow sense, fuzzy logic is a logical system, which is an extension of multivalued logic. However in a wider sense, fuzzy logic (FL) is almost synonymous with the theory of fuzzy sets, a theory which relates to classes of objects with unsharp boundaries in which membership is a matter of degree. In this perspective, fuzzy logic in its narrow sense is a branch of fuzzy theory. Even in its more narrow definition, fuzzy logic differs both in concept and substance from traditional multivalued logical systems (MATLAB Fuzzy logic toolbox user's guide).

The following is a list of general observations about fuzzy logic:

- *Fuzzy logic is conceptually easy to understand.* The mathematical concepts behind fuzzy reasoning are very simple. Fuzzy logic is a more intuitive approach without the far-reaching complexity.
- *Fuzzy logic is flexible.* With any given system, it is easy to add on more functionality without starting again from scratch.
- *Fuzzy logic is tolerant of imprecise data.* Everything is imprecise if you look closely enough, but more than that, most things are imprecise even on careful inspection. Fuzzy reasoning builds this understanding into the process rather than tacking it on to the end.
- *Fuzzy logic can model nonlinear functions of arbitrary complexity.* You can create a fuzzy system to match any set of input-output data. This process is made particularly easy by adaptive techniques like Adaptive Neuro-Fuzzy Inference Systems (ANFIS), which are available in Fuzzy Logic Toolbox software.
- *Fuzzy logic can be built on top of the experience of experts.* In direct contrast to neural networks, which use training data and generate opaque, impenetrable models, fuzzy logic lets you rely on the experience of people who already understand the system.
- *Fuzzy logic can be blended with conventional control techniques.* Fuzzy systems don't necessarily replace conventional control methods. In many cases fuzzy systems augment them and simplify their implementation.
- *Fuzzy logic is based on natural language.* The basis of fuzzy logic is human communication. This observation underpins many of the other statements about fuzzy logic. Because fuzzy logic is built on the structures of qualitative description used in everyday language, fuzzy logic is easy to use (MATLAB Fuzzy logic toolbox user's guide).

Generally, a fuzzy logic model is a functional relation between two multidimensional spaces. The relation between the input and output fuzzy spaces is known as fuzzy associative memories (FAM). Inside FAM, the linguistic variables and the attributes are specified and the associative rules between different fuzzy sets are elaborated in order to set up the following construction:

IF (premises) THEN (conclusions)
Every premise or conclusion consists of expressions as
(variable) IS (attribute)
connected through the fuzzy operator AND.

To implement a fuzzy system the following steps needs to be followed:

- *Fuzzification* is a coding process in which each numerical input of a linguistic variable is transformed in the membership function values of attributes.
- *Inference* is a process which is done in two steps: (i) The computation of a rule by intersecting individual premises, applying the fuzzy operator AND, (ii) Often, more rules drive to a same conclusion. To obtain the confidence level of this conclusion (i.e.,

the membership function value of a certain attribute of output linguistic variable) the individual confidence levels are joined by applying the fuzzy operator OR.

- *Defuzzification* is a decoding operation of the information contained in the output fuzzy sets resulted from the inference process, in order to provide the most suitable output crisp value. There are a number of methods which can be used for defuzzification presented by Paulescu et al. (2008).

2.3 Adaptive Network based Fuzzy Inference System (ANFIS)

The ANFIS model is a hybrid framework that is obtained by combining the concepts of fuzzy logic and neural networking into a unified platform. The model has a fuzzy inference system in the form of an adaptive network for system identification and a predictive tool that maps a given input space to its corresponding output space based on a representative training data set. The ANFIS inference system relies on both fuzzified human knowledge (human knowledge modelled in the form of fuzzy "if-then" rules) and a set of input-output data pairs (patterns) to accomplish the process of input-output mapping. The ANFIS modelling strategy is widely used in applications or systems that involve uncertainty or imprecision in the definitions of the variables constituting the system's behaviour. In other words, it has the ability to qualitatively model and represent human knowledge without the need for precise or quantitative definitions. Moreover, it is capable of modelling and identifying nonlinear systems as well as predicting chaotic time-dependant behaviour (Soyguder and Alli, 2009). There are mainly two approaches for fuzzy inference systems namely Mamdani and Sugeno. The difference is originated from the consequent part where fuzzy membership functions are used in Mamdani and linear or constant functions are used in Sugeno. One must have data at hand in order to apply Sugeno approach, whereas there is no such requirement for Mamdani approach (Ozger and Yıldırım, 2009). The architecture of ANFIS is shown in Fig. 3.

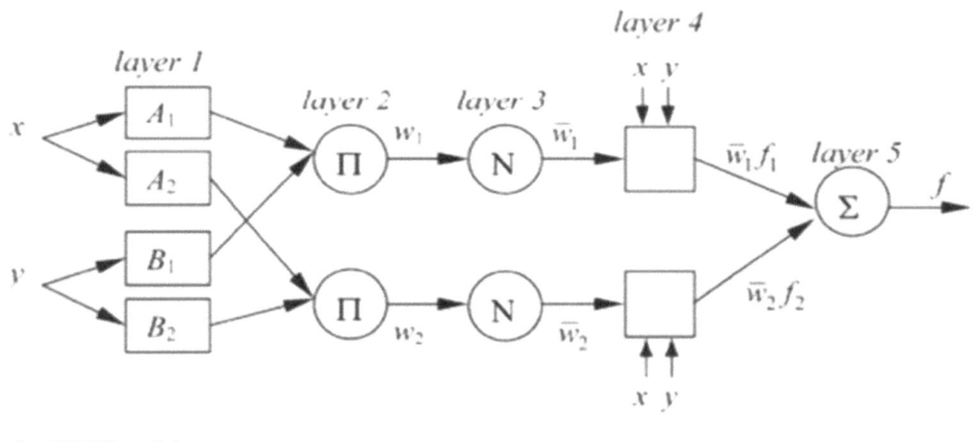

Fig. 3. ANFIS architecture

The functionality of nodes in ANFIS can be summarized as follows (Efendigil et al., 2009):

- *Layer 1*: Nodes are adaptive; membership functions (MFs) of input variables are used as node functions, and parameters in this layer are referred to as antecedent or premise parameters.
- *Layer 2*: Nodes are fixed with outputs representing the firing strengths of the rules.

- *Layer 3*: Nodes are fixed with outputs representing normalized firing strengths.
- *Layer 4*: Nodes are adaptive with node function given by Layer 1 for a first-order model, and with parameters referred to as defuzzifier of consequent parameters.
- *Layer 5*: The single node is fixed with output equal to the sum of all the rules' outputs.

2.4 Genetic Algorithms (GA)

Genetic algorithms are inspired by the way living organisms adapt to the harsh realities of life in a hostile world, i.e., by evolution and inheritance. The algorithm imitates the process of evolution of populations by selecting only fit individuals for reproduction. Therefore, a genetic algorithm is an optimum search-technique based on the concepts of natural selection and survival of the fittest. It works with a fixed-size population of possible solutions of a problem, called individuals, which are evolving in time. A genetic algorithm utilizes three principal genetic operators: selection, crossover and mutation (Kalogirou, 2004).

During each step (called a generation) in the reproduction process, the individuals in current generation are evaluated by a fitness-function, which is a measure of how well the individual solves the problem. Then each individual is reproduced in proportion to its fitness: the higher the fitness, the higher its chance to participate in mating (crossover) and to produce an offspring. A small number of newborn offspring undergo the action of the mutation operator. After many generations, only those individuals who have the best genetics (from the point of view of the fitness function) survive. The individuals that emerge from this "survival of the fittest" process are the ones that represent the optimal solution to the problem specified by the fitness function and the constraints (Kalogirou, 2004).

Genetic algorithms (GA) are suitable for finding the optimum solution in problems where a fitness function is present. Genetic algorithms use a "fitness" measure to determine which of the individuals in the population survive and reproduce. Thus, survival of the fittest causes good solutions to progress. A GA works by selective breeding of a population of "individuals", each of which could be a potential solution to the problem. The structure of the standard genetic algorithm is shown in Fig. 4.

```
                         Genetic Algorithm
Begin (1)
        t = 0 [start with an initial time]
        Initialize Population P(t) [initialize a usually random population of individuals]
        Evaluate fitness of Population P(t) [evaluate fitness of all individuals in population]
        While (Generations < Total Number) do begin (2)
                t = t + 1 [increase the time counter]
                Select Population P(t) out of Population P(t-1) [select sub-population for
                                                                  offspring production]
                Apply Crossover on Population P(t)
                Apply Mutation on Population P(t)
                Evaluate fitness of Population P(t) [evaluate new fitness of population]
        end (2)
end (1)
```

Fig. 4. The structure of a standard genetic algorithm (Kalogirou, 2004)

With reference to Fig. 4, in each generation, individuals are selected for reproduction according to their performance with respect to the fitness function. Actually, selection gives a higher chance of survival to better individuals. Subsequently, genetic operations are applied in order to form new and possibly better offspring. The algorithm is terminated either after a certain number of generations or when the optimal solution has been found (Kalogirou, 2004).

2.5 Data Mining (DM)

Data mining is a powerful technique for extracting predictive information from large databases. The automated analysis offered by data mining goes beyond the retrospective analysis of data. Data mining tools can answer questions that are too time-consuming to resolve with methods based on first principles. In data mining, databases are searched for hidden patterns to reveal predictive information in patterns that are too complicated for human experts to identify (Hoffmann & Apostolakis, 2003). Data mining is applied in a wide variety of fields for prediction, e.g. stock-prices, customer behaviour, and production control. In addition, data mining has also been applied to other types of scientific data such as astronomical and medical data (Li & Shue, 2004).

Data understanding starts with an initial data collection and proceeds with activities to get familiar with the data, to identify data quality problems, and to discover first insights into the data. Data preparation covers all activities that construct the final data set to be modelled from the initial raw data. The tasks of this phase may include data cleaning for removing noise and inconsistent data, and data transformation for extracting the embedded features (Li & Shue, 2004). Successful mining of data relies on refining tools and techniques capable of rendering large quantities of data understandable and meaningful (Mattison, 1996). The modelling phase applies various techniques, determines the optimal values of parameters in models, and finds the one most suitable to meet the objectives. The evaluation phase evaluates the model found in the last stage to confirm its validity to fit the problem requirements. No matter which areas data mining is applied to, most of the efforts are directed toward the data preparation phase (Li & Shue, 2004). The process of knowledge discovery in databases can be seen in Fig. 5.

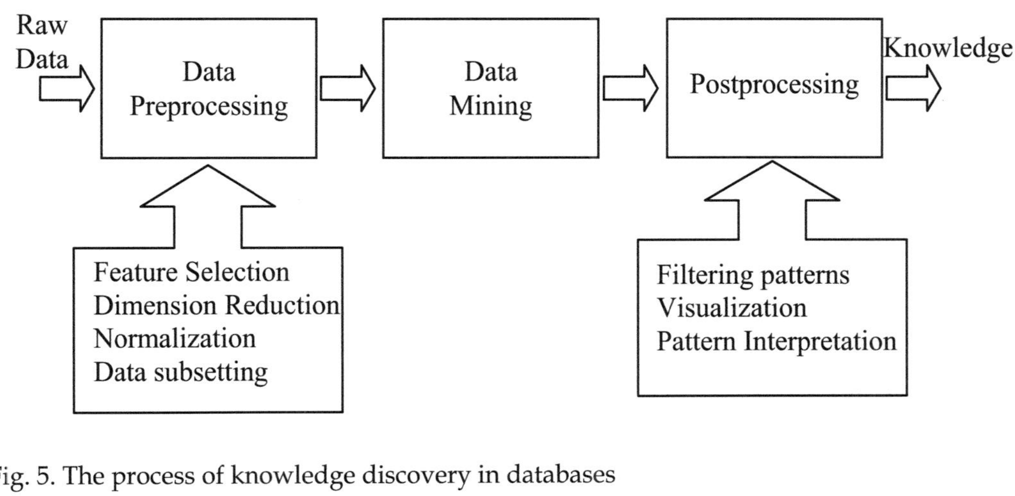

Fig. 5. The process of knowledge discovery in databases

3. Applications of Artificial Intelligence (AI) techniques in the solar energy applications

Artificial intelligence techniques have been used by various researchers in solar energy applications. This section deals with an overview of these applications. Some examples on the use of AI techniques in the solar energy applications are summarized in Table 1.

AI technique	Area	Number of applications
Artificial neural networks	Prediction of solar radiation	11
	Modelling of solar steam-generator	1
	Prediction of the energy consumption of a passive solar building	1
	Characterization of Si-crystalline PV modules	1
	Efficiency of flat-plate solar collectors	1
	Heating controller for solar buildings	1
	Modelling of a solar air heater	1
Fuzzy logic	Photovoltaic solar energy systems	2
	Sun tracking system	1
	Prediction of solar radiation	5
	Control of solar buildings	1
	Controller of solar air-conditioning system	2
Adaptive Network based Fuzzy Inference System	Prediction of solar radiation and temperature	3
Genetic algorithms	Photovoltaic solar energy systems	2
	Determination of Angström equation coefficients	1
	Solar water heating systems	2
	Hybrid solar–wind system	2
	PV-diesel hybrid system	2
	Solar cell	1
	Flat plate solar air heater	1
Data Mining	Solar cell	1

Table 1. Summary of numbers of applications presented in solar energy applications

3.1 Applications of artificial neural networks

Table 2 shows a summary of applications of artificial neural networks for solar energy applications.

Mellit and Pavan (2010) developed a Multi-Layer Perceptron (MLP) network for forecasting 24 h ahead solar irradiance. The mean daily irradiance and the mean daily air temperature are used as input parameters in the proposed model. The output was represented by the 24 h ahead values of solar irradiance. A comparison between the power produced by a 20 kWp Grid Connected Photovoltaic Plant and the one forecasted using the developed MLP-predictor shows a good prediction performance for 4 sunny days (96 h). As indicated by the authors, this approach has many advantages with respect to other existing methods and it can easily be adopted for forecasting solar irradiance values of (24-h ahead) by adding more input parameters such as cloud cover, pressure, wind speed, sunshine duration and geographical coordinates.

Authors	Year	Subject
Mellit and Pavan	2010	Prediction of solar radiation
Benghanem et al.	2009	
Rehman and Mohandes	2008	
Tymvios et al.	2005	
Mubiru and Banda	2008	
Sozen et al.	2004	
Soares et al.	2004	
Zervas et al.	2008	
Elminir et al.	2007	
Senkal and Kuleli	2009	
Moustris, K.	2008	
Kalogirou et al.	1998	Modelling of solar steam-generator
Kalogirou and Bojic	2000	Prediction of the energy consumption of a passive solar building
Almonacid et al.	2009	Characterization of Si-crystalline PV modules
Sözen et al.	2008	Efficiency of flat-plate solar collectors
Argiriou et al.	2000	Heating controller for solar buildings
Esen et al.	2009	Modelling of a solar air heater

Table 2. Summary of solar energy applications of artificial neural networks

Benghanem et al. (2009) have developed artificial neural network (ANN) models for estimating and modelling daily global solar radiation. They have developed six ANN-models by using different combination as inputs: the air temperature, relative humidity, sunshine duration and day of year. For each model, the output is the daily global solar radiation. For each of the developed ANN-models the correlation coefficient is greater than 97%. The results obtained render the ANN methodology as a promising alternative to the traditional approach for estimating global solar radiation.

Rehman and Mohandes (2008) used the air temperature, day of the year and relative humidity values as input in a neural network for the prediction of global solar radiation (GSR) on horizontal surfaces. For one case, only the day of the year and daily maximum temperature were used as inputs and GSR as output. In a second case, the day of the year and daily mean temperature were used as inputs and GSR as output. In the last case, the day of the year, and daily average values of temperature and relative humidity were used to predict the GSR. Results show that using the relative humidity along with daily mean temperature outperforms the other cases with absolute mean percentage error of 4.49%. The absolute mean percentage error for the case when only day of the year and mean temperature were used as inputs was 11.8% while when maximum temperature is used instead of mean temperature is 10.3%.

Tymvios et al. (2005) used artificial neural networks for the estimation of solar radiation on a horizontal surface. In addition, they used the traditional and long-utilized Angström's linear approach which is based on measurements of sunshine duration. The comparison of the performance of both models has revealed the accuracy of the ANN.

Mubiru and Banda (2008) used an ANN to estimate the monthly average daily global solar irradiation on a horizontal surface. The comparison between the ANN and empirical method has been given. The proposed ANN model proved to be superior over the empirical model because it is capable of reliably capturing the non-linearity nature of solar radiation. The empirical method is based on the principle of linearity.

Sozen et al. (2004) estimated the solar potential of Turkey by artificial neural networks using meteorological and geographical data (latitude, longitude, altitude, month, mean sunshine duration and mean temperature). The maximum mean absolute percentage error was found to be less than 6.74% and R^2 values were found to be about 99.89% for the testing stations. For the training stations these values were found to be 4.4% and 99.97% respectively. The trained and tested ANN models show greater accuracy for evaluating the solar resource possibilities in regions where a network of monitoring stations have not been established in Turkey. The predicted solar potential values from the ANN are given in the form of monthly maps.

Soares et al. (2004) used artificial neural networks to estimate hourly values of diffuse solar radiation at a surface in Sao-Paulo City, Brazil, using as input the global solar radiation and other meteorological parameters. It was found that the inclusion of the atmospheric long-wave radiation as input improves the neural-network performance. On the other hand traditional meteorological parameters, like air temperature and atmospheric pressure, are not as important as long-wave radiation which acts as a surrogate for cloud-cover information on the regional scale. An objective evaluation has shown that the diffuse solar radiation is better reproduced by neural network synthetic series than by a correlation model.

Zervas et al. (2008) used artificial neural networks to predict the daily global solar irradiance distribution as a function of weather conditions and each calendar day. The model was tuned using the meteorological data recorded by the "ITIA" Meteorological station of National Technical University of Athens, Zografou Campus, Greece. The model performed successfully on a number of validation tests. The future challenge is to extend the model, so that it can predict the output power of 50kWp PV arrays. This model will allow to take optimal decisions regarding the operation and maintenance of the PV panels. This work may prove useful for engineers who are interested in solar energy systems applications from both a general and a more detailed point of view.

Elminir et al. (2007) used an artificial neural network model to predict the diffuse fraction on an hourly and daily scale using as input the global solar radiation and other meteorological parameters, like long-wave atmospheric emission, air temperature, relative humidity and atmospheric pressure. A comparison between the performances of the ANN model with that of linear regression models has been given. The neural network is more suitable to predict diffuse fraction than the proposed regression models at least for the Egyptian sites examined.

Senkal and Kuleli (2009) also used artificial neural networks for the estimation of solar radiation in Turkey. Meteorological and geographical data (latitude, longitude, altitude, month, mean diffuse radiation and mean beam radiation) are used in the input layer of the network. Solar radiation is the output. The selected ANN structure is shown in Fig. 6. By using the ANN and a physical method, solar radiation was predicted for 12 cities in Turkey. The monthly mean daily total values were found to be 54 W/m^2 and 64 W/m^2 for the training cities, and 91 W/m^2 and 125 W/m^2 for the testing cities, respectively. According to the results of these 12 locations, correlation values indicate a relatively good agreement between the observed ANN values and the predicted satellite values.

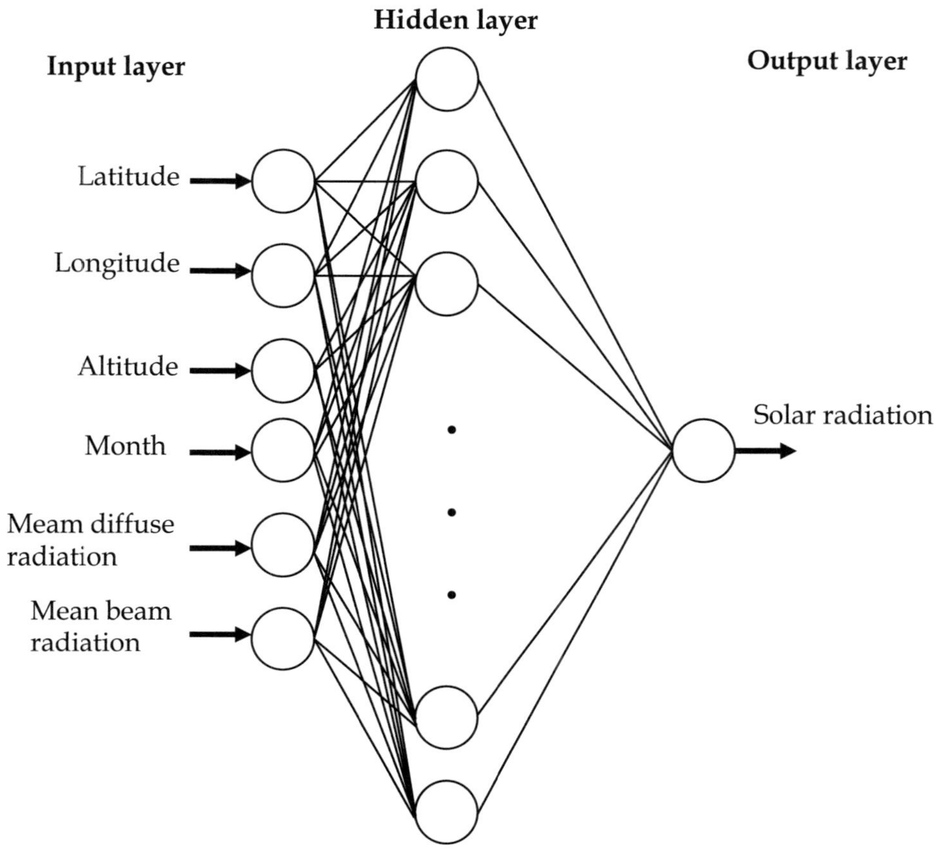

Fig. 6. ANN architecture used for the prediction of solar radiation with six neurons in the input layer by Senkal and Kuleli (2009)

Moustris et al. (2008) used neural networks for the creation of hourly global and diffuse solar irradiance data at representative locations in Greece. A very good agreement with a satisfactory outcome, is obtained between global and diffuse solar irradiance hourly data sets obtained by NNs (when trained with other, easy to find, weather and geographical parameters such as, air temperature, sunshine duration, cloud cover, latitude, etc.), and hourly solar irradiance values taken from pyranometer measurements, for the areas examined. Whenever solar data are missing, or in areas where meteorological stations do not measure and/or keep solar data, full solar irradiance time-series sets could be generated with a rather acceptable accuracy.

Kalogirou et al. (1998) used an artificial neural network to model the transient heat-up response of a solar steam-generation system. The input data are those that are easily measurable, i.e. environmental conditions and certain physical parameters (dimensions and sizes). The outputs are the measured temperatures, obtained over the heat-up period at different positions of the system. The architecture that was ultimately selected is shown in Fig. 7. The predictions of the neural network have been compared with the actual measured data (i.e. the learning set) and to the predictions from a computer program. The modelling, of the system presented, was able to predict correctly the profile of the temperatures at various points of the system within 3.9%.

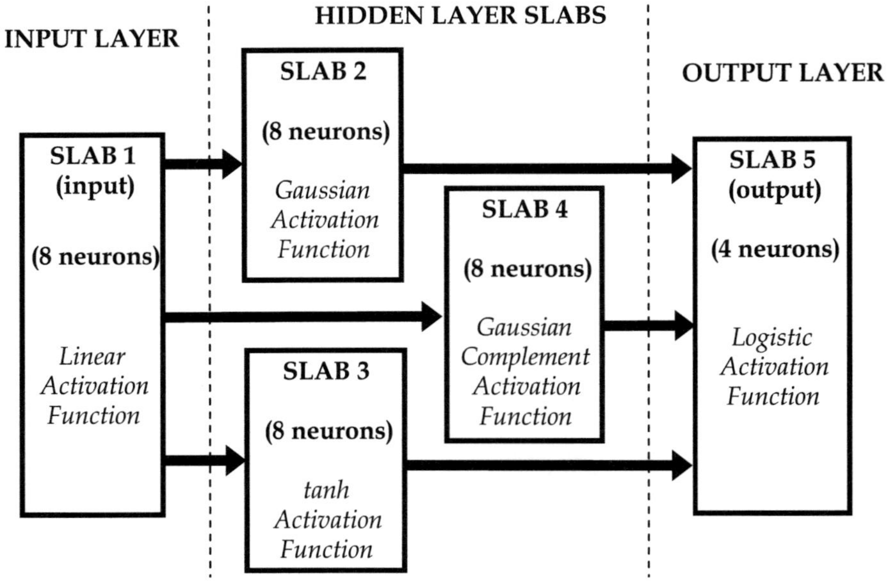

Fig. 7. The selected neural network architecture for modelling the transient heat-up response of a solar steam-generation system (Kalogirou et al., 1998)

Kalogirou and Bojic (2000) used artificial neural networks for the prediction of the energy consumption of a passive solar building. The building's thermal behaviour was evaluated by using a dynamic thermal building model constructed on the basis of finite volumes and time marching. The energy consumption of the building depends on whether all walls have insulation, on the thickness of the masonry and insulation, and on the season. Simulated data for a number of cases were used to train the artificial neural network. The ANN model proved to be much faster than the dynamic simulation programs.

Almonacid et al. (2009) used a neural network for predicting the electrical characteristics of Si-crystalline modules. I–V curves have been generated for Si-crystalline PV modules for a number of irradiance (G) and module temperature (T_m) combinations. The structure of the neural network is shown in Fig. 8. The input layer has two neurons or nodes (T_m and G), the

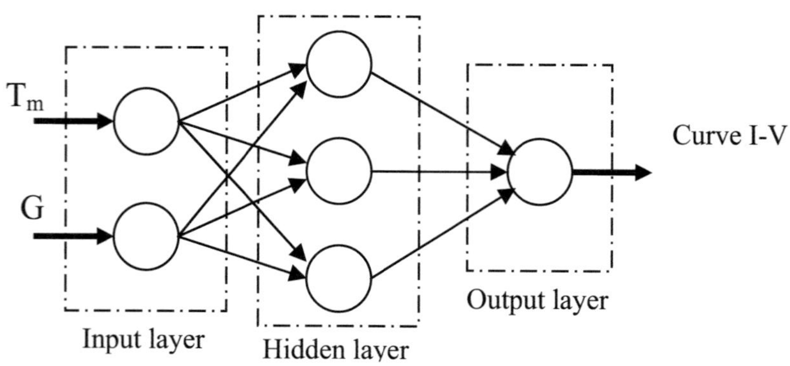

Fig. 8. Proposed neural network architecture for obtaining the I–V curves of PV modules (Almonacid et al., 2009).

second layer (hidden layer) has three nodes, and finally the last layer (output layer) has only one node: the points of the I–V curve. The results show that the proposed ANN introduces an accurate prediction for Si-crystalline PV modules' performance when compared with the measured values.

Sözen et al. (2008) developed a new formula based on artificial neural network techniques to determine the efficiency of flat plate solar collectors. The selected ANN architecture is depicted in Fig. 9.

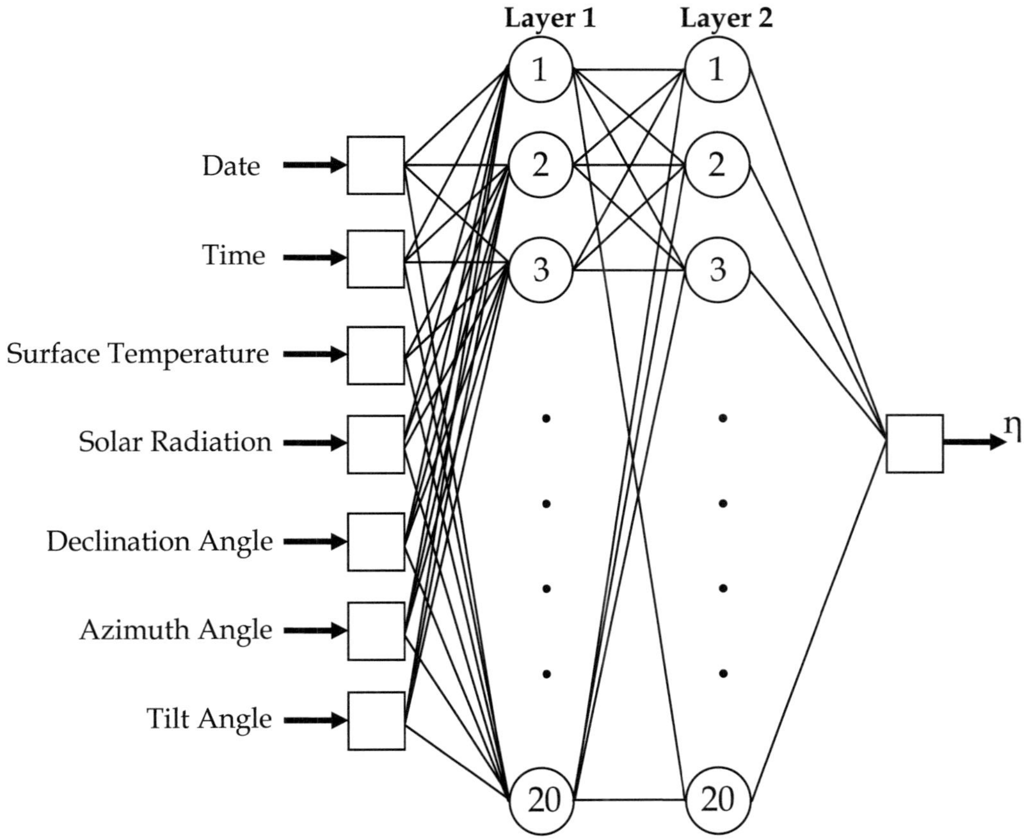

Fig. 9. ANN structure used by Sözen et al. (2008)

Date, time, surface temperature on collector, solar radiation, declination angle, azimuth angle and tilt angle are used as input to the network. The efficiency of flat-plate solar collector is in the output of the ANN. The results show that the maximum and minimum deviations were found to be 2.558484 and 0.001969, respectively. The advantages of the ANN model compared to the conventional testing methods are speed, simplicity and capacity of the ANN to learn from examples.

Argiriou et al. (2000) used ANN in order to control the indoor temperature of a solar building. The performance of the ANN controller has been tested both experimentally and in a building thermal simulation environment. The results showed that the use of the proposed controller can lead to 7.5% annual energy savings in the case of a highly insulated passive solar test cell.

Esen et al. (2009) proposed the modelling of a solar air heater system by using an artificial neural network and wavelet neural network. Two output parameters (collector efficiency and the air temperature leaving the collector unit) were predicted by the models. For this purpose, an experimental solar air heating system was set up and tested in clear day conditions. The data used as inputs to the model were obtained from measurements made on a solar air heater. A neural network-based method was intended to adopt solar air heater system for efficient modelling. Comparison between predicted and experimental results indicates that the proposed neural network model can be used for estimating the efficiency of solar air heaters with reasonable accuracy.

3.2 Applications of fuzzy logic

In recent years, the number and variety of applications of fuzzy logic have increased significantly. Table 3 shows a summary of fuzzy logic applications for solar energy systems.

Authors	Year	Subject
Altas and Sharaf Salah et al.	2008 2008	Photovoltaic solar energy systems
Alata et al.	2005	Sun tracking system
Şen Paulescu et al. Gomez and Casanovas Gomez and Casanovas Iqdour and Zeroual	1998 2008 2002 2003 2005	Prediction of solar radiation
Gouda et al.	2006	Control of solar buildings
Lygouras et al. Lygouras et al.	2007 2008	Controller of a solar air-conditioning system

Table 3. Summary of solar energy applications of fuzzy logic

Altas and Sharaf (2008) carried out a study of a stand-alone photovoltaic energy utilization system feeding a hybrid mix of electric loads which is fully controlled by a novel and simple on-line fuzzy logic-based dynamic search, detection and tracking controller that ensures maximum power point (MPP) operation under variations in solar insolation, ambient temperature and electric load fluctuations. The proposed MPP detection algorithm and dual fuzzy logic MPP tracking controller are tested using the Matlab/Simulink software environment by digitally simulating the PV array scheme feeding hybrid DC loads. Besides the MPP detector and dual fuzzy logic MPP tracking controller, the scheme includes two more control units, one for the voltage control of the common DC load bus, and the other for the speed control of the permanent magnet DC motor (PMDC) using DC/DC choppers. The MPP is detected and tracked with minimum error as the solar irradiation level change resulting in different maximum power operating points.

Salah et al. (2008) used a fuzzy algorithm for energy management of a domestic photovoltaic panel. The algorithm is validated on a 1kW peak (kWp) photovoltaic panel and domicile apparatus of different powers installed at the Energy and Thermal Research Centre in the north of Tunisia. Criteria are verified on the system behaviour during days covering different seasons of the year. The power audit, established using measures, confirms that the energy save during daylight reaches 90% of the photovoltaic panel available energy.

Alata et al. (2005) developed a multipurpose sun tracking system using fuzzy control. Sugeno fuzzy inference system was utilized for modelling and controller design. In addition, an estimation of the insolation incident on a two axis sun tracking system was determined by fuzzy IF-THEN rules. The simulations, along with the virtual reality 3-D, are regarded as powerful tools to investigate the behaviour of the systems prior to installation. Thus, the need for real values of the simulation parameters makes it closer to real applications. The step tracking that is considered in the design of multi-purpose sun tracking systems is taken every four minutes (one degree movement by the sun), and hence, less energy is needed for driving the sun trackers.

Şen (1998) used a fuzzy logic algorithm for estimating the solar irradiation from sunshine duration measurements. The fuzzy approach has been applied for three sites with monthly averages of daily irradiances in the western part of Turkey. The fuzzy algorithm developed herein does not provide an equation but can adjust itself to any type of linear or nonlinear form through fuzzy subsets of linguistic solar irradiation and sunshine duration variables. It is also possible to augment the conditional statements in the fuzzy implications used in this paper to include additional relevant meteorological variables that might increase the precision of solar irradiation estimation. The application of the proposed fuzzy subsets and rule bases is straightforward for any irradiation and sunshine duration measurements in any part of the world.

Paulescu et al. (2008) used fuzzy logic algorithms for atmospheric transmittances prediction for use in solar energy estimation. Two models for solar radiation attenuation in the atmosphere were presented. The first model encompasses self-dependent fuzzy modelling of each characteristic transmittance, while the second is a proper fuzzy logic model for beam and diffuse atmospheric transmittances. The results lead to the conclusion that developing parametric models along the ways of fuzzy logic is a viable alternative to classical parameterization. Due to the heuristic nature of the fuzzy model input-output map, it has lead to more flexibility in adapting to local meteo-climatic conditions.

Gomez and Casanovas (2002) considered solar irradiance as a case study for physical fuzzy modelling of a climate variable. The uncertainty of the solar irradiance is treated as a fuzzy uncertainty whilst other variables are considered crisp. The approach is robust as it does not rely on statistical assumptions, and it is a possible alternative to modelling complex systems. When compared with non-fuzzy models of solar irradiance, the fuzzy model shows an improved performance, and when compared with experimental data, the performance can be evaluated by fuzzy indices that take into account the uncertainty of the data and the model output.

A fuzzy model of solar irradiance on inclined surfaces has been developed by Gomez and Casanovas (2003). The fuzzy model includes concepts from earlier models, though unlike these, it considers non-disjunctive sky categories. The proposed model offers performance similar to that of the models with the best results in the comparative analysis of literature, such as the Perez model.

Iqdour and Zeroual (2005) used the Takagi-Sugeno fuzzy systems for modelling daily global solar radiation recorded in Marrakesh, Morocco. The results obtained from the proposed model have been compared with two models based on higher order statistics; the fuzzy model provides better results in the prediction of the daily solar radiation in terms of statistical indicators.

Gouda et al. (2006) investigated the development of a quasi-adaptive fuzzy logic controller for space heating control in solar buildings. The main aim of the controller is to reduce the lagging overheating effect caused by passive solar heat gain to a room space. The quasi-adaptive fuzzy logic controller is shown in Fig. 10. The fuzzy controller is designed to have two inputs: the first is the error between the set-point temperature and the internal air temperature and the second is the predicted future internal air temperature. The controller was implemented in real-time using a test cell with controlled ventilation and a modulating electric heating system. Results compared with validated simulations of conventionally controlled heating, confirm that the proposed controller achieves superior tracking and reduced overheating when compared with the conventional method of control.

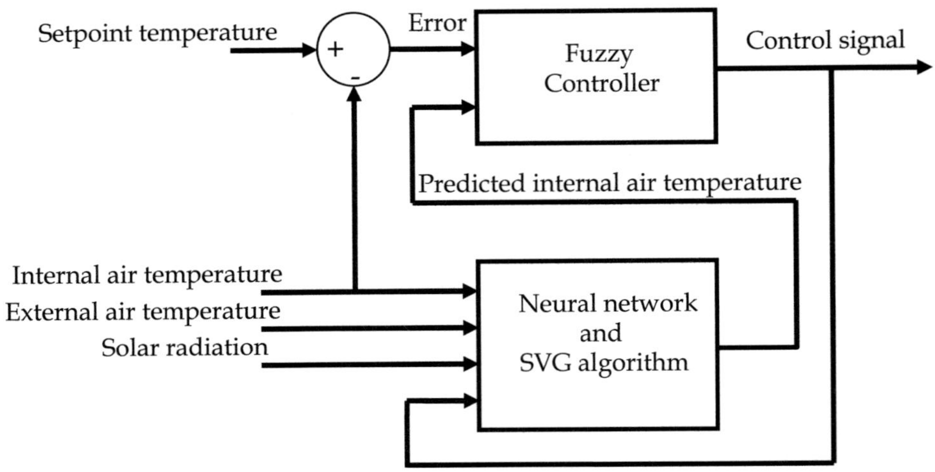

Fig. 10. Quasi-adaptive fuzzy logic controller developed by Gouda et al. (2006).

Lygouras et al. (2007) investigated the implementation of a variable structure fuzzy logic controller for a solar powered air conditioning system and its advantages. Two DC motors are used to drive the generator pump and the feed pump of the solar air-conditioner. Two different control schemes for the DC motors rotational speed adjustment are implemented and tested. The first one is a pure fuzzy controller, its output being the control signal for the DC motor driver. The second scheme is a two-level controller. The lower level is a conventional PID controller, and the higher level is a fuzzy controller acting over the parameters of the low level controller. Comparison of the two control schemes presented in this paper shows that the two-level controller behaves better in all situations.

Lygouras et al. (2008) used a fuzzy-logic controller to adjust the rotational speed of two DC motors of a solar-powered air-conditioner. Initially, a traditional fuzzy-controller has been designed; its output being one of the components of the control signal for each DC motor driver. Subsequently, according to the characteristics of the system's dynamics coupling, an appropriate coupling fuzzy-controller (CFC) is incorporated into a traditional fuzzy-controller (TFC) to compensate for the dynamic coupling among each degree of freedom. This control strategy simplifies the implementation problem of fuzzy control, but can also improve the controller performance. This mixed fuzzy controller (MFC) can effectively improve the coupling effects of the systems, and this control strategy is easy to design and implement.

3.3 Applications of Adaptive Network based Fuzzy Inference System (ANFIS)

Table 4 lists the applications of Adaptive Network based Fuzzy Inference System for solar energy systems.

Authors	Year	Subject
Chaabene and Ammar	2008	Prediction of solar radiation
Moghaddamnia et al.	2009	
Mellit et al.	2008	

Table 4. Summary of solar energy applications of ANFIS

Chaabene and Ammar (2008) used a neuro-fuzzy dynamic model for forecasting irradiance and ambient temperature. The medium term forecasting (MTF) gives the daily meteorological behaviour. It consists of a neuro-fuzzy estimator based on meteorological parameters' behaviour during the days before, and on time distribution models. As for the short term forecasting (STF), it estimates for a 5 min time step ahead, the meteorological parameters evolution. According to normalized root mean square error (NRMSE) and the normalized mean bias error (NMBE) computation, the meteorological estimator carries out satisfactory estimation of the meteorological parameters.

Moghaddamnia et al. (2009) estimated daily solar radiation from meteorological data sets with local linear regression (LLR), multi-layer perceptron (MLP), Elman, NNARX (neural network auto-regressive model with exogenous inputs) and adaptive neuro-fuzzy inference system (ANFIS). They used five relevant variables for estimating the daily solar radiation (extraterrestrial radiation, daily maximum temperature, daily mean temperature, precipitation and wind velocity). In general, they have concluded that the ANFIS model does not have the ability to estimate solar radiation precisely, but LLR and NNARX models are the most suitable models for the area under study.

Mellit et al. (2008) proposed a new model based on neuro-fuzzy for predicting the sequences of monthly clearness index and applied it for generating solar radiation, which has been used for the sizing of a PV system. The authors proposed a hybrid model for estimating sequences of daily clearness index by using an ANFIS; the proposed model has been used for estimating the daily solar radiation. An application for sizing a PV system is presented based on the data generated by this model. Fig. 11 shows the proposed ANFIS-based prediction for the monthly clearness index.

3.4 Applications of genetic algorithms

Table 5 summarizes various applications of genetic algorithms for solar energy systems.

Larbes et al. (2009) investigated the use of intelligent control techniques for maximum power point tracking in order to improve the efficiency of PV systems, under different temperature and irradiance conditions. Initially, the design and simulation of a fuzzy logic-based maximum power point tracking controller was proposed. Compared to the perturbation and observation controller, the proposed fuzzy logic controller has improved the transitional state and reduced the fluctuations in the steady state. To improve the design and further improve the performances of the proposed fuzzy logic-based maximum power point tracking controller, genetic algorithms were then used to obtain the best subsets of the membership functions as they are very fastidious to be achieved by the designer. The obtained optimized fuzzy logic maximum power point tracking controller was then simulated under different temperature and irradiance conditions. Compared to the fuzzy logic controller, this optimized controller showed much better performance and robustness. It has not only improved the response time in the transitional state but has also reduced considerably the fluctuations in the steady state.

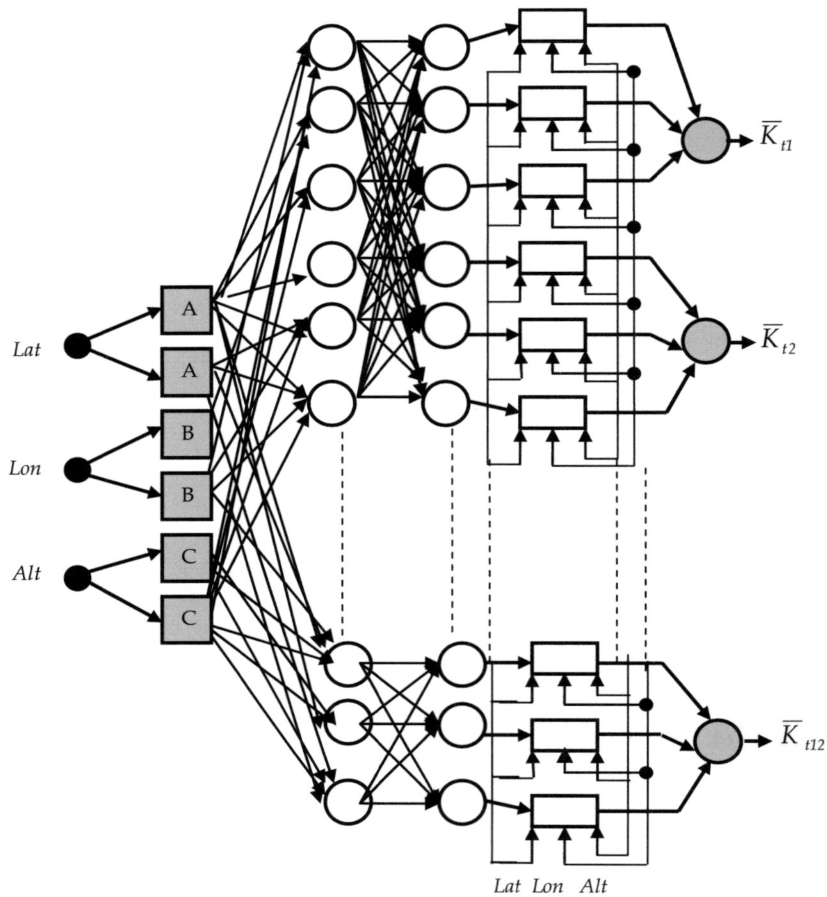

Fig. 11. The proposed ANFIS-based prediction for monthly clearness index proposed by Mellit et al. (2008)

Authors	Year	Subject
Larbes et al. Zagrouba et al.	2009 2010	Photovoltaic solar energy systems
Şen et al.	2001	Determination of Angström equation coefficients
Loomans and Vısser Kalogirou	2002 2004	Solar hot water systems
Koutroulis et al. Yang et al.	2006 2008	Hybrid solar-wind system
Bala and Siddique Dufo-Lopez and Bernal-Agustin	2009 2005	PV-diesel hybrid system
Lin and Phillips	2008	Solar cell
Varun	2010	Flat plate solar air heater

Table 5. Summary of solar energy applications of genetic algorithms

Zagrouba et al. (2010) proposed to perform a numerical technique based on genetic algorithms (GAs) to identify the electrical parameters of photovoltaic (PV) solar cells and modules. These parameters were used to determine the corresponding maximum power point from the illuminated current-voltage (I-V) characteristic. The one diode type approach is used to model the AM1.5 I-V characteristic of the solar cell. To extract electrical parameters, the approach is formulated as a non convex optimization problem. The GAs approach was used as a numerical technique in order to overcome problems involved in the local minima in the case of non convex optimization criteria. Compared to other methods, they found that the GAs is a very efficient technique to estimate the electrical parameters of PV solar cells and modules. The electrical parameters resulting from the use of the GA-based fitting procedure, with those given by the Pasan cell tester software is shown in Table 6.

Electrical parameters	Pasan software	Genetic algorithms
I_s (A)	Not performed	1.2170×10^{-2}
I_{ph} (A)	0.1360	0.1360
R_s (Ω)	0.2790	0.0363
R_{sh} (Ω)	99999	99050
n	Not performed	1.0196

Table 6. Comparison between the electrical parameters of the solar cell determined using GAs and those given by the Pasan software (Zagrouba et al., 2010)

Şen et al. (2001) used a genetic algorithm for the determination of Angström equation coefficients. Good correlation is obtained in all the cases, showing the validity of the Angström equation for Turkish locations. The authors have presented a new way of estimating the Angström equation parameters using GAs.

Loomans and Visser (2002) used a genetic algorithm for the optimization of large solar hot water systems. The genetic algorithm tool calculates the yield and the costs of solar hot water systems based on technical and financial data of the system components. The genetic algorithm allows for optimization of separate variables such as the collector type, the number of collectors, the heat storage mass and the collector heat exchanger area. The applicability of the genetic algorithm was tested for the optimization of large solar hot water systems. Among others, the sensitivity of the optimum system design to the tap water draw-off and the draw-off pattern has been determined using the optimization algorithm. As the genetic algorithm is a discrete optimization tool and is implemented in the design tool through the use of databases, the number of variables in principle is free of choice.

Kalogirou (2004) used artificial intelligence methods like artificial neural-networks and genetic algorithms, to optimize a solar-energy system in order to maximize its economic benefits. The system is modelled using a TRNSYS computer program and the climatic conditions of Cyprus, included in a typical meteorological year (TMY) file. An artificial neural-network is trained using the results of a small number of TRNSYS simulations, to learn the correlation of collector area and storage-tank size on the auxiliary energy required by the system from which the life-cycle savings can be estimated. Subsequently, a genetic algorithm is employed to estimate the optimum size of these two parameters, for maximizing life-cycle savings; thus the design time is reduced substantially. As an example, the optimization of industrial process heat-system employing flat-plate collectors is presented. The results are shown in Table 7, where the actual results of the genetic algorithm program are presented together with the results of the traditional method. The optimum solutions obtained from the present methodology give increased life-cycle savings of 4.9 and 3.1% when subsidized and non-subsidized fuel prices are used respectively, as compared to solutions obtained by the traditional trial-and-error method.

Fuel price	Parameter	Optimum system obtained from GA	Practical selection to that of GA (1)	Traditional method (2)	Percentage difference between (1) and (2)
29.6 €/L (Subsidized)	Area (m^2) Volume (m^3) LCS (€)	301.6 14.1 13,990	300 14 13,987	300 20 13,336	4.9
48.4 €/L (non-subsidized)	Area (m^2) Volume (m^3) LCS (€)	410 29.9 60,154	410 30 60,156	400 30 58,337	3.1

Table 7. Results of the solar-system optimization (Kalogirou, 2004)

Koutroulis et al. (2006) developed a methodology for the optimal sizing of stand-alone photovoltaic (PV)/wind-generator (WG) systems using genetic algorithms. The cost (objective) function minimization was implemented using genetic algorithms, which, compared to conventional optimization methods such as dynamic programming and gradient techniques, have the ability to attain the global optimum with relative computational simplicity. The proposed method has been applied for the design of a power generation system which supplies electricity to a residential household. The simulation results verify that hybrid PV/WG systems feature lower system cost compared to the cases where either exclusively WG or exclusively PV sources are used.

An optimal sizing method used to optimize the configurations of a hybrid solar-wind system employing battery banks is proposed by Yang et al. (2008). Based on a genetic algorithm, which has the ability to attain the global optimum with relative computational simplicity, an optimal sizing method was developed to calculate the optimum system configuration that can achieve the customers required loss of power supply probability (LPSP) with a minimum annualized cost of system (ACS). The decision variables included in the optimization process are the PV module number, wind turbine number, battery number, PV module slope angle and wind turbine installation height. The proposed method has been applied to the analysis of a hybrid system which supplies power to a telecommunication relay station, and good optimization performance has been found. Furthermore, the relationships between system power reliability and system configurations were also given. Although a solely solar or a wind turbine solution can also achieve the same desired LPSP, it represents a higher cost. The relationships between system power reliability and system configurations have been studied, and the hybrid system with 3–5 days' battery storage is found to be suitable for the desired LPSP of 1% and 2% for the studied case.

Bala and Siddique (2009) carried out the optimal sizing of PV array, storage battery capacity, inverter capacity, backup diesel generator set capacity and operational strategy of a solar-diesel mini-grid of an isolated island-Sandwip in Bangladesh using genetic algorithms. This study reveals that the major share of the costs is for solar panels and batteries. Technological development in solar photovoltaic technology and development in batteries production technology make rural electrification in isolated islands more promising and demanding.

Dufo-Lopez and Bernal-Agustin (2005) developed the HOGA (hybrid optimization by genetic algorithms), which is a program that uses a genetic algorithm (GA) to design a PV-diesel system (sizing, operation and control of a PV-diesel system). The program has been developed in C++. A PV-diesel system optimized by HOGA is compared with a stand-alone PV-only system that has been dimensioned using a classical design method based on the available energy under worst-case conditions. In both cases, the demand and solar irradiation are the same. The computational results show the economical advantages of the PV-hybrid system. HOGA is also compared with a commercial program for optimization of hybrid systems.

Lin and Phillips (2008) used a genetic algorithm to optimize the multi-level rectangular and arbitrary gratings. Solar cells with optimized multi-level rectangular gratings exhibit a 23% improvement over planar cells and 3.8% improvement over the optimal cell with periodic gratings. Solar cells with optimized arbitrarily shaped gratings exhibit a 29% improvement over planar cells and 9.0% improvement over the optimal cell with periodic gratings. The enhanced solar cell efficiencies for multi-level rectangular and arbitrary gratings are attributed to improved optical coupling and light trapping across the solar spectrum.

Varun (2010) used GAs for estimating the optimal thermal performance of a flat plate solar air heater having various system and operating parameters. The present work facilitates the domain of optimized values for different parameters which are decisive for ultimately finding the best performance of such a system. The basic values like number of glass covers, irradiance and Reynolds number are the key inputs on the basis of which the entire set of optimized values of parameters like wind velocity, panel tilt angle, emissivity of plate and ambient temperature are estimated by the proposed algorithm and finally the efficiency is calculated. Different optimized parameters for Reynold numbers ranging from 2000 to 20000 have been evaluated.

3.5 Applications of data mining

Table 8 summarizes various applications of data mining for solar energy systems.

Authors	Year	Subject
Kusama et al.	2007	Solar cell

Table 8. Summary of solar energy applications of data mining

Only one application is found in this area. This is by Kusama et al. (2007) who used data mining assisted by theoretical calculations for improving dye-sensitized solar cell performance. This method led to new knowledge about the influence of imidazole (crystalline heterocyclic compound used mainly in organic synthesis) derivatives as additives in an electrolytic solution on the cell performance. It was found that the solar energy conversion efficiency is strongly correlated to the Mulliken charge of the carbon atom at position 4 in the imidazole group. This result indicates that data mining assisted by theoretical calculations should facilitate the rate that cell performance is improved. Data mining combined with theoretical calculations successfully elucidated a new research direction for developing an improved electrolytic solution for dye-sensitized solar cell using base additives.

4. Conclusions

From the description of the various applications presented in this chapter, one can see that artificial intelligence techniques have been applied in a wide range of fields for modelling, prediction and control of solar energy systems. What is required for setting up such an AI system is data that represents the past history and performance of the real system and a selection of a suitable model. The selection of this model is usually done empirically and after testing various alternative solutions. The performance of the selected models is tested with the data of the past history of the real system.

In this chapter, various AI techniques used in a number of solar energy systems have been reviewed. Available literature summaries published in this area is also presented. AI techniques are becoming useful as alternate approaches to conventional techniques. AI have been used and applied in different areas, such as engineering, economics, medicine, military, marine, etc. They have also been applied for modelling, identification, optimization, prediction and control of complex systems. As can be seen from the applications presented, AI techniques have been applied successfully in a wide range of solar energy applications.

Surely, the number of applications presented here is neither complete nor exhaustive but merely a sample of applications that demonstrate the usefulness and possible applications of artificial intelligence techniques. Like all other approximation techniques, artificial intelligence techniques have relative advantages and disadvantages. There are no rules as to when this particular technique is more or less suitable for an application. Based on the works presented here it is believed that artificial intelligence techniques offer an alternative method, which should not be underestimated.

5. References

Akerkar, R. (2005). *Introduction to Artificial Intelligence, Prentice-Hall, ISBN 81-203-2864-7, New Delhi.*

Almonacid, F., Rus, C., Hontoria, L., Fuentes, M. & Nofuentes G. (2009). Characterisation of Si-crystalline PV modules by artificial neural Networks. *Renewable Energy*, Vol. 34, pp. 941–949.

Alata, M., Al-Nimr, M.A. & Qaroush, Y. (2005). Developing a multipurpose sun tracking system using fuzzy control. *Energy Conversion and Management*, Vol. 46, pp. 1229–1245.

Altas, I.H. & Sharaf, A.M. (2008). A novel maximum power fuzzy logic controller for photovoltaic solar energy systems. *Renewable Energy*, Vol. 33, pp. 388–399.

Argiriou, A.A., Bellas-Velidis, I. & Balaras, C.A. (2000). Development of a neural network heating controller for solar buildings. *Neural Networks*, Vol. 13, pp. 811-820.

Bala, B.K. & Siddique, S.A. (2009). Optimal design of a PV-diesel hybrid system for electrification of an isolated island-Sandwip in Bangladesh using genetic algorithm. *Energy for Sustainable Development*, Vol. 13, pp. 137–142.

Benghanem, M., Mellit, A. &Alamri S.N. (2009). ANN-based modelling and estimation of daily global solar radiation data: A case study. *Energy Conversion and Management*, Vol. 50, pp. 1644–1655.

Bourbakis, N.G. (1992). *Artificial Intelligence Methods and Applications, World Scientific Publishing Co., ISBN 981-02-1057-4,* Singapore.

Chaabene, M. & Ammar, M.B. (2008). Neuro-fuzzy dynamic model with Kalman filter to forecast irradiance and temperature for solar energy systems. *Renewable Energy*, Vol. 33, pp. 1435–1443.

Dufo-Lopez, R. & Bernal-Agustin, J.L. (2005). Design and control strategies of PV diesel systems using genetic algorithms. *Solar Energy*, Vol. 79, pp. 33–46.

Efendigil, T., Onut, S. & Kahraman, C. (2009). A decision support system for demand forecasting with artificial neural networks and neuro-fuzzy models: A comparative analysis. *Expert Systems with Applications*, Vol. 36, pp. 6697–6707.

Elminir, H.K., Azzam, Y.A. & Younes F.I. (2007). Prediction of hourly and daily diffuse fraction using neural network, as compared to linear regression models. *Energy*, Vol. 32, pp. 1513–1523.

Esen, H., Ozgen, F., Esen, M. & Sengur A. (2009). Artificial neural network and wavelet neural network approaches for modelling of a solar air heater. *Expert Systems with Applications*, Vol. 36, pp. 11240–11248.

Gouda, M.M., Danaher, S. & Underwood C.P. (2006). Quasi-adaptive fuzzy heating control of solar buildings. *Building and Environment*, Vol. 41, pp. 1881–1891.

Gomez, V. & Casanovas, A. (2002). Fuzzy logic and meteorological variables: a case study of solar irradiance. *Fuzzy Sets and Systems*, Vol. 126, pp. 121–128.

Gomez, V. & Casanovas, A. (2003). Fuzzy modelling of solar irradiance on inclined surfaces. *Solar Energy*, Vol. 75, pp. 307–315.

Hoffmann, D. & Apostolakis, J. (2003). Crystal Structure Prediction by Data Mining. *Journal of Molecular Structure*, Vol. 647, pp. 17-39.

Iqdour, R. & Zeroual, A. (2005). Prediction of daily global solar radiation using fuzzy systems. *International Journal of Sustainable Energy*, Vol. 26, No. 1, pp. 19–29.

Kalogirou, S.A., Neocleous, C.C. & Schizas, C.N. (1998). Artificial neural networks for modelling the starting-up of a solar steam-generator. *Applied Energy*, Vol. 60, pp. 89-100.

Kalogirou, S.A. & Bojic, M. (2000). Artificial neural networks for the prediction of the energy consumption of a passive solar building. *Energy*, Vol. 25, pp. 479–491.

Kalogirou, S.A. (2000). Applications of artificial neural-networks for energy systems. *Applied Energy*, Vol. 67, pp. 17-35.

Kalogirou, S.A. (2001). Artificial neural networks in renewable energy systems applications: a review. *Renewable and Sustainable Energy Reviews*, Vol. 5, pp. 373–401.

Kalogirou, S.A. (2004). Optimization of solar systems using artificial neural-networks and genetic algorithms. *Applied Energy*, Vol. 77, pp. 383–405.

Koutroulis, E., Kolokotsa, D., Potirakis, A. & Kalaitzakis, K. (2006). Methodology for optimal sizing of stand-alone photovoltaic/wind-generator systems using genetic algorithms. *Solar Energy*, Vol. 80, pp. 1072–1088.

Kusama, H., Konishi, Y. & Sugihara, H. (2007). Data mining assisted by theoretical calculations for improving dye-sensitized solar cell performance. *Solar Energy* Materials & Solar Cells, Vol. 91, pp. 76–78.

Larbes, C., Ait Cheikh, S.M., Obeidi, T. & Zerguerras, A. (2009). Genetic algorithms optimized fuzzy logic control for the maximum power point tracking in photovoltaic system. *Renewable Energy*, Vol. 34, pp. 2093–2100.

Li, S.T. & Shue L.Y. (2004). Data Mining To Aid Policy Making in Air Pollution Management. *Expert System with Applications*, Vol. 27, pp. 331-340.

Lin, A. & Phillips, J. (2008). Optimization of random diffraction gratings in thin-film solar cells using genetic algorithms. *Solar Energy Materials & Solar Cells*, Vol. 92, pp. 1689–1696.

Loomans, M. & Vısser, H. (2002). Application of the genetic algorithm for optimisation of large solar hot water systems. *Solar Energy*, Vol. 72, pp. 427–439.

Lygouras, J.N., Botsaris, P.N., Vourvoulakis, J. & Kodogiannis, V. (2007). Fuzzy logic controller implementation for a solar air-conditioning system. *Applied Energy*, Vol. 84, pp. 1305–1318.

Lygouras, J.N., Kodogiannis, V.S., Pachidis, Th., Tarchanidis, K.N. & Koukourlis, C.S. (2008). Variable structure TITO fuzzy-logic controller implementation for a solar air-conditioning system. *Applied Energy*, Vol. 85, pp. 190–203.

Mattison, R. (1996). *Data Warehousing: Strategies, Technologies and Techniques Statistical Analysis*, New York; London: McGraw-Hill.

MATLAB Neural Network Toolbox v. 4.0.4
http://www.mathworks.com/access/helpdesk/help/pdf_doc/nnet/nnet.pdf

Matlab Fuzzy logic toolbox user's guide. Natick: The Math Works Inc.,
http://www.mathworks.com/

Mellit, A., Kalogirou, S.A., Shaari, S., Salhi, H. & Hadj Arab, A. (2008). Methodology for predicting sequences of mean monthly clearness index and daily solar radiation data in remote areas: Application for sizing a stand-alone PV system. *Renewable Energy*, Vol. 33, pp. 1570-1590.

Mellit, A. & Pavan A.M. (2010). A 24-h forecast of solar irradiance using artificial neural network: Application for performance prediction of a grid-connected PV plant at Trieste, Italy. *Solar Energy*, Vol. 84, No. 5, pp. 807-821.

Moghaddamnia, A., Remesan, R., Kashani, M.H., Mohammadi, M., Han, D. & Piri, J. (2009). Comparison of LLR, MLP, Elman, NNARX and ANFIS Models-with a case study in solar radiation estimation. *Journal of Atmospheric and Solar-Terrestrial Physics*, Vol. 71, pp. 975–982.

Moustris, K., Paliatsos, A.G., Bloutsos, A., Nikolaidis, K., Koronaki, I. & Kavadias K. (2008). Use of neural networks for the creation of hourly global and diffuse solar irradiance data at representative locations in Greece. *Renewable Energy*, Vol. 33, pp. 928–932.

Mubiru, J. & Banda, E.J.K.B. (2008). Estimation of monthly average daily global solar irradiation using artificial neural networks. *Solar Energy,* Vol. 82, pp. 181–187.

Nilsson, N. (1980). *Principles of Artificial Intelligence,* Springer-Verlag, ISBN 3-540-11340-1, New York.

Ozger, M. & Yıldırım, G. (2009). Determining turbulent flow friction coefficient using adaptive neuro-fuzzy computing technique. *Advances in Engineering Software,* Vol. 40, pp. 281–287.

Paulescu, M., Gravila, P. & Tulcan-Paulescu, E. (2008). Fuzzy logic algorithms for atmospheric transmittances of use in solar energy estimation. *Energy Conversion and Management,* Vol. 49, pp. 3691–3697.

Rehman, S. & Mohandes, M. (2008). Artificial neural network estimation of global solar radiation using air temperature and relative humidity. *Energy Policy,* Vol. 36, pp. 571–576.

Salah, C.B., Chaabene, M. & Ammar, M.B. (2008). Multi-criteria fuzzy algorithm for energy management of a domestic photovoltaic panel. *Renewable Energy,* Vol. 33, pp. 993–1001.

Şen, Z. (1998). Fuzzy algorithm for estimation of solar irradiation from sunshine duration. *Solar Energy,* Vol. 63, pp. 39–49.

Sen, Z., Öztopal, A. & Sahin, A.D. (2001). Application of genetic algorithm for determination of Angström equation coefficients. *Energy Conversion & Management,* Vol. 42, pp. 217-231.

Senkal, O. & Kuleli, T. (2009). Estimation of solar radiation over Turkey using artificial neural network and satellite data. *Applied Energy,* Vol. 86, pp. 1222–1228.

Soares, J., Oliveira, A.P., Boznar, M.Z., Mlakar, P., Escobedo, J.F. & Machado, A.J. (2004). Modelling hourly diffuse solar-radiation in the city of Sao Paulo using a neural-network technique. *Applied Energy,* Vol. 79, pp. 201–214.

Soyguder, S. & Alli, H. (2009). An expert system for the humidity and temperature control in HVAC systems using ANFIS and optimization with Fuzzy Modelling Approach. *Energy and Buildings,* Vol. 41, No. 8, pp. 814-822.

Sozen, A., Arcaklioğlu, E. & Ozalp, M. (2004). Estimation of solar potential in Turkey by artificial neural networks using meteorological and geographical data. *Energy Conversion and Management,* Vol. 45, pp. 3033–3052.

Sözen, A., Menlik, T. & Unvar, S. (2008). Determination of efficiency of flat-plate solar collectors using neural network approach. *Expert Systems with Applications,* Vol. 35, pp. 1533–1539.

Tymvios, F.S., Jacovides, C.P., Michaelides, S.C. & Scouteli, C. (2005). Comparative study of Angström's and artificial neural networks' methodologies in estimating global solar radiation. *Solar Energy,* Vol. 78, pp. 752–762.

Varun, S. (2010). Thermal performance optimization of a flat plate solar air heater using genetic algorithm. *Applied Energy,* Vol. 87, pp. 1793–1799.

Zagrouba, M., Sellami, A., Bouaicha, M. & Ksouri, M. (2010). Identification of PV solar cells and modules parameters using the genetic algorithms: Application to maximum power extraction. *Solar Energy,* Vol. 84, No. 5, pp. 860-866.

Yang, H., Zhou, W., Lu, L. & Fang, Z. (2008). Optimal sizing method for stand-alone hybrid solar–wind system with LPSP technology by using genetic algorithm. *Solar Energy*, Vol. 82, pp. 354–367.

Zervas, P.L., Sarimveis, H., Palyvos, J.A. & Markatos, N.C.G. (2008). Prediction of daily global solar irradiance on horizontal surfaces based on neural-network techniques. *Renewable Energy*, Vol. 33, pp. 1796–1803.

16

Ray-Thermal-Structural Coupled Analysis of Parabolic Trough Solar Collector System

Yong Shuai, Fu-Qiang Wang, Xin-Lin Xia and He-Ping Tan
School of Energy Science and Engineering, Harbin Institute of Technology, No. 92, West Dazhi Street, Harbin 150001
P. R. China

1. Introduction

An effective approach to sustainable energy is the utilization of solar energy. The parabolic trough collector with central receiver is one of the most suitable systems for solar power generation. A type of concentrating solar collector that uses U-shaped troughs to concentrate sunlight onto a receiver tube, containing a working fluid such as water or oil, which is positioned along the focal line of the trough. Sometimes a transparent glass tube envelops the receiver tube to reduce heat loss. Parabolic troughs often use single-axis or dual-axis tracking. Temperatures at the receiver can reach 400°C. The heated working fluid may be used for medium temperature space or process heat, or to operate a steam turbine for power or electricity generation. As designed to operate with concentrated heat fluxes, the receiver will be subjected to the high thermal stresses which may cause the failure of receivers.

The thermal stress of receiver or tube heat exchangers has drawn many researchers' attention. Numerous studies have been carried out to investigate the temperature distributions and thermal stress fields of receiver or tube heat exchangers. A numerical analysis had been conducted by Chen [1] to study the effect on temperature distributions of using porous material for the receiver. Experiments were conducted by Fend [2] to research the temperature distributions on the volumetric receivers used two novel porous materials. A finite element analysis was conducted by Islamoglu [3] to study the temperature distribution and the thermal stress fields on the tube heat exchanger using the SiC material. To reduce the thermal stresses, Agrafiotis [4] employed porous monolithic multi-channeled SiC honeycombs as the material for an open volumetric receiver. Low cycle fatigue test of the receiver materials was conducted at different temperatures by Lata et al. [5], the results showed that the high nickel alloys had excellent thermo-mechanical properties compared to the austenitic stainless steel. Almanza and Flores [6, 7] proposed a bimetallic Cu-Fe type receiver, and the experimental test results showed that, when operated at low pressure, the bimetallic Cu-Fe type receiver had a lower thermal gradient and less thermal stress strain than the steel receiver. In Steven's study [8], the receiver is divided into 16 sections, and the average solar radiation heat flux of each section is calculated. The average heat flux is used as boundary condition for each corresponding section in the thermal analysis model. This method is fairly straightforward and simple, but the deviations generated during the heat flux transformation process are enormous.

In this section, the conjugate heat transfer and thermal stress analyses of tube receiver are carried out with concentrated solar irradiation heat flux conditions. A ray-thermal-structural sequential coupled method is adopted to obtain the concentrated heat flux distributions, temperature distributions and thermal stress fields of tube receiver. The concentrated solar irradiation heat flux distribution converged by solar parabolic collector is obtained by Monte-Carlo ray tracing method and used as boundary conditions for CFD analysis by fitting function method. Steady state conjugate heat transfer is performed to calculate temperature field using CFD system and the resulted temperature defined at the nodes of CFD mesh is interpolated as input data to the nodes in the thermal-stress analysis mesh.

2. Methodology

2.1 Radiative flux calculation

Monte Carlo (MC) method is a statistical simulation method for radiative transfer, which can be performed by tracing a finite number of energy rays through their transport histories. What a ray does at each interaction and where it goes is then determined by the probability for each process (refraction, reflection, absorption, diffraction, scatter and emission). Modest [9] and Siegel [10] have described the MC simulation in detail, respectively.

A Monte-Carlo ray tracing computational code [11], which is based on the radiative exchange factor (REF) theory, is developed to predict the heat flux distribution on the bottom surface of the tube receiver. The REF $RD_{i,j}$ is defined as the fraction of the emissive power absorbed by the jth element in the overall power emitted by the ith element. The jth element can absorb the emissive power within the system by the means of direct radiation, direct reflection and multiple reflections. The values of the $RD_{i,j}$ are determined by both the geometry and radiative characteristics of the computational elements.

The REF within the spectral band $\Delta\lambda_k$ ($k = 1,2,....,M_b$) can be expressed as follows:

$$RD_{i,j,\Delta\lambda_k} = N_{i,j} / N_i \tag{1}$$

where N_i is the total bundles emitted by the i th element, $N_{i,j}$ is the bundles absorbed by the j th element, and M_b is the total spectral bands of the wavelength-dependent radiation characteristics of the surface. As shown in Fig. 1, the concentrated heat flux distribution on the bottom surface of the tube receiver can be expressed as follows:

$$q_{r,j} = \frac{A_i}{A_j} \sum_{k=1}^{M_b} RD_{i,j,\Delta\lambda_k} E_{sun,\Delta\lambda_k} \tag{2}$$

where $q_{r,j}$ is the heat flux of the j th surface element of the tube receiver, A_i is the area of the imaginary emission surface, A_j is the area of the j th surface element of the tube receiver, and $E_{sun,\Delta\lambda_k}$ is the sun average spectral irradiance within the spectral band $\Delta\lambda_k$.

2.2 Thermal stress analyses

In order to analyze thermal stress, a ray-thermal-structural coupled method [12] is adopted to obtain temperature distribution and thermal stress field of tube receiver in the parabolic trough solar thermal collector system. At the first step, the concentrated solar radiation heat flux distribution q_c on the bottom half periphery of tube receiver, which is used as the input

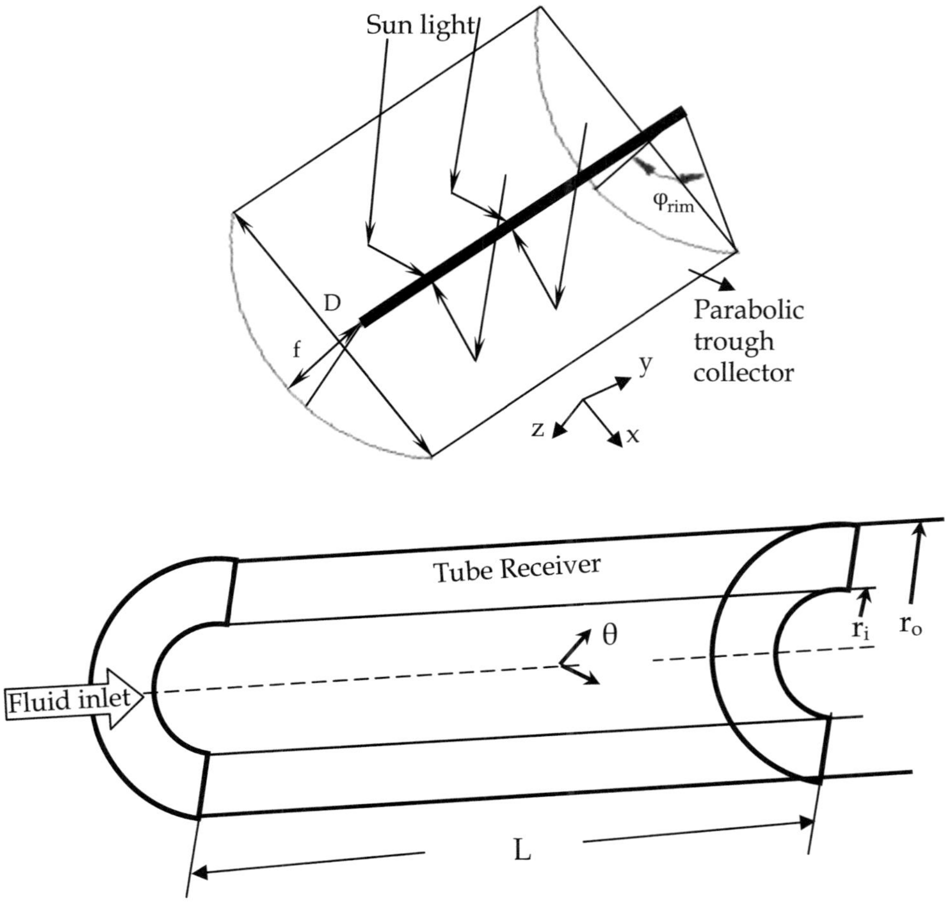

Fig. 1. Schematic diagram of the parabolic collector and receiver

data for the CFD analyses, will be calculated by the solar concentration system program with the Monte-Carlo ray tracing method. The thermal model proposed for the solar parabolic collector with tube receiver system is illustrated in Fig. 1. The geometrical parameters of the parabolic trough collector and tube receiver for this study are illustrated in Table 1. As seen from this table, the transmissivity of the glass envelop is highly close to 1, and the thickness of glass envelop is very thin, therefore, the values and distribution of heat flux are impacted very slightly when passing through the glass envelop. Therefore, this investigation doesn't consider the impact of glass envelop. During the heat flux distribution calculation process, the external cylinder surface of tube receiver will be discretized to 300 nodes along the circumference and 300 nodes along the tube length direction. Therefore, the solar concentration system program will obtain 300 × 300 heat flux values on the discrete nodes. No optical errors or tracking errors were considered for the solar concentration system program, and the calculation conditions are: the non-parallelism angle of sunlight is 16' and the solar radiation flux is 1,000 W/m^2.

At the second step, the concentrated heat flux distribution calculated by the Monte-Carlo ray tracing method will be employed as input data for the CFD analyses by means of using the boundary condition function in Ansys software. In this study, the fitting function

Parabolic trough collector and tube receiver	Value
Focal length of parabolic trough collector	2,000 (mm)
Length of parabolic trough collector	2,000 (mm)
Opening radius of parabolic trough collector	500 (mm)
Height of parabolic trough collector	1500 (mm)
Outer diameter of tube receiver (r_{out})	70 (mm)
Inner diameter of tube receiver (r_{in})	60 (mm)
Glass cover diameter	100 (mm)
Length of tube receiver	2,000 (mm)
Reflectivity of parabolic trough collector	0.95
Absorptivity of tube receiver	0.9
Transmissivity	0.965

Table 1. Geometrical parameters of the parabolic trough collector and tube receiver

method is introduced for the calculated heat flux distribution transformation from the Monte-Carlo ray tracing model to the CFD analysis model. The radiation heat flux distribution calculated by the Monte-Carlo ray tracing method along the bottom half periphery of tube receiver will be divided in to several sections, and the heat flux distribution of each section will be fitted by a polynomial regression function with highly fitted precision. The calculated heat flux distribution on the bottom half periphery of tube receiver is shown in Fig.2 and Fig. 3. Six polynomial regression functions are employed as the fitted functions and illustrated as follows:

$$\begin{cases} q = 12 & x \in [-35, \ \ -17.82] \\ q = 13740.23 + 770556.99 \times x & x \in [-17.82, \ \ -16.54] \\ q = 43418.96 + 2.57 \times x & x \in [-16.54, \ \ 0] \\ q = 43418.96 - 2.57 \times x & x \in [0, \ \ 16.54] \\ q = 13740.23 - 770556.99 \times x & x \in [16.54, \ \ 17.82] \\ q = 12 & x \in [17.82, \ \ 35] \end{cases} \tag{3}$$

The six fitted function curves are also drawn in Fig. 3. As seen from this figure, the fitted function curves can match the calculated heat flux distribution well with high precision.

At the third step, the CFD analyses will obtain the temperature distributions. Thermal oil (Syltherm 800) and stainless steel are used as the heat transfer fluid and the material of tube receiver respectively. The thermal-physical properties of the thermal oil and four different materials are presented in Table 2. The boundary conditions applied on the tube receivers are illustrated as follows:

- The flow has a uniform velocity u at atmosphere temperature at the tube receiver inlet;
- The top half periphery of tube receiver is subjected to a uniform heat flux distribution which is the sun average radiation in the air (the value is 1,000 W/m^2);
- The bottom half periphery of tube receiver is subjected to the concentrated heat flux distribution calculated by the Monte-Carlo ray tracing method which is fitted by six polynomial regression functions;

- Zero pressure gradient condition is employed across the fluid outlet boundary.

At the forth step, the finite element analysis (FEA) will obtain the Von-Mises thermal stress fields, which is a synthesis stress of radial stress, axial stress and circumferential stress. According to the Von-Mises stress theory [13], the formulation to calculate the Von-Mises stress σ_{eff} is:

$$\sigma_{\text{eff}} = \sqrt{\sigma_r^2 + \sigma_z^2 + \sigma_\theta^2 - (\sigma_r\sigma_z + \sigma_r\sigma_\theta + \sigma_\theta\sigma_z)} \tag{4}$$

where σ_r , σ_z , σ_θ are the radial stress, axial stress and circumferential stress respectively. The resulted temperature fields defined at the nodes of CFD analysis meshes are interpolated as input data to the nodes of the thermal stress analysis meshes. This simulation approach is fairly straightforward and has been adopted by many investigators.

Property	Fluid	Tube receiver			
	Thermal Oil	Stainless steel	Aluminum	Copper	SiC
Density (kg m^{-3})	938	7900	2698	8930	3210
Specific Heat (J kg^{-1} K^{-1})	1970	500	879	386	2540
Viscosity (10^{-6} Pa s)	15.3	48	247	384	42
Thermal Conductivity (W m^{-1} K^{-1})	`0.118	220	70	128	427
Poisson Ratio	–	0.25	0.32	0.31	0.17
Young's Modulus (Gpa)	–	17.2	23.6	17.1	4.8
Thermal expansion coefficient (10^{-6} K^{-1})	–	450	130	270	400

Table 2. Thermal-physical properties of heat transfer fluid and tube receiver

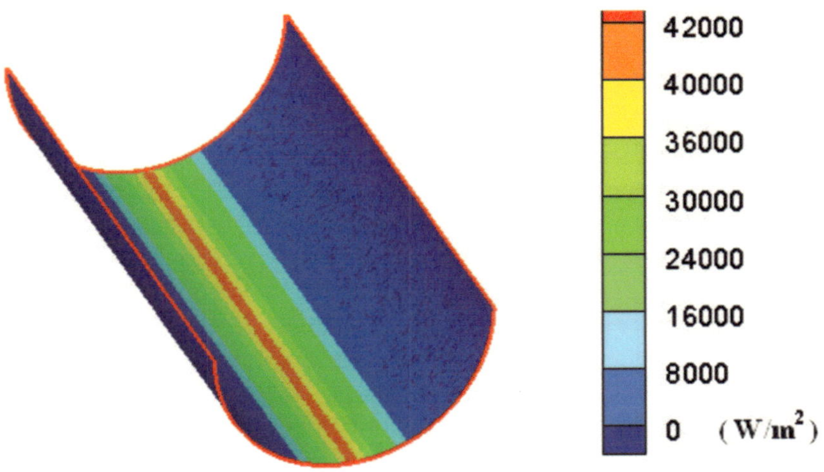

Fig. 2. Concentrated solar irradiation heat flux distribution on the bottom surface of tube receiver.

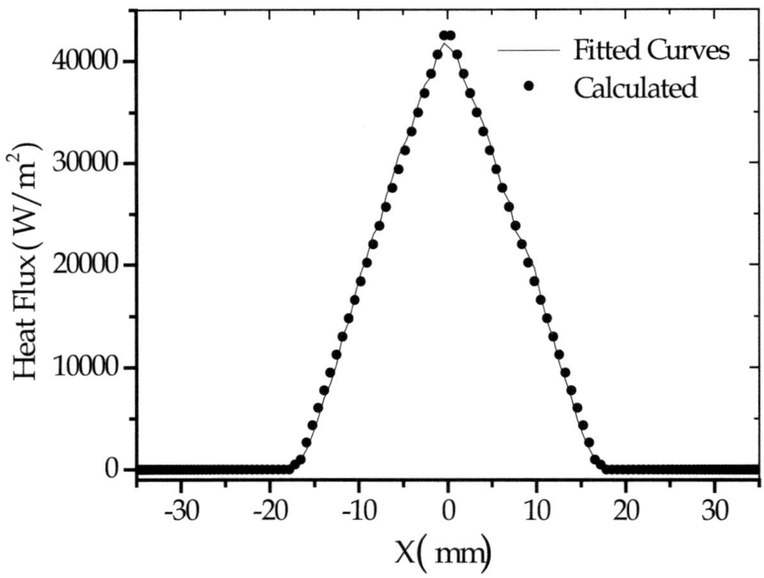

Fig. 3. Calculated heat flux distribution on the bottom half periphery of tube receiver and the fitted function curves.

The validation of this simulation approach has been described in [14], and the comparison between the simulation results and the experimentations reveals a high level of compliance. The detail of the computational meshes is presented in Fig. 4. All of the meshes are generated with O-grid method by Ansys Workbench software. In this study, a finer solid part mesh is used in thermal stress analysis to produce a reasonably accurate degree of freedom solution. There are 24,000 mesh elements in solid part and 62,000 mesh elements in fluid part for the CFD analysis and 123,280 mesh elements in the finer solid part mesh for thermal stress analysis.

3. Ray-thermal-structural analysis of concentric tube receiver

3.1 Comparisons between uniform and concentrated heat flux conditions

The temperature distribution and thermal stress field of the tube receiver with uniform and concentrated solar irradiation heat flux conditions are obtained. Fig. 5 and Fig. 6 show the temperature contours on the outlet surface and outer surface of the tube receiver respectively both for uniform and concentrated solar irradiation heat flux conditions. The maximal temperature for concentrated solar irradiation heat flux condition is 21 K higher than the maximal temperature for uniform heat flux condition. For concentrated solar irradiation heat flux condition, there are five temperature contour sections at the outlet surface of tube receiver, compared to three temperature contour sections for uniform heat flux conditions. As seen from Fig. 5 and Fig. 6, compared to uniform heat flux condition, the temperature gradients varying with θ and L are higher for concentrated solar irradiation condition, this is caused by the highly concentrated heat flux on bottom surface of tube receiver.

The maximal effective thermal stresses are found at the outlet surface of tube receiver both for uniform and concentrated solar irradiation heat flux conditions. Fig. 7 shows the effective thermal stress contours on the outlet surface. As we expected, due to the higher

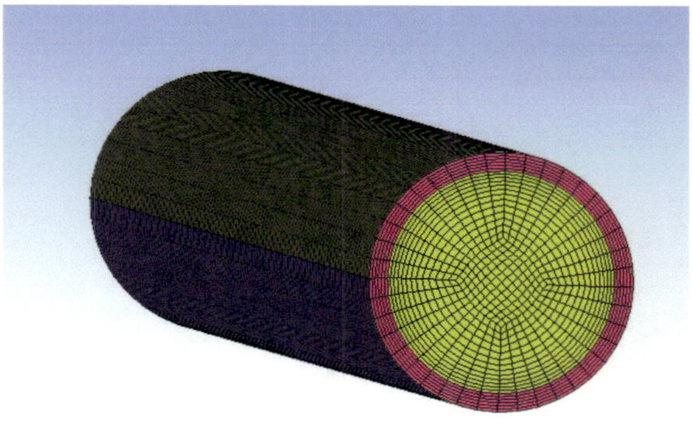

(a) Meshes for CFD analysis

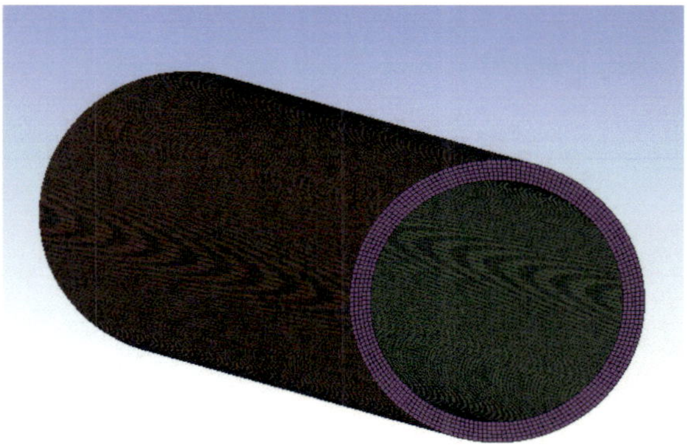

(b) Finer meshes for thermal stress analysis

Fig. 4. Computational meshes for CFD and thermal stress analysis

temperature gradient, the concentrated solar irradiation heat flux condition has a much higher effective thermal stress. The maximal effective thermal stress for concentrated solar irradiation heat flux condition is 73.6 Mpa, which is 4.2 times of the maximal effective thermal stress for uniform heat flux condition.

3.2 Comparisons between different materials

Fig. 8 shows the temperature profiles across the circumference on the tube inner surface at the tube outlet section. Among the four different material conditions, the SiC condition has the highest maximum temperature. Due to the low conductivity of SiC and stainless steel compared with the conductivity of aluminum and copper, as seen from this figure, the temperature gradients of the stainless steel and SiC conditions are much higher than those of the aluminum and copper conditions, which can cause higher thermal stress and reduce the durability of tube receiver.

The numerical result shows that the maximum effective stresses are found at the circumference on the tube inner surface at the tube outlet section at θ=270° for all the four different material conditions. Fig. 9 shows the effective stress profiles on the tube inner surface along the length direction at θ=270° for the four different material conditions. As

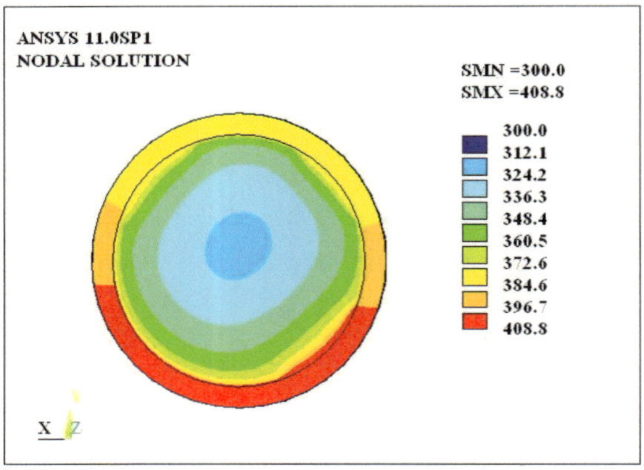

(a) Uniform heat flux

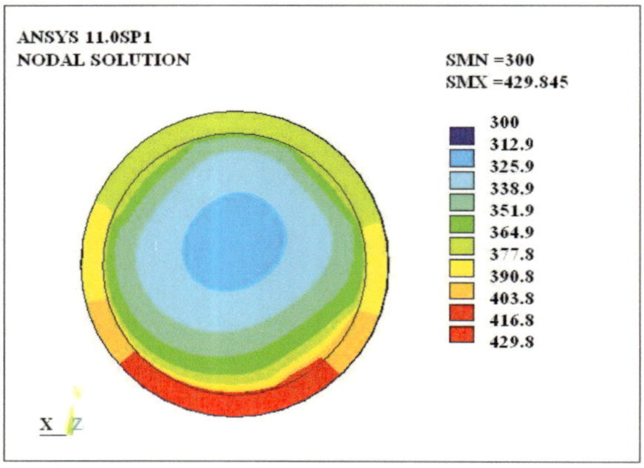

(b) Concentrated solar irradiation heat flux

Fig. 5. Temperature contour on the outlet surface.

seen from the figure, the four profiles have almost the same trend line. At the two free ends of the tube, the effective stress values are much higher than the effective stress values at other positions. This phenomenon may_be caused by the bending movement of the tube receiver at the two free ends due to the outward defection of the tube. From the tube inlet end to z=0.1m, the effective stress values decrease sharply to the lowest of each profile. From z=0.1m to z=1.9m, the tangential stresses of each profile almost keep constant. From z=1.9m to the tube outlet end, the compressive tangential stresses increase sharply to the maximum. Among the four different material conditions, the stainless steel condition has the highest maximum effective stress and the copper condition has the lowest maximum effective stress which is only 4.9 MPa.

In this study, the stress failure ratio F_c ($F_c = \delta_{\text{eff}} / \delta_b \times 100\%$) is introduced to assess the thermal stress level of each material condition. Fig. 10 presents the stress failure ratio profiles on the tube inner surface along the length direction at $\theta = 270^{\circ}$ for the four different material conditions. The copper condition has the lowest stress failure ratio and the stainless steel condition has the highest stress failure ratio which is about 6 times of the copper condition. Therefore, from the standpoint of thermal stress, copper is recommended as the material of tube receiver.

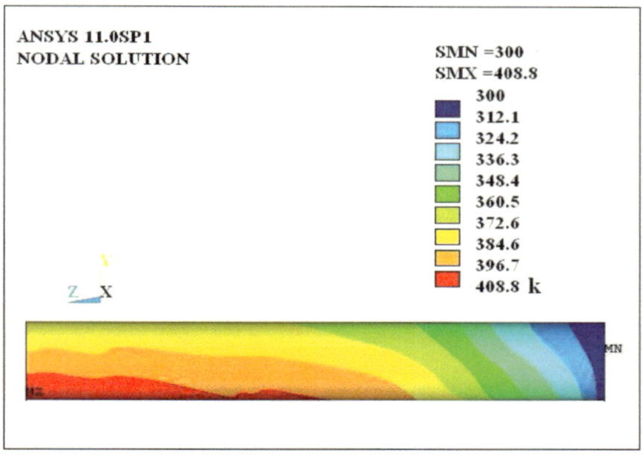

(a) Uniform heat flux

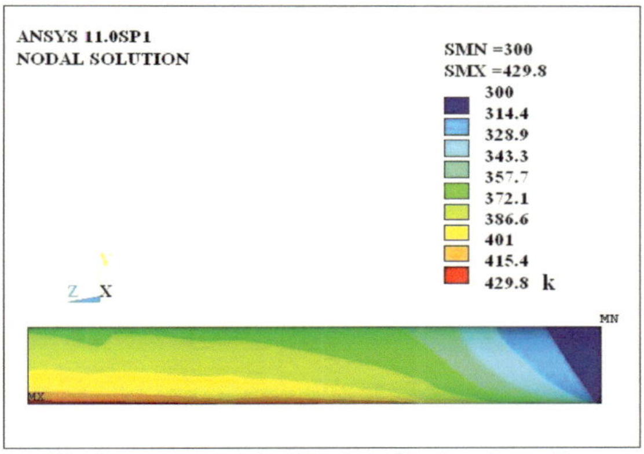

(b) Concentrated solar irradiation heat flux

Fig. 6. Temperature contour on the outer surface.

4. Ray-thermal-structural analysis of eccentric tube receiver

As mentioned in the previous section, the tube receivers are designed to operate under extremely nonuniform heat flux, cyclic weather and cloud transient cycle's conditions, which in turn will produce high temperature gradients and large deflection of tube receiver. The high temperature gradients will generate the large thermal stresses which may cause the failure of tube receiver, and the deflection of tube receiver will induce the rupture of glass envelop which will result in the increase of heat loss. Therefore, it is necessary to seek some new approaches to reduce the thermal stresses and deflection of the tube receiver.

Hitherto, mainly three methods have been proposed to reduce the thermal stresses or deflection of receiver:

- Optimizing the size of tube receivers or operation parameters, such as, employing small diameter tubes; or controlling the fluid flow rate.
- Receivers with homogenous solar radiation heat flux distribution on the surface. Generally, these kinds of receivers are designed using ray tracing methods to obtain the isosurface of solar radiation. At present, the literature survey indicates that the research on receivers with homogenous solar radiation heat flux distribution remains at the theory stage, and a large amount of manufacturing problems wait to solve further.

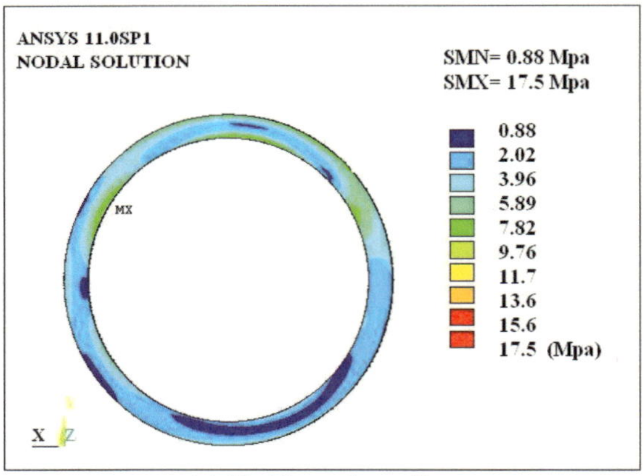

(a) Uniform heat flux

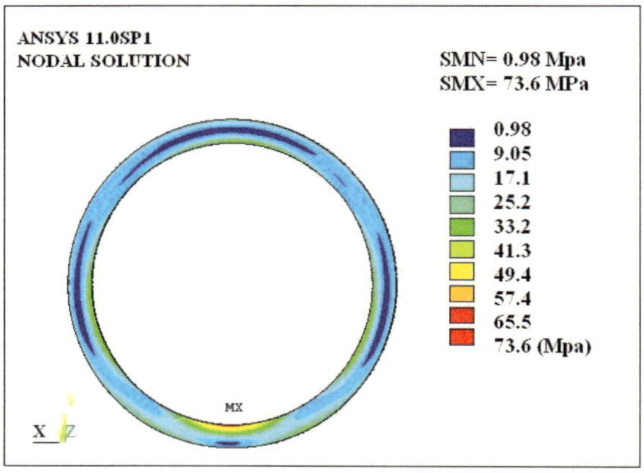

(b) Concentrated solar irradiation heat flux

Fig. 7. Effective stress contour on the outlet surface.

- Compound wall copper-steel receiver. The compound wall receiver is composed of two parts: the internal tube stratified is made of copper to obtain an excellent heat transfer performance to reduce the temperature gradients, and the external tube stratified is made of steel to strengthen the intensity of the tube receiver. The compound wall copper-steel tube receivers have been applied to the Solar Power Plant of the National University of Mexico. Though the compound wall copper-steel receiver can reduce the deflection of tube receiver, it will introduce the contact resistance if the two stratifications can not contact well and the solar radiation absorption efficiency will be affected.

With the aim to reduce the thermal stresses of tube receiver during application, an eccentric tube receiver for the parabolic trough collector system is introduced. The aim of the new-type receiver is to:

- Reducing the thermal stresses effectively
- Without adding the mass of tube receiver
- Easy to manufacture

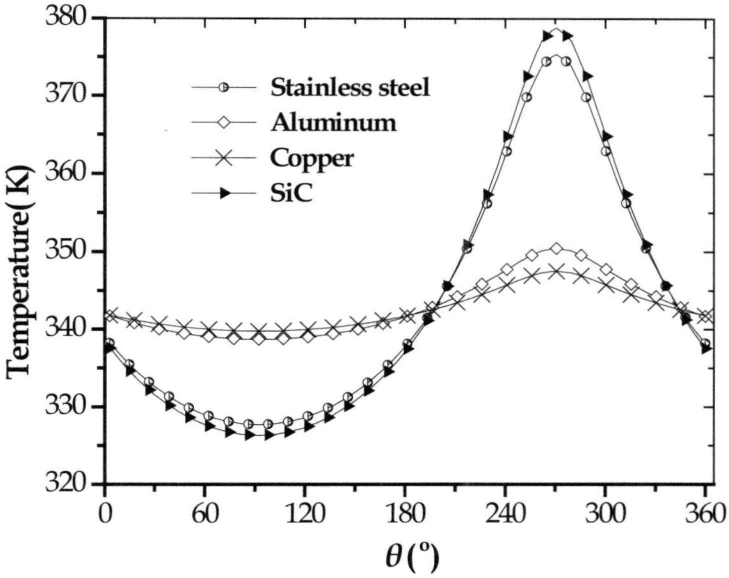

Fig. 8. Temperature profiles across the circumference on the tube inner surface at the tube outlet section

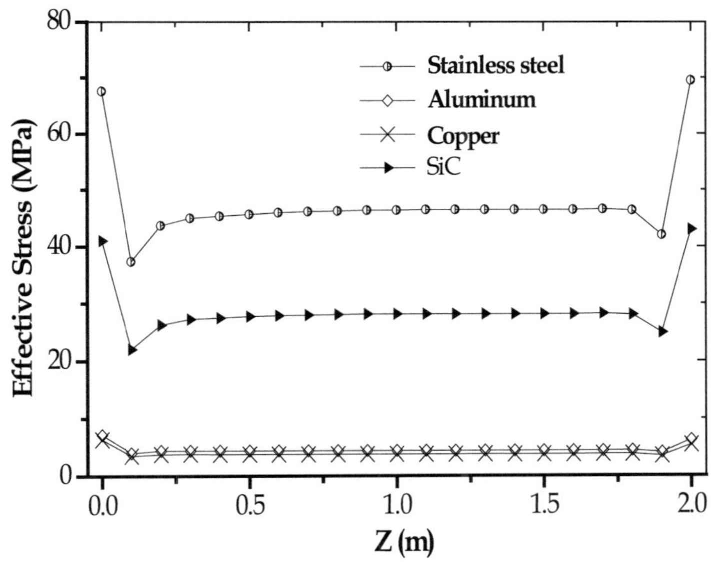

Fig. 9. Effective stress profiles on the tube inner surface along the length direction at θ=270°

4.1 Construction of eccentric tube receiver

To meet the above requirements of the new type receiver, the eccentric tube receiver for parabolic trough collector system is introduced. Fig. 11 shows the diagram of the eccentric tube receiver. The eccentric tube receiver is proposed on the basis of concentric tube receiver. As seen from this figure, the center of internal cylinder surface of concentric tube

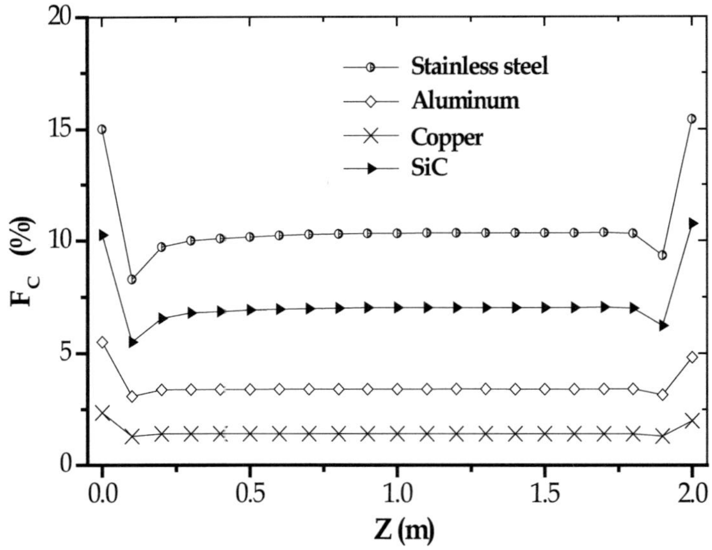

Fig. 10. Stress failure ratio profiles on the tube inner surface along the length direction at θ=270°

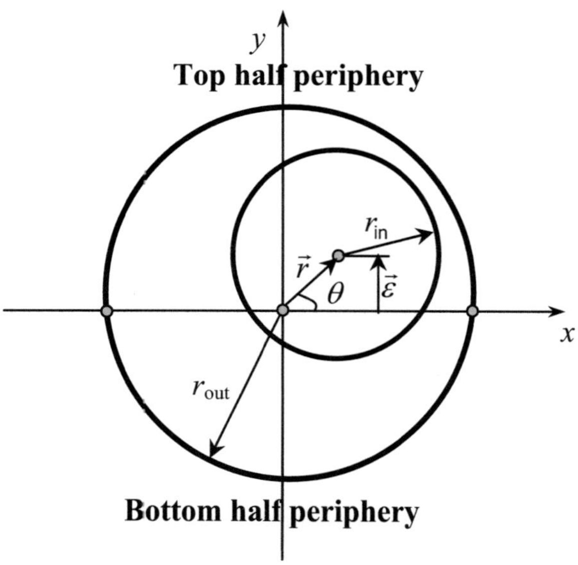

Fig. 11. Schematic diagram of physical domain and coordinate system for the eccentric tube receiver.

receiver is moved upward (or other directions), which is not located at the same coordinate position with the center of external cylinder surface. Therefore, the wall thickness of the bottom half section of tube receiver will increase without adding any mass to the entire tube receiver. With the same boundary conditions for numerical analyses, the increase of wall thickness will not only strengthen the intensity to enhance the resistance of thermal stress, but also can increase the thermal capacity, which in turn will be benefit to alleviate the extremely nonuniform temperature distribution situation.

As seen from Fig. 11, the origin of coordinate system is placed at the center of the external cylinder surface. In this study, the vector eccentric radius $\vec{r}$ (the origin of coordinate system points to the center of the internal cylinder surface); the vector eccentricity $\vec{\varepsilon}$ (the projection of vector $\vec{r}$ on the y-axis); and the oriented angle θ (the angle between the vector $\vec{r}$ and the x-axis) are introduced to describe the shape of eccentric tube receiver.

4.2 Comparison between the concentric and eccentric tube receiver

The eccentric tube receiver with the center of internal cylinder surface 3 mm moved upward along the y-axis (the magnitude of vector eccentricity $\vec{r}$ is 3 mm, and the oriented angle θ is 90°) is chosen for the comparison research. The temperature distributions and thermal stress fields of eccentric tube receiver are compared with those of concentric tube receiver under the same boundary conditions and material physical properties.

Fig. 12 shows the temperature distributions along the internal circumference at the outlet section for both the concentric and eccentric tube receivers. As seen from this figure, the concentric tube receiver has a higher value of peak temperature which is about 5 °C higher than that of eccentric tube receiver. Along the bottom half internal circumference (the θ is between 180° and 360°) where the peak temperatures of both the concentric and eccentric tube receivers are found, the temperature gradients of concentric tube receiver are higher than those of eccentric tube receiver which can lead to the higher thermal stresses, the cause of this phenomenon should be attributed to the thermal capacity increase on the bottom section of tube receiver due to the wall thickness increase on this section.

The thermal stress fields along the internal circumference at the outlet section for both the concentric and eccentric tube receivers are presented in Fig. 13. The peak thermal stress

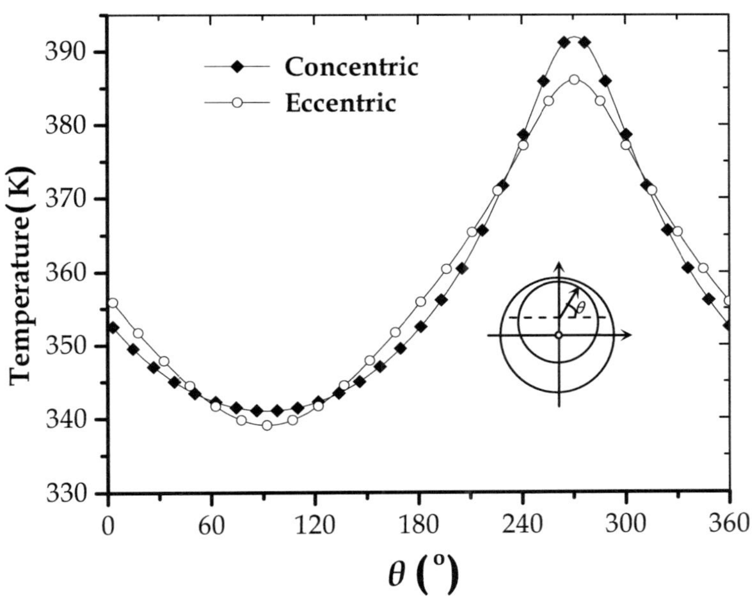

Fig. 12. Temperature profiles along the internal circumference at the outlet section for both the concentric and eccentric tube receivers.

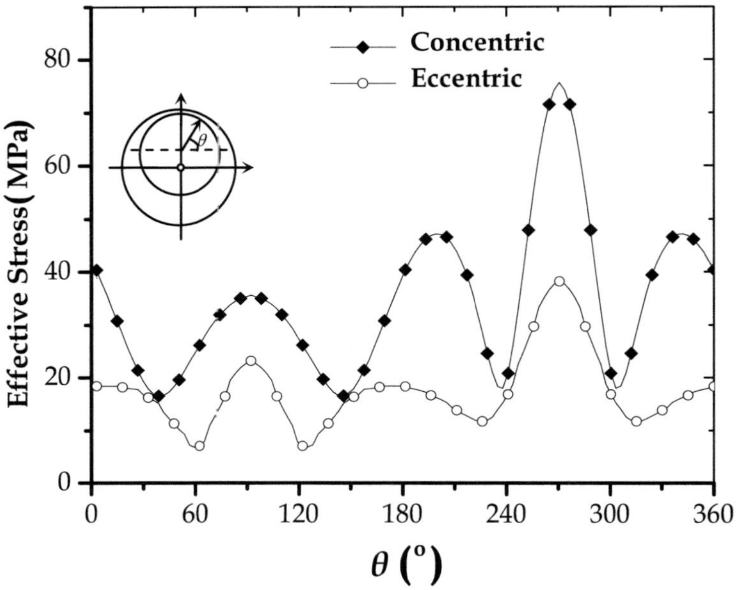

Fig. 13. Thermal stress profiles along the internal circumference at the outlet section for both the concentric and eccentric tube receivers.

values of the two profiles are both found at θ=270° where the peak temperature values are also located at. Attributed to the lower temperature gradients and intensity strengthen on the bottom half section of tube receiver, the peak thermal stress value of the eccentric tube receiver which is only 38.2 MPa is much lower compared to that of the concentric tube receiver which is 71.5 MPa. Therefore, adopting eccentric tube receiver as the tube receiver for parabolic trough collector system can reduce the thermal stresses effectively up to 46.6%, which means the eccentric tube receiver can meet the requirements of the new type receiver.

5. Conclusions

The ray-thermal-structural sequential coupled method is adopted to obtain the concentrated heat flux distributions, temperature distributions and thermal stress fields of both the eccentric and concentric tube receivers. Aiming at reducing the thermal stresses of tube receiver, the eccentric tube receiver is introduced in this investigation. The following conclusions are drawn.

1. For concentrated solar irradiation condition, the tube receiver has a higher temperature gradients and a much higher effective thermal stress.
2. The radial stresses are very small both for uniform and concentrated heat flux distribution conditions due to the little temperature difference between the inner and outer surface of tube receiver. The maximal axial stresses are found at the outer surface of tube receiver both for uniform and concentrated solar irradiation heat flux conditions. The axial stress has more impact on thermal stress compared to radial stresses.
3. The temperature gradients and effective stresses of the stainless steel and SiC conditions are significantly higher than the temperature gradients and effective stresses

of the aluminum and copper conditions. The stainless steel condition has the highest stress failure ratio and the copper condition has the lowest stress failure ratio.

4. Adopting eccentric tube as the tube receiver for parabolic trough collector system can reduce the thermal stress effectively up to 46.6%. The oriented angle has a big impact on the thermal stresses of eccentric tube receiver. The thermal stress reduction of tube receiver only occurs when the oriented angle is between 90° and 180°.

6. Acknowledgements

This work was supported by the National Key Basic Research Special Foundation of China (No. 2009CB220006), the key program of the National Natural Science Foundation of China (Grant No. 50930007) and the National Natural Science Foundation of China (Grant No. 50806017).

7. References

C.F. Chen, C.H. Lin, H.T. Jan, Y.L. Yang, Design of a solar collector combining paraboloidal and hyperbolic mirrors using ray tracing method, Opt. Communication 282 (2009) 360-366.

T. Fend, R.P. Paal, O. Reutter, J. Bauer, B. Hoffschmidt, Two novel high-porosity materials as volumetric receivers for concentrated solar radiation, Sol. Energy Mater. Sol. Cells 84 (2004) 291-304.

Y.S. Islamoglu, Finite element model for thermal analysis of ceramic heat exchanger tube under axial concentrated solar irradiation convective heat transfer coefficient, Mater. Design 25 (2004) 479–482.

C.C. Agrafiotis, I. Mavroidis, A.G. Konstandopoulos, B. Hoffschmidt, P. Stobbe, M. Romero, V.F. Quero, Evaluation of porous silicon carbide monolithic honeycombs as volumetric receivers/collectors of concentrated solar radiation, Sol. Energy Mater. Sol. Cells 91 (2007) 474-488.

J.M. Lata, M.A. Rodriguez, M.A. Lara, High flux central receivers of molten salts for the new generation of commercial stand-alone solar power plants, ASME J. Sol. Energy Eng. 130 (2008) 0211002/1–0211002/5.

R.F. Almanza. DSG under two-phase and stratified flow in a steel receiver of a parabolic trough collector, ASME J. Sol. Energy Eng. 124 (2002) 140–144.

V.C. Flores, R.F. Almanza, Behavior of compound wall copper-steel receiver with stratified two-phase flow regimen in transient states when solar irradiance is arriving on one side of receiver, Sol. Energy 76 (2004) 195–198.

Steven, G., Macosko, R.P., 1999. Transient thermal analysis of a refractive secondary solar collector. SAE Technical Paper, No. 99–01–2680.

M.F. Modest. Radiative heat transfer. 2nd ed. California: Academic Press; 2003.

R. Siegel, J.R. Howell. Thermal radiation heat transfer. 4th ed. New York/London: Taylor & Francis; 2002

Y. Shuai, X.L. Xia, H.P. Tan, Radiation performance of dish solar collector/cavity receiver systems, Sol. Energy 82 (2008) 13–21.

F.Q Wang, Y. Shuai, G. Yang, Y. Yuan, H.P Tan. Thermal stress analysis of eccentric tube receiver using concentrated solar radiation. Solar Energy, 2010, Accepted.
J.H. Fauple, F.E. Fisher, Engineering design–a synthesis of stress analysis and material engineering, Wiley, New York, 1981.
Y.F. Qin, M.S. Kuba, J.N. Naknishi, Coupled analysis of thermal flow and thermal stress of an engine exhaust manifold, SAE Technical Paper 2004-01-1345.

17

Some Techniques in Configurational Geometry as Applied to Solar Collectors and Concentrators

Reccab M Ochieng and Frederick N Onyango
Department of Physics and Materials Science, Maseno University, P.O. Box 333, Maseno 40105, Kenya

1. Introduction

All systems, which harness and use the sun's energy as heat, are called solar thermal systems. These include solar water heaters, solar air heaters, and solar stills for distilling water, crop driers, solar space heat systems and water desalination systems.

This chapter presents analysis based on configurational geometry of solar radiation collectors and concentrators using system models that have the same dimensions, material structure and properties. The work shows that different elements added to concentrators of well known configurations increase the geometric concentration ratio.

The need to develop effective solar thermal systems is not only to reduce the effects of global warming but also to reduce the overall costs and risks of climate change. Therefore, it is paramount to develop technologies for utilizing clean and renewable energy on a large scale.

Solar energy being the cleanest source of renewable energy free of Green House Gas (GHG) emission has seen the development of many gadgets and new technologies which include power generation (e.g., photovoltaic and solar thermal), heating, drying, cooling, ventilation, etc.

Development of the technologies utilizing solar energy focuses on improving the efficiency and reducing the cost. The objective of this book chapter is to present an analysis based on configurational geometry of solar radiation collectors and concentrators using system models that have been used to demonstrate the technique of configurational geometry in design and applications of a number of systems.

Geometry configuration plays an important role in most if not all solar collectors and concentrators. A number of collectors and concentrators have symmetries which allow them to collect and concentrate solar thermal energy. Since solar collector and concentrator surfaces are normally planes or curves of specific configurations, the analysis of system processes can be carried out through the use of the laws and rules of optics. Because of the known geometries and symmetries found in the collectors and concentrators, analysis of the collection and reflection of light, hence radiation analysis can also be done using configurational geometries of the systems. We shall discuss the general principles of operation of solar collectors and concentrators then show in a number of ways that it is possible to design collectors and concentrators innovatively using the method of configurational geometry. By use of some examples, we shall show the importance and effect of configurational geometry on the Geometric Concentartion Ratio, CR_g, of a concentrator, defined as the area of the collector aperture A_a, divided by the surface area of the receiver, A_r (Garg & Kandpal, 1978). We show that for given dimensions of a specific solar collector and concentrator system, (a modified cone concentrator and a modified inverted cone concentrator), the configurational geometries give different concentration ratios unless certain conditions are prescribed. We also demonstrate that different new elements and components can be incorporated in well known configurational geometries to improve the performance of collectors and concentrators. In this chapter, we first give a brief discussion on the general aspects of concentrators and collectors which is then followed by

a. a mathematical procedure in concentrators and collectors with respect to configurational geometry,
b. a technique of generating cone concentrators and collectors from hyperbloid configurations,
c. a discussion of configurational geometry in straight cone concentrators and inverted cone concentrators and collectors and.

2. General theoretical considerations

A typical flat plate collector consists of an absorber plate, one or more transparent cover(s), thermal insulation, heat removal system and an outer casing.

An absorber plate is generally a sheet of metal of high thermal conductivity like copper which is normally coated with black paint or given a special coating (called selective coating) so that it absorbs the incident solar radiation efficiently and minimizes loss of heat by radiation from the collector plate.

In the flat plate solar collector, a glass plate of good quality, which is transparent to incoming solar radiation to act as cover, is fixed about 2-4 cm above the absorber plate. This prevents convective heat loss from the absorber plate and prevents infrared radiation from the plate escaping to the atmosphere. If the plate temperature under normal operation is expected to be higher than 80^0C, two glass plates separated from each other may be used.

The absorber plate rests on a 5-15 cm thick bed of glass wool or any other good thermally insulating material of adequate thickness, which is also placed along the sides of the collector plate to cut down heat loss by conduction.

The most common method of removing heat from the collector plate is by fixing tubes, called risers at spacing of about 10-25 cm. Good thermal contact between the tube and plate is very important for efficient operation of the collector hence the tubes could be soldered, spot welded, tied with wires or clamped to the plate. These risers are connected to larger pipes called headers at both ends so that heat removal fluid can enter from the lower header and leave from the upper header. This configuration of absorber plate is called the fin type and is most commonly used. The heat removal fluid, usually water or oil, flows through these tubes to carry away the heat received from the sun. In another type of collector, heat removal fluid flows between two sheets of metal sealed at the edges, the top acting as the absorber plate.

All parts of the collector are kept in an outer case usually made of metal sheets. The case is made air tight to avoid considerable loss of heat from the collector plate to the ambient.

The collector is finally placed on a stand so that the absorber plate is correctly inclined to the horizontal and receives maximum amount of heat from the sun during a particular season or the entire year.

Flat plate solar collectors may be divided into two main classifications based on the type of heat transfer fluid used. Either liquid or gases (most often air) is used in collectors. Liquid heating collectors are used for heating water and non-freezing aqueous solutions and occasionally for non-aqueous heat transfer liquids such as thermal oils, ethylene glycol e.t.c,. Air-heating collectors are used for heating air used for solar dying or space heating (such as rooms).

Many advanced studies both experimental and theoretical have been carried out on flat plate solar collectors. Accurate modelling of solar collector system using a rigorous radiative model applied for the glass cover, which represents the most important component, has been reported by (Maatouk & Shigenao, 2005).

A different category of solar thermal systems known as solar concentrators are also used in solar thermal systems. Solar concentrators are the collection of devices which increase solar radiation flux on the absorber surface as compared to the radiation flux existing on the entrance aperture. Figure 1 show schematic diagrams of the most common conventional configurations of concentrating solar collectors. Optical concentration is achieved by the use of reflecting or refracting elements positioned to concentrate the incoming solar radiation flux onto a suitable absorber. Due to the apparent diurnal motion of the sun, the concentrating surface, whether reflecting or refracting will not be in a position to redirect the solar radiation on the absorber throughout the day if both the concentrator surface and absorber are stationary. This requires the use of a tracking system.

Ideally, the total system consisting of mirror/lens and absorber should follow the sun's apparent motion so that the sun rays are always captured by the absorber. In general, therefore, a solar concentrator consists of (i) a focusing device (ii) a blackened metallic absorber provided with a transparent cover and (iii) a tracking device for continuously following the sun. Temperatures as high as 3,000^{0}C can be achieved with solar concentrators which find applications in both photo-thermal and photovoltaic conversion of solar energy.

The use of solar concentrators may lead to advantages such as increase energy delivery temperatures, improved thermal efficiency due to reduced heat loss, reduced cost due to replacement of large quantities of expensive material(s) for constructing flat plate solar collector systems by less expensive reflecting and/or refracting elements and a smaller absorber tube. Additionally there is the advantage of increased number of thermal storage options at elevated temperatures thus reducing the storage cost. Earlier works by (Morgan 1958), (Cornbleet, 1976), (Basset & Derrick, 1978), (Burkhard & Shealy, 1975), (Hinterberger & Winston, 1968a), (Rabl 1976a, 1976b, 1976c), (Rabl & Winston, 1976), provide some important information and ideas on the development and design of solar collectors and concentrators as employed in this work.

The use of optical devices in solar concentrators makes it necessary that some of the parameters characterizing solar concentrators are different than those used in flat plate solar collectors. Several terms are used to specify concentrating collectors. These are:

i. Aperture area
ii. Acceptance angle
iii. Absorber area
iv. Geometric concentration ratio
v. Local concentration ratio
vi. Intercept factor
vii. Optical efficiency
viii. Thermal efficiency.

The aperture area, A_a, is defined as the plane area through which the incident solar radiation is accepted whereas the acceptance angle $(\theta_{\max})$ defines the limit to which the incident ray path may deviate from the normal drawn to the aperture plane and still reach the absorber. A concentrator with large acceptance angle needs only seasonal adjustments while one with small acceptance angle must track the sun continuously.
The absorber area (A_{abs}), is the total area that receives the concentrated solar radiation. It is the area from which useful energy can be removed and the geometric concentration ratio (CR_g), or the radiation balance concentration ratio of a solar concentrator is defined as the ratio of the collecting aperture area (A_{Ap}), to the area of the absorber (A_{abs}). Mathematically this is given by

$$(CR_g) = \frac{A_{Ap}}{A_{abs}} \tag{2.1}$$

The brightness concentration ratio or the local concentration ratio is a quantity that characterizes the nonuniformity of illumination over the surface of the absorber.
It is the ratio of the radiation flux arriving at any point on the absorber to the incident radiation flux at the entrance aperture of the solar concentrator. In some literature, the brightness ratio is called optical concentration ratio (CR_o) and is defined as the average irradiance (radiant flux) (I_r) integrated over the receiver area (A_r) divided by the insolation incident on the collector aperture. Mathematically, this takes the form

$$CR_o = \frac{\frac{1}{A_r}\int I_r dA_r}{I_a} \tag{2.2}$$

The intercept factor (γ) for a concentrator-receiver system is defined as the ratio of energy intercepted by the absorber of a chosen size to the total energy reflected/refracted by the focusing device, that is,

$$\gamma = \frac{\int_{-\omega/2}^{+\omega/2} I(x)dx}{\int_{-\infty}^{+\infty} I(x)dx} \tag{2.3}$$

where $I(x)$ is the solar flux at a certain position (x) and ω is the width of the receiver. For a typical concentrator-receiver design its value depends on the size of the absorber, the surface area of the concentrator and solar beam spread.
The optical efficiency (η_0), of a solar concentrator-receiver system is defined as the ratio of the energy absorbed by the absorber to the energy incident on the concentrator's aperture. It includes the effect of mirror/lens surface shape and reflection/transmission losses, tracking accuracy, shading, receiver cover transmittance of the absorber and solar beam incidence effects.
In a thermal conversion system, a working fluid may be a liquid, a vapour or gas is used to extract energy from the absorber. The thermal performance of a solar concentrator is characterized by its thermal efficiency, which is defined as the ratio of useful energy delivered to the energy incident on the aperture of the concentrator.

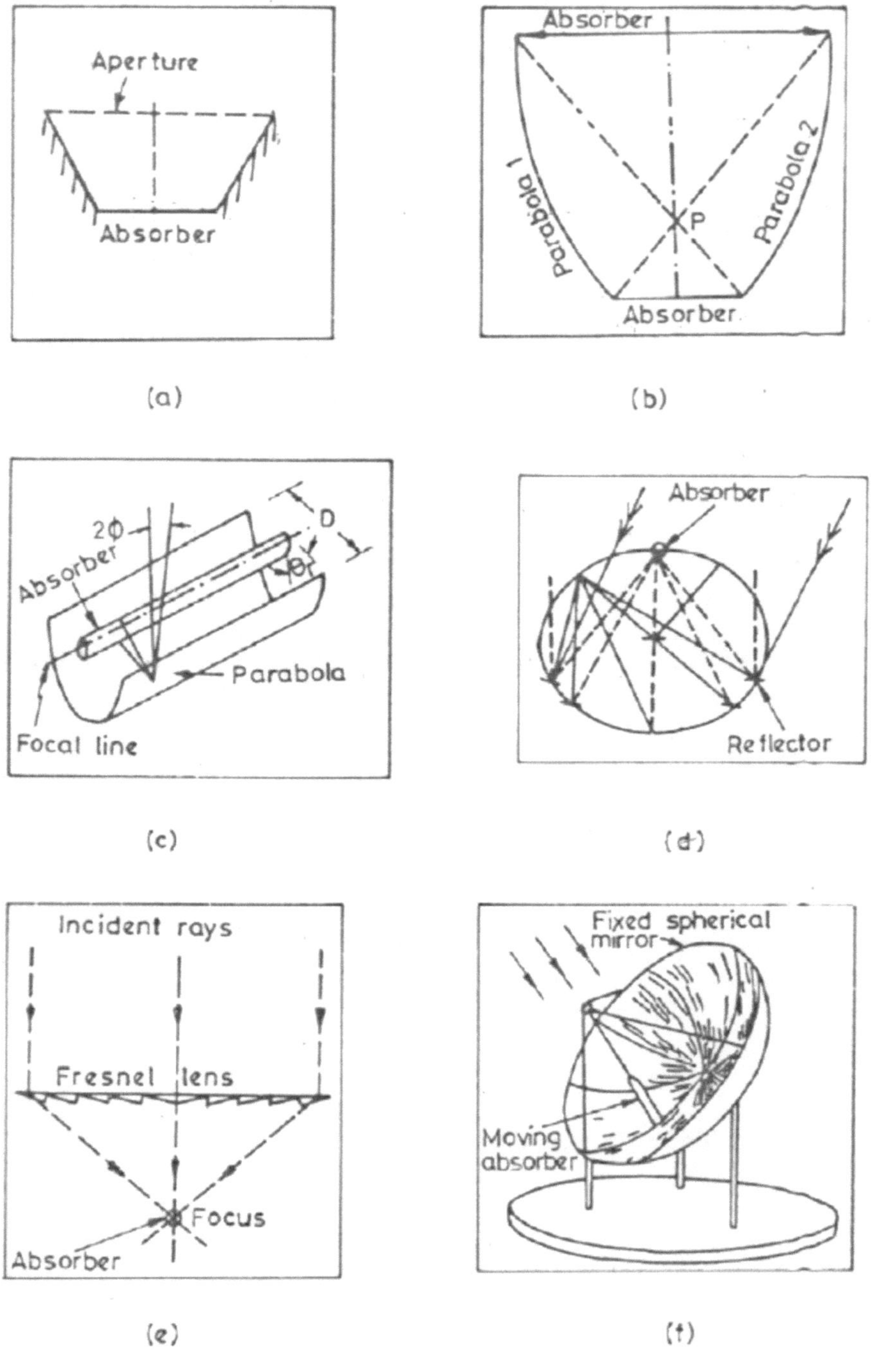

Fig. 1.1. *Schematic diagrams of the most common solar concentrators: (a) Flat plate absorber with plane reflectors (V trough), (b) compound parabolic concentrator, (c) Cylindrical parabolic trough, (d) Russel's fixed mirror solar concentrator, (e) Fresnel lens, (f) Hemispherical bowl. (Adopted from Garg and Kandpal, 1999).*

The instantaneous efficiency of a solar concentrator may be calculated from an energy balance on the absorber. The useful energy delivered by a concentrator is given by

$$Q = \eta_0 I_b A_a - U_L (T_{abs} - T_a) A_{abs} \tag{2.4}$$

where I_b is the direct beam on the concentrator, U_L is called the overall heat loss coefficient for the collector of the concentrator and is the sum for the heat loss from the bottom, U_b, the sides, U_s, and the top, U_t, i.e.,

$$U_L = U_b + U_s + U_t \,. \tag{2.5}$$

The other symbols have their usual meanings as previously defined. In situations where the receiver is not protected by a transparent cover, the useful heat collected by the receiver Q can be calculated as,

$$Q = A_{Ap} \cdot \left(\alpha \cdot C \cdot E^S - \varepsilon \cdot \sigma \cdot T_A^4 - U_L \left(T_A - T_a \right) \right) \tag{2.6}$$

with $\left(A_{Ap} \right)$ being the entrance aperture area, α being the absorptivity of the absorber with respect to the solar spectrum, (C) the concentration factor, E^S, the radiation density of the direct solar radiation and ε the average emissivity of the absorber with respect to the black body radiation at the absorber temperature T_A. σ stands for the Stefan-Boltzmann constant whereas U_L is the heat loss coefficient due to convection and conduction. In Eq. (2.6), thermal radiation input from the ambient (with the ambient temperature T_a) to the receiver is neglected.

Taking into account that for the heat transfer from the absorber to the heat transfer fluid a temperature difference is required, the following expression is also valid for the useful energy:

$$Q = A_{Abs} \cdot U_I \cdot \left(T_A - T_F \right) \tag{2.7}$$

with U_I being the inner heat transfer coefficient from the absorber to the fluid, T_F being the average temperature of the heat transfer fluid and A_{Abs} being the absorber area. Using Eq. (2.6) and Eq. (2.7), the energy balance equation can be rewritten replacing the absorber temperature by the fluid temperature:

$$Q = A_{Ap} \cdot \left(F \cdot \alpha \cdot C \cdot E^S - F \cdot \varepsilon \cdot \sigma \cdot T_F^4 - F \cdot U_L \cdot (T_F - T_a) \right) \tag{2.8}$$

The parameter F is the heat removal factor and is defined from the energy balance of flat plate solar collectors as

$$F = \frac{A_{Abs} \cdot U_I}{A_{Abs} \cdot U_I + A_{Ap} \cdot U_L + A_{Ap} \cdot 4 \cdot \sigma \cdot T_F^3} \tag{2.9}$$

The thermal efficiency of the receiver, η_{th}, is defined by the ratio of the useful heat to the incoming solar radiation in the aperture. The resulting expression for the efficiency is

$$\eta_{th} = \frac{Q}{A_{Ap} \cdot C.E^S} = F \cdot \alpha - \frac{F \cdot \varepsilon \cdot \sigma \cdot T_F^4}{C \cdot E^S} - \frac{F \cdot U_L \cdot (T_F - T_a)}{C \cdot E^S} \tag{2.10}$$

Using Eq. (2.4) and Eq. (2.5), the instantaneous efficiency of a concentrator having a top cover may be written as

$$\eta = \frac{Q}{I_b A_a} = \eta_0 - \frac{U_L\left(T_{abs} - T_a\right)}{I_b C}. \quad (2.11)$$

The linear approximation of heat loss factor made in Eq. (2.5) for a concentrator with top cover is valid for small operating temperatures only. At high operating temperatures, where the radiation loss term dominates the convective losses, energy balance may be expressed as

$$Q = \eta_0 I_b A_a - U_L (T_{abs}^4 - T_a^4) A_{abs} \quad (2.12)$$

where U_L now takes into account the accompanying convective and conduction losses also, hence Eq. (2.11) may now be modified as

$$\eta = \eta_0 - \frac{U_L\left(T_{abs}^4 - T_a^4\right)}{I_b C}. \quad (2.13)$$

Since the absorber surface temperature is difficult to determine, it is convenient to express the efficiency in terms of the inlet fluid temperature T_i by means of the heat removal factor F as

$$\eta = F\left[\eta_0 - \frac{U_L\left(T_i - T_a\right)}{I_b C}\right]. \quad (2.14)$$

Comparing Eq. (2.10) and Eq. (2.14) one sees that there is a parallel between the "static" efficiency (η_0), the emissivity and the absorptivity of the concentrator. Eq. (2.14) is a first order steady state expression for the instantaneous efficiency of a solar concentrator having a top cover. The instantaneous efficiency of a solar concentrator receiver system is dependent on two types of quantities, namely the concentrator receiver design parameters and the parameters characterizing the operating conditions. The optical efficiency, heat loss coefficient and heat removal factor are the design dependent parameters while the solar flux, inlet fluid temperature and the ambient temperature define the operating conditions.

Geometric optics is used as the basic tool in designing almost any optical system, image-forming or not. Intuitive ideas of a ray of light, roughly defined as the path along which light energy travels together with surfaces that reflect or transmit light are often used in solar collector and concentrator designs. When light is reflected from a smooth surface it obeys the well-known law of reflection which states that the incident and reflected rays make equal angles with the normal to the surface and that both rays lie in one plane.

When light is transmitted, the ray direction is altered according to the law of refraction, Snell's law which states that the sine of the angle between the normal and the incident ray gives a constant ratio to the sine of the angle between the normal and the refracted ray, all the three directions being coplanar.

A major part of design and analysis of solar collectors and concentrators involves ray tracing, i.e., following the paths of rays through a system of reflecting and refracting surfaces. The result of such processes may or may not create images of the source of the ray. Depending on the surface structure, properties and materials used, two types of systems;

image-forming concentrators and non image-forming concentrators arise. The process of ray tracing is used extensively in lens design, but the requirements are somewhat different for concentrators. In conventional lens design, the reflecting or refracting surfaces involved are almost always portions of spheres and centers of spheres lie in one straight line (axisymmetric optical system), so the special methods that take advantage of the simplicity of forms of surfaces and symmetry can be applied.

Nonimaging concentrators do not, in general, have spherical or symmetric surfaces. In fact, sometimes, there are no explicit analytical forms for the surfaces, although there is usually an axis or a plane of symmetry and ray-tracing schemes are conveniently based on vector formulations. Detailed analyses are often dealt with in computer programs on the basis of each different shape.

In principle, the use of ray tracing tells us all there is to know about the geometric optics of a given optical system, image forming or not. However, ray tracing alone is often little or no use for inventing new systems having properties for a given purpose. We need to have ways of describing the properties of optical systems in terms of general performance, using parameters such as, for example, the concentration ratios. A primitive form of nonimaging concentrator, the light cone has been used for many years (see for example, (Hotler *et.al.* 1962), (Witte, 1965), (Williamson, 1952), (Welford & Winston, 1978).

The option to integrate cost effective storage systems directly into solar thermal facilities represents a significant advantage of solar thermal systems over other concepts using renewable energy sources. This idea shall also be discussed with reference to configurational geometry of cone cylinder combination concentrators and collectors.

In the evaluation or calculation of the geometric concentration ratio of most concentrators, standard methods have been employed. This work departs from the traditional approach and outlines the mathematical foundation for such calculations. It will be shown that using the mathematical technique, for a straight cone with a collector area A_{coll}, situated a distance H_2 from the apex and an absorber area, A_{abs}, at a distance H_1 from the apex, the ratio of the squares of H_2 to H_1 give the geometric concentration ratio of the cone concentrator.

3. Mathematical procedures in concentrators and collectors with respect to configurational geometry

In the evaluation or calculation of the geometric concentration ratio of most concentrators, standard methods have been employed. This work departs from the traditional approach and outlines the mathematical foundation for such calculations. We then proceed to determine the concentration ratio of a modified cone concentrator.

The work shows that for a straight cone with a collector area A_{coll}, situated a distance H_2 from the apex and an absorber area, A_{abs}, at a distance H_1 from the apex, the ratio of the squares of H_2 to H_1 give the geometric concentration ratio of the cone concentrator (Figure 3.1).

Figure 3.2 shows a mall elemental volume of a cone that has been generated from a CPC. If the cone subtends an angle $\delta\theta$ and $\delta\phi$ at the origin, its cross-sectional area at a distance r from the apex is $r^2 \sin\theta\delta\theta\delta\phi$. Let us cut a cross section of the cone a distance H_1 from the origin so that the elemental area given by

$$dA_{abs} = H_1^2 \sin\theta\delta\theta\delta\phi \tag{3.1}$$

acts as the absorber area or the exit aperture for a cone concentrator.
Extending the length a distance H_2 from the apex we obtain an elemental collector area or entrance aperture, dA_{coll}, given by

$$dA_{coll} = H_2^2 \sin\theta\delta\theta\delta\phi \tag{3.2}$$

Where $H_2 > H_1$.
From Eq. (3.1) and (3.2) the "elemental" geometric concentration ratio is given by

$$\Delta C_g = \frac{dA_{coll}}{dA_{abs}} = \frac{H_2^2}{H_1^2} \tag{3.3}$$

We shall now explore the calculation of the geometric concentration ratio from the point of view of the relation between the area of a surface of revolution and the length of the curve that generates it.

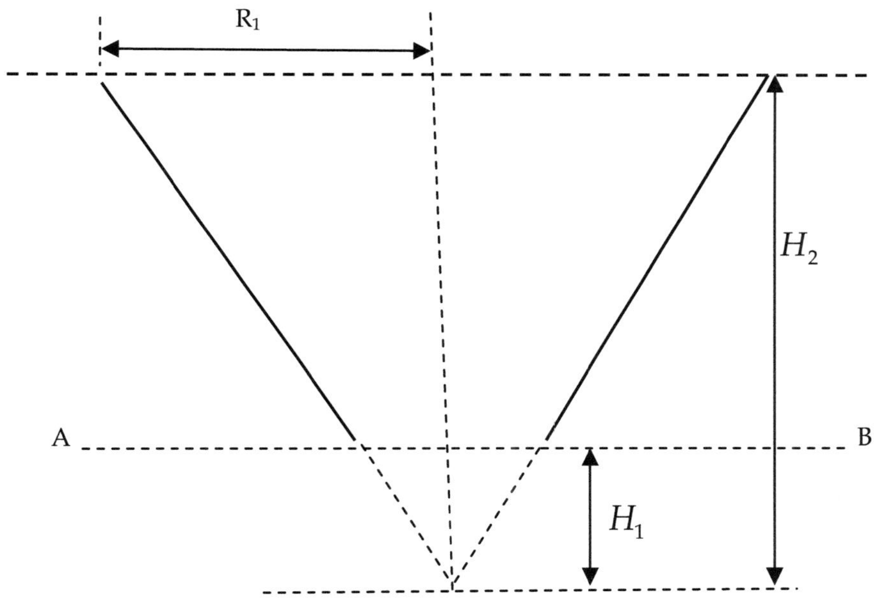

Fig. 3.1. *Schematic diagram of a cone showing the distance H₁ from the apex to the cross section AB (absorber area) and the distance H₂ from the apex to the collector area*

Suppose that a curve *AB* in the xy – plane like the one shown in Figure 3a is revolved about the x – axis to generate a surface. If *AB* is approximated by an inscribed polygon, then each segment *PQ* of the polygon will sweep out part of a cone whose axis lies along the x – axis (magnified view in Figure 3b). If the base radii of the *frustrum* of the cone are r_1 and r_2, as shown in Figure 3c, and its slant height is L, then its lateral surface area, A_r, is given as (Grant & Phillips, 1978)]
The total of the *frustrum* areas swept out by the segments of the inscribed polygon from *A* to *B* will give an approximate area *S* of the surface swept out by the curve *AB*. The approximation leads to an integral for *S* as follows.

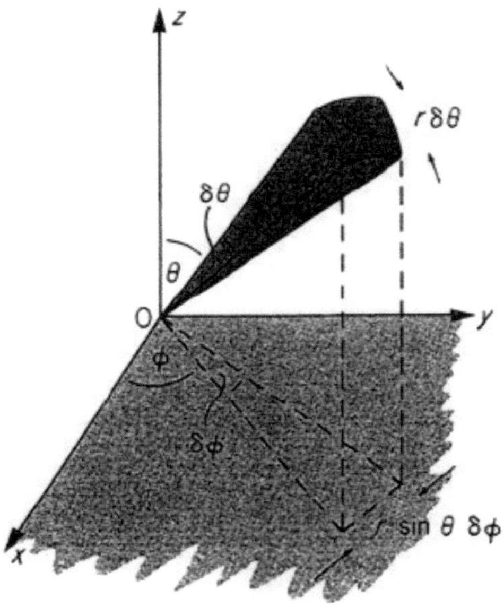

Fig. 3.2. *The cone subtends angles $\delta\theta$ and $\delta\phi$ at the origin, and its cross sectional area at a distance r from the apex is $r^2 \sin\theta\delta\theta\delta\phi$.[1].*

If we let the coordinates *P* be (x, y) and *Q* be $(x+\Delta x, y+\Delta y)$, then the dimensions of the *frustrum* swept out by the line segment *PQ* are

$$r_1 = y\ ,\ r_2 = y + \Delta y \quad L = \sqrt{(\Delta x)^2 + (\Delta y)^2} \tag{3.5}$$

The lateral area of the *frustrum,* from Eq. (4), is

$$A = \pi(r_1 + r_2)L = \pi(2y + \Delta y)\sqrt{(\Delta x)^2 + (\Delta y)^2} \tag{3.6}$$

Adding the individual *frustrum* areas over the interval $[a.b]$ from left to right, we obtain

$$\text{Cone frustrum sum} = \sum_{x=a}^{b} \pi(2y + \Delta y)\sqrt{(\Delta x)^2 + (\Delta y)^2} \tag{3.7}$$

Eq. (3.7) can be rewritten as

$$\text{Cone frustrum sum} = \sum_{a}^{b} 2\pi\left(y + \frac{1}{2}\Delta y\right)\sqrt{1 + \left(\frac{\Delta y}{\Delta x}\right)^2} \tag{3.8}$$

Assuming y and dy/dx to be continuous functions of x, then the sum in Eq.(3.8) approaches a limit given as

$$S = \int_{a}^{b} 2\pi y\sqrt{1 + \left(\frac{dy}{dx}\right)^2}\, dx \tag{3.9}$$

We therefore define the area of the surface to be the value of this integral. Eq. (3.9) is easily remembered if we write

$$\sqrt{1+\left(\frac{dy}{dx}\right)^2}\,dx = ds \tag{3.10}$$

and take

$$S = \int 2\pi y ds \tag{3.11}$$

We interpret $2\pi y$ as the circumference and ds as the slant height. Thus $2\pi yds$ gives the lateral area of a *frustrum* of cone of slant height ds if the point (x, y) is the midpoint of the element of arc length ds.

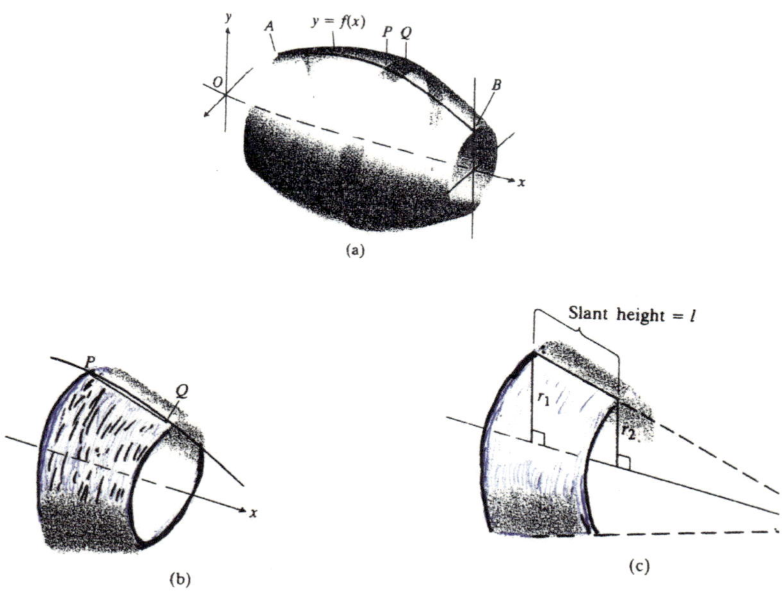

Fig. 3.3. *Generation of a cone from a curve AB by using frustrums as parts of the cone.*

4. Generating cone concentrators and collectors from hyperbloid configurations

The starting point of our analysis is based on the principles and operations of the cone concentrator. These principles are used to build on new systems when additional elements and components are added to modify the cone giving it two different configurations.

Figure 4.1 shows a parabolic compound concentrator (CPC) with axes of the two parabolas passing through the foci of the parabolas. We use these axes to generate a cone which embeds the parabola as shown in Figure 4.2. By drawing two lines parallel and passing through the foci of the first and second parabola, the lines meet at a point just below the foci. The two lines are then rotated along the axis of the compound parabola in order to form a

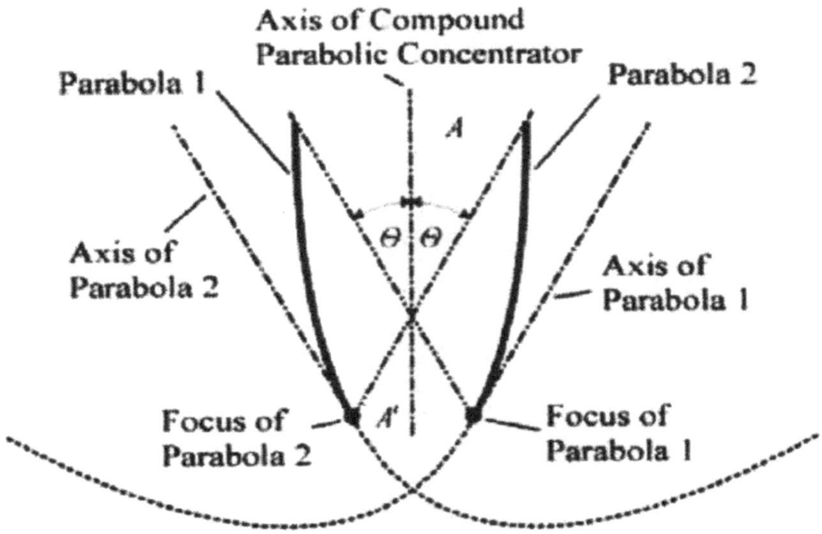

Fig. 4.1. *Compound parabolic concentrator (CPC).*

three dimensional straight cone. If the parabola embedded in Figure 2 is removed we remain with a cone having the same extreme angle as the original parabola provided the dimensions are not changed. A two dimensional schematic of such a cone generated in this manner is shown in Fig. 4.2.

If the cone as shown in Fig. 4.2 has a semi angle θ and θ_i is the extreme input angle, then the ray which enters at the extreme input angle will just pass through the aperture after one reflection if $2\theta = (\pi/2 - \theta_i)$. For a given entry diameter, an expression for the length of the cone can be obtained. It can be seen that some other rays such as the one indicated by the double arrow which enter at some angle less than the extreme input angle will be turned back by the cone. If a longer cone is used which has more reflections, some rays will still be turned back.

The cone is far from being an ideal concentrator. (Williamson, 1952) and (Witte, 1965) attempted some analysis of the cone concentrator by restricting their analysis to the meridian rays. The meridian ray analysis gives a very optimistic estimate of the concentrator; however, it does not address the problem of back reflections. Nevertheless, the cone is very simple compared to the image forming concentrators and its general form suggest a new direction in which to look for better concentrators. Modifications of the straight cone led to a Winston cone (http://scienceworld.wolfram.com/physics /WinstonCone.html), Hilderband and Winston, 1982), (Winston 1970), (Welford and Winston, 1989). A schematic of the Winston cone is shown in Fig. 4.3.

If an attempt is made to improve on the cone-concentrator by applying the edge-ray principle which stated loosely require that all extreme rays should form sharp image points and should emerge from the rim of the exit aperture, then one is led to the description of the Compound Parabolic Concentrator (CPC). These are prototypes of a series of nonimaging concentrators that approach very close to being ideal and having the maximum theoretical concentration ratio. Cones are much easier to manufacture than CPCs. Parabloids of revolutions (which of course CPCs are not) seem a more natural choice to conventional physicists as concentrators.

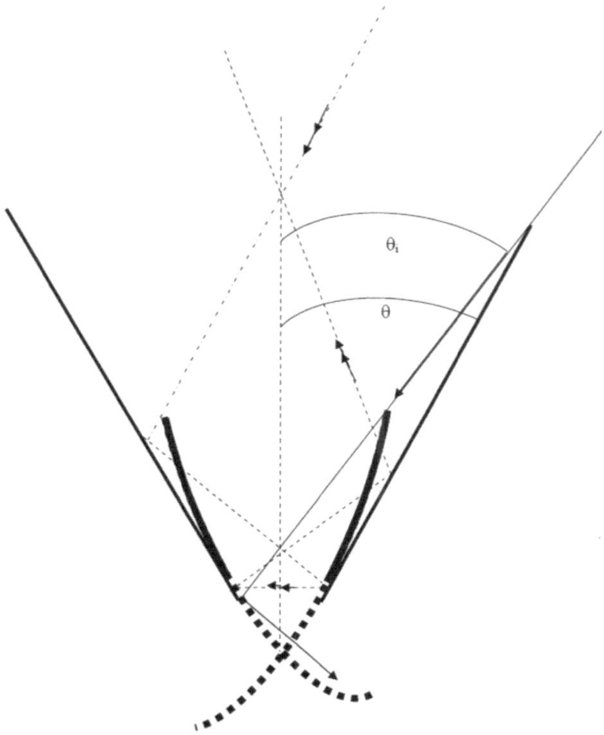

Fig. 4.2. *The cone concentrator with an "embedded" parabolic concentrator.*

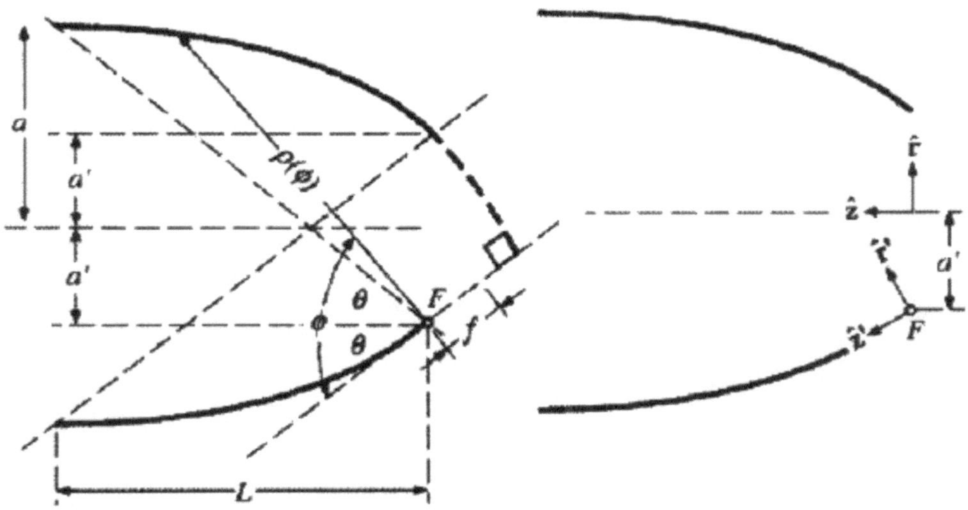

Fig. 4.3. *Winston cone concentrator.*

As a result of the foregoing, we consider two configuration models of modified cone concentrators as shown in Figure 5.1 and Figure 5.4 in which we analyze the systems by considering the heat exchange processes. We introduce a "reverse modelling technique" by creating certain components and elements in the design that will not only reduce the "back reflections" but also increase the geometric concentration ratios of the systems. The introduction of an outer cylinder limits the extent of the entrance aperture to a certain value just like in the case of a fixed lens, which has a specific radius of curvature. The smaller cylinder and the bigger cylinder serve two other purposes. The first purpose is to guide the radiation on to the absorber and the second purpose is to reduce the size of the exit aperture, thus increasing the concentrations ratio.

In the first model of the concentrator shown in Figure 5.1, the system is made of polyvinyl (PVC) pipes of two different diameters that form the radiation guide to the copper ring which acts as an absorber. The collector is a cone also made of polyvinyl and has a base that fits exactly on top of the inner cylinder. The cone is glued on top of the smaller cylinder. The system of cone and two cylinders form the collector and concentrator. The system is placed on top of one of the copper rings depicted in Figure 5.2. The copper rings are made from a copper tube having very small thickness.

The copper rings are painted black to allow for maximum absorption of radiation from a halogen lamp for example, that is used to simulate solar radiation. The inside surface of the larger cylinder and the external surface of the smaller cylinder are painted with white barium sulphate paint to allow for maximum reflection of radiation.

Let I_D be the direct radiation from a halogen lamp for example, falling directly on one of the rings and let I_R be the radiation falling on the ring as a result of being collected on the surface of the cone and guided to the ring by the walls of the cylinder forming the guide. If I_T is the total radiation from the halogen lamp, then

$$I_T = I_D + I_R - L \tag{4.1}$$

where $L = l_c + l_r$. Here l_c represents the loss due reflections and radiation from the cone and the guide whereas l_r represents the loss of radiation from the ring.

Using Eq. (4.1), we can write the heat transfer / balance equations for the two systems (the open system and the concentrator) as:

$$I_D - l_r = m_w c_w \frac{dT_{open}}{dt} + m_c c_c \frac{dT_{open}}{dt} \tag{4.2}$$

where m_w is the mass of the water in the ring and c_w is the specific heat capacity of water while m_c is the mass of the copper ring and c_c is the specific heat capacity of copper. For the concentrator we can write the heat balance equation as

$$I_D + I_R - (l_c + l_r) = m_w c_w \frac{dT_{conc}}{dt} + m_c c_c \frac{dT_{conc}}{dt} \tag{4.3}$$

For the first configuration of the solar collector and concentrator (inverted cone) shown in Figure 5, one of the important relations is

$$S_1^2 = R_1^2 + H_1^2 , \tag{4.4}$$

The surface area of the cone according to Eq. (3.9) is therefore given by

$$A_s = \pi R_1 S_1 \tag{4.5}$$

where R_1 is the base radius, H_1 the vertical height and $\sqrt{R_1^2 + H_1^2} = S_1$ the slant height of the cone.
Equation (4.4) is true if we assume that the surface area of the cylinder enclosing the surface area of the cone does not participate in collecting the radiation, however, it does to some extent, therefore the total surface area of the cone and cylinder concerned with the collection of radiation may be written in the form

$$A_s = \pi R_1 S_1 + \pi (R_1 + d_1)^2 \mu H_1 \tag{4.6}$$

with the second term in Eq. (4.6) representing the surface area of the cylinder involved in collection of radiation. μ, whose value lies between 0 and 1, that is, $0 \le \mu \le 1$. $\mu = 0$ corresponds to the case in which the outer cylinder physically exists but only provides guiding of solar radiation down on to the absorber.
The surface area of the ring on which radiation falls is given as

$$A_{ring} = \pi (R_1 + d_1)^2 - \pi R_1^2, \tag{4.7}$$

and can be rewritten as

$$A_{ring} = \pi \left(d_1^2 + 2R_1 d_1 \right). \tag{4.8}$$

The geometric concentration ratio in this configuration calculated by substituting Eq. (4.6) and Eq. (4.8) in to Eq. (2.1), results in

$$\left(CR_g \right)_1 = \frac{R_1 S_1 + (R_1 + d_1)^2 \mu H_1}{\left(d_1^2 + 2R_1 d_1 \right)}. \tag{4.9}$$

which for $\mu = 0$ reduces to

$$\left(CR_g \right)_1 = \frac{R_1 S_1}{\left(d_1^2 + 2R_1 d_1 \right)} \tag{4.10}$$

5. Experiment

Six (6) thermocouples T1, T2, T3, T4, T5, and T6 (Fig. 7) to measure temperatures of water in the rings with the concentrator and that without the concentrator are interfaced to a Fluke-2286/5 data logger through a temperature-measuring card attached to the data logger. Three of the thermocouples T1 T2 and T3 measure the temperature of the water in the ring with the concentrator while the other three thermocouples T4 T5 and T6 measure the temperature of the water in the ring without the concentrator.
In this work, both rings separated by a distance of 1.5 meters to avoid any effect of one ring on another are filled with water at the same temperature and the valves closed. The water remaining in the inlet is then drained and the two identical halogen lamps rated 90 watts powered by a 12 volt power source placed at the same height of 25cm from the rings are switched on simultaneously to heat the rings and thus, the water.

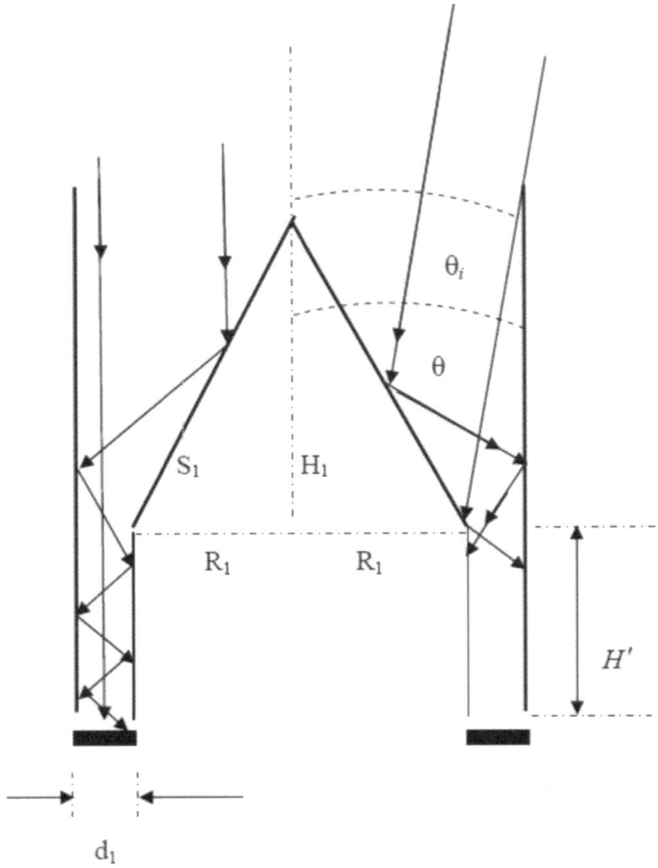

Fig. 5.1. *Vertical cross –section of the concentrator showing the path of both direct radiation collected by the cone absorber surface and guided to the ring by the radiation guide.*

The readings of the first three thermocouples are automatically averaged at pre-set time intervals and the resulting average value of the temperature stored in a specified register in the data logger. The other three thermocouples inserted into the ring not covered by the concentrator measure the temperature of water contained in the ring. The average temperature of the water is taken at the same pre-set time interval as that of the concentrator. The data logger (programmed using machine language) display on its output screen and print both the average temperature of the water in the ring with the concentrator and the average temperature of water in the ring without the concentrator.

Both copper tubes used to make the rings in the experiment had diameters of $0.01\,\text{m}$. The height, H', of the radiation guide formed by the two cylinders was $10\,\text{cm}$. The vertical height, H, of the cone used in this work was 0.2 cm.

Calculations done on the amount of heat reflected on the surface of the cone and inside the cylindrical guide show that if the initial radiation is reflected only once on the surface of the cone in to the radiation guide, then the guide will receive 98% of the original heat energy (http://www.oceanoptics.com/Products/ispref.asp, 2006)). Continuing the analysis on the 98% limit, a second reflected radiation inside the guide will posses 96.04% of the original radiation and a third beam reflected from the guide will have 94.12% of the original energy.

A fourth reflected beam possesses 92.24% of the initial radiation and a fifth reflected beam has 90.39% of the original radiation. Further analysis and simulation show that after several reflections from the surface of the cone through the radiation guide, the amount of radiation which will fall on the ring in this configuration is still more than 50% of the original radiation. Figure 5.5 shows the curve obtained from simulation of the remaining energy versus the number of reflections (Ochieng *et. al.* 2007). It is thus reasonable to assume that

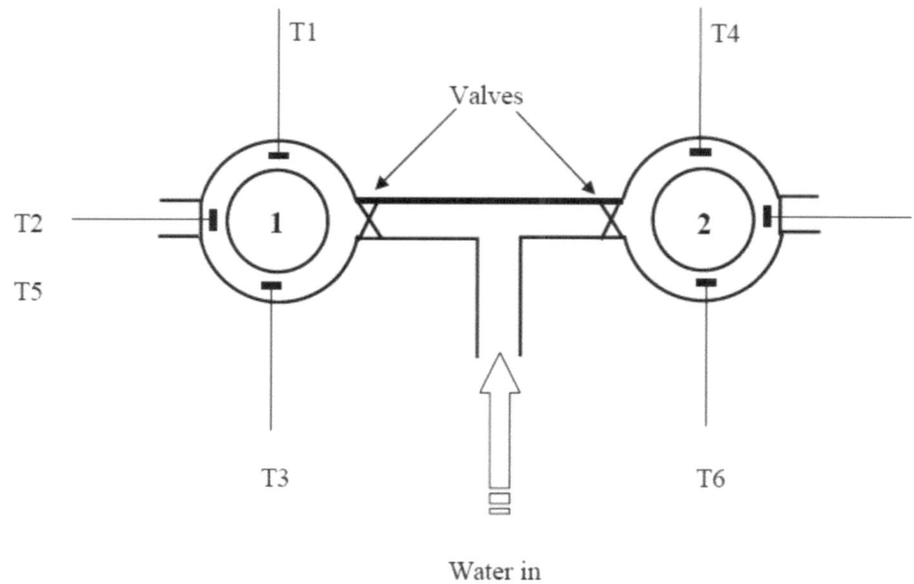

Fig. 5.2. *Arrangement of the rings and thermocouples for solar water heater experiment.*

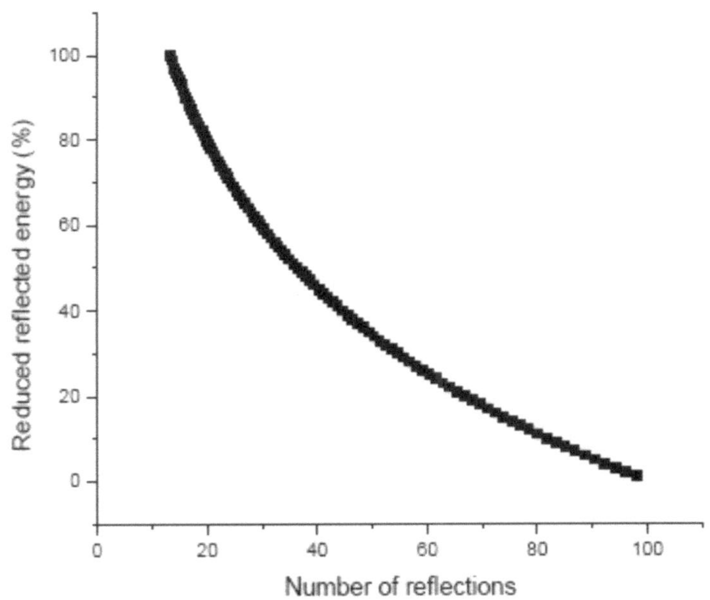

Fig. 5.3. *Reflected energy versus the number of reflections.*

the ring in the concentrator is heated by the radiation coming directly from the halogen lamp as well as by some radiation collected on the surface of the cone and guided to the ring by the walls of the cylinders forming the guide.
In this section we analyze and calculate the concentration ratio of a modified cone shown in Figure 8. This cone has an axial cylinder whose exterior surface area is considered to be involved in solar radiation collection.
If the modified cone concentrator has an axial cylinder as shown in Fig. 5.4, then the axial cylinder can also be taken to aid in radiation collection. Hence, the total area of solar collector in a modified cone concentrator can be written as

$$A_{coll} = \pi R_x \sqrt{\left[R_2 - \left(l_2 + d_2\right)\right] + H_2^2} + 2\pi l_2 H_2 \beta \tag{5.1}$$

where $R_x = R_2 - (l_2 + d_2)$ and $\sqrt{\left[\left(R_2 - (l_2 + d_2)\right)\right]^2 + H_2^2}$ is the slant height, S_2, of the modified cone concentrator

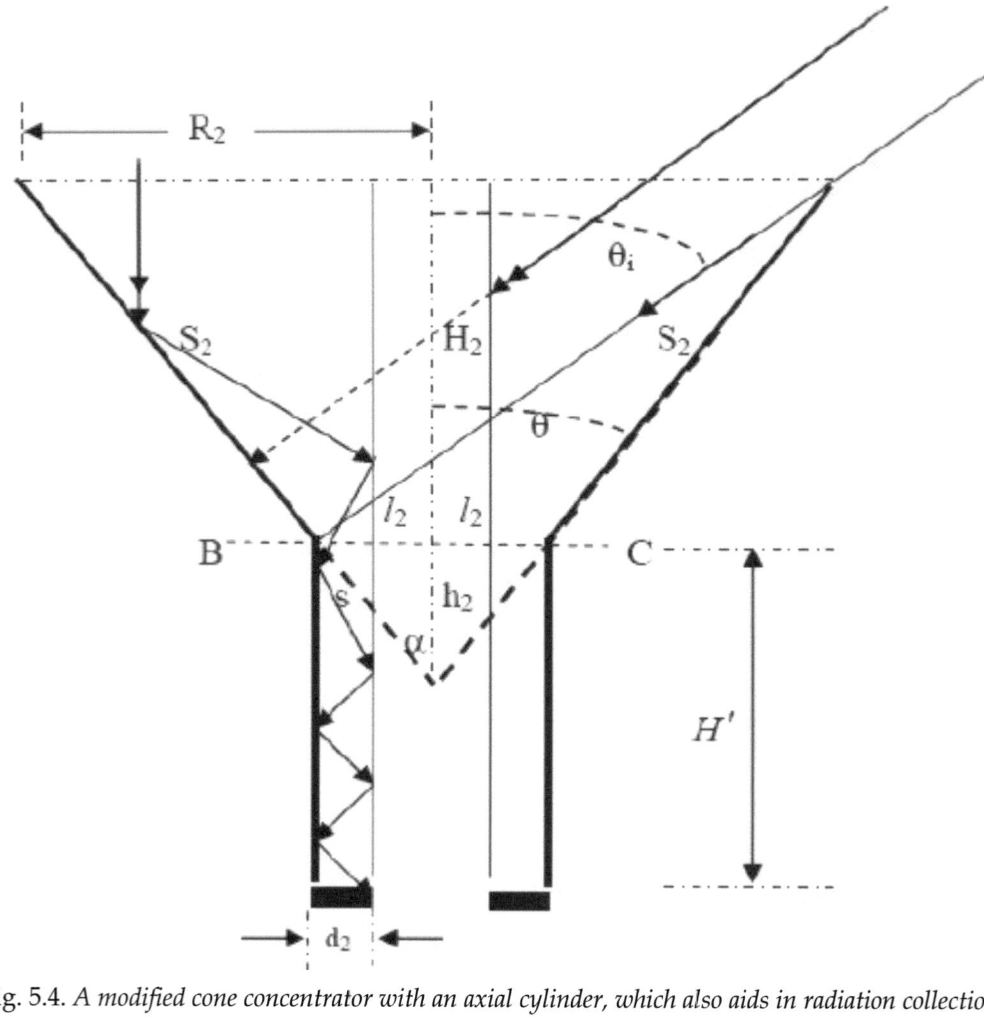

Fig. 5.4. *A modified cone concentrator with an axial cylinder, which also aids in radiation collection.*

The absorber surface area of the modified cone concentrator is then given by

$$A_{abs} = \pi\left(2l_2 d_2 - 2d_2^2\right) \tag{5.2}$$

Thus, the expression for the geometric concentration ratio of the modified cone concentrator according to the configuration of Fig. 8 is given by

$$\frac{A_{coll}}{A_{abs}} = \frac{R_x\left(\sqrt{\left(R_2 - \left(l_2 + d_2\right)\right)^2 + H_2^2}\right) - l_2\sqrt{l_2^2 + h_2^2} + 2\left(l_2 - d_2\right)^2 H_2 \beta}{\left(\frac{d_2}{2}\right)^2} \tag{5.3}$$

It has also been assumed in this configuration that the axial cylinder is involved in the collection of radiation, hence the factor β indicate the percentage of its surface area involved in the collection of solar radiation. $0 \le \beta \le 1$, where $\beta = 0$ has the physical meaning that the external surface area of the axial cylinder is not part of the collection area of solar radiation.

6. Results

Fig.6.1 show results obtained for the experimental set up shown in Fig. 5.2 in which the concetrator configuration of Fig. 5.4 was used.

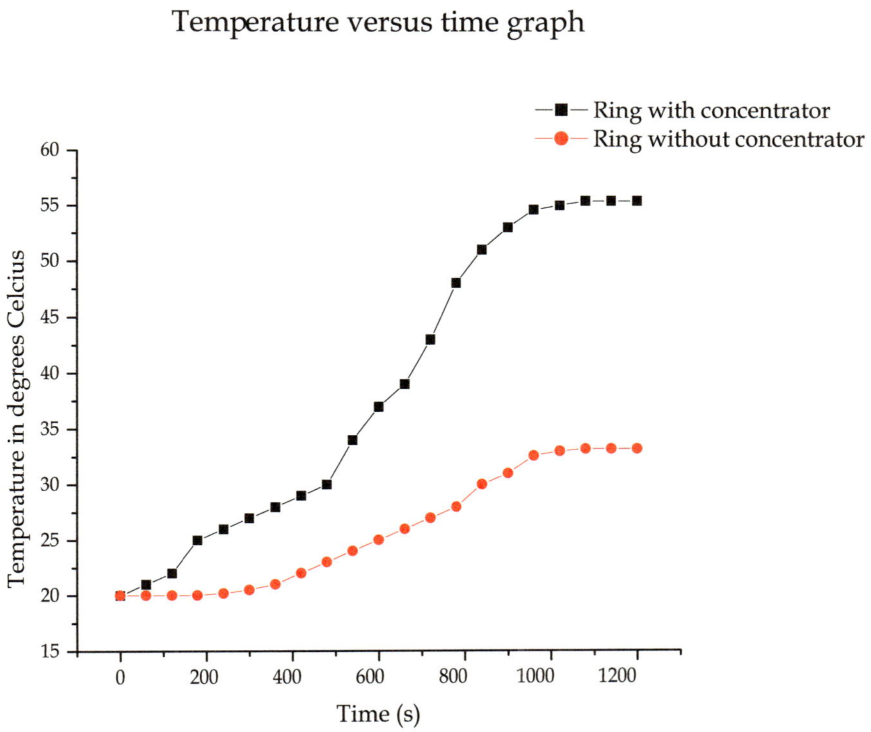

Fig. 6.1. *Curves for heating water in an open ring and in a cone-cylinder concentrator.*

The figure show curves obtained for the temperature of water in the ring with the concentrator and the ring without the concentrator as functions of time using the model of Fig.8. It was also found that the water in the ring with the concentrator rose to more than 55.4°C after being exposed to the radiation for a period of 1,200 seconds while water in the ring without the concentrator reached a temperature of 33.2°C for the same exposure time of 1,200 seconds to radiation. The water in the ring with the concentrator thus attained a higher temperature than the water in the ring without the concentrator (open ring).
Using models in which the dimensions given hold, theoretical geometric concentration ratios of different configurations of cones and cone-cylinder combinations concentrators with modifications were calculated: $H_2 = 2\text{ m}$, $H_1 = h_2 = 0.5\text{ m}$, $d_2 = 0.01\text{ m}$ $R_1 = R_2 = 1\text{ m}$, $d_1 = 0.04\text{ m}$, $S_1 = S_2 = 2.69\text{m}$, $l_2 = 0.228\text{ m}$. For the given dimensions, our results show that for a straight cone as shown in Fig. 3.1, the geometric concentration ratio is found to be 16 by employing Eq. (3.3). The value for $S_1 = S_2 = 2.69\text{m}$ is calculated from the values of H_1, H_2 R_1 and R_2. The geometric concentration ration of the concentrator shown in Fig. 5 calculated using Eq. (4.10) is found to be 33.33. With $\beta = 1$ in Eq. 5.1, the theoretical concentration ratio of the configuration works out to be 90,400.

Concentrator	Description of concentrator	Theoretical value of geometric concentration ratio
Figure 3.1	Cone concentrator (Using Eq. 3.3) (Using Eq. 2.1)	16.00 19.10
Figure 5.1	Inverted cone-cylinder concentrator with a radiation guide and outer cylinder	33.33
Figure 5.4	Straight cone concentrator with radiation guide and axial cylinder	90,400
Figure 5.4	Straight cone concentrator with radiation guide but no axial cylinder $(\beta = 0)$	90,000

Table 1. Summary of results of theoretical geometric concentration ratios for different configurational geometries of cone, cone-cylinder concentrators.

7. Discussion

The results obtained in this work show that a mathematical procedure can be used to calculate the effective collector and absorber surface areas of cone concentrators. The technique can be extended to a modified cone concentrator and the values obtained for the ratio of the collector to the absorber area of the concentrator give the geometric concentration ratio. Though this method gives the correct values of concentration ratios, just as the inverse sine of the acceptance angle method, both do not present any idea on the size or configurational geometry of the system.
It can be seen from this work that configurational geometry of cone and cone cylinder-combinations of concentrators give different theoretical geometric concentration ratios for the same dimensions. Additional components are seen to improve the geometric concentration ratio.

8. Conclusion

The work shows that for a straight cone with a collector area A_{coll} ,situated a distance H_2 from the apex and an absorber area, A_{abs} , at a distance H_1 from the apex, the ratio of the squares of H_2 to H_1 give the geometric concentration ratio of the cone concentrator. The geometric concentration ratio has been calculated using configurational geometry giving the same value when the dimensions of the cone are used. The method of using the ratio of the squares of the heights can therefore be used as a quick way of approximation of the concentration ratio of a straight cone without taking many measurements.

No literature exist for the solar concentrator shown in Fig. 5.1 that have been reported in this work except for the work by (Ochieng and Onyango, 2009). However, variants of this concentrator such as the straight cone type in which some literature can be found, (Williamson, 1952), (Witte, 1965), (Smith, 1966), (Welford and Wilson, 1978). The inclusion of cylindrical component is a completely a new addition. There is therefore need to build prototypes and test them with actual solar radiation. The values of the theoretical geometric concentrations can be raised by making the denominator in Eq. (4.10), and Eq. (5.1) smaller. Two possible ways of doing this are: (1) to reduce the diameter of the copper ring but keep the other dimensions of the concentrator constant and (2) to increase the surface area of the cone but keep the other parameters of the concentrator constant.

As can be seen from Figure 6.1, the rise in temperature of water in the open ring seems to be gradual while that of the ring with the concentrator is steeper. A possible explanation of this effect is that the water in the ring with the concentrator is heated by direct radiation from the halogen lamp as well as by the radiation collected on the surface of the cone which is guided on to the ring by the cylinder walls whereas in the open ring, the heating only takes place by the direct radiation from the lamp. It can also be noted that in the open ring, energy loss is due to two processes, namely radiation and convection while the ring in the concentrator may loose little energy by convection because it is almost a closed system.

As in the straight cone, both configurations discussed in this work have semi angles denoted by θ . The extreme angle is denoted by θ_i . It can be seen from Figure 5.4 that some of the radiation enter through the angle less than θ_i that would have been "back-reflected" are multiply reflected from the surface of the axial cylinder and the surface of the cone in to the radiation guide and end on the absorber. The amount of radiation that will not be "back-reflected" can be calculated by determining the shading by the axial cylinder on to the surface of the cone.

All the radiation, axial and off axis in the inverted cone concentrator of Figure 5.1 end up on the absorber. However, it is easy to see that by making the cone angle bigger; the cone becomes flatter and tends to a flat plate collector in which case not all the radiation falling on its surface will be channeled in to the guide.

One of the advantages of using the configuration of Figure 5.4 rather than that of Figure 5.1 for a solar collector concentrator is that the surface area of the cone can be increased easily without affecting the other dimensions of the system just by increasing the slant height S_2. If an attempt is made to do the same thing with the other configuration (Figure 5.1) by keeping the cone angle constant, bigger cylinders forming the radiation guide must be used. This, however, has a counterproductive effect of increasing the surface area of the absorber and thus reducing the geometric concentration ratio.

Since actual systems have not been constructed and tested, there are many parameters that can be varied in order to find out appropriate dimensions for such systems which maximize

the geometric concentration ratio. These parameters include the height H' of the radiation guide, the base radius R, the slant height S, and the vertical height H of the cone.
The reason for suggesting changing H' to a smaller value say by an amount δ, is that the area on which radiation will be lost through the guide will change from $\pi H\left(d^2 + 2Rd\right)$ to $\pi\left[H - \delta\right]\left(d^2 + 2Rd\right)$. Since $H - \delta < H$, a smaller area will be available for radiation loss inside the guide. PVCs were preferred in this work because they are poor conductors of heat.

9. References

Basset, I.M., and Derrick, G.H. (1978). The collection of diffuse light into an extended absorber. *Optical and Quantum Electronics* 10, 61-82.

Burkhard, D.G., and Shealy, D.L. (1975). Design of reflectors which distribute sunlight in a specific manner. *Sol. Energy* 17, 221-227.

Cornbleet, S. (1976). *Microwave Optics*. Academic Press, New York.

Garg, H.P., and Kandpal, T.C. (1999), *Laboratory Manual on Solar Thermal Experiments.*, Narosa, New Delhi, ISBN 81-7319-340-1

Grant, I.S., and Phillips, W.R., (1978). *Electromagnetism,* John Wiley and Sons, ISBN 0 471 99707 2, Great Britain.

Hildebrand, R.H., and Winston, R. (1982), Throughput of Diffraction-Limited Field Optics System for Infrared and Millimetric Telescope. *Appl. Opt.* 21, 1844-1846.

Hinterberger, H., and Wilson, R. (1968a). Use of a solid light funnel to increase phototube aperture without restricting angular acceptance. *Rev. Sci. Instr.* 39, 1217-1218.

Hotler, M.L., Nudelman, S., Suits, G.H., Wolf, W.L., and Zissis, G.J. (1962), *Fundamentals of Infrared Technology,* Macmillan, New York.

Maatouk, K., and Shigenao, M. (2005). Theoretical approach of a flat plate solar collector with clear and low-iron glass covers taking into account the spectral absorption and emission within glass cover layers. *Renewable Energy.* 30, 1177-1194. http://www.oceanoptics.com/Products/ispref.asp (2006)).

Morgan, S.P. (1958). General solutions of the Luneburg lens problem. *J. Appl. Phys.* 29, 1358-1368.

Ochieng R M and Onyango F N. A new type of solar concentrator employing a cone-cylinder combination, International Journal of Global Energy Issues, Vol. 31, No. 2, 169-182. 2009, ISSN (Print) 0954-7118, (Online) 1741-5128.

Ochieng, R.M., Onyango, F.N., and Owino, A.J. (2007), Determination of global absorptivity and emissivity of some opaque bulk materials using an integrating sphere calorimeter without ports. *Meas. Sci. Technol.* 18, 2667-2672, ISSN 0975-0233. dio:10.1088/0957-0233/18/8/043

Rabl, A., and Winston,R., (1976), Ideal concentrators for finite sources and restricted exit angles, *Appl. Opt.* 15, 2880-2883.

Rabl, A., (1976a), Solar concentrators with maximal concentration for cylinder absorbers, *Appl. Opt.* 15, 1871-1873.

Rabl, A., (1976b), Comparison of solar thermal concentrators, *Sol. Energy* 18, *93-111*.

Rabl, A., (1976c), Optical and thermal properties of compound parabolic concentrators, *Sol. Energy* 18, 497-511.

Welford, W.T., and Winston, R (1978), *The Optics of Nonimaging Concentrators-Light and Solar Energy,* Academic Press, New York.

Williamson, D. E., (1952), Cone channel condensed optics, *J. Opt. Soc. Amer.* 42, 712-715.

Witte, W., (1965), Cone channels optics, *Infrared Phys.* 5, 197-185.

18

Applications Oriented Research on Solar Collectors at the "Politehnica" University of Timişoara

Ioan Luminosu, Aldo De Sabata and Coleta De Sabata,
"Politehnica" University of Timişoara
Romania

1. Introduction

Solar energy, as a free and clean energy source, is successfully used in the present in the European Union countries for individual and social buildings climatization, water heating and cleaning, electricity and motor force production, heating in swimming pools etc. Examples are numerous so that we briefly present a selection.

The Self-Sufficient Solar House (SSSH) project has been started within the Fraunhofer Institute for Solar Energy Systems in Freiburg, Germany in 1988. The construction of the building started in 1991 and has been accomplished in October 1991. The aim of the project has been to demonstrate that the needs of thermal and electric energy of a residence may be satisfied exclusively by solar means. Measurements and calculation showed that, for a primary equivalent energy of 600 MWh invested in the construction of the residence, the system returns that energy in a period of 10-20 years and its life time is about 80 years, meaning that SSSH provides an economy of primary energy for about 60 years and diminishes the pollution of the ambient (Sthal et al., 1994).

The Northern Solar Heating Demonstration Project (NSHDP) has been started in Denmark in 1985 with the aim of domestic hot water production for a block with 150 flats. The system works with an efficiency of 25-33% and provides an annual thermal energy of 485-585 kWh for every square meter of collecting surface (Pedersen, 1993).

Through the Thermosyphoning Air Panels project (TAP), an elementary school situated 26 km north of London, Great Britain, with a living surface of 1631 m^2, has been architecturally adapted to the use of solar energy in 1989 (Lo et al., 1994). The school is used from Monday to Friday by 181 pupils and 11 teachers between 8 am and 3.30 pm.

The solar collector is the active part of the solar to thermal energy conversion chain. The importance of the role played by the solar collector in solar applications explains the high amount of research performed for increasing its efficiency and for decreasing its price and impact on the ambient. Such research goals are fulfilled by devising new materials and innovating geometries. Simple, efficient, cheap and ambient-friendly solar installations are the attractive ones (Haberl et al., 2008). Some examples are cited below.

The energy autonomy of mountain meteorological stations and huts in the Italian Alpes is achieved by means of thermal, photovoltaic and biogas systems. The accumulation boiler is placed 3 m below the collector. The liquid agent circulates by gravity and thermosyphoning (De Beni et al., 1994).

The use of a small photovoltaic (PV) module for the generation of the electricity supply of the fan attached to a thermal collector increases the overall efficiency by 3-5% and improves energy autonomy (Eicker & Steeberger, 1996).
Insulating transparent materials reduce thermal losses and allow reaching temperatures by 80-120 K above ambient (Folkerts et al., 1996).
Polymer based materials, immune to the action of hot, salted water and to freezing enter the structures of sheets used for covering swimming pools (Rommel et al., 1996).
The research in the field of Solar Energy performed within the Physics Department from the "Politehnica" University of Timişoara, Romania has had the goal to evaluate and exploit the solar potential of the western part of the country. The results are relevant for the Danube-Kris-Mures-Tisza region. Procedures, equipment and installations for hot air production, drying of ceramic products, bitumen melting, home climatization, stock raising, hot water production, solar radiation measurement and environment protection have been devised. An air solar collector, a Trombe wall and solar collectors built from recyclable materials are treated in this chapter. Next, some thermal applications of Solar Energy that rely on the solar collectors are presented.

2. Air solar collector

The plane solar collector represented in Fig. 1 has been devised, constructed and studied. The elements in Fig. 1 are: transparent plate - 1, absorber - 2, pipe for air circulation - 3, insulating sheet - 4, support - 5, chassis - 6. The elevation angle is denoted by s and the overheight by h. The pipe for air circulation is represented in Fig. 2.

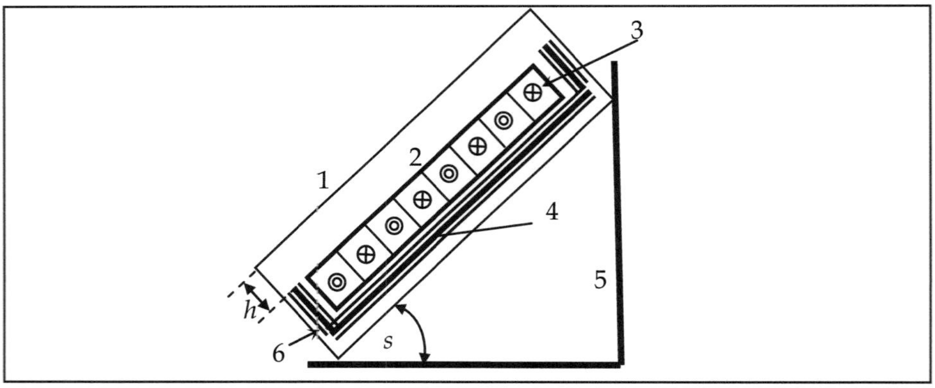

Fig. 1. Air solar collector

The absorbing plate plays also the role of a wall for the air circulation pipe. Two neighbouring corridors share a common wall, so that the transport factor is unity, $F_t = 1$. The air flow rate may be in the interval 25 - 135 m³/h and the length of the pipe is $L_{cd} = 12.4\ \text{m}$. The air admission and evacuation holes are square shaped, with a side $a = 0.25\ \text{m}$. The collector surface is $A = aL_{cd} = 3.1\ \text{m}^2$. The surface of the collector is tilted by an angle $s = 45\ \text{deg}$. The air is circulated by a low power fan, having a power of 30 W for an air flow rate of 135 m³/h, while the power of the installation is about 800 W. A photovoltaic panel may be used for powering the fan. The absorption – transmission equivalent product is $(\tau\alpha)_{eff} = 0.844$ and the thermal insulation is characterized by $U = 4.73\ \text{Wm}^{-2}\text{K}^{-1}$.

Some regions of the absorbing surface are shadowed by the window supports and by the walls of the chassis. The first effect has a daily variation, while the second one may be considered to have a hourly variation. A cross section through the lateral window support is represented in Fig. 3. The fluid crosses n times the shadows created by the central and lateral supports , with n = 5 the number of pipes. The total length of the shadow may be expressed as

$$L_1 = n\left[h\tan(\theta) + b - d\right], \tag{1}$$

where h is the overheight (Fig. 1), b is the width of the central support (Fig. 2), d is width of the insulation (Fig. 3) and θ is the angle between the incoming sun ray and the normal to the absorber. For example, at equinox, $\theta = \omega\Delta\tau$, with ω – the apparent angular speed of the Sun and $\Delta\tau$ - time from noon.

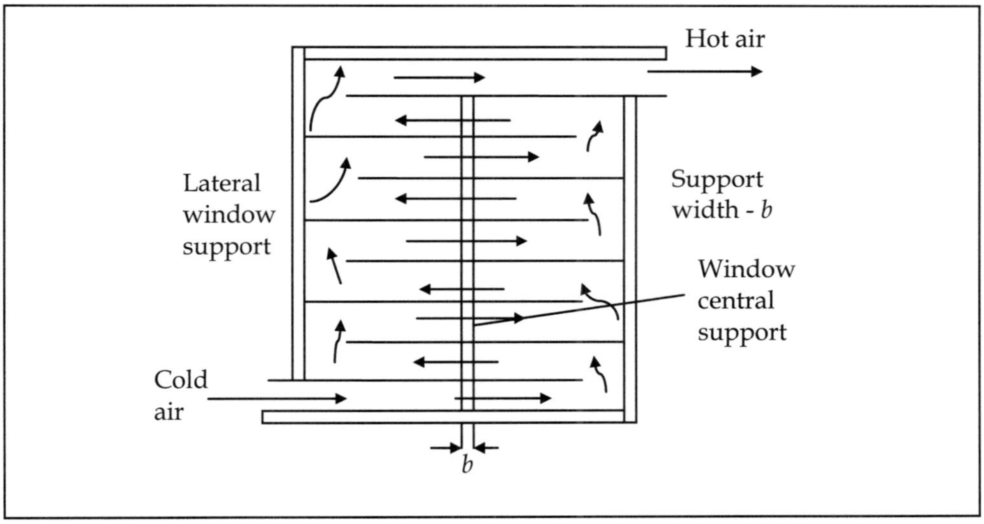

Fig. 2. Pipe for air circulation.

The length of the pipe that is irradiated allowing for the heat to be absorbed is

$$L = L_{cd} - L_1 . \tag{2}$$

The surface of the fluid that is irradiated, $A_c = aL$, results:

$$A_c = a\left[L_{cd} - n\left(h\tan(\theta) - d + b\right)\right] \tag{3}$$

so that the fraction of surface that is effectively used is

$$f = 1 - \frac{n\left[\tan(\theta) + b - d\right]}{L_{cd}} . \tag{4}$$

The variation of the fraction f with the hour is represented in Fig. 4. It may be seen that $f \approx 0.85$ for a time interval of 4 – 6 hours centred at noon.

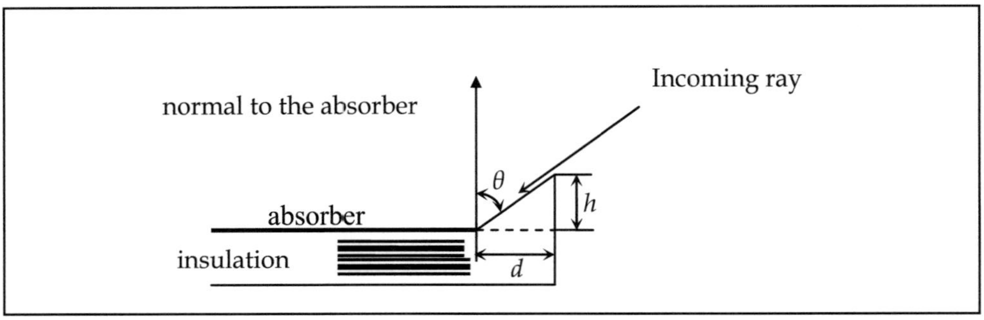

Fig. 3. Shadowing of the surface.

In order to find the equations that characterize the system, we note that the heat obtained by thermal conversion is transferred to the working agent. The fluid enters the collector at a temperature T_{fi} and exits at a temperature T_{fe}. The energy balance for the fluid that flows through a small segment of pipe, of length Δy, is

$$\dot{m}C_pT_f\Big|_y - \dot{m}C_pT_f\Big|_{y+\Delta y} + q_u'\Delta y = 0\,, \tag{5}$$

where $\dot{m}$ is the mass flow rate, C_p is the isobar specific heat of the fluid, q_u' is the heat flux absorbed by the unit length of a current tube and T_f is the temperature of the fluid.

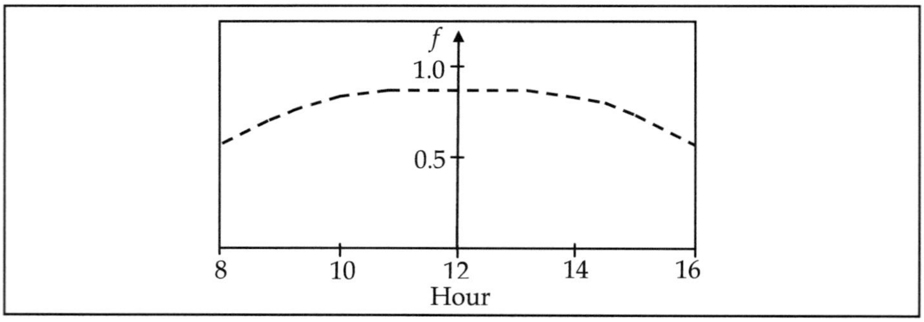

Fig. 4. Irradiated fraction of surface versus hour.

The flux absorbed per unit length may be expressed as

$$q_u' = aF'\left[S - U\left(T_f - T_a\right)\right] \tag{6}$$

where $S = (\tau\alpha)_{eff}G_c$ is the total flux density absorbed by the black plate, G_c is the solar flux density in the plane of the collector, F' is an efficiency factor and U is the coefficient of heat loss in the ambient.

By manipulating (5) and (6), the equation of the temperature may be obtained:

$$T_f = T_a + \frac{S}{U} + \left(T_{fi} - T_a - \frac{S}{U}\right)\exp\left(-\frac{F'AU}{\dot{m}C_p}y\right). \tag{7}$$

By setting $y = L$, the temperature at the collector output T_{fe} may be obtained.

If the collector is functioning in an open regime, the input temperature is equal to the ambient one $T_{fi} = T_a$, which, substituted into (7) yields (Luminosu, 1983)

$$T_f = T_a + \frac{S}{U}\left[1 - \exp\left(-\frac{F'aU}{\dot{m}C_p}y\right)\right]. \tag{8}$$

For $y = L$, and by using $A_c = aL$, the temperature at the output of the collector results:

$$T_{fe} = T_a + \frac{S}{U}\left[1 - \exp\left(-\frac{F'U}{\dot{m}C_p}A_c\right)\right]. \tag{9}$$

The temperature rise $\Delta T = T_{fe} - T_a$ versus the radiant power density absorbed by the black plate S is represented in Fig. 5. The curves are linear and start from the origin. Temperature rises as high as 50°C may be achieved.

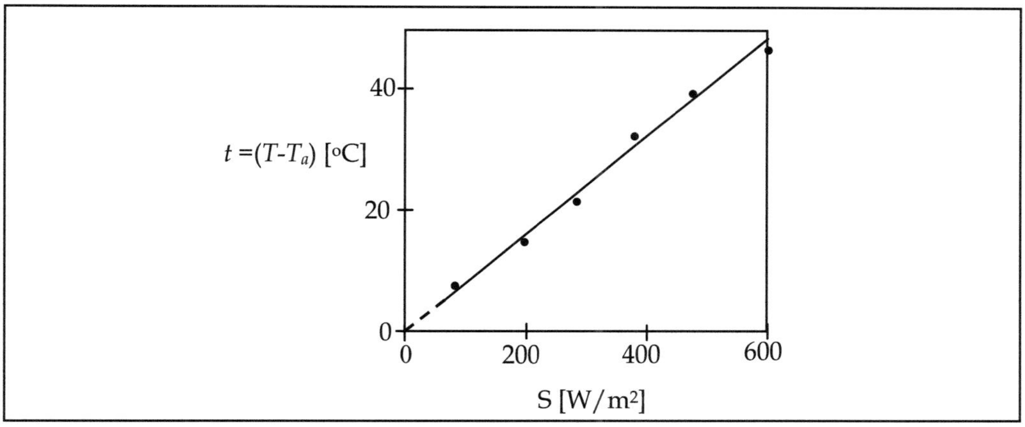

Fig. 5. Temperature rise versus absorbed power density.

The energy flow for the air collector in open state (heat per time unit or power), $\dot{Q}_u = \dot{m}C_p\left(T_{fe} - T_a\right)$ is (De Sabata & al. 1983):

$$\dot{Q}_u = \dot{m}C_p\frac{S}{U}\left[1 - \exp\left(-\frac{F'UA_c}{\dot{m}C_p}\right)\right]. \tag{10}$$

The collector power versus the density of the flux absorbed by the black plate is represented in Fig. 6, at various mass flow rates of the fluid. The power increases with the incoming radiation and the flow rate. At large flow rates, at noon, the power may increase up to 800 W.

The specific power is the ratio of the energy flow to the collecting surface

$$\dot{q}_u = \frac{\dot{Q}_u}{A_c}. \tag{11}$$

The values of the specific power are listed in Table 1. Measurements have shown that this quantity reaches larger values in the afternoon than before noon for similar values of the incident flux. This result may be explained by the fact that the carcass of the device provides additional heat to the fluid when the radiation intensity decreases.

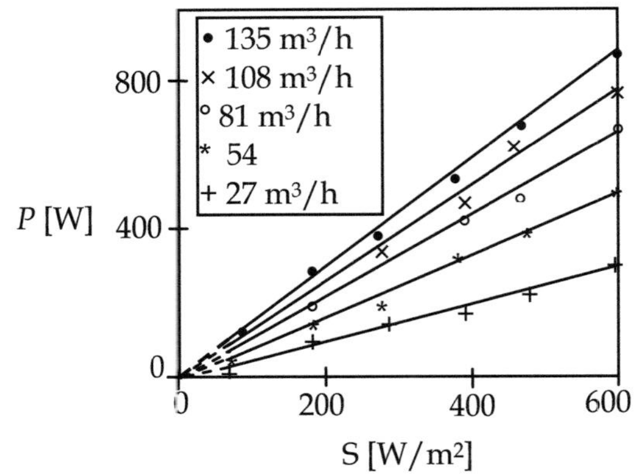

Fig. 6. Collector power versus absorbed radiation, parameterized by the flow rate

S[W/m²]	100	200	300	400	500	600
$\dot{q}_u$ [W/m²]	33	74	124	152	200	218

Table 1. Absorbed flux density and specific power.

The instantaneous efficiency of the collector is

$$\eta_i = \frac{\dot{Q}_u}{A_c G_c} . \tag{12}$$

Equations (10) and (12) imply

$$\eta_i = \frac{\dot{m} C_p S}{U A_c G_c} \left[1 - \exp\left(-\frac{F' U A_c}{\dot{m} C_p} \right) \right] . \tag{13}$$

The variation of the efficiency with the absorbed flux, for various values of the flow rate is represented in Fig. 7.

The long term performance of the collector is given by the average efficiency in the considered time interval

$$\tilde{\eta} = \frac{\dot{Q}_{u,average}}{A_c \tilde{G}_c} \tag{14}$$

where $\dot{Q}_{u,average}$ is the average value of the power provided by the collector and $\tilde{G}_c$ is the average value of the incident radiant power density in the considered time interval.

The hourly variation of the average efficiency is represented in Fig. 8, parameterized by the flow rate.

The curves presented in Fig. 8 show that efficiencies are high around noon, when the incidence angles are small and the absorption - transmission products are high. The time variation of the incidence angle determines changes of the absorption-transmission product which, at its turn, determines the variation of the efficiency. The curves present maxima at noon, but they are asymmetric with respect to noon: the slopes of the curves are smaller in the afternoon when the fluid is additionally heated by the metallic support. At high flow rates (135 m^3/h), the efficiency of the collector approaches 40%. This reasonably high efficiency and the unsophisticated design recommend this solar collector for home climatization and for drying applications in industry.

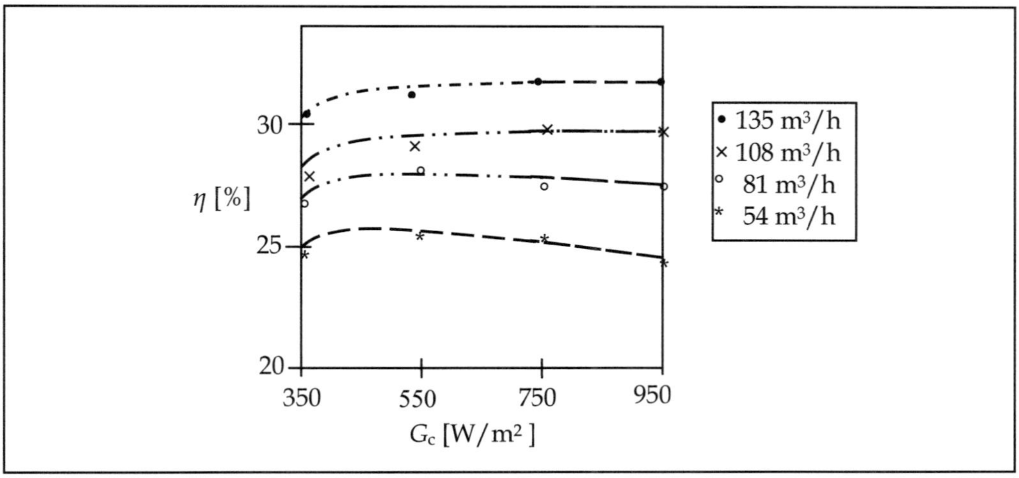

Fig. 7. Efficiency versus irradiation.

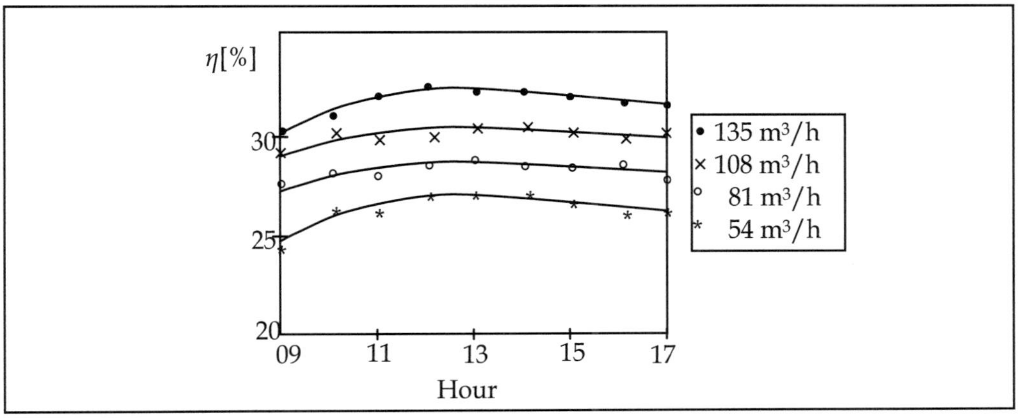

Fig. 8. Hourly variation of the average efficiency.

3. Trombe wall

The Trombe wall is the main element of heating systems for buildings based on passive solar gain. For an outside temperature t_{ext}=0°C and an inside temperature t_{int}=20°C, a simple wall (without solar installations) transfers heat towards the interior if the normal solar irradiation is greater than 465.2 W/m^2 (Athanasouli & Massouporos, 1999). Such conditions are met in Timişoara, Romania during transition months, between 11 am and 1 pm. In order to increase the contribution of the wall to the energy required for heating the room and in order to decrease energy losses during night time, the wall may be covered with a glass plate during daytime and additionally with a curtain at night fall (Ohanesion & Charteres, 1978). The solar panels mounted on the eastern and southern walls of a school supplied each year a thermal energy of 2469 kWh during classes (Lo et al., 1994).

An experimental setup with Trombe wall has been built at the "Politehnica" University of Timişoara in order to evaluate the opportunity of implementing passive solar installations in the region. The installation has been used for heating a living room, complementary to electric power, during transition months (March, April, September and October). The Trombe wall has been placed on the southern wall of an ordinary building with four rooms at the ground floor, otherwise heated by classical means. The three rooms that were not heated by solar means have been maintained at a temperature of $21 \pm 1\ ^\circ\mathrm{C}$, so that the heat lost through the door of the target room could be neglected (De Sabata et al., 1986a, 1986b).

The dimensions of the solar heated room were $2.80 \times 4.75 \times 1.75$ m and the dimensions of the window on the southern wall were 1.0×0.75 m. The walls made with bricks were 0.39 m thick and were plastered with lime and mortar. The concrete foundation was $h = 1$ m deep and 0.49 m thick. The underground water layer is situated at a depth smaller than four meters and it has a temperature $t_f = 10\ ^\circ\mathrm{C}$. The surface of the Trombe wall was $A_T = 8.8\ \mathrm{m}^2$ (Fig. 9). The curtain from *I* covered the wall during night time. The air dampers $L_{1,2,3}$ controlled the direction of the air flow. A water container *C* was attached to the passive wall in order to raise the inside air humidity. The small power fan *F* ($P = 10$ W) contributed to the uniformity of the thermal field.

The heating of the room has been provided by a radiator with thermostat *R* and the Trombe wall. The heat supplied by the two devices balanced the thermal losses of the room through the eastern wall, the floor and the window (Luminosu, 2003a). Temperatures at points 1..12 have been measured with the thermometer *V*, having an error of $\pm 0.1\ ^\circ\mathrm{C}$. The global radiation intensity *G* has been measured with an error of $\pm 5\%$ by means of an instrument built in our laboratory (Luminosu et al., 2010), the electric power with an aem1CM4a instrument (*N* on Fig. 9), with an error of ± 5 Wh and the air humidity has been measured with the hygrometer *H*, having an error of $\pm 5\%$. Additionally, the velocity of the air current has been measured with the anemometer FEET (*A*, Fig. 9), error $\pm 10\%$ and the illumination at the centre of the room has been measured with a Lux PU150 light meter.

The average air velocity has been found to be $\overline{v} = 0.15$ m/s, which corresponds to the upper comfort limit and, due to the additional water container, the humidity has been kept in between the limits 35..70%, a range well inside the comfort limits. The lighting at the centre of the room has been in the range 50..70 lx in the horizontal plane; these values have been achieved by operating the blinds and by turning on the 12 W ECOTONE light bulbs for about 4 hours a day.

The measured values of the solar radiation (1), temperature at the upper air damper (2), temperature at the centre of the room (3) and ambient temperature versus hour are presented in Fig. 10. Measurements have been performed in autumn (October and November) and spring (mid February and March). Temperature ranges of 14..17.5 °C at the centre of the room, 21..31°C at the upper air damper and 18..22 °C near a wall shared with an adjacent room have been obtained.

The daily average radiant energy has been $\bar{H}_d = 99.1\ \text{MJ}$. Adding up the hourly measured heats resulted in the following average daily heats: the heat lost by the room $\bar{Q}_{dL} = 22.4\ \text{MJ}$, the heat supplied by the passive wall $\bar{Q}_{dT} = 10.26\ \text{MJ}$ and the electric energy for heating $\bar{Q}_{del} = 12.31\ \text{MJ}$

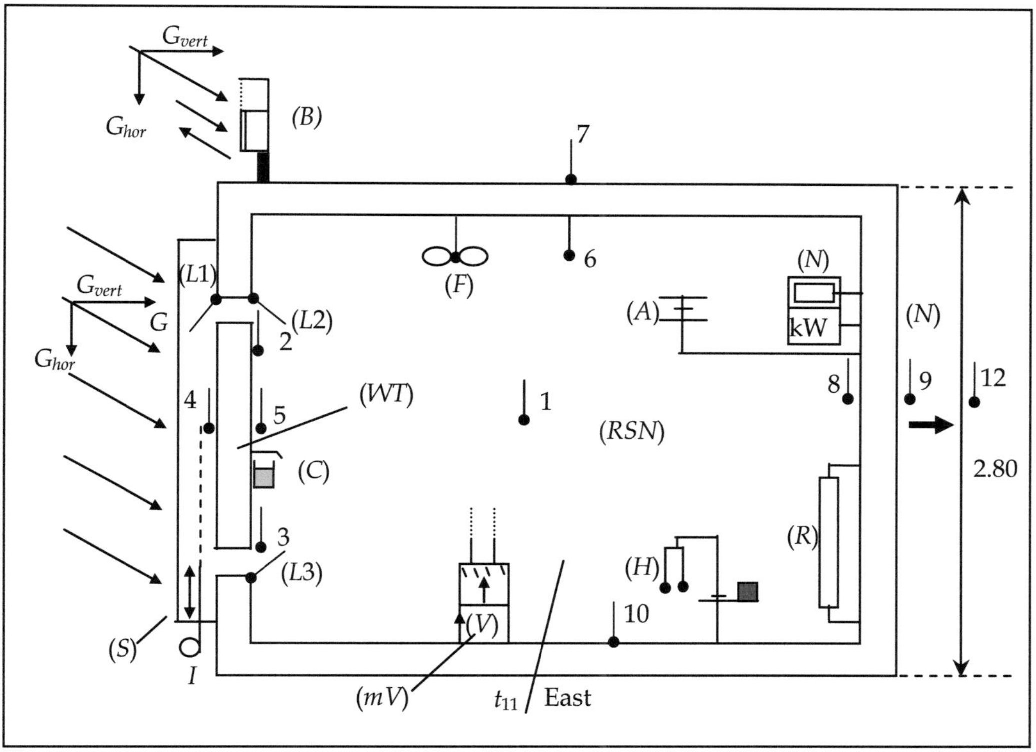

Fig. 9. Room with Trombe wall and measuring points.

The power of the Trombe wall has been $\bar{P}_T = 237.5\ \text{W}$. As the average number of days with clear sky during the transition months is $N = 46$, the annual average heat supplied by the wall is $Q_{yT} = NQ_{dT} = 131\ \text{kWh}$. The daily efficiency of the passive wall is $\bar{\eta}_T = 100 \times \frac{\bar{Q}_{dT}}{\bar{H}_d}$. Depending on the season, the efficiency of the considered wall varied between 7.8 and 10.4%. The specific annul heat of the wall is $\bar{q}_{yT} = \frac{\bar{Q}_{yT}}{A_T} = 14.9\ \text{kWh/m}^2$.

The sensation of thermal comfort is determined by the inside temperature and the temperatures of the walls and objects the human body establishes a radiant energy exchange with. According to hygienists (Săvulescu, 1984), the radiant temperature (°C) is given by

$$t_{rad} = \sum_{j=1}^{n} f_j t_j \tag{15}$$

and the room temperature by

$$t_{room} = \frac{t_{int} + t_{rad}}{2}, \tag{16}$$

where t_{int} is the inside room temperature, n is the number of elements the body exchanges radiant energy with and f_j are the shape factors $f_j = \frac{A_j}{A}$ (A_j – area of the j'th element, A – exchange area).

The level of comfort is optimal when the room temperature is equal to the comfort temperature prescribed by hygienists. According to Bradke (in Săvulescu, 1984), an inside air temperature $t_{int} = 21°C$ must have a radiant temperature correspondent $t_{rad,adm} = 16.3°C$ and a comfort temperature one of $t_{comf} = 18.7°C$.

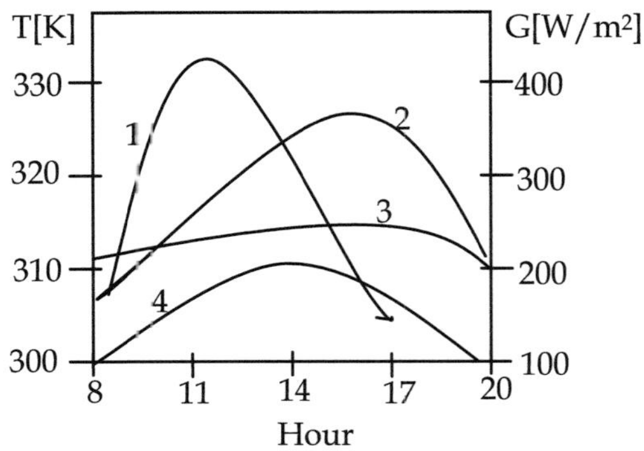

Fig. 10. Temperature of the passive wall and global solar radiation versus hour.

The shape factors f_j and the average temperatures $\overline{t}_j$ of the walls of the room heated by the passive wall, the average radiant temperature $\overline{t}_{rad}$ and the room temperature $\overline{t}_{room}$ are given in Table 2.

Radiant element	f_j	$\overline{t}_j$ (°C)	$\overline{t}_{rad}$ (°C)	$\overline{t}_{room}$ (°C)
Eastern wall	0.09	16	17.9	19.5
Southern wall	0.24	26		
Western wall	0.09	18		
Northern wall	0.24	18		
Ceiling	0.16	14		
Floor	0.16	13		

Table 2. Thermal comfort inside the room.

The Trombe wall produces a room temperature by 0.8°C higher than the comfort temperature prescribed by hygienists.

The thermal comfort factor, according to Van Zuilen (in (Săvulescu, 1984)), is given by

$$B = C + 0.25\left(\overline{t}_{int} - \overline{t}_{rad}\right) + 0.1\overline{x} - 0.1\left(37.8 - \overline{t}_{int}\right)v^{1/2}, \tag{17}$$

with x - absolute humidity inside, $\overline{x} = 12\ g/kg$; C - constant depending on the season, $C = -10.6$ in this case; v - velocity of the air.
Depending on B, the thermal sensation of comfort may be optimal $B = 0$, satisfactory $B = \pm 1$, or discomforting $B = \pm 3$. In our case we have $B = -0.325$, meaning that comfort reaches an optimal state.

4. Solar collectors from recyclable materials

Applications of Solar Energy in urban areas are facilitated by the existing infrastructure. However, in isolated locations, additional preparations that raise the costs of installations are necessary. Therefore, the possibility of using waste materials, resulted from demolishment of old buildings and from old appliances for devising low cost, small size solar collectors has been studied in our laboratory (Luminosu, 2007a). Transforming waste into raw material for a useful application has both a favorable impact on prices and on ambient. The main mechanisms of this impact are: decrease in the quantity of polluting waste; decrease in the demand for metal and glass from industry; decrease of energy consumption from classical sources; raise in the quality of life by the availability of low cost and ambient friendly energy in isolated locations; economy in transportation costs, as discarded materials are often available at the place were the collectors are built (e.g. following demolishments of old buildings); and economy in fabrication costs, as materials are often preprocessed and already cut into usable shapes, so that the collectors may be realized in modest mechanical workshops.

4.1 Solar collector from old glass plates

A first solar collector has been realized from glass plates, Fig. 11. The represented elements are: metallic frame - 1; vertical glass plates oriented towards south - 2; heated water - 3; cold water tank - 4; taps - 5, 6; mechanical support - 7; expansion bowl - 8; solarimeter - 9. Water is stored between the glass plates. One plate is transparent, while the other plate is painted in black, in order to absorb the solar radiation. The hot water is removed through

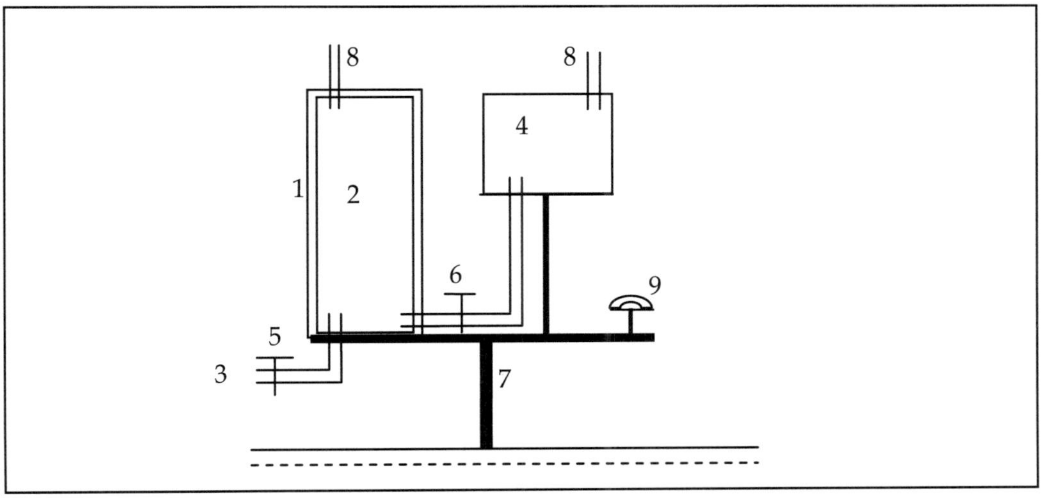

Fig. 11. Collector with glass plates.

the tap 5. The collector is filled with water contained in the tank 4, by the principle of communicating vessels, through the tap 6. The collector is positioned vertically in order to avoid breaking of the glass plates. The dimensions of the plates are 40 × 70 cm. The dimensions of the collector and the quantity of water stored between the glass plates must be kept reasonably low, by mechanical reasons related to the resistance to bending of the glass. The thickness of the water layer is 1.5 cm and the mass of water is m=4.2 kg.

The collector has been experimentally tested. Solar radiations has been measured with a solar wattmeter built in our laboratory (Luminosu et al., 2010). The water temperature T_w and the ambient temperature T_a have been monitorized. The water has been heated in time intervals comprised between 0.5 and 5.5 hours, symmetrically placed around noon. Measurements have been taken every 0.5 hours. It has been found that, under clear sky conditions, the water temperature raised by approximately 32°C with respect to the ambient temperature so that the water could be used for domestic purposes. The obtained average efficiency of the collector has been $\bar{\eta} = 33.3\%$.

4.2 Solar collector based on the heat exchanger of an old refrigerator

A second design consisted of a solar collector built around some parts of an old refrigerator. These parts are frequently available following the current replacement of old, heavy energy consuming refrigerators with modern, ecological ones. The disclosed heat exchangers and polystyrene sheets from the old refrigerators may be used for building small sized solar collectors, with favourable effects on the ambient.

The design of a collector that uses parts from an old "Arctic" refrigerator is presented in Fig. 12.

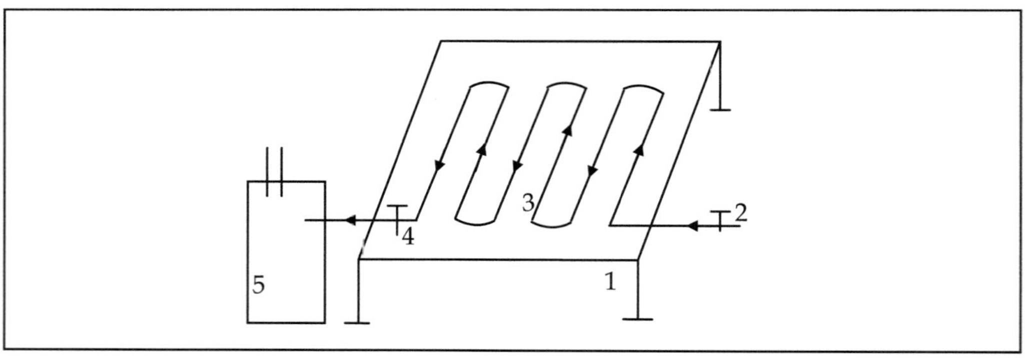

Fig. 12. Collector with pipes from an old refrigerator.

The elements in Fig. 12 are: mechanical support – 1; tap for cold water – 2; heat exchanger – 3; tap for hot water – 4; container with warm water – 5. The heat exchanger is 0.90 m long and 0.45 m wide, the pipes circulating the working fluid are spaced by 6 cm and the collecting area is 0.405 m². The collector is oriented towards south, at a tilt angle of 45 deg. A greenhouse effect is created by means of a glass plate, 3 mm thick. The hot water is accumulated in a Dewar pot. A coefficient of thermal losses $U = 6.453\ \mathrm{Wm^{-2}K^{-1}}$ and an absorbtion – transmission equivalent product $(\tau\alpha) = 0.847$ have been determined. The collector has been studied in open circuit.

For large flow rates of the water, of up to 3.60 kg/h and for densities of the solar flux of 500..600 W/m², the raise of the water temperature may reach up to 30°C and the efficiency

may be larger than 50%. In this way, the temperature of the water in the Dewar pot reaches 50..60°C, a temperature that allows the domestic use.
In conclusion, the use of recyclable materials for devising small sized thermal solar collectors has favourable impacts both on the way of life in isolated places and on the ambient.

5. The "Politehnica" solar house

Solar houses are equipped with thermal solar systems that maintain the inside temperature at a comfortable level and produce hot water for domestic use. As maximum solar radiation and energy need are not synchronous events, several types of thermal solar installations, which complement the classical ones, have been conceived. Some examples from the literature include: a hybrid solar system with heat pump, plane collectors and storage tank with $CaCl_2 \cdot 6H_2O$ (Çomakli, 1993); thermal solar system with heat pump that relies on the heat accumulated in the roof of the building (Loveday & Craggs, 1992); and thermal solar system with plane collectors complementary to the gas installation (Pedersen, 1993). Close to our laboratory, an experimental Solar House has been built and experimented with.

5.1 The solar house and measuring devices

The building has two rooms, a lobby and an access hall. A "minimal thermal loss enclosure", situated at the first floor has been defined and provided with a double layered door and a triple layered window. The dimensions of the room are $3.5 \times 3.5 \times 2.8$ m, giving a total volume $V_r = 35\,\text{m}^3$ and a total thermal exchange area $A_r = 63.7\,\text{m}^2$. The technical room is situated at the ground floor. A bedrock thermal accumulator, in the shape of a parallelopiped of dimensions $1.5 \times 1.5 \times 4$ m and filled with river stone $(C = 16.6\,\text{MJ/Kg})$ is deposited in the basement. The concrete walls are 40 cm thick and insulated with mineral wool. The main side of the building is south oriented.
The energy system shown in Fig. 13 includes the plane collectors - 1, the heat exchanger - 2, the thermal accumulator - 3, the heated room - 4 and the technical room - 5. The collecting field consists of twelve "Sadu 1" solar collectors connected in parallel. Each of the plane collectors is provided with aluminium pipes with inner diameter of 20 mm, facing south and tilted by an angle $s = 45\,\text{deg}$ from horizontal. The dimensions of the collectors are $2.0 \times 1.0 \times 0.12$ m and they are insulated with a 50 mm thick layer of mineral wool. The case is made of 0.8 mm steel plates. The heat-transfer fluid is water, activated by a 40 W Riello TF108 pump at a mass flow rate $m_w = 300\,\text{kg/h}$. The total collecting surface is $A_c = 24\,\text{m}^2$ and the thermal and optical parameters are $U_c = 3.7\,\text{W/m}^2$ and $(\tau\alpha)_{eff} = 0.81$.
The heat exchanger is of air-water type with copper coil and it provides a power of 60 W and a mass flow rate $m_a = 1154\,\text{kg/h}$. The heat carried by the hot water from the collectors to the coil of the heat exchanger is transferred to the air and carried to the bedrock. The direction of the air flow between the heat exchanger, tank and heated room through the nozzles *C*, *D* and *H* is determined by the slide dampers mounted at points *a*, *b*, *c* and *d* (Fig. 13). The heated room (minimum loss enclosure 4) may be heated either by solar means (the hot airflow comming fron the accumulator through nozzle *H*) or electrically from the radiator *R* equipped with a thermostat. The temperatures at points *A*, *B*, *C*, *D*, *H* (heat carrying fluid), *F* (hall), *I* (tank), *G* (exterior) and *T* (technical room) are read on the electric thermometer *V* with an error of $\pm 0.5°\text{C}$. The thermometer is equipped with 1N4148 diode

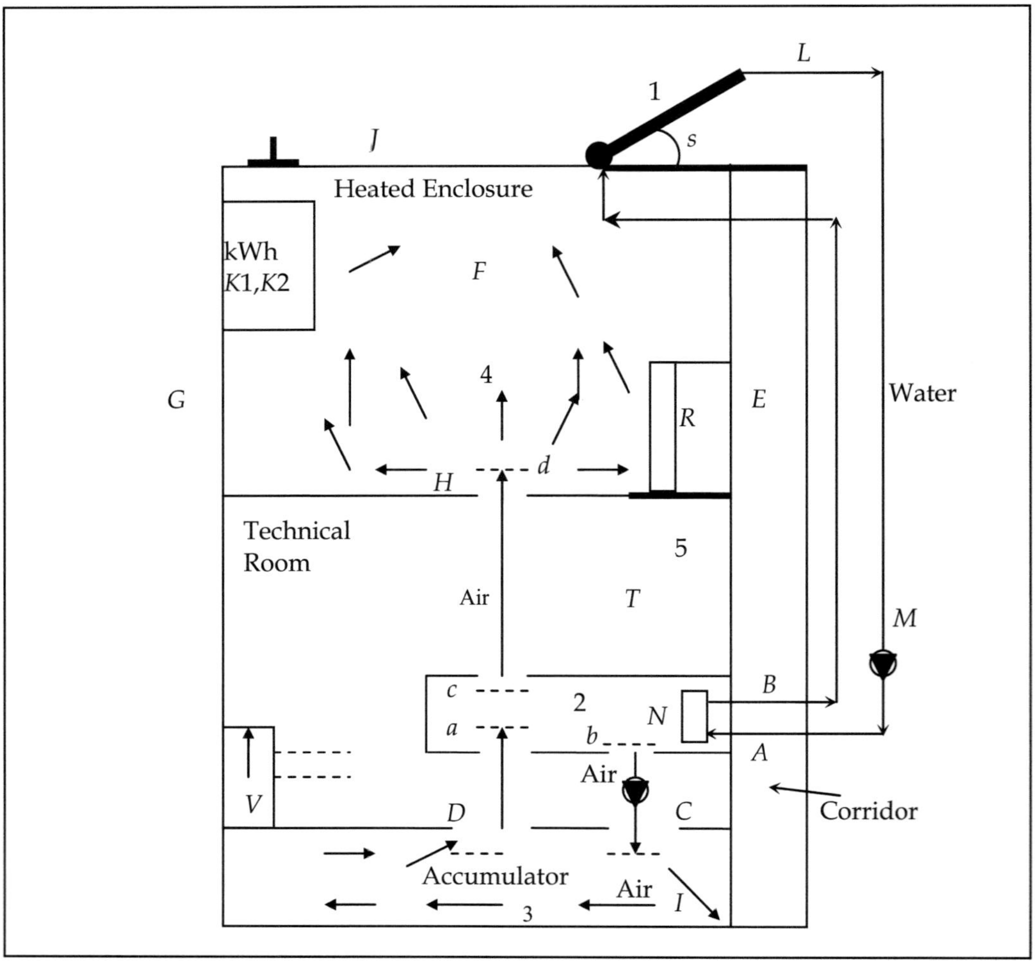

Fig. 13. Simplified chart of the energy system of the Solar House

sensors. The intensity of the solar radiation *G* is read on the pyrheliometer *J* with an error of $\pm 1\ \mathrm{W/m^2}$. The flow rate is obtained by dividing the volume recorded with the AEM BN5 water gauge, with an error of $\pm 25\ \mathrm{cm^3}$, at point *M*, by the recording period of time. The air velocity is measured with a FEET anemometer at point *N*, with an error of $\pm 0.5\ \mathrm{m/s}$, so that the air flow rate may be evaluated from $V_a = A_a v_a = 895\ \mathrm{m^3/h}$ (A_a is the area of the orifice of the nozzle). The electric energy used by the radiator *R* for heating is read on the AEM 1CM4A meter at point *K*1, with an error $\Delta Q_{el} = \pm 5 \times 10^{-3}\ \mathrm{kWh}$ and the energy used by the pumps is read on a similar meter at point *K*2. A βM135 temperature detector is mounted at point *L*. The detector triggers a control circuit that starts the pumps if the water at the collectors output has a temperature over 50ºC.

5.2 Analytic model for the solar house

The heat loss per time unit through walls, ceiling, and through window and door openings is given by (De Sabata & Luminosu, 1993)

$$\dot{Q}_1 = \sum_{i=1}^{2} m_i A_i \frac{\Delta T_i}{R_i} \tag{18}$$

where $\Delta T_1 = T_F - T_G$ and $\Delta T_2 = T_F - T_E$; m - thermal mass coefficient, $m_1 = 0.90$ for the walls and $m_2 = 1.2$ for the window and the door; A_i - the corresponding surface areas and R_i - global thermal resistances.

The heat per time unit required to warm up the air infiltrated through the shutters of the window and the door may be expressed as

$$\dot{Q}_2 = E(iL)v^{4/3} + \dot{Q}_{door} \tag{19}$$

where E=1 (first floor), i - air infiltration coefficient i=0.081 $Ws^{4/3}m^{1/3}K^{-1}$, L - lengths of the shutters, L_{door}=5.4 m, L_{window}=4.4 m; v - wind velocity, v=3.4 m/s (typical value).

The thermal resistance is given by

$$R = \frac{1}{\alpha_{int}} + \sum_{j=1}^{3} \frac{d_j}{k_j} + \frac{1}{\alpha_{ext}} \tag{20}$$

where $\alpha_{int,ext}$ - surface thermal exchange coefficients, $\alpha_{int} = 8\ \text{Wm}^{-2}\text{K}^{-1}$, $\alpha_{ext} = 22.8\ \text{Wm}^{-2}\text{K}^{-1}$; d_j - thicknesses of the successive layers of materials that forms the walls; k_j - heat conductivity of the layers [$Wm^{-1}K^{-1}$].

The heat loss per time unit for the room is the sum

$$\dot{Q}_L = \dot{Q}_1 + \dot{Q}_2 \, . \tag{21}$$

The hourly heat loss is $Q_{hL} = 3600\dot{Q}_L$ and the daily heat loss $Q_{dL} = \sum_{1}^{n} Q_{hL}$ (n - number of hours).

The heat lost by the room is compensated through solar and electric gains:

$$\overline{Q}_{dL} = \overline{Q}_{H \to F} + \overline{Q}_{el} \tag{22}$$

Hourly measurements have been carried out over several series of 3-4 days during spring (March, April, May) and autumn (September, October, November), 2000. In order to obtain average insolation characteristics, the experimental data have been statistically processed as described below.

The measurement period has been split into 12 h intervals, successively numbered 1, 2, ..., n; then, $n = n_1 + n_2$, n_1 - number of intervals with significant insolation, n_2 - number of intervals without solar radiation (night time and days with overcast sky). The hourly and daily average energy have been calculated with:

$$H_h = 3600 G_{hc} A_c, \; \overline{H}_h = \frac{1}{n_1} \sum_{n_1} H_h \tag{23}$$

$$\overline{H}_d = \sum_{i=1}^{p} \overline{H}_{hj} \tag{24}$$

p - number of 1 h intervals in an insolation day, $p = 1..8$.

The hourly average temperatures at points shown in Fig. 13 have been calculated using the equation

$$\overline{t}_{hq} = \frac{1}{n_{1,2}} \sum_{i=1}^{n_{1,2}} t_{hqi},\ q = A,B,C,D,E,F,G,H,I\ . \tag{25}$$

The elements of the energy system have been labeled as follows (Fig. 13): j=0 – collecting area; j=1 – collectors, between A and B; j=2 – heat exchanger, between C and D; j=3 – accumulator, between I and H; j=4 – room, between H and F. The hourly and daily average heat have been calculated for each segment using

$$\overline{Q}_{hj} = 3600 m_x C_x \Delta\overline{t}_{hj},\ \ \overline{Q}_{dj} = \sum_{(p)} \overline{Q}_{hj} \tag{26}$$

(e.g. $\Delta t_{h1} = \overline{t}_{hA} - \overline{t}_{hB}$); the subscript x identifies the nature of the fluid: $x = a$ – air, $x = w$ – water.
The average efficiencies of the successive links have been calculated with

$$\overline{\eta}_j = \frac{\overline{Q}_{d,j+1}}{\overline{Q}_{d,j}}\ . \tag{27}$$

For example, for the collectors we have $\overline{Q}_{d,0} = \overline{H}_d$, $\overline{\eta}_1 = \frac{\overline{Q}_{d,1}}{\overline{H}_d} = \frac{\sum_{1}^{8} 3600 m_w C_w \left(\overline{t}_{hA} - \overline{t}_{hB}\right)}{\overline{H}_d}$.

The average efficiency of the system is given by:

$$\overline{\eta}_{syst} = \prod_{j=1}^{3} \overline{\eta}_j\ . \tag{28}$$

5.3 Experimental results

The hourly variation of the quantity $\overline{H}_h$ versus hour of the average day is represented in Fig. 14 (Luminosu, 2003b).
The daily average of the radiant energy has been $\overline{H}_d = 389.8\ \mathrm{MJ}$. The average hourly temperatures at points A, B, C, D and I versus hour are represented in Fig. 15.
The average temperature at A, at noon has been 83°C. The highest temperature at A, i.e. 87°C, has been reached during May and September. During March and November, the same point has reached the lowest temperature, 61°C.
The maximum average temperature of the air in the heat exchanger has been of 52°C. The temperature of the accumulator has been carefuly maintained above 30°C all throughout the measurement period $\left(t_{min,st} = 30^\circ \mathrm{C}\right)$. The average increase in the temperature of the tank during the daily loading period has been $\Delta t = 11^\circ \mathrm{C/day}$. The average decrease in temperature during the extraction of heat from the bedrock has been of 4.5°C/day. The average temperature inside the heated room has been kept at $(20 \pm 1)^\circ \mathrm{C}$ for an ambient (exterior) temperature variation between 4 and 15°C. The average daily heat transferred by the collectors to the heat exchanger has been $\overline{Q}_{d,A\to B} = \overline{Q}_{d1} = 291.6\ \mathrm{MJ}$. An efficiency $\overline{\eta}_1 = 0.75$ for the collecting field has been obtained.

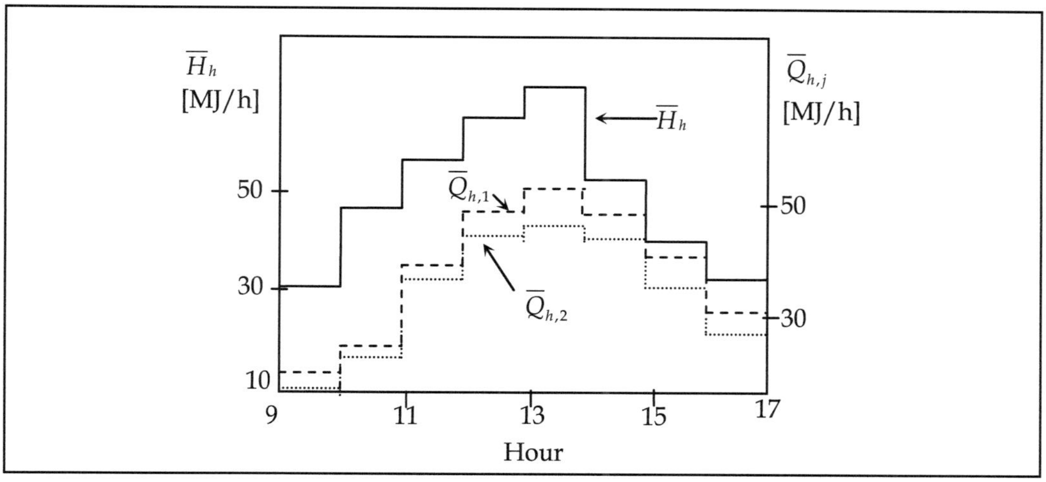

Fig. 14. Hourly averaged parameters $\overline{H}_h$, $\overline{Q}_{h,1}$ and $\overline{Q}_{h,2}$ versus hour.

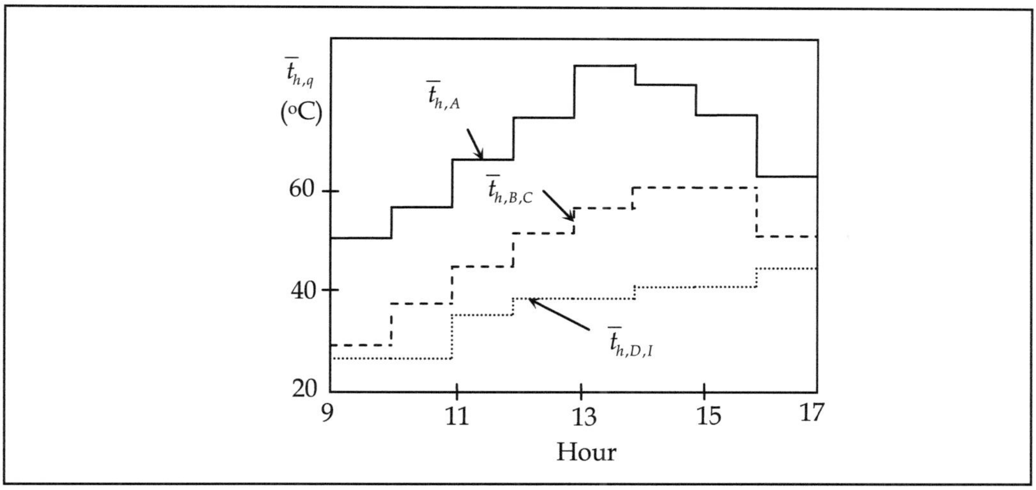

Fig. 15. Hourly averaged temperatures at points *A*, *B*, *C*, *D* and *I*.

The daily average heat swept away by the air curent from the coil conected between *A* and *B* (Fig. 13) has been $\overline{Q}_{d,D\to C} = \overline{Q}_{d2} = 239.4$ MJ, so that the average efficiency of the heat exchanger resulted as $\overline{\eta}_2 = 0.82$. The average quantity of heat transferred from the air current to the bedrock has been $\overline{Q}_{d,I\to H} = \overline{Q}_{d3} = 183.7$ MJ and the corresponding efficiency $\overline{\eta}_{3ld} = 0.77$. The room had a solar gain $\overline{Q}_{d,H\to F} = \overline{Q}_{d4} = 115.7$ MJ, so that the efficiency of the heat extraction from the storage environment resulted as $\overline{\eta}_{3ds} = 0.63$. The global efficiency of accumulation and storage of heat could then be calculated: $\overline{\eta}_3 = \overline{\eta}_{3ld}\overline{\eta}_{3ds} = 0.49$.
By using (28), one gets for the efficiency of the system $\overline{\eta}_{syst} = 0.30$.
The daily power consumption of the pumps is $\overline{Q}_{el,pumps} = 5.2$ MJ.
The average heat lost daily in the heated enclosure has been $\overline{Q}_{dl4} = 186.6$ MJ, which is compensated by solar energy $\overline{Q}_{d4}$ given above and by the energy provided by the electric radiator $\overline{Q}_{d,el,heat} = 70.9$ MJ. The solar energy ratio for room heating is

$$p = \frac{\bar{Q}_{d4} - \bar{Q}_{el,pump}}{Q_{dL4}} \times 100 = 59\% \ . \qquad (29)$$

5.4 Discussion

The solar system has an efficiency of 30% with respect to the incident solar energy. The thermal energy produced by the energy chain of the residence could provide 60% of the needs of the minimum loss enclosure. As the global efficiency is the product of individual ones, a possibility to increase the efficiency is o decrese the number of elements in the series conection.

A typical value for the southern side of the roof of an average residence is $A' = 40 \text{ m}^2$. This collecting area would give each year, at the location with solar conditions similar to those considered above, a quantity of heat as high as $Q_u' = \bar{q}_{yu} A' = 3977 \text{ kWh}$.

The present study might be extrapolated to thermal systems that do not contain heat exchangers. In this case, the water collector has to be replaced with air collectors. The hot air may be directed both towards the room and towards the thermal storage tank.

As a conclusion, the development of passive and active solar architecture in the Euroregion might be beneficial for both private residences and institutional buildings.

6. Thermal system for drying ceramic blocks

Solar collectors may be used with good results as complementary sources of heat in technological processes that take place at moderate thermal levels. Such applications lead to the reduction of conventional fuel consumption and have favourable impact on the environment.

Air solar collectors are used worldwide in complex installations for the climatization of buildings and for drying industrial and agricultural products. In the case of plane solar collectors with air and bedrock between the absorbing and transparent plates, the rocks in the current tube increase the turbulence of the air flow, so that the coefficient of thermal transfer and consequently the efficiency are also increased (Choudhury & Garg, 1993). Air solar collectors with thermosyphoning and rocks in the fluid current tube are used for heating social buildings during daytime (Lo et al., 1994). Solar installations optimized through exergetic analysis are used in Mexico for drying mango fruits (Torres-Reyes et al., 2001).

At the Physics Department from the "Politehnica" University of Timişoara, a thermal system with plane collectors designed for drying ceramic blocks has been realized. The system relies on hot air from the collectors during the daytime and on heat accumulated in water tanks in the night time.

6.1 Description of the system

The thermal system has been placed on the roof of an industrial hall belonging to the Plant for Ceramics Products from Jimbolia, near Timişoara. The hall was 12 m long and had a volume $V_h = 312.5 \text{ m}^3$, Fig. 16, (De Sabata et al., 1994).

A common practice for drying ceramic blocks relies on the Johnson burner with fuel oil. At the place, the power was P=770 kW. Hot gases resulted after the burning process are blowed with a ventilator over the drying hall.

Fig. 16. Hall for drying ceramic products.

The hall was divided into 10 corridors; each corridor contained $n_1 = 8000$ (hollow) bricks posed on mobile shelves. The drying process consists of removing a quantity of water $m_2 = 0.5\text{ kg}$ from each brick, such the humidity decreases below 5% (Luminosu, 1993).
The minimum quantity of heat needed for a drying cycle is $Q_{cycle} = 25 \times 10^{10}\text{ J}$. The average quantity of water that must be evacuated in a 10 day cycle is $M = 167\text{ kg/h}$. The variation in humidity of the air is $\Delta x = 5 \times 10^{-3}$ kg of water per kg of air. The air in the hall must be renewed N=9 times per hour. As the working temperature varies between 40 and 60°C, a fraction of the heat Q_{cycle} may be obtained by solar conversion.
The longitudinal axis of the hall was oriented in the E-W direction. On the south oriented roof, a plane solar collector with air has been posed. The collectors were tilted by an angle $s = 30\text{ deg}$ and the total collecting surface was $A_C = 600\text{ m}^2$.
The path of the air current is presented in Fig. 17. The air was blown with fans placed in each corridor, having a power of 100 W.
The quality of the ceramic products is determined by the uniformity of the drying process. Consequently, the storage of thermal energy of solar origin for subsequent use during periods without sun is important.

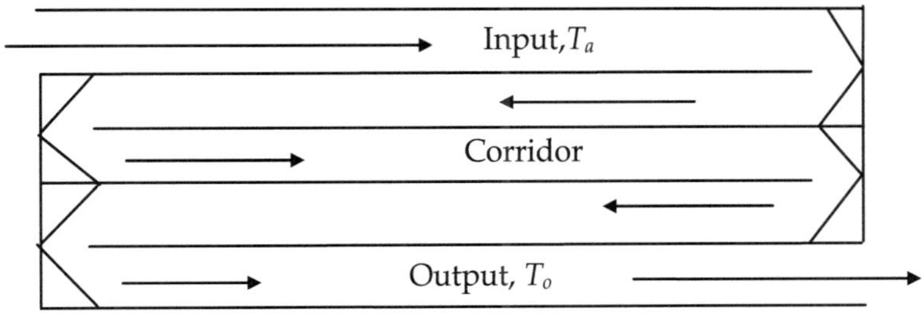

Fig. 17. Air current tube

A storage system of thermal energy as sensible heat has been designed to supply the requirements of the drying process during nighttime or for one or two days with low solar radiaton. Water collectors of type Sadu1 have been mounted on the roof of a neighboring hall, having a collecting area $A_C' = 360\text{ m}^2$. The average hourly specific power of the collectors has been $q = 2.09\times10^3\ \mathrm{kJ\cdot m^{-2}\cdot h^{-1}}$. The average hourly captured thermal energy has been $Q_h' = 7.524\times10^5\ \mathrm{kJ\cdot h^{-1}}$. The flow rate of the water through the storage installation has been $\dot{m}_w = 900\ \mathrm{kg/h}$. The thermal energy has been stored as sensible heat in a storage tank having a volume $V' = 54\ \mathrm{m^3}$. The hot water has been circulated with a pump at a flow rate of $800\ \mathrm{l/h}$ through radiators with horizontal pipes during nighttime. The air has been blown over the radiators by using fans and then the heated air was directed to the drying hall.

6.2 Experimental results

The drying procedure has been applied to ceramic blocks of dimensions $24\times12\times8$ cm. The density of the burnt material was 1300 kg/m^3.

Experiments at an industrial scale have been performed in the period April - September 1999. The drying process has been divided into cycles and it consisted of 3 cycles per month with a duration of 6..8 days/cycle. Several physical quantities have been monitorized: the solar radiation intensity G had a variation in the interval 460..920 W/m^2; the ambient temperature t_a (23..33°C); the air temperature at the hall entrance t_{in} (40..60°C); the water temperature at the output of the water collector t' (40..73°C); the flow rate of the working fluid v $(2.2..2.6)\times10^3\ \mathrm{m^3/h}$; the relative humidity of the air in the hall (30..35%).

The solar system has been used for heating the circulated air for 8 hours per day, in the interval 8 am – 4 pm. From the accumulation tank, heat has been extracted for time periods comprised between 8 and 16 hours per day. The Johnson burner has been used in parallel to the solar system in order to provide an air temperature at the input of the drying system of 40..60°C and the prescribed air humidity of 30..35%.

Cylindrical samples with radii of 2 cm and heights of 8 cm have been periodically extracted from the blocks. The samples have been weighted and compared with the burnt material in order to determine the mass of water from the ceramic block. The drying process was considered completed when the mass of water from the block was below 150 g.

The drying periods have been found as follows: n=6 days in June and July; n=7 days in May and August; n=8 days in April and September.

The following quantities have been determined:

a. the heat injected in the drying hall by the air collectors:

$$Q_{air} = \sum_{j=1}^{8} \dot{m}_{air} C_{air} \Delta t_j \Delta\tau \tag{30}$$

where $\Delta t_j = t_j - t_a$, t_j is the temperature of the air heated by the collectors and $\Delta\tau$ corresponds to the 8 hours interval when the air solar collectors were used;

b. the heat injected into the hall from the storage system:

$$Q_{storage} = \sum_{i=1}^{16} \dot{m}_{air} C_{air} \Delta t_i \Delta\tau \tag{31}$$

where $\Delta t_i = t_i - t_a$, t_i is the temperature of the air heated by the storage system and $\Delta\tau$ corresponds to the 16 hours interval when the storage system was used for heating;

c. the heat provided by the thermal solar system

$$Q_{syst} = Q_{air} + Q_{storage}\,; \tag{32}$$

d. the heat provided by the Johnson burner

$$Q_J = m_J q \tag{33}$$

where m_J is the mass and $q = 42\ \mathrm{MJ/kg}$ is the calorific power of the fuel oil;

e. the total heat used for heating the hall:

$$Q_{nec} = Q_{syst} + Q_J\,; \tag{34}$$

f. the total energy cosumed for the hall

$$W = Q_{nec} + W_{electric} \tag{35}$$

where $W_{electric}$ is the electric energy that could be read on a meter;

g. the fraction of heat of solar origin from the total energy used for the hall:

$$f = \frac{Q_{syst}}{W}\,. \tag{36}$$

h. The efficiencies of the solar installations have been calculated by dividing the heat they provided by the solar energy incident on the collecting surfaces.

The monthly averages of these quantities are presented in Table 3.

Month	April	May	June	July	August	September
n (days/cycle)	8	7	6	6	7	8
G (W/m^2)	741	833	864	849	780	656
$\langle Q_{syst} \rangle$ (GJ/cycle)	80	91	84	82	81	72
$\langle Q_J \rangle$ (GJ/cycle)	173	161	166	170	173	190
$\langle Q_{nec} \rangle$ (GJ/cycle)	253	252	250	252	254	262
$\langle W_{electric} \rangle$(GJ/cycle)	0.51	0.45	0.38	0.38	0.45	0.51
$\langle f \rangle$ (%)	32	36	34	33	32	27
η_{air} (%)	53	56	60	60	57	54
$\eta_{storage}$ (%)	34	37	40	41	38	35

Table 3. Monthly averaged quantities that characterize the drying process.

The results presented in Table 3 for one year show that the solar thermal system may provide approximately one third of the thermal energy needed for the process of industrial drying of ceramic blocks. The calculated efficiencies might change from year to year following solar radiation and weather variability.

Experiments revealed that the presented system provided a uniform distribution of temperature so that a reduction by 10% of the number of blocks broken during the burning process with respect with other drying systems used within the same plant resulted.

The energy chain could be built with inexpensive and readily available materials and parts, produced by the local industry.

7. Solar heater for bitumen melting

7.1 Experimental installation

The extension of the applications field of solar energy is possible by identifying new industrial activities for which the thermal solar conversion is appropriate, efficient and cheap. Low and medium temperature thermal solar installations (50-80°C) have the largest efficiencies (40-50%).

Bitumen has many applications in civil engineering industry and road and highway construction. In industry, bitumen is heated by classical means in a three-phase process: heating up to 50-65°C for melting; heating up to 100-125°C for the asphalt mixture; maintaining the thermal level during inactive periods.

The D80/100 bitumen used in road construction has the following physical properties: penetration at 25°C of 0.0085 m, a melting point at 47.5°C, a ductility at –25°C of 1.30 m and a density at 25°C of 1050 kg/m³. As the melting temperature is sufficiently low, it is possible to use low and medium temperature thermal solar installations in the first phase of the heating process.

At the present time, the literature on this subject is rare. At the Physics Department of the "Politehnica" University of Timişoara, an experimental setup has been devised in an outdoor laboratory in order to test the possibility of using solar energy for bitumen preheating (De Sabata & Nicoara, 1984; Mihalca & al., 1988; De Sabata, 1986c). The results have been encouraging, although the thermal conductivity of the bitumen $\lambda_{bi} = 0.174\ \mathrm{Wm^{-1}K^{-1}}$ is much smaller than the thermal conductivity of water $\lambda_w = 0.651\ \mathrm{Wm^{-1}K^{-1}}$ (at 60°C). Further research in this direction is still necessary in order to find the optimal solution.

The experimental installation is presented in Fig. 18. The elements are: cylinder of iron plate – 1; mechanical support for the envelope – 2; insulating support for the cylinder – 3; envelope made of glass plates – 4; thermometer – 5, indicating the temperature in the collector, T_c and the ambient temperature T_a; device for the variation of the tilt angle of the axis of the cylinder with respect to the horizontal – 6. The cylindrical tank has a length of 0.30 m, a diameter of 0.15 m a mass of 1.17 kg and it contains 6.4 kg of bitumen. The installation is facing south and the axis of the tank is tilted by an angle of 30 deg with respect to the horizontal (Luminosu & al., 2007b).

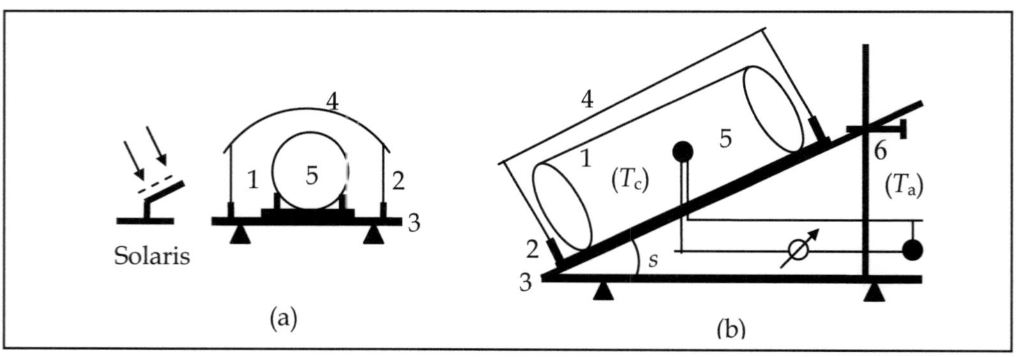

Fig. 18. Installation with semicylindrical glass envelope; (a) front view; (b) side view.

The results reported in Table 4 below have been obtained by measurements performed in 2003. The following quantities are considered: $t_{bi,k,av}$ – the hourly average temperature of the

bitumen at hour k; $t_{a,k,av}$ - the ambient hourly average temperature at hour k; $G_{k,av}$ – average hourly irradiance; $G_{d,av}$ - average daily irradiance; p – specific power; η – efficiency.
The experimental results show that the bitumen may be heated by solar means up to a temperature of 50-65°C. The thermal field in the bitumen mass is influenced by the solar radiation, the geometry of the installation and the ambient. The achieved efficiency of the laboratory installation for bitumen heating has been between 8.1 and 9.1%. The results have been favourable enough to suggest trying industrial applications.

Hourly interval $\Delta\tau$[h]	08-09	09-10	10-11	11-12	12-13	13-14	14-15	15-16	16-17	17-18	18-19
$t_{bi,k,av}$ [°C]	21,0	24,5	30,0	36,0	42,5	47,5	50,5	54,0	56,5	56,5	53,5
$t_{a,k,av}$ [°C]	18,5	19,5	21,0	24,5	26,5	28,5	30,0	32,5	32,5	31,0	28,5
$G_{k,av}$ [W/m²]	344	438	760	863	978	960	747	684	431	386	297
p [W/m²]	65,6										
$G_{d,av}$ [W/m²]	721										
η [%]	9,1										

Table 4. Hourly values of the quantities $t_{bi,k,AV}$, $t_{a,k,AV}$, $G_{k,AV}$, $I_{d,AV}$, p and η.

7.2 Industrial thermal solar system for bitumen preheating

The diagram of the solar system for bitumen preheating that has been realized at Săcălaz, near Timişoara, in cooperation with the Roads and Highways Regional Direction is presented in Fig. 19 (Luminosu et al., 2007b).
The solar installation has been placed on an existing construction. The elements in Fig. 19 are: solar collector, with a surface of 300 m²; roof made of iron plates – 2; pipes penetrating the bitumen – 3; compartment with bitumen preheated at 90-100°C - 4; heat exchanger with oil – 5; tank for bitumen heating at 100-150°C - 6; metallic meshes distanced by 0.5 m (mounted in order to homogenize the temperature in the solar trap) – 7; thermometers – 8; fire place – 9. An iron plate, having a thickness of 0.75 mm is placed between the glass plate

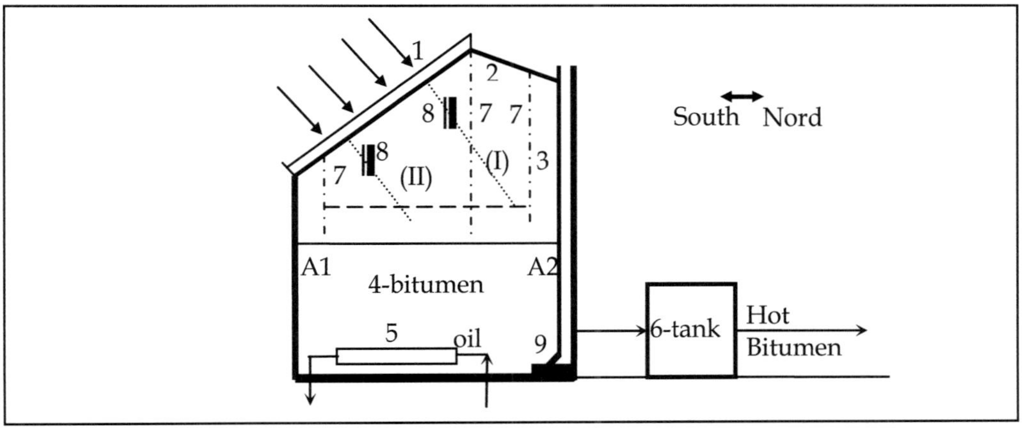

Fig. 19. Diagram of the industrial installation for bitumen preheating.

and the surface of the bitumen. The solar installation accomplishes the bitumen heating up to 50-55°C, with the favourable consequence of saving conventional fuel.
Financing conditions allowed only for preliminary measurements. The temperature has been measured in the volume in between the surface of the bitumen and the roof (the solar trap). We present as examples, in Table 5, the hourly averages of the temperatures of the bitumen t_{Bi} and ambient t_a.
The maximum average temperature of the bitumen, 54-56°C has been obtained around 14h30. In the daytime when measurements have been performed, the maximum average temperature in the solar trap has been larger than the ambient temperature by 27°C. It has been evaluated a saving of approximately 80 kg conventional fuel for 1 m² collecting surface per year. A further saving of fuel is obtained if the bitumen extraction is made around 4-5 pm from the upper portion of the tank.

Hour	9h30min	10h30min	12h30min	14h30min	16h30min	18h30min
$\langle t_a \rangle$ [°C]	27,5	28,5	34,0	35,0	33,5	31,0
$\langle t_{Bi} \rangle$ [°C]	38.0	47,5	55,5	56,5	55,0	52,5

Table 5. Average temperatures in the solar trap.

8. Conclusion

Research in solar energy has been approached at the "Politehnica" University of Timişoara in 1976, motivated by economical and ecological problems related to classical fuels.
Solar collectors have been conceived and realized and several thermal solar installations for producing hot air and water have been devised and applied in industry. Solar energy technology has also been applied to waste water cleaning and to building climatization. A part of this experience has been presented in this chapter. The installations have been realized and tested in Timişoara, Romania. The obtained results are relevant for the south-eastern part of Europe.
The experimentally determined efficiencies of the solar installations have been comparable with efficiencies of similar installations produced in other European countries. This proves the possibility of implementing solar energy applications in the region based on the local industry and on locally devised solutions. However, a further involvement of the local industry in the field of solar energy in particular and of renewable energy in general, as well as the education of the population in this spirit are actions to be considered in the near future.

9. References

Athanasouli, G. & Massouporos, P. (1999). A model of the thermal restoration transient state of an opaque wall after the interruption of solar radiation. *Solar Energy*, Vol. 66, No. 1, (May 1999) pp. (21-31), ISSN 0038-092X.

Chouhury, C. & Garg. H. P. (1993). Performance of air-heating collectors with packed air-flow passage. *Solar Energy*, Vol. 50, No. 3, (March 1993), pp. (205-221), ISSN 0038-092X.

De Beni, G.; Friesen, R. & Olmo, M. (1994). Utilization of solar thermal energy in mountain refuges through an innovative system. *Solar Energy*, Vol. 52, No. 2, (February, 1994), pp. (221-224), ISSN 0038-092X.

De Sabata C.; Marcu, C. & Luminosu, I. (1994). Some industrial utilization of solar energy in South-West Romania. *Renewable Energy*. Vol. 5, No. 1-4, (August 1994), pp. (387-389), ISSN 0960-1481.

De Sabata, C. & Luminosu, I. (1993). Complex experimental base for the study of heat conversion and solar energy storage. *Solar Energy in Romania*. Vol. 2, No. 1-2, (July 1993), pp. (115-116).

De Sabata, C.; Luminosu, I.; Mihalca, I. & Ercuţa, A. (1986a). Thermal phenomena in the active part of some T-M wall models. *Scientific Bulletin of the "Politehnica" University of Timişoara, Trans. Math. - Phys.* Vol. 31(45), No. 1, (May, 1986), pp. (81-84), ISSN 1224-6069.

De Sabata, C.; Luminosu, I.; Ercuţa. A. & Baea, R. (1986b). Experimental study on a Trombe wall efficiency concerning dwelling climatization. *Scientific Bulletin of the "Politehnica" University of Timişoara, Trans. Math. - Phys.* Vol. 31(45), No. 1, (May, 1986), pp. (117-120), ISSN 1224-6069.

De Sabata, C. (1986c). Contributions to the use of nonconventional energy in construction of roads and highways; *Ptischea (Roads)*, Vol. 25, No. 2, (February, 1986), pp. (1-4), ISSN n/a (in Bulgarian).

De Sabata, C. & Nicoară, L. (1984). Use of solar energy for bitumen preheating in tanks. *Revue générale des routes et des aerodromes*. No. 614, (December, 1984), pp. (69-71), ISSN 0035-3191 (in French).

De Sabata, C.; Marcu, C. & Luminosu, I. (1983). On the energy flow rate in a plane solar collector that produces industrial hot air. *Scientific Bulletin of the "Politehnica" University of Timişoara, Trans. Math. - Phys.* Vol. 28(42), No. 2, (November, 1983), pp. (99-102), ISSN 1224-6069 (in Romanian).

De Sabata, C.; Mihailovici, D.; Baea, R.; Luminosu, I. & Gangăl, M. (1981). Use of solar energy for bitumen heating in high capacity cylindrical tanks. *Scientific Bulletin of the "Politehnica" University of Timişoara, Trans. Math.-Physics*. Vol. 26(40), No. 2, (November, 1981), pp. (65-72), ISSN 1224-6069 (in Romanian).

Folkerts, L.; van Orshoven, D.; Pavic, D. & Mack, M. (1996). A new design tool for collectors with transparent insulation (TIM), *Proceedings of EuroSun '96*, Book 1, pp. 171-175, ISBN n/a, Frieburg, October 1996, DGS Sonenenergie GmbH Pub., Frieburg

Haberl, R.; Frank, E. &Vogelsanger, P. (2008). Cost/Benefit ratio analysis of a maximum lean solar combisystem. *Proceedings of EuroSun 2008*. Part: Domestic and Services Water Heating, (USB - 8 pages), ISBN n/a, Lisabona, (October 2008), ISES Pub., Lisabona

Çomakli, Ö.; Kaygusuz, K. & Ayhan, T. (1993). Assisted heat pump and energy storage for residential heating. *Solar Energy*. Vol. 51, No. 5, (November 1993), pp. (357-366), ISSN 0038-092X

Lo, S., N., G.; Deal, C., R. & Norton, B. (1994). A school building reclad with thermosyphoning air panels. *Solar Energy*. Vol. 52, No. 1, (January 1994), pp. (49-58), ISSN 0038-092X

Loveday, D., L. & Craggs, C. (1992). Stochastic modelling of temperatures affecting the in situ performance of a solar - assisted heat pump: the univariate approach. *Solar Energy*. Vol. 49. No. 4, (October 1992), pp. (279-287), ISSN 0038-092X

Luminosu, I.; De Sabata, C. & De Sabata, A. (2010). Research in Solar Energy at the "Politehnica" University of Timişoara". *Thermal Science.* Vol. 14, No. 1 , (January 2010), pp. (157-169), ISSN 0354-9836.

Luminosu, I. (2007a). *Thermal Phenomena and Applications of Thermal Solar Conversion.* "Politehnica" Pub., ISBN 978-973-625-423-9, Timişoara (in Romanian).

Luminosu, I.; De Sabata, C. & But, A. (2007b). Solar equipment for preheating bitumen. *Thermal Science.* Vol. 11, No. 1, (January 2007), pp. (127-136), ISSN 0354-9836.

Luminosu, I. (2003a). Experimental studies and economic considerations on a living space heated through a passive solar gain and through electric power. *Thermal Science.* Vol. 7, No. 1, (January 2003), pp. (47-61), ISSN 0354-9836.

Luminosu, I. (1993). Raising of plane solar collectors efficiencies by study of physical phenomena that occur in thermal conversion. *PhD Thesis,* "Politehnica" University of Timişoara, 1993.

Luminosu., I. (1983). Study upon temperature performance of a flat plane solar collector in view of obtaining technological hot air. *Scientific Bulletin of the "Politehnica" University of Timişoara, Trans. Math. - Phys.* Vol. 28(42), No. 1, (May, 1983), pp. (79-82), ISSN 1224-6069.

Mihalca, I.; Luminosu, I.; Ercuţa, A. & Damian, I. (1988). Thermal field in solar energy heated bitumen mass. *Scientific Bulletin of the "Politehnica" University of Timişoara, Trans. Math. - Phys.* Vol. 28(42), No. 1, (May, 1988), pp. (116-118), ISSN 1224-6069.

Ohanession, P. & Charteres, W. (1978). Thermal simulation of a passive solar house using a Trombe Michelle wall structure. *Solar Energy.* Vol. 20, No. 3, (March 1978), pp. (275-281), ISSN 0038-092X

Pedersen, P., D. (1993). Experience with a large solar DHW system in Denmark – The Nordic Solar Heating Demonstration Project. *Solar Energy.* Vol. 50, No. 3, (March 1993), pp. (259-266), ISSN 0038-092X

Rommel, M.; Kohl, M.; Graf, W.; Brucker, F. & Lustig, K. (1996), Development of flat-plate collectors with selectively coated polymer. *Proceedings of EuroSun '96,* Book 1, pp. 330-335, ISBN n/a, Frieburg, October 1996, DGS Sonenenergie GmbH Pub., Frieburg.

Săvulescu, T., D. (1984). *Ventilation and Heating Installations.* "Ed. Tehnică" Pub., ISBN n/a, Bucureşti (in Romanian).

Stahl, W.; Voss, K. & Goetzberger, A. (1994). The self-sufficient solar house in Frieburg. *Solar Energy,* Vol. 52, No. 1, (January 1994), pp. (111-127), ISSN 0038-092X

Torres-Reyes, E.; Cervantes de Gortari, J., G.; Ibarra-Salazar, B., A. & Pico-Nunez, A. (2001). A design method of flat-plate solar collectors based on minimum entropy generation. *Exergy Int. J.,* Vol. 1, No. 1, (January 2001), pp. (46-52), ISSN 1742-8297

19

Thermal Performance of Photovoltaic Systems Integrated in Buildings

D. Bigot, F. Miranville, A. H. Fakra, I. Ingar, S. Guichard and H. Boyer
University of Reunion
France

1. Introduction

1.1 History of photovoltaic systems ...

Photovoltaics is one of the leading chains of "sustainable development". Indeed, when one observes the development programs of energy systems in the countries or nations that move towards sustainable development, we find that the solar (and through it the production of energy through photovoltaics) represents the main axis of development.

One might at first believe that knowledge of the photovoltaic effect is recent. In fact, we must go back to 1839 with the French physicist Edmund Becquerel who first discovered the photovoltaic effect. It was during the period between the second half of the 19th and the Second World War (1945) that scientific knowledge related to solar phenomena were mastered. Thus, in 1875, Werner von Siemens presented to the Academy of Sciences in Berlin an article on the photovoltaic effect in semiconductors and it was Albert Einstein who first was able to explain the photovoltaic principle, thereby won the Nobel Prize for Physics in 1923.

After the Second World War, when the world gets in another war called "cold war" between the East Block in the West Block, the simmering conflict reached its apogee in the arms race and especially in the space conquest. The space industry is now rapidly finding new and innovative solutions that would power satellites into space. This was a boon for the photovoltaic sector and will help structure an industry.

Thus, in 1954, with the developed of a high efficiency photovoltaic cell for the time (6%) and in 1958, the rise of the yield to 9% and above, VANGUARD, the first satellite equipped with photovoltaic cells was sent to the space.

The oil shocks of the 1970s allowed the industry to begin its development in civilian applications in 1973 with the construction of the first house powered by solar cells at the University of Delaware. The next step was the construction of the first car equipped with a photovoltaic energy, which in 1983 covered a distance of 4000 km in Australia.

Yet in 1980, while the industry is launched commercially, the following years have seen its development focus mainly on rural electrification as well as some isolated houses for professional use (refuges, measuring stations, etc.) and for many villages in developed countries.

Since 1990, awareness of the phenomenon of global warming induced the development of the concept of sustainable development, with effect of boosting the photovoltaic and allows it to pass a critical level.

With the advent of power electronics, the use of PV systems connected to the network exploded in 2007, to represent over 90% of PV capacity installed. Then programs of grid-connected photovoltaic roofs appeared in 1995 in Japan and Germany, with a generalization from 2001.

1.2 ... Integrated in buildings

Integrating the frame represents the substitution of traditional building elements of homes and buildings with PV systems. This type of system offers the advantage of improving the profitability of a construction project from the substitution by the photovoltaic modules of traditional materials or equipment.
This may seem surprising at first, because of the cost of a photovoltaic system, but in fact, given the fact that the cost of redeeming a photovoltaic system integrated into the frame is low, the overall calculation of a construction project it is in fact improved.
The types of integration are built:

- Sloped roofs
    - Photovoltaic tiles
    - Systems integration of conventional modules
    - PV steel bins
    - PV membranes (in some cases)
- Flat roofs
    - PV membranes
    - Photovoltaic steel trays (in some cases)
- Facades
    - Conventional modules mounted on a metal specific structure
    - Systems integration of conventional modules (for some systems)
- Sun visors
    - Conventional modules mounted on a specific metal structure
- Windows
    - Semi transparent modules, translucent or only on specific structure

In France, the systems integrated into buildings can receive a higher purchase price with the "premium built integration". This bonus is awarded when the systems met the eligibility criteria.
The base rates and the premium rates are adjusted each year so that by 2009, rates for purchases are as follows:

Metropolitan France	Corsica & Overseas French territories
Basic price : 0.32823 € / kWh Grant for integration: € 0.27353 / kWh Integrated tariff: € 0.60176 / kWh	Basic price: € 0.43764 / kWh Grant for integration: € 0.16412 / kWh Integrated tariff: € 0.60176 / kWh

Table 1. Rates of photovoltaic energy in France in 2009

The solar integrated building is a French particularity.
The architectural integration of photovoltaics promotes a healthier growth of the sector, because avoids speculative effects related to the explosion of large solar parks.
Since July 2006, France has thus feed-in tariffs of PV highest in the world with a significant premium to the frame integration, thereby stimulating the market to maturity.

Financial incentive in addition to the context of fiscal incentive (tax credit, VAT at 5.5%, aid communities) and regulatory objectives are clear and ambitious (23% renewable by 2020, nine buildings positive energy around 2020).
Investment in building integrated photovoltaics is now a 20-year guaranteed investment, which generates an average return rate of 8 to 10%.

1.3 From global warming ...

Global Warming is one of the most controversial sciences from the 21 st century, challenging the very structure of our global society. The problem is that global warming is not just a scientific concern, but encompasses economics, sociology, geopolitics, local politics, and individuals' choice of lifestyle. Global warming is caused by the massive increase of greenhouse gases, such as carbon dioxide, in the atmosphere, resulting from the burning of fossils fuels and deforestation. There is clear evidence that we have already elevated concentrations of atmospheric carbon dioxide to their highest level for the last half million years and maybe even longer. Scientists believe that is causing the Earth to warm faster than at any other time during, at the very least, the past one thousand years. The most recent report by the Intergovernmental Panel on Climate Change (IPCC) states that there is a clear evidence for 0.6°C rise in global temperature and 20 cm in sea level during the 20 th century. The IPCC also predicts that the global temperatures could rise by between 1.4°C and 5.8°C and sea level could rise by between 20 cm and 88 cm by the year 2100.

1.4 ... to energetic optimization of buildings

The electrical energy consumed in a building is divided on such items as lighting, equipments (fridge, Hi-Fi, TV ,...) production of hot water, electric heating, air conditioning ..., but the position still remains the majority consumer of heating or cooling (depending on whether it is summer or winter). However, the introduction of PV plant on the roofs of buildings can impact the use of air conditioning because the plant acts as a double skin on the roof. Thus, there is a direct relationship between PV installation on a building with its power consumption, because of the central role played by the insulation in reducing the operating time of air conditioning, so the power consumption of the building. The challenge is thus to be able to model the coupling PV production - profile of energy consumption in a building to control the operation of energy consuming equipments that they operated during hours of plant output PV. This is the approach that, combined with a master's program of energy in the building, ultimately achieves what is called a positive energy building, meaning a building that produces, much less energy than it consumes.

2. Methodology

2.1 Context and objectives

The purpose of this study is to establish a physical model able to describe, transiently, the thermal phenomena taking place in a Building Integrated Photovoltaic systems (BIPV) or Building Added Photovoltaic system namely BAPV (see Fig 1). The model should be able to take into account the different positions of photovoltaic panels (inclined, horizontal and vertical) that are integrated into the architecture of the building (roof, facades, awnings, closeroofs, etc.).

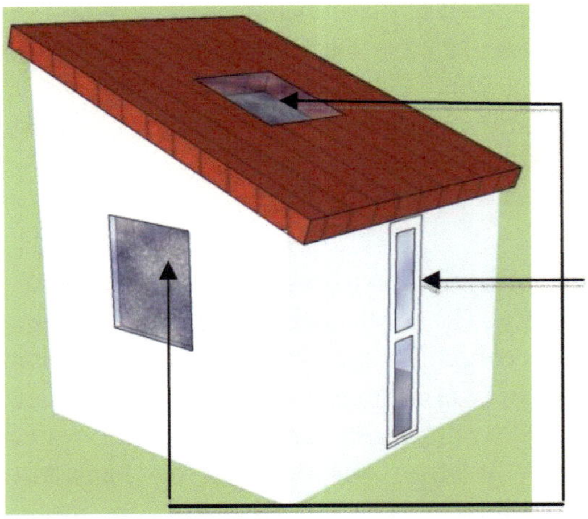

Fig. 1. Illustration of BIPV and BAPV

Moreover, it is important for the results to be validated, in order to allow their use when dealing with studies about renewable energy and building energy consumption and optimisation. For this, worldwide recognized procedures exist and can be applied to implemented models to test their general behaviour under given conditions.

2.2 Literature review

In the early 1990s, BIPV started to become more important. Many studies have been conducted to describe the evolution of the temperature of BIPV products and systems. These studies were conducted in different fields of scientific research:

- The development of physical models ;
- The design of experimental devices ;
- Finally, the conception of softwares capable to give the evolution of this temperature in the space and the time. Other research has been conducted to ensure the reliability of the software to simulate correctly the temperature.

This paragraph presents an inventory of all the work developed in the field of BIPV.

2.2.1 Existing models

In 2001, Zondag did a numerical study on the BIPV roof (Zondag, 2001). Kropf (Kropf, 2003) developed a simulation model for dynamical calculation of the heat gain of BIPVT. Many models have been conducted as for BIPV-Thermal system included facade- integrated and roof-integrated PVT (Photovoltaic and Thermal collector), system ventilation BIPV (Zondag, 2008); (Bazilian & Prasad, 2001); (Bazilian, 2002.); (Kondratenko, 2003.). Generally, there are three models types for characterize dynamic and steady state aspects: experimental, numerical (3D, 2D and 1D), and analytical models of performance of BIPV systems.

In 2006, Wang (Wang et al, 2006) compared four thermal models, one-dimensional, transient, for different roofs to evaluate the impacts of BIPV on the building's heating-and-cooling loads: ventilated air-gap BIPV, non-ventilated (closed) air-gap BIPV, close roof mounted BIPV, and conventional roof with no PV and no air gap. An objective of this study was to evaluate the photovoltaic performances and building cooling-and-heating loads across the different roofs in order to select the appropriate roof BIPV system.

Jie (Jie et al, 2007) developed a two-dimensional thermal model of PV glass panel and model of the PV-Trombe wall system. These models can also be used to predict the temperature distribution of a room at any time.
Guiavarch and Peuportier (Guiavarch & Peuportier, 2006) calculated the thermal yield of a BIPVT (Building Integrated Photovoltaic and thermal collector system) for climate of Paris and Nice (in France).
Tian (Tian et al., 2007) introduced a PTEBU model. This model described the thermal influence of BIPV on microclimate of urban canopy layer.
In 2008, Fung (Fung & Yang, 2008) presented the Semi-transparent Photovoltaic module Heat Gain (SPVHG) model for evaluating the heat gain of semi-transparent photovoltaic modules of BIPV applications on one-dimensional transient heat transfer.
In the same year, Chow (Chow et al., 2007) have taken over the work of Jiméner and presented a numerical modelling (based on the multi-node scheme) of a building integrated photovoltaic and water-heating (BIPVW) system. The authors have gathered the various existing models in the field of thermal building, photovoltaic and thermal collector system in a single dynamic model of the BIPVW system.
In 2008, Jiménez (Jiménez et al., 2008) worked on the linear and non linear continuous time modelling of physical systems using discrete time data (stochastic models), in particularly, on BIPV systems. Continuous-discrete stochastic state space model consists of a set of Stochastic Differential Equations (SDE's) describing the dynamics of the system in continuous time and a set of discrete time measurement equations.
In 2009, Skoplaki (Skoplaski, & Palyvos, 2009) established the important role of the operating temperature in relation to electrical efficiency of a BIPV array. Authors developed an implicit correlation for the PV operating temperature and compared the empirical modelling with other implicit and explicit equations for Tc (cell/module operating Temperature) in the scientific literature review. The author recommends being careful in applying a particular expression to BIPV installation because the available equations have been developed with a specific mounted geometry or building integration level in mind.
In 2009, Nynne (Nynne et al., 2009) presented a new mathematical modelling of the heat transfer of BIPV modules for a stochastic non-linear physical system. This model takes into account ambient wind velocity and the PV module temperature fluctuations.
In 2010, Steven and Benjamin (Steven & Benjamin, 2010) developed a Finite Element Model (F.E.M) applied into a double pane glazing system. The model is capable to study the thermal and electrical performance for an opaque Active Thermal Insulator (A.T.I.) glazing system. A.T.I.- systems represent a new thermal control technology that uses solar energy to compensate for passive heat losses or gains in building envelopes.

2.2.2 Experiment tests cases and validations

Many Scientific studies were conducted, in the world, on BIPV and BAPV systems. Most significant and recent studies that were treated in the thermal aspect of BIPV systems are as follows.
In 2001, Cherruault (Cherruault & Wheldon, 2001) evaluated the performance of BIPV system installed in the refurbishment of the roof at the University of Reading in UK. The report contained many references of thermal measurement database of the BIPV.

In 2008, Xu (Xu & Dessel, 2008) worked on the technology of Active Building Envelope (ABE) in particularly, the experimental ABE window-system for the testing room in USA. Test set up and protocol of measuring errors of BIPV was described.
Trinuruk developed other experimental test cases in 2009 (Trinuruk et al., 2009). Indeed, Authors developed Scale Models that represented structure envelopes of building in Thailand and integrated BIPV systems (Photovoltaic in the façade of the wall). Experimental facility and measuring equipment set up was presented in this work. The PV module installed can be positioned with different inclination angles relative to the test room (Scale Model).
Bigot has worked on Isotest cells (scale models tests rooms) and compared the experimental thermal performance of BIPV in tropical region. Results of experimental measurements have shown that building added photovoltaic systems might reduce the temperature inside the building around 3-6 degrees depending on the chosen configurations (Bigot et al. 2009).
In 2009, Mei (Mei et al., 2009) presented the results of a laboratory based experimental investigation undertaken to determine the potential for high temperature operation in such a BIPV installation in UK.
In this same year, Park (Park et al., 2010) studied the thermal performance of a semi-transparent PV module that was designed as a glazing component. The experiment was performed under both Standard Test Condition (STC) and outdoor conditions.

2.3 A combined approach (M.E.V.)

To determine both the thermal behaviour and the performances of BIPV and BAPV installed according to the state of the art, it is important to combine three parts: Modelling, Experimentation and Validation (what we'll indicate as M.E.V.). The first one consists of the detailed modelling of the energetic system composed of the whole building, including the added PV panels. For this, it is necessary to define the correct level of description in order to be able to use the results during practical studies. This part is illustrated on the left side of fig 2.
The detailed model then goes through a validation step, including comparisons with measurements and also two important analyses: parameters sensitivity and optimization. At the end of the process, the model can be used to determine, in several conditions, the general thermal behaviour of the building as well as its performances.
The corresponding steps indicated on the right side of fig 2., concern the experimental part. Indeed, to be able to run the validation process, it is necessary to have at least one measurement database, for comparisons. Such data have to be of high quality and correspond to realistic conditions. It also has to be constituted such that direct comparisons of equivalent variables in both the model and the experimentation can be done. Parameters as time step of measurements, location of the sensors, errors calculations of the whole data acquisition system have to be determined to ensure the accuracy of the results.
The last part combine the two preceding ones, in the sense that it uses the predictions of the model, and also the measurements included in the database. The validation step follows a worldwide known procedure, known as the BESTEST procedure (Judkoff et al., 1995). It consists of verifying the calculation code of the model, through analytical tests and inter-software comparisons, and also of validating the model, through predictions/measurements comparisons and also a detailed analysis of the model, with sensitivity analysis, optimisation and corroboration.

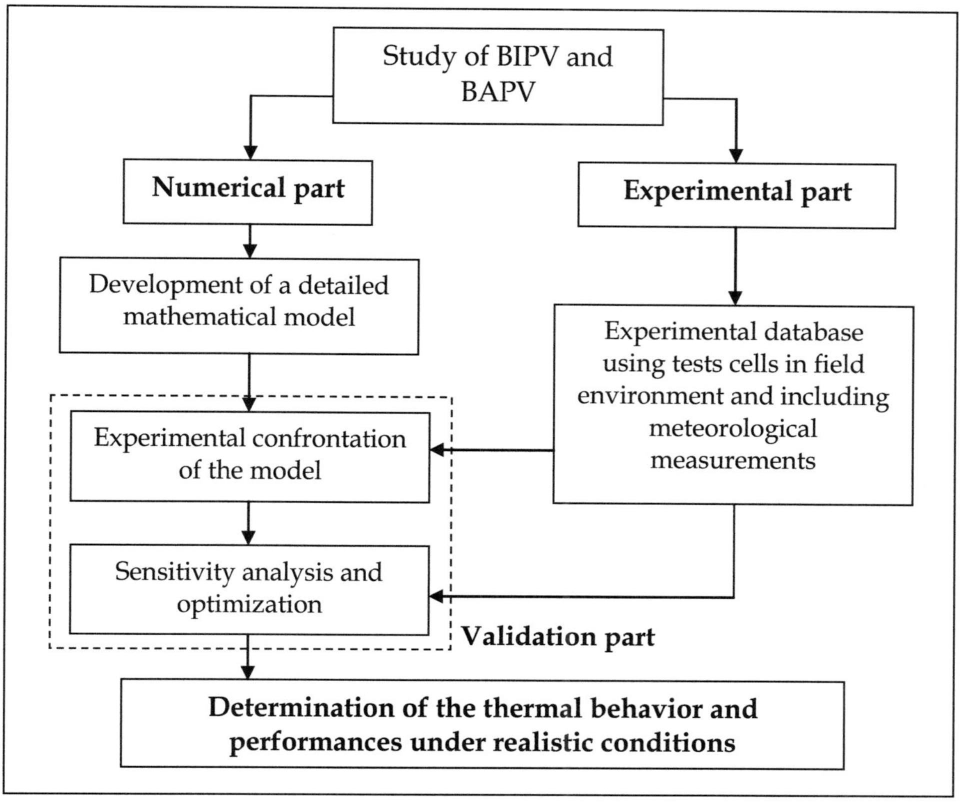

Fig. 2. General overview of the methodology

2.4 Numerical and experimental tools

To apply the above methodology, numerical and experimental tools are needed. In our case, they have been totally developed and dedicated to the present study and constitute an original contribution to international studies about complex walls, especially including PV systems. Many publications have involved these tools, for example (Miranville, 2003) and (Bigot, 2009).

The numerical code used to predict the thermal response of the whole building envelope is part of the thermo-hygro-aeraulic simulation codes and is based on a multizone description of the physical system (here composed of the building and its very specific wall with PV). Specifics developments have been done to allow the correct modelling of the system, with a very special focus on radiative exchanges in semi-transparent layers. The corresponding model is described further and constitutes the main addition to the building simulation code that is necessary for predicting the temperature field.

In terms of experimental equipment, a dedicated platform has been set up, build in field environment, constituting a unique case for the French overseas departments. It is composed of several test cells, as it will be described further, allowing the collection of experimental databases, needed for comparisons with code predictions. Combining the two tools give a powerful mean to analyse the adequacy between models and measurements and thus go further in the knowledge about building physics.

2.5 Performance indicators

Once a model is validated, it can be used to evaluate the thermal performance of the building; if the aim of the study is to calculate the thermal performance of a wall, several performance indicators can be used:

- The R-value
- The percentage of reduction of the heat flux

The R-value is the most known performance indicator for walls, as it is part of heat transfer theory, in particular for steady state conditions. In field environment, with measurements, it is possible to calculate the R-value, using dynamic values. The used method to reach this objective is called the average method and is well known among performance materials researchers. Restrictions for the obtaining of correct values are imposed. If well used, it is possible to determine a R-value which is very near from the indicator in steady-state conditions.

The average method is precisely described in (ISO-9869, 1994) and is based on an evaluation of the thermal resistance R of a wall with the following mathematical expression:

$$R = \frac{\sum_{i=1}^{n}(T_{se,i} - T_{si,i})}{\sum_{i=1}^{n}\varphi_i} \quad [m^2.K / W]$$

With:

$T_{se,i}$: outer surface temperature of the wall [K]

$T_{si,i}$: inner surface temperature of the wall [K]

φ_i : heat flux density through the wall [W/m²]

Another well-used indicator, when dealing with performance of complex walls, is the percentage of reduction of the heat flux. Its application requires comparative experimental or numerical studies, one set with the specific wall, another set equiped with a reference wall. The calculation is simply done according to the following equation:

$$\text{percent reduction} = \frac{\int_{\text{evaluation period}} \varphi_{\text{wall with PV}} \cdot dt - \int_{\text{evaluation period}} \varphi_{\text{wall without PV}} \cdot dt}{\int_{\text{evaluation period}} \varphi_{\text{wall without PV}} \cdot dt}$$

These two indicators are often used to demonstrate the thermal performance of building walls, and are usually evaluated in the post-processing step of models results.

3. Modelling of Building Integrated PV (BIPV)

3.1 Physical and structural description

In this study, interest has focused on photovoltaic systems installed on buildings. Specifically, on systems that are installed on the walls of a building, either in front or on the roof. Such systems are generally integrated into the architecture of the building; they are designated by the term "BIPV" i.e. "Building Integrated Photovoltaics". These systems can be installed on the roof of a building, like sun protection in front, in walls, Trombe walls, or embedded in glass windows.

In this context, and in order to approach the building simulation code that will be subsequently used, it was decided to consider these systems as a particular type of wall. The walls of a building are generally opaque except glasses of windows. So the photovoltaic wall system has been considered like an assembly of the photovoltaic panel and the wall that supports it.

The characteristic of a photovoltaic system, compared to other types of walls encountered in a building, is that a part of its component layers is semitransparent. Semitransparent layers are mainly those of the panel that produce electricity. These layers form an assembly of materials, generally glass, and the silicium under it (or other semiconductor material that can produce electricity when exposed to radiation). In addition, silicium is typically encapsulated in two layers of material in order to ensure mechanical protection (see Fig 3).

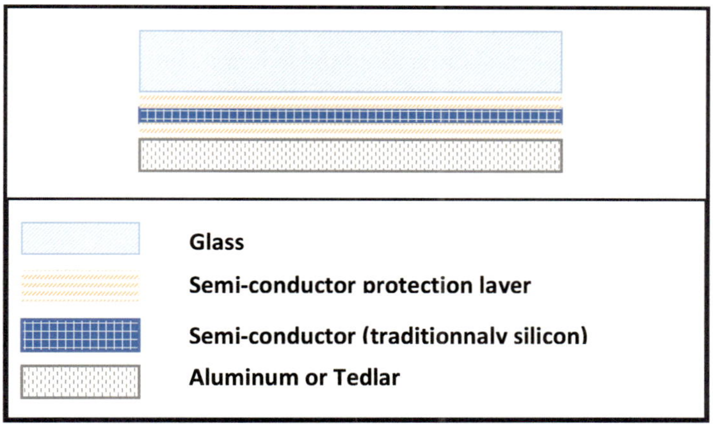

Fig. 3. Cross section of a typical photovoltaic panel

In these semitransparent layers, complex radiative phenomena occur. Indeed, the multiplicity of layers causes complex reflection phenomena in the semitransparent medium. This is shown in Fig 4. A ray of light that reach the surface of a layer of material will be decomposed into three fluxes: absorbed, reflected and transmitted to deeper layer.

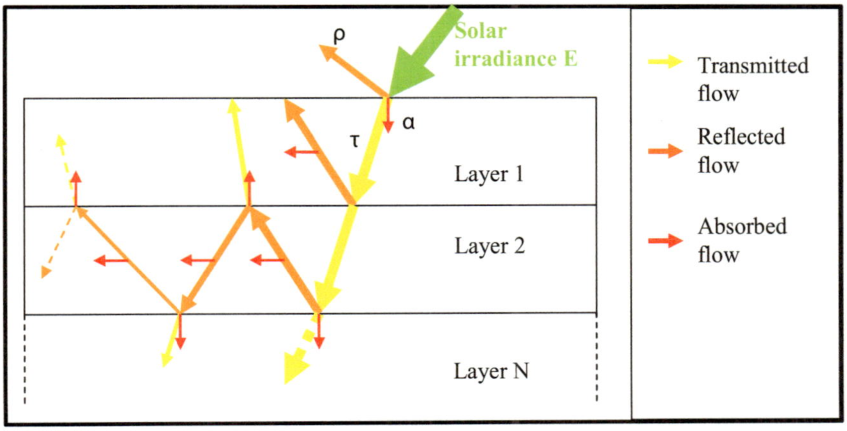

Fig. 4. Section view of the multiple reflections phenomena in a semi-transparent multilayer material

Furthermore, another feature of the system is that it may contain air or water gaps. These air gaps may be contained in the wall where the panel is installed or between the wall and the photovoltaic panel (as in the case of Trombe walls or on some photovoltaic roofs). The blades of water are present in hybrid PV systems. These layers of fluid are complex to model, and are host of phenomena due to different ventilation or fluid circulation system integration in the building. They may be influenced by conditions outside the system (such as wind in the case of opened air gaps in roof installations).

3.2 Thermal phenomena and assumptions

The walls are modelled layer by layer. The goal is to find the energy transfer across the solar system and its coupling with the building, it is not necessary to model finely phenomena. In addition, the coupling of the wall model with the PV will be done with an existing code, named ISOLAB (Miranville, 2003). This code models each type of walls in the same manner, by reducing the thermal problem at the scale of the material layer.

ISOLAB is a building simulation code able to predict the heat and mass transfer in buildings according to a nodal 1D description of the building and its corresponding thermo-physical and geometrical parameters. The resolution is based on a finite difference numeric scheme and the system of differential equations, written in a matrix form, is solved numerically for each time step.

In the version of ISOLAB that was used as the basis for this work, the walls are described by using heat balance equation. This equation is discretized by finite difference method dynamically according to a nodal 1D description in the thickness of each wall.

The heat transfer equation takes classically into account the conduction phenomena in different layers. It is to be noticed that the phenomena occurring in convective fluid layers and radiative semitransparent layers must be described specifically.

Regarding the fluid layers, the choice was made to use empirical models. These models can characterize the convective heat flux by determining the coefficient of convective heat exchange between the fluid and the considered wall. This coefficient will depend on the flow regime in the fluid layer and the temperature of the fluid. Several models have been chosen to perform the tests; they were chosen to meet the most technical configurations of the panel (Bigot, 2009). Note that the chosen models are not necessarily the most appropriate in some cases. The goal here is to test the ability of these models to describe our system. It will be necessary in the future to choose other models as appropriate, and to validate them. These models were implemented directly in the PV model code. They are chosen automatically by the program as needed (cavity vertical, inclined, horizontal, or depending on the configuration of the air layer in terms of opening to the outside, and thus ventilation).

To model the radiative phenomena in the semitransparent medium, the model chosen follows the "ray tracing" method. It is presented in the next section.

3.3 Derivation of the problem

The « ray tracing » method is a model that can describe radiative exchanges in semitransparent mediums. In this work, the model was inspired of Robert Siegel works (Siegel, 1992). This model consists on a net radiative balance of fluxes at each layer of material. As its name suggests, a ray of light will be followed and dispatched every time it will meet a new material surface (see Fig 4). With each new surface it encounters, the ray will be divided into three parts until meeting an opaque layer: the flux absorbed by the layer encountered, the flux transmitted through this layer, and the flux reflected by this layer to the layer where the ray comes from. These phenomena are reproduced until encounter an opaque layer (the layer N where $\tau > 0$ on Fig. 4).

A system describing radiative flux exchanges can be defined for such a problem:

$\Phi_{abs}(i,1,j)$ is the flow absorbed by the layer i at the iteration j on its exterior face ($\Phi_{abs}(i,2,j)$ corresponds to the inside); $\Phi_{trans}(i \rightarrow k,j)$ is the flux transmitted on the layer k by the layer i in the iteration j, and $\Phi_{ref}(i \rightarrow k,j)$ is the reflected flux by the layer i on the layer k for the iteration j. In the below relations, the indicated physical parameters are the following:

α_i : absorption coefficient of the layer i

τ_i : transmission coefficient of the layer i

ρ_i : reflectivity coefficient of the layer i

ε_i : emissivity coefficient of the layer i

F_{pe} : view factor between the panel and the environment

F_{pi} : view factor between layers i and j

E : incident shortwave radiation

T_i : temperature of the layer i

Φ_{abs} : absorbed radiation flux

Φ_{trans} : transmitted radiation flux

Φ_{ref} : reflected radiation flux

In terms of equations, the physical phenomenon can be described as indicated below :

- Initial condition:

$$\Phi_{abs}(1,1,1) = E \cdot S \cdot \alpha_1 \cdot F_{pe}$$

$$\Phi_{trans}(1 \rightarrow 2,1) = E \cdot S \cdot \tau_1 \cdot F_{12}$$

$$\Phi_{trans}(N+1 \rightarrow N,1) = \varepsilon_{N+1} \cdot \sigma \cdot S \cdot T_{N+1}^4 \cdot F_{N+1,N}$$

- Boundary conditions: for $2 \leq j \leq I$:

$$\Phi_{abs}(1,2,j) = \Phi_{abs}(1,j-1,2) + \left(\Phi_{ref}(2 \rightarrow 1,j-1) + \Phi_{trans}(2 \rightarrow 1,j-1)\right) \cdot \alpha_1 \cdot F_{21}$$

$$\Phi_{trans}(N+1 \rightarrow N,j) = 0 \ ; \ \Phi_{trans}(1 \rightarrow 2,j) = 0 \ ; \Phi_{ref}(1 \rightarrow 2,j) = 0$$

$$\Phi_{ref}(N+1 \rightarrow N,j) = \left(\Phi_{trans}(N \rightarrow N+1,j-1) + \Phi_{ref}(N \rightarrow N+1,J-1)\right) \cdot \rho_{N+1} \cdot F_{N+1,N}$$

- System description: for $2 \leq j \leq I$ and $2 \leq i \leq N$:

$$\Phi_{abs}(i,1,j) = \Phi_{abs}(i,j-1,1) + \left(\Phi_{ref}(i-1 \rightarrow i,j-1) + \Phi_{trans}(i-1 \rightarrow i,j-1)\right) \cdot \alpha_i \cdot F_{i-1,i}$$

$$\Phi_{abs}(i,2,j) = \Phi_{abs}(i,j-1,2) + \left(\Phi_{ref}(i+1 \rightarrow i,j-1) + \Phi_{trans}(i+1 \rightarrow i,j-1)\right) \cdot \alpha_i \cdot F_{i+1,i}$$

$$\Phi_{ref}(i \rightarrow i-1,j) = \left(\Phi_{trans}(i-1 \rightarrow i,j-1)\right) \cdot \rho_i \cdot F_{i-1,i}$$

$$\Phi_{ref}(i \rightarrow i+1,j) = \left(\Phi_{trans}(i+1 \rightarrow i,j-1)\right) \cdot \rho_i \cdot F_{i+1,i}$$

$$\Phi_{trans}(i \rightarrow i-1,j) = \left(\Phi_{trans}(i+1 \rightarrow i,j-1) + \Phi_{ref}(i+1 \rightarrow i,j-1)\right) \cdot \tau_i \cdot F_{i,i-1}$$

$$\Phi_{trans}(i \rightarrow i+1,j) = \left(\Phi_{trans}(i-1 \rightarrow i,j-1) + \Phi_{ref}(i-1 \rightarrow i,j-1)\right) \cdot \tau_i \cdot F_{i,i+1}$$

The absorbed flux by the layer situated after the PV system and the absorbed flux by each layer are known:

$$\Phi_{abs}(N+1,1) = \alpha_{N+1} \cdot F_{N,N+1} \cdot \sum_{j=1}^{j=I} \Phi_{trans}(N \rightarrow N+1,j)$$

$$\Phi_{abs}(i,1)=\sum_{j=1}^{j=I}\Phi_{abs}(i,j,1) \qquad \Phi_{abs}(i,2)=\sum_{j=1}^{j=I}\Phi_{abs}(i,j,2)$$

Iterations can be stopped when the residual energy of the system is lower than a threshold value (*erreur*):

$$\sum_{i=1}^{i=N}\left(\Phi_{trans}(i,j)+\Phi_{ref}(i,j)\right)-\sum_{i=1}^{i=N}\left(\Phi_{trans}(i,j-1)+\Phi_{ref}(i,j-1)\right)\le erreur$$

The integration of the PV module to the building simulation is done according to the synoptic of Fig. 5. Once the thermal model of the considered building without PV panels is generated, a test is done in order to detect the inclusion of PV panels; if PV panels are detected, the PV module generates the corresponding system of equations and solves the whole model. Results can then be analysed.

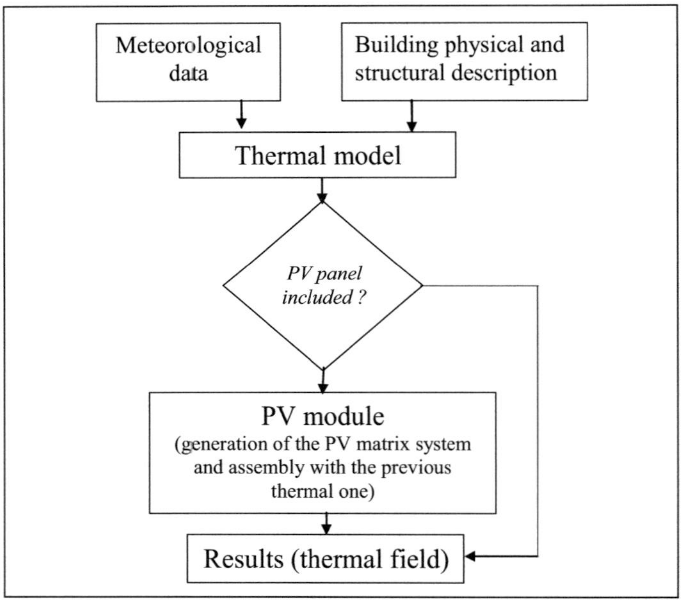

Fig. 5. Integration of the PV calculation module to the existing ISOLAB code.

3.4 Numerical resolution

By discretizing the heat equation below as described above, we obtain a system describing the evolution of the temperature in each building wall. This system of equations can be written in matrix form to facilitate its handling and resolution.

$$\frac{\partial^2 T}{\partial x^2}=\frac{1}{a}\cdot\frac{\partial T}{\partial t}-P \quad \text{where} \quad a=\frac{\lambda}{\rho\cdot C_p}$$

In the case where the material is a semi-transparent layer, P is the volumic heat power absorbed by the semi-transparent layer. P is null in other cases.

We solve this equation by discretizing with a finite difference method. Each layer of material is cut in many nodes. Three types of equations are obtained:

- A first for nodes inside the layer:

$$T_c^t = -\frac{\Delta t}{\tau_c} \cdot T_{c+1}^{t+1} + \left(1 + 2 \cdot \frac{\Delta t}{\tau_c}\right) T_c^{t+1} - \frac{\Delta t}{\tau_c} \cdot T_{c-1}^{t+1} + P$$

- A second for nodes on extremity of the wall or near a fluid layer (c is the number of the node in the wall):

$$T_c^t = \left(1 + 2 \cdot \frac{\Delta T}{\tau_c}\right) \cdot T_c^{t+1} - \frac{2\Delta t}{\tau_c} \cdot T_{c-1}^{t+1} - \frac{2\Delta t}{C_c} \cdot \varphi_{inc}$$

- A third for nodes of the surface between two conductive materials:

$$T_c^t = \frac{k_{c+1}}{k_c + k_{c+1}} \cdot T_{c+1}^t + \frac{k_{c+1}}{k_c + k_{c+1}} \cdot T_{c-1}^{t+1}$$

Where: $\tau_c = \frac{C_c}{k_c}$; $C_c = \rho_c \cdot C_{pc} \cdot \Delta x$; $k_c = \frac{\lambda_c}{\Delta x}$

For surface node temperatures, the φ_{inc} corresponds to the sum of convective and radiative exchange fluxes.

These equations are applied to all nodes of the building system, and we obtain an equation system that describes the evolution of each temperature. It can be expressed in a numeric form by the following matrix equation:

$$[A]_i \cdot [T]^t = [A]_e \cdot [T]^{t+1} + B$$

Matrixes $[A]_i$ and $[A]_e$ describe the composition of the various materials constituting the building, while [B] corresponds to outside or internal solicitations of the system. Matrixes $[T]^t$ and $[T]^{t+1}$ contain all nodes temperatures of all walls.

Finally, a matrix system is obtained that describes the temperature evolution of the PV wall. It is included like a traditional wall by ISOLAB to the matrix building system. Function of the surfaces, the PV wall is partly or totally substituted to the wall where the PV panel is installed.

4. Experimentation of BIPV

4.1 A dedicated experimental platform

In order to apply the preceding combined methodology, a dedicated experimental platform was set up, in field environment. It is indeed very important to be able to determine the physical behaviour of the whole building equipped with the BIPV or the BAPV, under realistic conditions. For this, the experimental platform includes several cells, facing north, and fully instrumented. A meteorological station is also integrated, to allow the measurement of the climatic conditions of the location. The cells are of two types. A large scale test cell, named LGI, is used to represent typical conditions of a real building and its thermal response. Four other cells (ISOTEST cells) are installed on the platform, reduced size and dedicated to the simultaneous comparison of different types of walls installed on buildings. An overview of the platform is presented on fig 6 and the two types of cells are illustrated on fig 7.

Fig. 6 & 7. The experimental platform and the test cells

The study undertaken here is made with ISOTEST test cells in order to compare directly the cases between the buildings which are equipped with a PV panel and those which are not (see fig 8 and 9). These experimental cells have indeed been set up to allow a comparison between the several types of roof components, all in the same conditions. Each of them is equipped with a specific roof component and is fully instrumented to allow the physical observation of the energetic behaviour. It has an interior volume of about $1m^3$ and is conceived from a modular structure, which means that with the same cell we can study different configurations and phenomena. This is why the walls are movable. It constitutes a basis for the thermal studies of building components, with the advantage of flexibility and easy-to-use, especially when several products must be tested. It is installed in-situ, which allows us a better observation of the actual behaviour of the cell. Thanks to this method, we are able to know the temperature of each part of the system in different configurations but in the same environmental conditions. Comparisons between the test cells have been made. Before this, a calibration step has been done to make sure that the four cells had the same thermal behaviour.

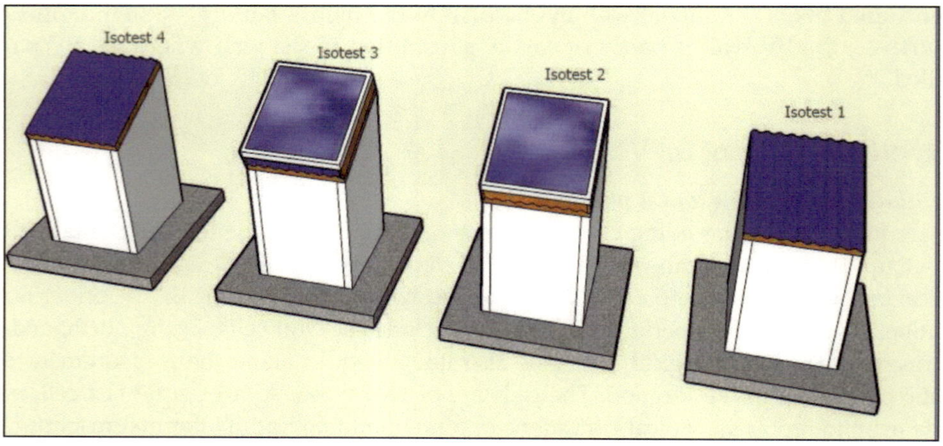

Fig. 8. Current aerial view of Isotest Cells

Fig. 9. Photography of Isotest cell without and with PV panel.

4.2 Data acquisition sensors and errors

The data measured in this experiment are inside surface temperatures of walls and roof, air temperatures, and heat flux through each roofs (see fig 10). The global error of these measurement equipments (sensors and data acquisition system) is about one degree Celsius (±1°C) for the temperature and ±10% for the heat flux (Miranville, 2002). The last study made with this equipment dating for one year, it was necessary to calibrate the equipment. This was done by running a calibration procedure consisting in determining the calibration coefficient allowing the correct inter-comparison of the response of the cells.

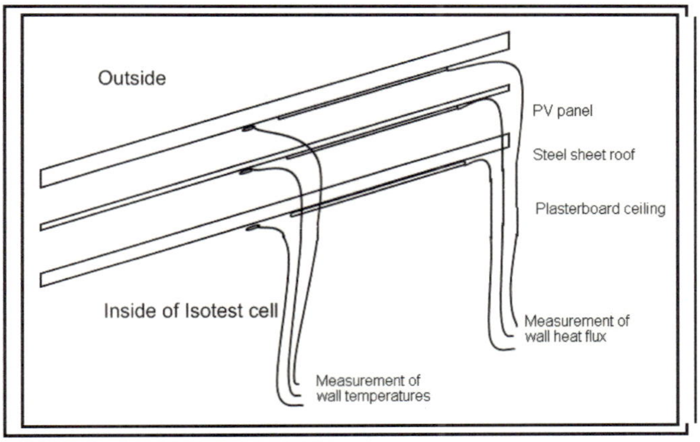

Fig. 10. Sensors installation in the roof wall.

5. Validation

5.1 Overview

Building simulation codes are useful to point out the energetic behaviour of a building as a function of given inputs. The steps involved in this process depend on a mathematical

model, which is considered a global model because it involves several so-called elementary models (conductive, convective, radiative, etc.). Therefore the validation procedure will involve verifying not only the elementary models, but also their coupling, as the building model can be seen as the coupling of a given combination of elementary models.
For several years a common international validation methodology has been developed, which, among others, has led to Anglo-French cooperation. This latter brought to fruition a common validation methodology, involving two test categories, as indicated in table 2.

Verification of the basic theory Verification of good numerical behaviour Comparison of software Analytic verification of elementary models	'Pre-Tests'
Parametric sensitivity analysis Empirical validation	'Post-Tests'

Table 2. Global validation methodology

The first, generally called 'a priori' or 'pre-' tests, involves the verification of the programming code, from the under-lying theory of the elementary models, to software comparisons, and finally to analytic verifications. The objective is to ensure the correct implementation of the elementary models and the correct representation of their coupling at the level of the global model.
This important step of validation justifies the development of dedicated software tools, such as the BESTEST procedure (Judkoff et al., 1995). This latter is essentially based on the comparison between the programming code predictions with so-called reference software results, for a range of different configurations. As a result it includes aspects of verification of correct numerical behaviour and of cross-software comparison, and allows us to compare the program to analogue tools. If the results compare well with those found during this procedure, the programming code is considered acceptable.
The second part of the validation methodology, known as the 'a posteriori' or 'post-'tests, involves two main steps, the parametric sensitivity analysis and, most important, the empirical validation. This second step is fundamental, because it compares the program's predictions with the physical reality of the phenomena, using measurements. It therefore requires an experiment to be set-up, with the aim of obtaining high quality measurements.
The sensitivity analysis of the model consists of finding the set of parameters with most influence on a particular output. It is also used when seeking the cause of any difference between the model and measurements, and allows us to focus this search on a restricted set of parameters, which control the considered output.
Further, the empirical validation methodology is a function of the given objective and of the type of model under consideration; in our case, the empirical validation must allow us to demonstrate the correct thermal behaviour of the building envelope, in particular at the level of the complex wall including a PV panel.

5.3 Empirical validation

In order to improve the PV model, a comparison has been made between measurements and simulation data (see fig. 11) for the case of the PV panel with a confined air layer. In previous articles, the ISOLAB code has already been validated in many cases by comparisons with other building simulation codes, as well as experimental validations. This comparisons can show advantages and disadvantages of the model. In figures presented below, the main temperatures are compared for the previous cell.

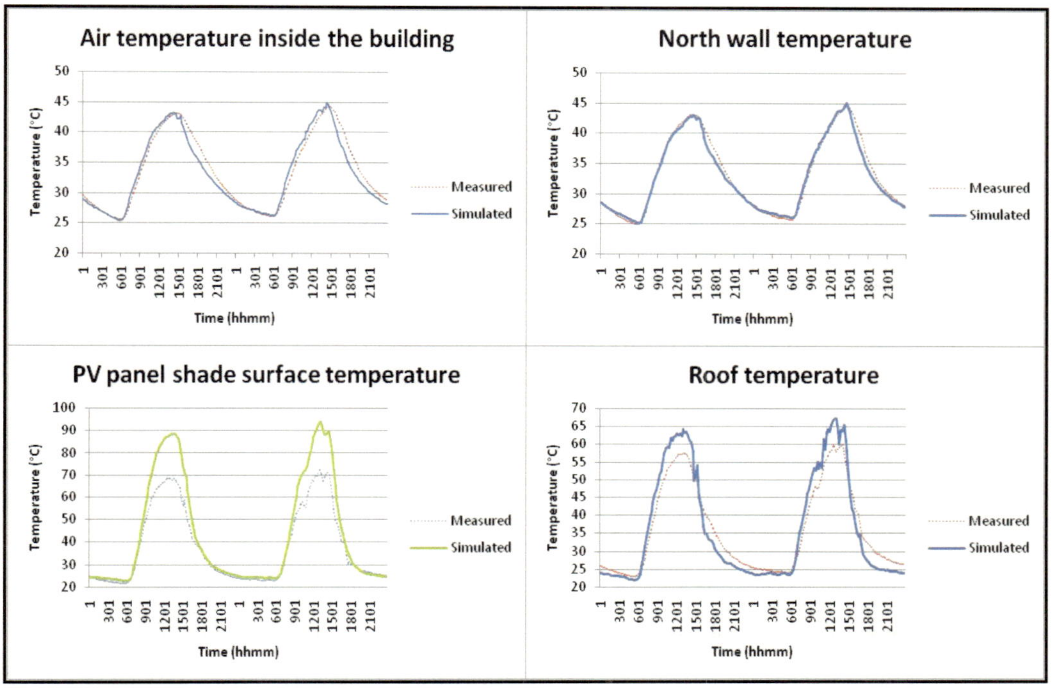

Fig. 11. Temperatures of the PV installation with a confined air layer.

For the temperatures obtained for the body of the cell, a good agreement is obtained, the average difference of temperature being weak, of the order of 1°C. Nevertheless Figure 11 shows, although the PV model has a good dynamic behaviour in the case of a confined air layer, noticeable differences between the model and the reality of measurements. These differences can be related to:

- Thermo-physical properties (conduction, thermal capacity, transmitivity, absorptivity...) of each PV panel material, which are not exactly known. Industrials did not give details of those properties in order to protect their copyright.
- The precision of the radiative model (of PV panels) or convective model (of air layer) in the PV modelling.

To give elements of answers for these differences between predictions and measurements, a sensitivity analysis was made, as explained in the following paragraph.

5.4 Sensitivity analysis

The sensitivity analysis consists in performing several simulation runs by oscillating each parameter according to a sinusoid over its range of interest. Analyzing the spectrum (Fourier transform or power spectral density) of the output, identification of the most influential factors can be easily derived (Mara, 2000); (Mara et al., 2000); (Mara, 2002).

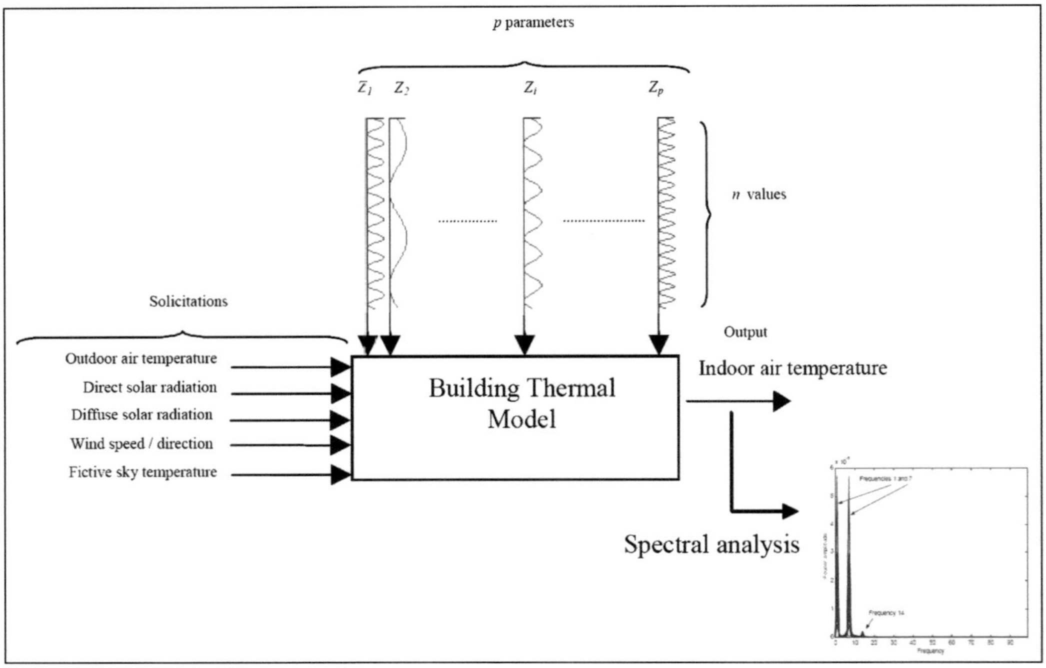

Fig. 12. Procedure of sensitivity analysis.

The proposed FAST method (Fast Fourier Amplitude Transform) uses a sinusoidal sampling of parameters around their base value, each parameter having its own frequency, the variation being applied to a simulation on the other as shown in fig. 12. Thus, the process is analogous to the use of an experimental design where the parameters are varied in each test according to a predetermined pattern, so to sweep the best surface model response.
The sensitivity analysis is composed of three steps:

- The first step that put in evidence the most influential parameters, shown on the figure 13 (Fourier spectrum). For each frequency that corresponds to each parameter it can be shown if it has an effect on the outputs.
- The second step presents principal effects of each parameter on the outputs. It represents the linear effect of each parameter.
- The third step presents non linear effects of parameters on the outputs. Contrary to principal effects, it takes into account the effect of a parameter in interaction with other parameters.

In this study, only principal effects are presented, because non linear effects are negligible compare to principal effects (the maximum interactional effect is about 0.1°C).
The sensitivity analysis was run with a thermal simulation of the building during two days in January 2009. A variation of 10% was applied to all parameters contained in the building and PV panel descriptions.
In a First run, the inside air temperature of the building was chosen has the output. Results show that several parameters of the PV thermal model are influential on this temperature (see fig. 13 an fig. 14).
The fig. 13 shows the Fourier spectrum, and also parameters of influence. Fig. 14 shows parameter effects, and the magnitude of influence of each parameter, described by a frequency number (see Table 3).

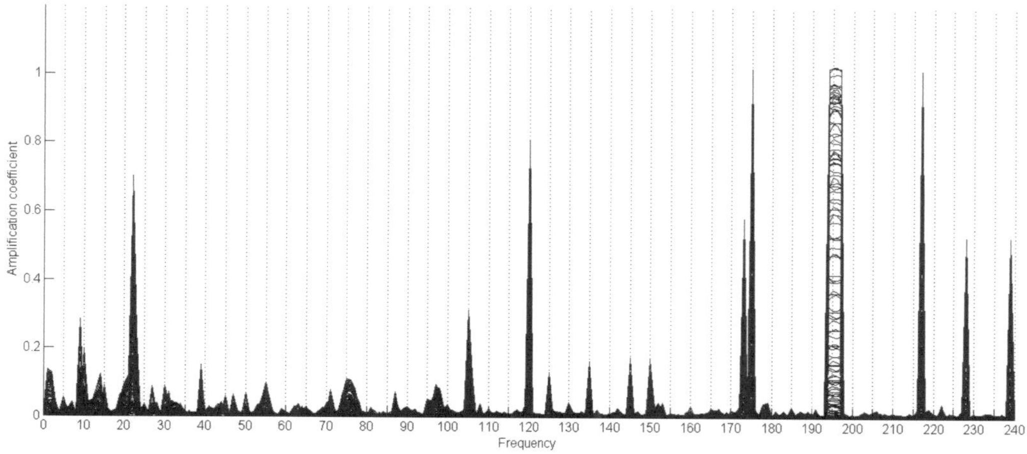

Fig. 13. Fourier spectrum of the sensitivity analysis for the inside air temperature of the building

Parameter numbers	Signication of the parameter
120	Roof azimuth
125	Roof tilt
135	Galvanized steel thickness
145	Galvanized steel density
150	Galvanized steel heat capacity
175	Aluminum thermal conductivity
194	EVAinf transmissivity
195	Silicon transmissivity
196	EVAsup transmissivity
197	Glas transmissivity
217	Aluminum absorptivity
228	Convective heat transfer of the air gap under PV panel

Table 3. Designation of influential parameters of PV model on temperatures of the building

Because the inconsistency seems to come from the modelling of the PV system (ie the assembly of the PV panel and the roof wall), the sensitivity analysis was made for temperatures of all layers of the PV panel system and for the building inside air temperature.
The analysis emphases thermo-physical parameters like thermal conductivity, heat capacity or transmitivity. These results show that three types of thermal transfer must be described more precisely or in a different way, because they are very influential on the air temperature inside the building:

- the transmission of solar irradiation through the semi-transparent system in the PV panel, and the absorption of solar irradiation by the first opaque layer,
- the thermal conduction through all opaque layers after semi-transparent complex system,
- the convection transfer in air gaps in the PV complex wall (like the air gap besides the PV panel).

Furthermore, optical properties of semi transparent layers and characterization of the flow in inclined air gaps are not easy to visualize or describe. These phenomena have been described by commonly accepted parameters, but it is not sure it corresponds exactly to reality. So these results seem quite realistic.

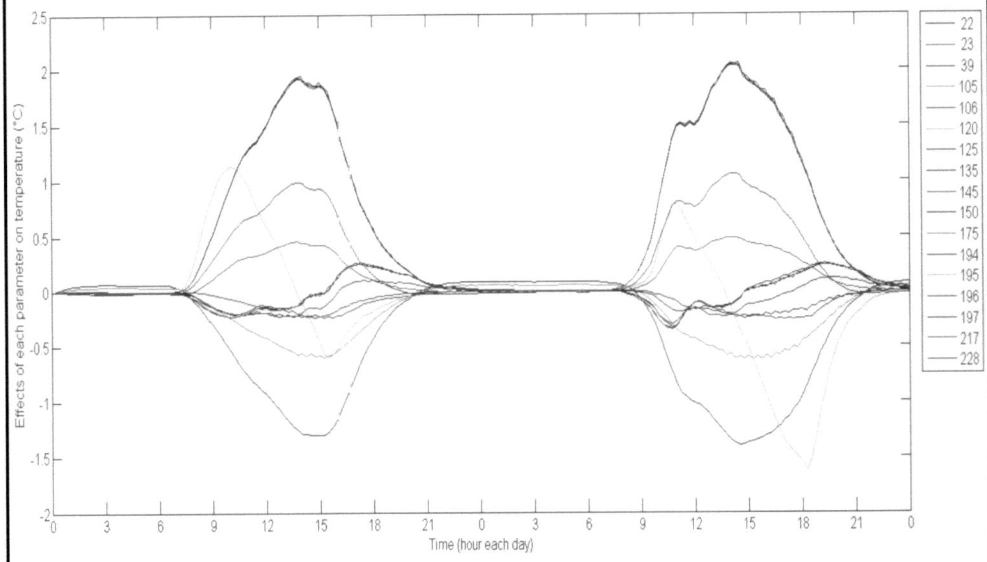

Fig. 14. Principal effect of sensitivity analysis of the inside air temperature of building

Focusing the sensitivity analysis on different layers of the PV complex wall, it can be shown that the most influential parameters are those presented above. Basically, it depends on the transmitivity of all semi-transparent layers through which solar irradiation is transferred, on the conductivity of all opaque layers, and on convective heat transfer coefficients of air gaps of the system.

The next step consists in optimizing parameters of the thermal modelling, as it is introduced as following.

5.5 Optimization

The optimization is the step where the model can be improved and validated. It can be made by using optimization algorithms. In this chapter, we present the use of a free optimization program called GENOPT (Wetter, 2001). This program was set up to allow anyone to use it with his own simulation code. It has been coupled with many building simulation codes like EnergyPlus, TRNSYS, SPARK, IDA-ICE or DOE-2.

GENOPT make the optimization by running simulations of the studied code. It changes values of parameters in the inputs of the program and notes the variation induced on the outputs. As it is shown on fig. 14, it needs only three files to run: the input file, the output file and also the program it has to run. Furthermore, it needs information about the optimization algorithm, studied parameters and the cost function.

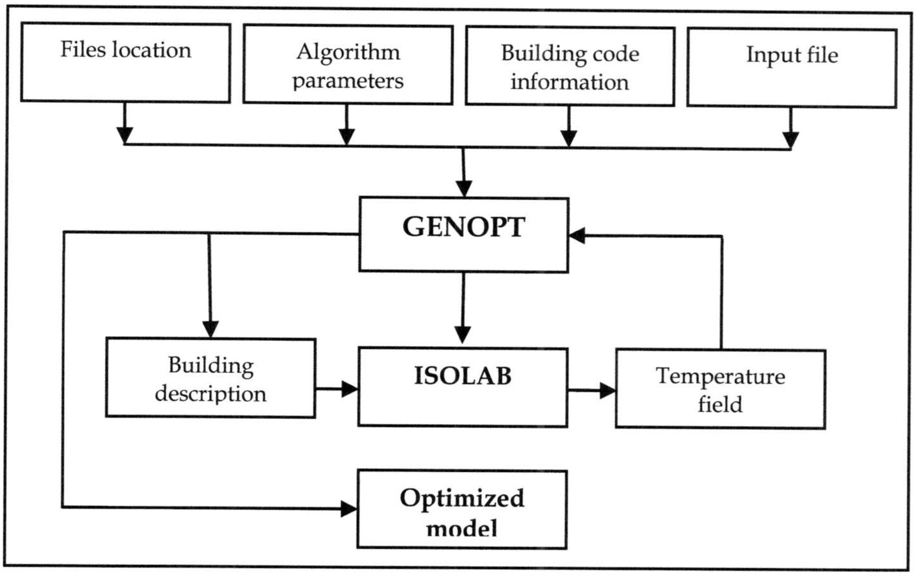

Fig. 14. Synoptic of the coupling of the building simulation code ISOLAB with GENOPT.

To use GENOPT as it is presented in fig. 14, it is necessary to create a complete standalone simulation code; i.e. a program that does not need the human intervention to run a simulation. This step is particularly complex in our case, because ISOLAB was made to be used with the presence of a human kind in all steps of the simulation process.

The interfacing between GENOPT and ISOLAB is in the last test phase. The next step will be the optimisation procedure of the PV system, with the precise determination of the best set of parameters, including conductive, convective and radiative aspects.

Finally, the corroboration of the optimised model will terminate the validation procedure, and allow the generalised use of the model for precise building design.

6. Conclusion

6.1 Thermal Performance of BIPV

The review on BIPV has demonstrated that not only a unique physical model exists, capable of predicting the thermal evolution of the building envelope with the influence of photovoltaic systems in various configurations (integrated-façade, integrated-roof, integrated-glazing, etc.). This chapter has presented a semi-detailed model of a fully coupled PV model, integrated in a building simulation code. The model was used to predict the temperature field in the complex wall constituted by the PV system and its support wall. A global validation procedure (including a sensitivity analysis) has been conducted to determine the precision level of the results and has shown that the performance of the BIPV was greatly dependant on the radiative heat transfer within the semi-transparent layers and the convective heat transfer in the fluid layers. Moreover, the opaque layer included in the system plays also, according to its radiative properties, an important role on the whole behaviour of the system. The main problem is the modelling of convective air-gaps, in which coupled heat transfers arise, the intensity of the coupling being function of the configurations of the photovoltaic installation (angles, thickness and distribution of air spaces in the panel, etc).

6.2 Model validity

Experimentation data was compared to simulation data. This comparison shows that the thermal model has a good dynamic. However, there are some fairly large differences in amplitude for temperatures of the PV complex wall. To provide some answers to this problem, a sensitivity analysis was run and brought to light the most important parameters on the behaviour of the system. An optimisation procedure is planned, to determine the best set of parameters to lead to the best performance of the BIPV. Adjusting these parameters will considerably reduce the observed difference between measurements and predictions, and lead to the validation of the building envelope model. This important step is in progress and will be presented in future works.

6.3 Coupling with PCMs

One possible perspective is to couple the BIPV with MCPs (phase change materials). These are materials capable of changing of physical state within wide ranges of temperatures according to desired applications (building insulation, passive cooling, thermal energy storage, textile industry, etc.).
These materials have the ability to store or to release a large amount of energy as latent heat during phase change liquid-solid. They can be classified into three broad categories:

- The MCP organic (paraffin and fatty acid)
- The MCP inorganic (hydrated salt)
- The MCP eutectic (organic-organic, organic-inorganic, inorganic-inorganic)

The choice of MCPs is based on a number of factors such as latent and sensible heat, thermal conductivity in liquid and solid phases but also the impact on the overall thermal performance of the entire system and its cost.
The coupling of the PCM with BIPV could be considered as liquid-solid phase change to reduce the temperature rise within the BIPV but also increase their performance and their life.

6.4 Toward zero net energy buildings

The building simulation code used for this study henceforth includes a generic model, fully coupled, for the complete modelling of the integration of PV panels in buildings. More and more used in the world, as a means of electricity production using renewable energy, PV systems are of great potential and are subject to numerous research programs. Their inclusion in building envelopes opens the way for zero net energy constructions, whose potential in terms of energy consumption and reduction of global warming is more and more recognised. In a near future, with constant developments and improvements, our building simulation code will be able to predict the energetic behaviourof zero net energy buildings and thus the evaluation and optimisation of their performances.

7. References

Bazilian M. D., Prasad D. Thermal and electrical performance monitoring of a combined BIPV array and modular heat recovery system. In: ISES Solar World Congress Adelaide, Australia, 2001

Bazilian M. D. Australia's first BiPV/thermal test facility (an ACRE funded research project). In: PV in Europe, Rome, 2002

Bigot D., Miranville F., Fakra A., H. Boyer H. (2009). A nodal thermal model for photovoltaic systems: impact on building temperature field elements of validation for humid climatic conditions. Energy and Buildings, Vol. 41, June 2009, 1117-1126

Cherruault J., Wheldon A. (2001). Evaluation of a BIPV roof, designed for expandability and using coloured cells. DTI Substainable Energy Programmes. DTI Pub/URN 01/1395, ETSU S/P2/00297/REP, 80 p. University of Reading, Renewable Energy Helpline

Chow T., He W., Chan A. L. S., Fong K. F., Lin Z., Li J. (2008). Computer modelling and experimental validation of a building-integrated photovoltaic and water heating system. Applied Thermal Engineering, Vol. 28, October 2007, 1356-1364

Fung T. Y., Yang H. (2008). Study on thermal performance of semi-transparent building-integrated photovoltaic glazing's. Energy and Buildings, Vol. 40, February 2007, 341-350

Guiavarch A., Peuportier B. (2006). Photovoltaic collectors efficiency according to their integration in buildings. Solar Energy, January 2006, Vol. 80 issue 1, 65–77

Jie J., Hua Y., Wei H., Gang P., Jianping L., Bin, J. (2007). Modeling of a novel Trombe wall with PV cells. Building and Environment, Vol. 42, January 2006, 1544-1552

Jiménez M. J., Madsen H., Bloem J., Dammann B. (2008). Estimation of non-linear continuous time models for the heat exchange dynamics of building integrated photovoltaic modules. Energy and Buildings, Vol. 40, February 2007, 157-167

Judkoff R. D., Neymark J. S. A Procedure for Testing the Ability of Whole Building Energy Simulation Programs to Thermally Model the Building Fabric. Journal of Solar Energy Engineering, Transactions of ASME, Volume 117, pp. 7-15, 1995

Kondratenko IV. Urban retrofit building integrated photovoltaics [BIPV] in Schotland, with particular reference to double skin facades. PhD thesis, University of Glasgow, 2003

Kropf S. PV/T Schiefer, Optimierung der Energieeffizienz von Gebaüden durch gegenseitige Erga¨nzung von Simulation und Messung am Beispiel der Hinterlu¨ ftung geba¨ udeintegrierter Photovoltaik. PhD report, ETH Zurich, 2003

Mara T., Boyer H., Garde F. and Adelard L. Présentation et Application d'une Technique d'Analyse de Sensibilité Paramétrique en Thermique du Bâtiment, Société Française de Thermique SFT 2000, Lyon, France. p.795-800. 2000.

Mara T., Garde F., Boyer H., Mamode M., Empirical validation of the thermal model of a passive solar test cell. Energy and Buildings. Vol.1320, p.1 - 11. 2000.

Mara T., Boyer H., Garde F. Parametric Sensitivity Analysis of test cell thermal model using spectral analysis. ASME Journal of Solar Energy Engineering. Vol.124, p.237 - 242. (2002)

Mei L., Infield D. G., Gosttschalg R., Loveday D. L., Davies D., Berry M. (2009). Equilibrium thermal characteristics of building integrated photovoltaic tiled roof. Solar Energy, Vol. 83, July 2009, 1893-1901

Miranville F. Contribution à l'Etude des Parois Complexes en Physique du Bâtiment. Thesis. University of La Reunion, La Reunion (France). 2002.

Miranville F., Boyer H., Mara T., Garde F. On the thermal behaviour of roof-mounted radiant barriers under tropical and humid climatic conditions. Energy and Buildings, Volume 35, Issue 10, November 2003, Pages 997-1008

Norme ISO-9869-1994, Isolation thermique - Elements de construction - Mesures in-situ de la resistance thermique et de la transmittance thermique

Nynne F., Maria, J., Hans B., Henrik M. (2009). Modelling the heat dynamics building integrated and ventilated photovoltaic modules. Energy and Buildings, Vol. 41, May 2009, 1051-1057

Park K. E., Kang H. G., Kim H. I., Yu G. J., Kim J. T. (2010). Analysis of thermal and electrical performance of semi – transparent photovoltaic module. Energy, Vol. 35, July 2009, 2681 – 2687

Siegel R. 1992. Thermal Radiation Heat Transfer. Hemisphere, Washington.

Skoplaski E., Palyvos J. A. (2009). Operating temperature of photovoltaic modules: Asurvey of pertinent correlations. Renewable Energy, Vol. 34, june 2008, 23-29

Steven V. D., Benjamin F. (2010). Active thermal insulators: finite elements modelling and parametric study of thermoelectric modules integrated into a double pane glazing system. Energy and Buildings, Vol. 42, February 2010, 1156-1164

Tian W., Wang Y., Xie Y., Wu D., Zhu L., Ren J. (2007). Effect of building integrated photovoltaic on microclimate of urban canopy layer. Building and Environment, Vol. 42, February 2006, 1891-1901

Trinuruk P., Sorapipatana C., Chenvidhya D. (2009). Estimating operating cell temperature of BIPV modules in Thailand. Renewable Energy, Vol. 34., February 2009, 2515-2523

Wang Y., Tian W., Ren J., Zhu L., Wang Q. (2006). Influence of a building's integrated-photovoltaic on heating and cooling loads. Applied Energy, Vol. 83, December 2005, 989-1003

Wetter M. GENOPT - A generic optimization program. In R. Lamberts, C. O. R. Negrao, and J. Hensen, editors, Proc. of the 7th IBPSA Conference, volume I, pages 601-608. Rio de Janeiro, 2001.

Xu X., Dessel V. S. (2008). Evaluation of a prototype active building envelope window – system. Energy and Buildings, Vol. 40, February 2007, 168-174

Zonda H. A. Combined PV-air collector as heat pump air preheater. Staffelstein, 2001

Zondag H. A. (2008). Flate-Plate PV-Thermal collectors and systems: A review. Renewable and Sustainable Energy Reviews, Vol. 12, December 2005, 891-959

20

Working Fluid Selection for Low Temperature Solar Thermal Power Generation with Two-stage Collectors and Heat Storage Units

Pei Gang, Li Jing, Ji Jie
Department of Thermal Science and Energy Engineering, University of Science and Technology of China, Jinzhai Road 96#, Hefei City, Anhui Province, People's Republic of China

1. Introduction

Organic Rankine Cycle (ORC) is named for its use of an organic, high molecular mass fluid that boils at a lower temperature than the water. Among many well-proven technologies, the ORC is one of the most favorable and promising ways for low-temperature applications. In comparison to water, organic fluids are advantageous when the plant runs at low temperature or low power. The ORC is scalable to smaller unit sizes and higher efficiencies during cooler ambient temperatures, immune from freezing at cold winter nighttime temperatures, and adaptable for conducting semi-attended or unattended operations [1]. Simpler and cheaper turbine can be used due to the limited volume ratio of organic fluid at the turbine outlet and inlet [2]. In the case of a dry fluid, ORC can be employed at lower temperatures without requiring superheating. This results in a practical increase in efficiency over the use of the cycle with water as the working fluid [3]. ORC can be easily modularized and utilized in conjunction with various heat sources. The success of the ORC technology is reinforced by high technological maturity of majority of its components, spurred by extensive use in refrigeration applications [4]. Moreover, electricity generation near the point of use will lead to smaller-scale power plants, and thus the ORC is particularly suitable for off-grid generation.

The selection of the working fluid is of key importance in ORC applications. This is because the fluid must have not only thermophysical properties that match the application but also adequate chemical stability at the desired working temperature. There are several optimal characteristics of the working fluid:

1. Dry or isentropic fluid to avoid superheating at the turbine inlet, for the sake of an acceptable cycle efficiency;
2. Chemical stability to prevent deteriorations and decomposition at operating temperatures;
3. Non-fouling, non-corrosiveness, non-toxicity and non-flammability;
4. Good availability and low cost.

However, not all the desired general requirements can be satisfied in a practical ORC. In the previous research, numerous theoretical and experimental studies have focused on ORC fluid selection with special respect to thermodynamic properties. Hung et al. studied waste heat recovery of ORC using dry fluids. The results revealed that irreversibility depended on the type of heat source. Working fluid of the lowest irreversibility in recovering high-temperature waste heat fails to perform favorably in recovering low-temperature waste heat [5]. Liu et al. presented a performance analysis of ORC subjected to the influence of working fluid. It was revealed that thermal efficiency for various working fluids is a weak function of critical temperature [6]. Saleh et al. conducted a thermodynamic screening of 31 pure component working fluids for ORC using Backone equation of state. It was suggested that should the vapor leaving the turbine be superheated, an internal heat exchanger may be employed [7]. Madhawa et al. presented a cost-effective optimum design criterion for ORC utilizing low-temperature geothermal heat sources. Results indicated that ammonia possesses minimum objective function because of a better heat transfer performance, but not necessarily a maximum cycle efficiency [8]. Drescher et al. proposed a new heat transfer configuration with two thermal oil cycles to avoid the constriction of the pinch point between the organic fluid and thermal oil at the beginning of vaporization in biomass power and heat plants. Based on the new design, the influence of working fluids was analyzed and the family of alkyl benzenes showed highest efficiencies [9].

It should be noted that the majority of the previous research on ORC fluid selection was concerned in fields of waste heat recovery, geothermal and biomass applications. Integration of ORC and solar collectors has attracted limited attention. Wang et al. designed, constructed, and tested a prototype low-temperature solar Rankine system. With a 1.73 kW rolling-piston expander overall power generation efficiency is estimated at 4.2% or 3.2% for evacuated or flat plate collectors (FPC) respectively [10]. Ormat supplied a 1 MW power plant, based on ORC technology, to the new power facility of Arizona Public Service. It represented the first parabolic trough plant constructed since 1991 [11].

This paper combines ORC with compound parabolic concentrator (CPC). The feasibility and advantage of CPC application in solar thermal electric generation have been outlined [12, 13, 14]. In particular, FPCs are employed in series with CPC collectors. Three considerations should be made to understand the advantage of two-stage collectors. First, although CPC collectors offer relatively low overall heat loss when operated at high temperatures, efficiency may be lower than that of FPCs in low temperature ranges. Reflectivity of CPC reflectors and difference between the inner and outer diagram of the evacuated tube result in lower intercept efficiency. Thus, overall collector efficiency may be improved when FPCs are employed to preheat the working fluid prior to entering a field of higher-temperature CPC collectors. Second, FPC can absorb energy originating from all directions above the absorber (both beam and diffuse solar irradiance). Third, FPC currently costs less than CPC collector. Part of the reason is that production of FPC is considerably larger. Many excellent models of FPC are available commercially for solar designers [15]. Similarly, collector efficiency may be improved when two-stage heat storage units are employed with phase change material (PCM) of a lower melting point as the first stage, and PCM of a higher melting point as the second stage. Details are provided in the sections below.

Due this innovative design the working fluid selection criteria are different from that for a solo ORC or ORC plants in waste heat recovery, geothermal and biomass fields. The collector efficiency will be influenced directly by the thermophysical properties of the working fluid e.g. the enthalpy-temperature diagram in the isobaric heating process. Furthermore, the optimal proportion of FPC area to overall collector area for the two-stage collectors is determined by both the operation condition and selection of working fluid.

The low-temperature solar thermal electric generation with two-stage collectors and heat storage units is first designed. Subsequently, fundamentals of heat transfer and thermodynamics are illustrated. A mathematical model is established and a numerical simulation is carried out. Five widely or newly used fluids are considered in this study. The influences of working fluids on heat collection, ORC and global electricity efficiency are investigated. Performance comparison among R113, R123, R245fa, pentane and butane is presented.

2. Design and fundamentals

Figure 1 presents the diagram of low-temperature solar thermal electric generation with two-stage collectors and heat storage units. The system consists of FPC and CPC collectors, heat storage, and ORC subsystem. FPCs offer the advantage of accepting high pressure without leakage. The organic fluid flows through FPCs directly and is heated indirectly by CPC collectors with the intermediate of conduction oil. The ORC subsystem consists of evaporator (E), organic fluid/heat storage tank with PCM, turbine (T), generator (G), regenerator (R), condenser, and pumps. The first-stage heat storage is filled with PCM (1), while the second heat storage is filled with PCM (2). Melting point of PCM (1) is lower than that of PCM (2).

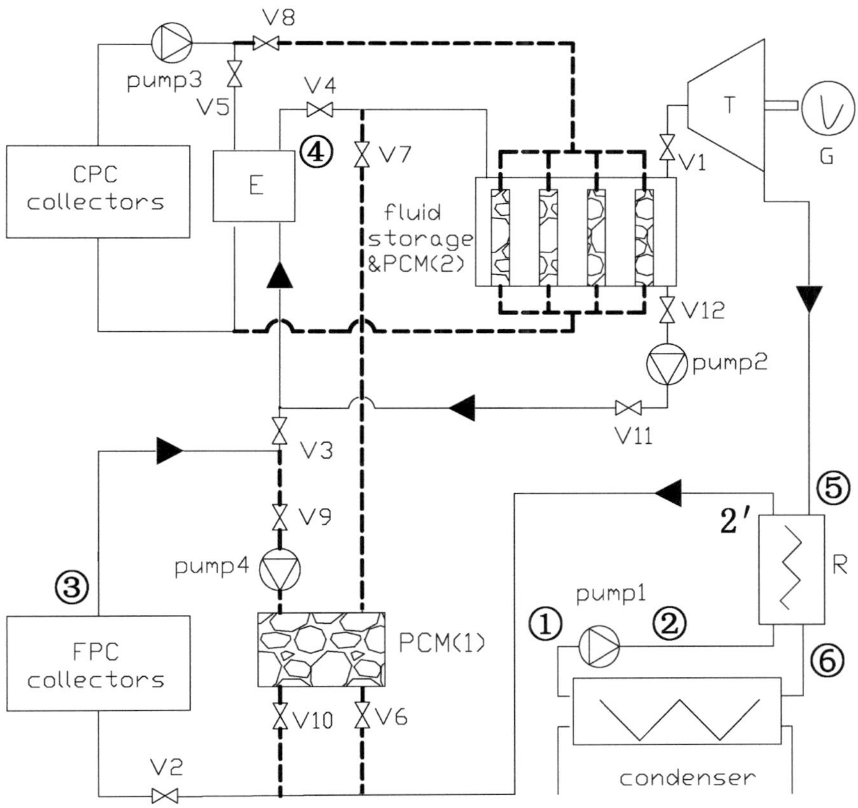

Fig. 1. Low-temperature solar thermal electric generation with two-stage collectors and heat storage units

There are three basic modes of the low-temperature solar thermal electricity system in the practical operating period. In Mode I, the system requires generation of electricity and irradiation is available. In this mode, Valves 1, 2, 3, 4, and 5 are open. Pumps 1 and 3 are running. Valves 11 and 12 may be open while Pump 2 may run to prevent superheating in the evaporator when irradiation is strong. Flow direction of the organic fluid is illustrated by arrows. Organic fluid is preheated in FPCs and subsequently vaporized in the evaporator under high pressure. In the event that organic fluid is not totally vaporized, liquid will drop into the fluid storage tank; it will not harm the turbine. Vapor flows into the turbine and expands, exporting power in the process because of enthalpy drop. The outlet vapor is cooled down in the regenerator and condensed to a liquid state in the condenser. Meanwhile, the liquid is pressurized by Pump 1 and warmed in the regenerator. Subsequently, organic fluid is sent back to the first stage collectors and is circulated. On the use of Pump 2, the system can run steadily in a wide irradiation range. Without any complicated controlling device, the process of heat storage or heat release can occur while electricity is being generated.

In Mode II, the system does not require generation of electricity but irradiation is sound. Valves 2, 8, 9, and 10 are open. Pumps 3 and 4 are running. The dashed lines in Fig.1 represent pipes for heat storage, with the exception of the line that passes through Valves 6 and 7. FPCs are connected with PCM (1) and CPC collectors are connected with PCM (2).

In Mode III, the system requires generation of electricity; however, irradiation is either extremely weak or unavailable. Valves 1, 6, and 7 are open, and Pump 1 is running. Organic fluid is preheated by the first-stage heat storage of PCM (1) and further heated by the second-stage heat storage of PCM (2).

Mode I is described as the simultaneous processes of heat collection and power conversion and is under special investigation in this work.

3. Working fluid properties

The ORC fluid can be classified into three categories according to the temperature-entropy ($T-s$) diagrams. It is noteworthy that for some kinds of fluids, the derivative of temperature with respect to entropy on the saturation vapor curve may change from positive value to negative value, e.g. $\frac{dT}{ds}$ of R123 on the saturation vapor curve is positive when T is smaller than 150°C while negative at higher temperature ranges. In this case, dry fluids are generally named for the positive $\frac{dT}{ds}$ in practical operation temperature range from the cold side to the hot side. And wet fluids would have negative $\frac{dT}{ds}$ on the saturation vapor curve. Meanwhile, isentropic fluids have approximately infinite value of $\frac{dT}{ds}$ (nearly vertical curve).

The working fluids of dry or isentropic type are more appropriate for ORC systems. The reason is that dry or isentropic fluids are superheated after isentropic expansion, thereby eliminating the concerns of impingement of liquid droplets on the turbine blades and making the superheated apparatus unnecessary [6]. Based on this consideration, five dry fluids are selected in the analysis. They are R113, R123, R245fa, pentane and butane. Some of properties of these fluids are listed in table 1. The optimal FPC proportion and the overall collector efficiency are related to the latent heat and heat capacity in saturation liquid states as discussed in Section 5.3.

	R123	R113	R245fa	pentane	butane
Critical pressure /Mpa	3.66	3.39	3.65	3.37	3.79
Critical temperature /°C	183.7	214.1	154.1	196.6	152.0
Boiling point /°C	27.8	47.5	15.1	36.1	-0.5
Latent heat, 120/°C kJ/kg	120.52	116.61	111.77	271.13	213.35
Heat capacity in saturation liquid state, $kJ/(kg{\cdot}°C)$	1.20	1.04	1.78	2.91	3.52

Table 1. Thermodynamic properties of the working fluids

4. Thermodynamics and heat transfer

4.1 Calculation of thermodynamic cycle

Figure 2 presents the scheme of thermodynamic cycle of a typical dry fluid. Point 1 illustrates the state of fluid at the condenser outlet; Point 2 at the Pump 1 outlet; Point 2′ at the regenerator outlet; Point 3 at the FPC collectors outlet; Point 4 at the evaporator outlet (on the normal condition of irradiation); and Point 5 at the turbine outlet. The points being referred to in Fig. 2 are placed in Fig. 1 with circles outside the numbers (with the exception of 2′). The reversible process of pressurization or expansion are described by 2 s or 5s respectively. Formulas for heat transfer and power conversion are developed below.
Enthalpy at Point 2′ is calculated by the following:

$$h_{2'} = h_2 + [h_5 - h_{6(T_6=T_2)}] \cdot \varepsilon_r \tag{1}$$

Where ε_r is the regenerator efficiency. Enthalpy at Point 6 is assigned by assuming $T_6 = T_2$. Total heat transferred to organic fluid from the collectors is calculated by the following:

$$Q = h_4 - h_{2'} \tag{2}$$

Power generated by the turbine (Eq.3) and that consumed by Pump 1 (Eq.4) are calculated by the following:

$$\begin{aligned} W_t &= (h_4 - h_5) \\ &= \varepsilon_t (h_4 - h_{5s}) \end{aligned} \tag{3}$$

$$\begin{aligned} W_{p,1} &= (h_2 - h_1) \\ &= v_1 (p_2 - p_1) / \eta_p \end{aligned} \tag{4}$$

Meanwhile, net power is calculated by the following:

$$W_{orc} = W_t \cdot \varepsilon_g - W_{p,1} - W_{p,2} \tag{5}$$

In case the negative effect of Pump 2 is considered, calculation of required power $W_{p,2}$ is presented in the following section. Practical ORC efficiency is calculated by the following:

$$\eta_{orc} = \frac{W_{orc}}{Q} \tag{6}$$

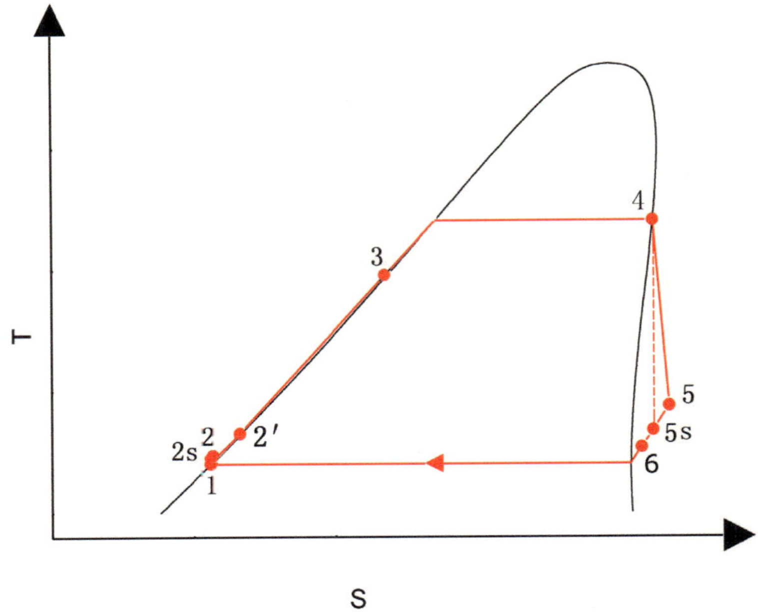

Fig. 2. Thermodynamic cycle of a typical dry fluid

4.2 Equations developed for total thermal efficiency of the collector system

The FPC or CPC collector module available in the market has an effective area of approximately 2.0 m^2. Its thermal efficiency can be expressed by the following equation:

$$\eta = \eta_0 - \frac{A}{G}(T - T_a) - \frac{B}{G}(T - T_a)^2 \tag{7}$$

Solar thermal electric generation system may demand tens or hundreds of collectors in series, and the temperature differences between neighboring collectors will be small. Thus, it is reasonable to assume the following: 1) the average operating temperature of the collector changes continuously from one module to anther module; and 2) the function of the simulated area of the collector system is integrable.

With inlet temperature T_i and outlet temperature T_o, the required solar collection area is obtained by the following [12]:

$$S = \int_{T_i}^{T_o} \frac{mC_p(T)}{\eta(T)G} dT \tag{8}$$

Temperature of conduction oil in the CPC changes within a small range. This is discussed further in Section 5.2. Heat capacity can be well approximated by the following [16]:

$$C_p(T) = C_{p,0} + \alpha(T - T_0) \tag{9}$$

In the case of FPCs, organic fluid is preheated in low temperature ranges and the first-order approximation of heat capacity can be used as well.

With $c_1 = A/G$, $c_2 = B/G$, the collection area according to Eqs. 8 and 9 is integrated by the following:

$$S = \frac{m}{c_2 G(\theta_2 - \theta_1)}\left[(C_{p,a} + \alpha\theta_1)\ln\frac{T_o - T_a - \theta_1}{T_i - T_a - \theta_1} + (C_{p,a} + \alpha\theta_2)\ln\frac{\theta_2 - T_i + T_a}{\theta_2 - T_o + T_a}\right] \tag{10}$$

where θ_1 and θ_2 are the arithmetical solutions of the following equations ($\theta_1 < 0$, $\theta_2 > 0$).

$$\eta_o - c_1\theta - c_2\theta^2 = 0 . \tag{11}$$

$$C_{p,a} = C_{p,0} + \alpha(T_a - T_0) \tag{12}$$

Subsequently, total thermal efficiency of the collector system is calculated using the following:

$$\eta_c = \frac{m}{GS}\int_{T_i}^{T_o} C_p(T)dT \tag{13}$$

Combining Eq.13 with Eqs.9 and 10, the following is obtained:

$$\eta_c = \frac{c_2(\theta_2 - \theta_1)[C_{p,0}(T_o - T_i) + 0.5\alpha(T_o - T_i)(T_o + T_i - 2T_0)]}{(C_{p,a} + \alpha\theta_1)\ln\frac{(T_o - T_a - \theta_1)}{T_i - T_a - \theta_1} + (C_{p,a} + \alpha\theta_2)\ln\frac{\theta_2 - T_i + T_a}{\theta_2 - T_o + T_a}} \tag{14}$$

Effect of c_1 is expressed by Eq.11 There are two inlet temperatures, as well as two outlet temperatures in the two-stage collectors. Total collector efficiency is calculated by the following:

$$\eta_c = \frac{Q}{GS} = \frac{\Delta H_1 + \Delta H_2}{\frac{\Delta H_1}{\eta_{FPC}} + \frac{\Delta H_2}{\eta_{CPC}}} \tag{15}$$

where η_{FPC} or η_{CPC} is the first- or second-stage collector efficiency, and ΔH_1 or ΔH_2 is the enthalpy increment of working fluid in the first- or second-stage collectors. The value of $C_{p,0}$ or α or collector heat loss coefficient varies when the fluid or the collector is different.

4.3 Heat transfer between conduction oil and working fluid

Thermal efficiency of FPCs can be calculated directly by the inlet and outlet temperatures of working fluid, according to Eq.14. On the other hand, thermal efficiency of CPC collectors is determined by the heat transfer process in the evaporator. The temperature relationship between working fluid and conduction oil must be established.

This section focuses on heat transfer in the evaporator, and the developed equations can easily be extended to the case of the condenser. Counter-current concentric tubes are adopted, and the parameters are listed in Table 2.

Parameters	Value	Parameters	Value
Outer diameter D_o mm	45	Generator efficiency ε_g	0.95
Inner diameter D_i mm	25	Regenerator efficiency ε_r	0.85
Turbine efficiency ε_t	0.80	Pump efficiency ε_p	0.75
Optical conversion of CPC η_0	0.644	Optical conversion of FPC η_0	0.857
First heat loss coefficient of collectors of CPC A $W/m^2 {}^oC$	0.749	First heat loss coefficient of collectors of FPC A $W/m^2 {}^oC$	3.157
Second heat loss coefficient of collectors of CPC B $W/m^2 {}^oC^2$	0.005	Second heat loss coefficient of collectors of FPC B $W/m^2 {}^oC^2$	0.014

Table 2. Specifications of the proposed low-temperature solar thermal electricity system

The following preconditions are assumed: 1) the influence of pressure drop on the saturated temperature arising from flow resistance in the evaporator is negligible; and 2) the two-phase flow is one-dimensional, that is, all parameters change only in the flow direction (Y).

4.3.1 Liquid-phase region of working fluid

The controlling equations for the energy balance of working fluid and conduction oil are as follows:

$$\frac{dT_f}{dY} = \frac{U\pi D_i(T_h - T_f)}{m_f C_{p,f}} \tag{16}$$

$$\frac{dT_h}{dY} = \frac{U\pi D_i(T_h - T_f)}{m_h C_{p,h}} \tag{17}$$

Total heat transfer coefficient is calculated by the following:

$$U = 1\Big/\left(\frac{1}{\bar{h}_i} + \frac{1}{\bar{h}_o}\right) \tag{18}$$

where

$$\bar{h}_i = Nu_f \frac{k_f}{D_i}$$

$$\bar{h}_o = Nu_h \frac{k_h}{(D_o - D_i)}$$

The convectional heat transfer coefficient can be calculated using the Dittus-Boelter equation [17]. When flow of the outer fluid is laminar, the concentric tube is considered isothermal at the inner annulus of the cross-section; it is insulated at the outer annulus, thus obtaining the heat transfer coefficient, according to the *Handbook of Heat Transfer* [18].

4.3.2 Binary-phase region of working fluid

Energy balance of the conduction oil remains to be controlled by Eq.17. However, energy balance (dryness) of organic fluid is controlled by the following:

$$\frac{dx}{dY} = \frac{U\pi D_i(T_h - T_f)}{m_f(h_{f,v} - h_{f,l})} \tag{19}$$

Convection heat transfer coefficient of two-phase flow can be obtained in Rohsenow's handbook [19].

4.4 Calculation of frictional resistance

Viscosity of the oil is generally larger as compared with that of working fluid or water. For precise simulation, flow frictional resistance of oil should be evaluated.
With N lines of the parallel concentric tubes, the required pump power is obtained by the following [17]:

$$W_{oil} \approx \int_{Y1}^{Y2} \frac{128\upsilon\dot{m}^2 v}{\pi N(D_o - D_i)^3(D_o + D_i)\eta_p} dY \tag{20}$$

where $\dot{m}$ is the total mass flow rate of oil through the tubes; υ is the viscosity of oil, $m^2 \cdot s^{-1}$; v is the specific volume of oil, $m^3 \cdot kg^{-1}$; and $Y2 - Y1$ is the length of a single tube. It is noted that Eq.20 may easily be extended to the case of organic fluid when the negative effect of Pump 2 is considered due to increased flow rate in the evaporator. Subsequently, the properties and diagrams in Eq.20 would be refined.

4.5 Overall thermal efficiency

Net electricity output W is obtained by subtracting oil pump power from net output of the ORC.

$$W = W_{orc} - W_{oil} \tag{21}$$

Global electricity efficiency is defined by the proportion of net electricity output to the total irradiation as follows:

$$\eta = \frac{W}{GS} \tag{22}$$

5. Results and discussion

The parameters for simulation are listed in table 2. The collectors and turbine are key issues of the low temperature solar thermal power system and the performance is proposed according to market available product [20, 21, 22]. The second-stage heat storage medium appropriate for the low temperature solar thermal electric system could be erythritol, which has melting point 120°C and heat of fusion 339.8 kJ/kg. Magnesium chloride hexahydrate ($MgCl_2 \cdot 6H_2O$) would be appropriate as well, which has melting point 117°C, heat of fusion 168.6 kJ/kg and thermal conductivity 0.694 $W/m \cdot K$ (solid). The evaporation temperature considered in this paper is 120°C, which would be well correlated with the above PCMs.

5.1 Comparison of ORC efficiencies

The global efficiency of the proposed system is determined by both heat collection and power conversion processes. In this section, influences of working fluids on the ORC efficiency are investigated. Performance of working fluids in the ORC is compared in table 3. The environment temperature is 20°C. In order to obtain the same dryness of 1.0 at the evaporator outlet, the collector area for each fluid is different. The state points are referred to as those in the thermodynamic cycle (figure 2). $(h_l - h_{2'})/(h_4 - h_{2'})$ represents the ratio of heat required in the sub-cooled heating process to the total heat absorbed by fluid in the ORC process. The relationship between this ratio and optimal FPC proportion will be analyzed in Section 5.3.

Table 3 shows that in the case of dry fluids, the regenerator can significantly warm working fluids from the condenser and complement the heat supplied from outside. The temperature arisen in the regenerator for R123, R113, R245fa, pentane or butane is 14.7°C, 23.3°C, 14.2°C, 25.5°C or 15.3°C respectively. The ORC efficiencies of the fluids are close, though R113 has a maximum value of 0.161 and butane has a minimum value of 0.147.

State point		R123	R113	R245fa	pentane	butane
1	t /°C	25	25	25	25	25
	h kJ/kg	225.14	222.67	232.46	-25.93	259.46
2s	t /°C	25.38	25.19	25.60	25.28	25.85
	h kJ/kg	225.89	223.07	233.78	-24.58	262.90
2	t /°C	25.62	25.33	25.93	25.47	26.32
	h kJ/kg	226.14	223.20	234.22	-24.13	264.05
2′	t /°C	40.34	48.46	40.61	50.96	41.58
	h kJ/kg	241.25	244.67	253.79	36.41	301.81
4	t /°C	120	120	120	120	120
	h kJ/kg	449.67	431.71	484.39	490.59	740.69
5s	t /°C	38.52	50.99	40.20	54.81	40.68
	h kJ/kg	406.00	391.45	437.25	392.84	649.49
5	t /°C	50.78	62.71	50.09	65.42	50.48
	h kJ/kg	414.73	399.50	446.68	412.39	667.73
6 assumed	t /°C	25.62	25.33	25.93	25.47	26.32
	h kJ/kg	396.95	374.24	423.66	341.16	623.31
6 real	t /°C	29.46	31.06	29.55	31.69	30.00
	h kJ/kg	399.62	378.03	427.11	351.84	629.97
$(h_l - h_{2'})/(h_4 - h_{2'})$		0.422	0.377	0.515	0.403	0.514
ORC efficiency		0.154	0.161	0.148	0.160	0.147

Note: h_l is the enthalpy of saturation liquid at 120°C.

Table 3. Comparison of working fluids performance in the ORC

5.2 Heat collection efficiency of single-stage collectors

For the purpose of a better understanding of the advantage of two-stage collectors on heat collection efficiency, a prior study on single-stage collectors is necessary. The collectors in single-stage system and the second stage collectors in two-stage system are CPC collectors connected with evaporator. And single-stage collectors could be interpreted as a special case of two-stage collectors with FPC proportion equal to 0. Heat transfer in the evaporator is simulated in order to establish the relationship between ORC operation temperature and

CPC efficiency. The organic fluid is heated from sub-cooled to binary phase conditions in the evaporator.

Table 4 shows the single-stage collectors efficiency and the specific distribution of thermodynamic parameters. The subscript of f or h represents organic fluid or conduction oil respectively. $x_{f,o}$ is the dryness of organic fluid at evaporator outlet. $T_{f,i}$ is the organic fluid inlet temperature. The evaporator inlet temperatures of the working fluids are different due to the use of regenerator. Since the fluids and conduction oil flow in a contrary direction, the inlet of fluids means location next to the outlet of conduction oil. In order to heat the organic fluid from sub-cooled to saturation vapor state in the evaporator, heat transfer irreversibility between the conduction oil and fluids is large. The average operating temperature of CPC collectors is higher than ORC evaporation temperature, regardless of the much lower inlet temperature of working fluid in the evaporator. A smaller mass flow rate m_h can reduce the outlet temperature of conduction oil $T_{h,o}$, but $T_{h,i}$ will be increased. The reason is that the fluid temperature is constant in the binary phase region and heat is required for evaporation. A smaller m_h would lead to a larger difference between inlet oil temperature and ORC evaporation temperature according to the law of conservation of energy. As collector efficiency declines more steeply at higher operating temperature, small m_h is not preferable.

The heat collection efficiency of single-stage collectors on condition of irradiation 750 W/m^2 for R123, R113, R245fa, pentane or butane is 46.47%, 46.05%, 47.02%, 46.37% or 47.01% respectively.

parameter	Organic fluid				
	R123	R113	R245fa	pentane	butane
m_f kg/s	1.23	1.38	1.12	0.57	0.59
m_h kg/s	7.00	7.00	7.00	7.00	7.00
$T_{f,i}$ °C	40.34	48.46	40.61	50.96	41.58
$x_{f,o}$	1.0	1.0	1.0	1.0	1.0
$T_{h,o}$ °C	116.94	118.61	114.51	117.21	114.43
$T_{h,i}$ °C	133.37	135.15	131.16	133.94	131.32
η_c %	46.47	46.05	47.02	46.37	47.01

Note: irradiation is 750 W/m^2

Table 4. Heat collection efficiency of single-stage collectors

5.3 Influences of working fluids on two-stage collectors

Economical and technological performances as well as collector efficiency have to be taken into consideration to evaluate the low temperature solar thermal power system with two-stage collectors. However, this work is concerned about influences of working fluids on heat collection and power conversion efficiency. In addition to key factors such as irradiation and environmental temperature that affect the efficiency of single-stage collectors, the proportion of FPC area to the total collector area plays an important role in both the overall heat collection efficiency and cost-effectiveness of the two-stage collectors. Figure 3 displays the heat collection efficiency of two-stage collectors varying with the proportion of FPC area to the total collector area y for each of the working fluids. The environment temperature is 20°C and irradiation is 750 W/m^2. The mass flow rates of the fluids are the same as those presented in table 4. The total collector area for each fluid is 690 m^2 and keeps constant when FPC proportion varies. The overall collector efficiency climbs when the FPC area increases in the lower proportion range. However, it drops with further increment of FPC area in the higher proportion range. There exists an optimal FPC proportion y_{opt}, at which the overall collector efficiency reaches the maximum for each fluid. The optimal FPC proportion alters when the working fluid is different.

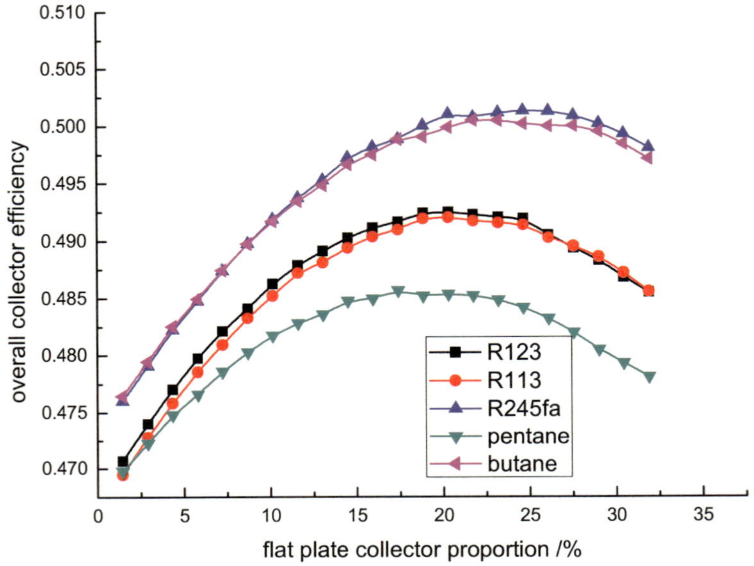

Fig. 3. Heat collection efficiency of two-stage collectors varying with the proportion of FPC area to the total collector area

Table 5 reveals the optimal FPC proportion and the maximum heat collection efficiency variation with working fluids and irradiation. Mass flow rate of each fluid through Pump 1 is equal to that through Pump 2. Thus, the dryness of the working fluids at the evaporator outlet should be 0.5 under normal condition without heat storage or heat release. On the use of Pump 2, superheating is avoidable even if irradiation is strong. Electricity is generated in a wide range of irradiation, and heat transfer between conduction oil and organic fluid is strengthened.

For each fluid, the optimal FPC proportion becomes larger when irradiation is weaker. The decrement of y_{opt} for R123, R113, R245fa, pentane or butane is about 4.2%, 5.5%, 3.9%, 3.1% or 3.0% respectively when irradiation changes from 850 W/m^2 to 650 W/m^2.

Among the five fluids, R245fa exhibits the highest heat collection efficiency accompanied with the largest FPC proportion. The ratio of heat required in the sub-cooled heating process to the total heat absorbed by fluid in the ORC process $(h_l - h_{2'})/(h_4 - h_{2'})$ for R245fa is the highest as shown in table 3. It seems that the preheating concept of FPC is especially suitable for fluids that have a large heat proportion in the sub-cooled heating process.

Irradiation W/m^2		Organic fluid				
		R123	R113	R245fa	pentane	butane
650	opt. y %	22.1	23.2	25.6	19.5	23.2
	max. $x_{f,o}$	0.347	0.326	0.320	0.334	0.309
	max. η_c %	47.37	47.30	48.33	46.45	48.18
750	opt. y %	18.9	18.8	24.0	17.6	21.7
	max. $x_{f,o}$	0.494	0.459	0.493	0.473	0.479
	max. η_c %	49.23	49.18	50.12	48.56	50.04
850	opt. y %	17.9	17.7	21.7	16.4	20.2
	max. $x_{f,o}$	0.649	0.604	0.672	0.623	0.658
	max. η_c %	50.70	50.56	51.51	50.13	51.41

Table 5. Performance analysis of working fluids on the two-stage collectors

On condition of irradiation of 750 W/m^2, the maximum heat collection efficiency for R123, R113, R245fa, pentane or butane is about 49.23%, 49.18%, 50.12%, 48.56% or 50.04% respectively. And the relative increment of heat collection efficiency is 5.94%, 6.80%, 6.60%, 4.73% or 6.45% respectively as compared with that of single-stage collectors (table 4).

6. Conclusion

Heat transfer irreversibility between conduction oil and organic fluids will be large if single-stage collectors are adopted. The low temperature solar thermal electric generation with two-stage collectors and heat storage units gives a flexible system which can react to different operation conditions. Besides, this kind of system displays superior heat collection efficiency as well as cost-effectiveness.

The regenerator can significantly warm working fluids and complement the heat supplied from outside. On the condition of evaporation temperature 120°C, environment temperature 20°C and irradiation 750 W/m^2, the ORC efficiency for R123, R113, R245fa, pentane or butane is 0.154, 0.161, 0.148, 0.160 or 0.147 respectively. Although R113 and pentane have the best ORC performance the highest collector efficiency is obtained on the use of R245fa and butane. And the heat collection efficiency is 49.23%, 49.18%, 50.12%, 48.56% or 50.04% respectively. The proportion of FPC area to the total collector area plays an important role in both the overall heat collection efficiency and cost-effectiveness of the two-stage collectors. And the optimal FPC proportion for R123, R113, R245fa, pentane or butane is 18.9%, 18.8%, 24%, 17.6% or 21.7% respectively. In consideration of frictional resistance of conduction oil as discussed in Section 4.4, the global electricity would be about 7.49%, 7.83%, 7.31%, 7.68%, 7.25% respectively.

7. Acknowledgments

This study was supported by the National Science Foundation of China [Project Numbers: 50974150, 50978241 and 50708105] and the National High Technology Research and Development Program of China (863 Program) [Project Number: 2007AA05Z444].

8. References

[1] Prabhu E. Solar trough ORC. Subcontract report NREL/SR-550-39433 (2006)

[2] Rogers G, Mayhew Y. Engineering thermodynamics, work and heat transfer, 4th ed. Harlow: Longman Scientific & Technical; (1992)239–243.

[3] Andersen WC, Bruno TJ. Rapid Screening of Fluids for Chemical Stability in Organic Rankine Cycle Applications. Industrial & Engineering Chemistry Research 44(2005)5560-5566

[4] Sylvain Quoilin, Vincent Lemort. Technological and economical survey of Organic Rankine Cycle systems, 5th European Conference Economics and Management of Energy in Industry, 14-17 April 2009

[5] Tzu-Chen Huang. Waste heat recovery of organic Rankine cycle using dry fluids. Energy Conversion & Management 42(2001)539-553.

[6] Bo-Tau Liu, Kuo-Hsiang Chien, Chi-Chuan Wang. Effect of working fluids on organic Rankine cycle for waste heat recovery. Energy 29(2004)1207-1217.

[7] Bahaa Saleh, Gerald Koglbauer, Martin Wendland, Johann Fischer. Working fluids for low-temperature organic Rankine cycles. Energy 32(2007)1210-1221.

[8] H.D. Madhawa Hettiarachchi, Mihajlo Golubovic, William M. Worek, Yasuyuki Ikegami. Optimum design criteria for an Organic Rankine cycle using low-temperature geothermal heat sources. Energy 32(2007)1698-1706

[9] Drescher and Brueggemann. Fluid selection for the Organic Rankine Cycle (ORC) in biomass power and heat plants. Applied Thermal Engineering 27 (2007) 223–228

[10] X.D. Wang, L. Zhao, J.L. Wang, W.Z. Zhang, X.Z. Zhao, W. Wu. Performance evaluation of a low-temperature solar Rankine cycle system utilizing R245fa. Solar Energy 84 (2010) 353–364

[11] S. Canada, G. Cohen, R. Cable, D. Brosseau, H. Price, Parabolic trough organic Rankine cycle solar power plant, NREL/CP-550-37077, Presented at the 2004 DOE Solar Energy Technologies, Denver, USA, 2004.

[12] Pei Gang, Li Jing, Ji Jie. Analysis of low temperature solar thermal electric generation using regenerative Organic Rankine Cycle. Applied Thermal Engineering 2010; 30: 998–1004.

[13] Li jing, Pei gang, Ji jie. Analysis of key factors in low temperature solar thermal electric power generation with Organic Rankine Cycle. CIESC Journal 60(2009)826-892.

[14] Optimization of low temperature solar thermal electric generation with Organic Rankine Cycle in different areas. Applied Energy (2010), doi:10.1016/j.apenergy.2010.05.013

[15] William Stine, Michael Geyer. Power from the sun, Solar Energy Research Institute, Solar Technical Information Program (U.S.), http://www.powerfromthesun.net/Chapter6/Chapter6.htm

[16] http://www.fiz-chemie.de/infotherm/servlet/infothermSearch

[17] Incropera FP, Dewitt DP, Bergman TL, Lavine AS. Fundamentals of Heat and Mass Transfer. Ge Xinshi; Ye Hong, trans.6th ed. Chemistry Industry Press (Chinese). 2007

[18] Kays W.M, Perkins H.C. Handbook of Heat Transfer. Chapter 7, New York, 1972.

[19] Warren M. Rohsenow, J.P. Hartnett. Handbook of heat transfer, McGraw-Hill, c1973, 14-1

[20] http://www.infinityturbine.com/ORC/ORC_Waste_Heat_Turbine.html

[21] Ritter Solar product CPC 16w OEM, http://www.rittersolar.de
[22] NAU FLATLINE BE Ultra, http://www.ecocalc.com/manufacturer_col/346/Nau+GmbH/FLATLINE+BE+Ultra?ep=1&prid=

Nomenclature

A	First heat loss coefficient, $W \cdot m^{-2} \cdot {}^oC^{-1}$
B	Second heat loss coefficient $W \cdot m^{-2} \cdot {}^oC^{-2}$
C_p	Heat capacity, $J \cdot kg^{-1} \cdot {}^oC^{-1}$
D	Diameter, m
G	Insolation, $W \cdot m^{-2}$
h	Enthalpy, $J \cdot kg^{-1}$
m	Mass ratio, $kg \cdot s^{-1}$
Nu	Nusselt number
p	Pressure, Pa
Q	Heat, $J \cdot kg^{-1}$
S	Collector area, m^2
T	Temperature, ° C
$\bar{h}$	Heat transfer coefficient, $W \cdot m^{-2} \cdot {}^oC^{-1}$
v	Specific volume, $m^3 \cdot kg^{-1}$
U	Total heat transfer coefficient, $W \cdot m^{-2} \cdot {}^oC^{-1}$
W	Power, $J \cdot kg^{-1}$
x	Dryness
Y	Length, m
y	FPC proportion
α	Heat capacity coefficient, $J \cdot kg^{-1} \cdot {}^oC^{-2}$
ε	Machine efficiency
η	Efficiency
κ	Conductivity, $W \cdot m^{-1} \cdot {}^oC^{-1}$
υ	Viscosity, $m^2 \cdot s^{-1}$
Subscripts	
1-5	State point
a	Environment
c	Collector
f	Organic fluid
g	Generator
h	Conduction oil
i	Inlet
o	Outlet

p	Pump
r	Regenerator
t	Turbine